G. Bastert
D. Wallwiener
(Eds.)

# Lasers in Gynecology

## Possibilities and Limitations

In Collaboration with
D. Pollmann and S. Rimbach

With 198 Figures, 91 in Color and 105 Tables

Springer-Verlag
Berlin Heidelberg GmbH

Prof. Dr. med. Dr. h. c.
GUNTHER BASTERT
Universitäts-Frauenklinik
Voßstr. 9
W-6900 Heidelberg, FRG

Priv.-Doz. Dr. med.
DIETHELM WALLWIENER
Universitäts-Frauenklinik
Voßstr. 9
W-6900 Heidelberg, FRG

With the Support of the Federal Ministry of Research and Technology of the Federal Republic of Germany (BMFT)

Copy-Editing: C. Hellwig, Heidelberg
Drawings in the contributions from the Universitäts-Frauenklinik Heidelberg by G. F. Duncan M. S., Oronoco, USA
Photographs in the contributions from the Universitäts-Frauenklinik Heidelberg by M. Liedtke, Heidelberg

ISBN 978-3-642-45685-5 ISBN 978-3-642-45683-1 (eBook)
DOI 10.1007/978-3-642-45683-1

Library of Congress Cataloging-in-Publication Data. Lasers in gynecology : possibilities and limitations / G. Bastert, D. Wallwiener (eds.). p. cm. Includes bibliographical references and index. 1. Generative organs, Female – Laser surgery. I. Bastert, G. B. (Gunther B.) II. Wallwiener, D. [DNLM: 1. Genital Diseases, Female – therapy. 2. Laser Surgery. 3. Lasers – therapeutic use. 4. Pregnancy Complications – therapy. WP 660 L3435] RG 109. L37L37 1992 618.1'059 – dc20 DNLM/DLC


Softcover reprint of the hardcover 1st edition 1992



Typsetting,
21/3130-5 4 3 2 1 0 – Printed on acid-free paper

# Contents

**Technical and Experimental Research**

# List of Contributors

The addresses of the first authors are given at the beginning of each contribution

# Laser in Gynecology

# Laser in Gynecology: An Overview

D. WALLWIENER [1], D. POLLMANN, W. STOLZ, S. RIMBACH, and G. BASTERT

The evolution of laser technology in medicine has progressed rapidly over the past years, and these devices are at present widely employed in surgical treatment. The use of the laser has become pervasive in gynecology as well. Of the surgical laser systems currently available for clinical application, the carbon dioxide, the neodymium : YAG, and the argon laser have proven their utility in gynecology [10, 15–17].

The potential user of laser techniques is, however, faced with controversial opinions concerning the indication spectrum for laser application on the one hand, and with the large variety of available lasers on the other. These difficulties sometimes seem insurmountable. The choice of the power range best suited is a further problem, since higher initial power means higher costs. We have thus attempted to make a synopsis of laser techniques and their application in operative gynecology.

In order to use lasers efficiently, a knowledge of the specific laser properties is essential. Though lasers operate on similar principles, they utilize thermal energy generated by light and tissue interaction, the different thermal effects depend on the wavelength absorbed preferentially.

[1] Universitäts-Frauenklinik, Abt. für Geburtshilfe und Gynäkologie, Voßstr. 9, W-6900 Heidelberg, FRG.

## The Carbon Dioxide Laser

The carbon dioxide laser (Fig. 1 a, b) emits light in the far infrared band of the electromagnetic spectrum at a wavelength of 10.6 μm. The carbon dioxide laser light is applied in the noncontact technique, the beam being transmitted to the target over a set of mirrors. Depth of laser light penetration into tissue is of primary importance. It depends on input power density (W/cm$^2$, or energy density, J/cm$^2$) and local tissue energy absorption characteristics. By varying power output, exposure time, and focusing, the user may influence penetration depth.

The cutting effect of the carbon dioxide laser beam is the result of thermal tissue reaction due to absorption of laser energy. One can distinguish various zones of tissue alteration: The carbon dioxide lasers create a maximal thermal effect at their foci, with surrounding gradients of decreasing injury (Fig. 2). At the focal point, power intensity is maximal and complete cell evaporation is achieved. The size of these areas of tissue damage is largely dependent on the energy applied. Almost complete absorption of the carbon dioxide beam by water is a major advantage. The $CO_2$ light penetrates only fractions of millimeters into the target, causing minimal thermal damage to the tissue beyond the vaporization area. By defocused application in

Lasers in Gynecology
Edited by G. Bastert and D. Wallwiener

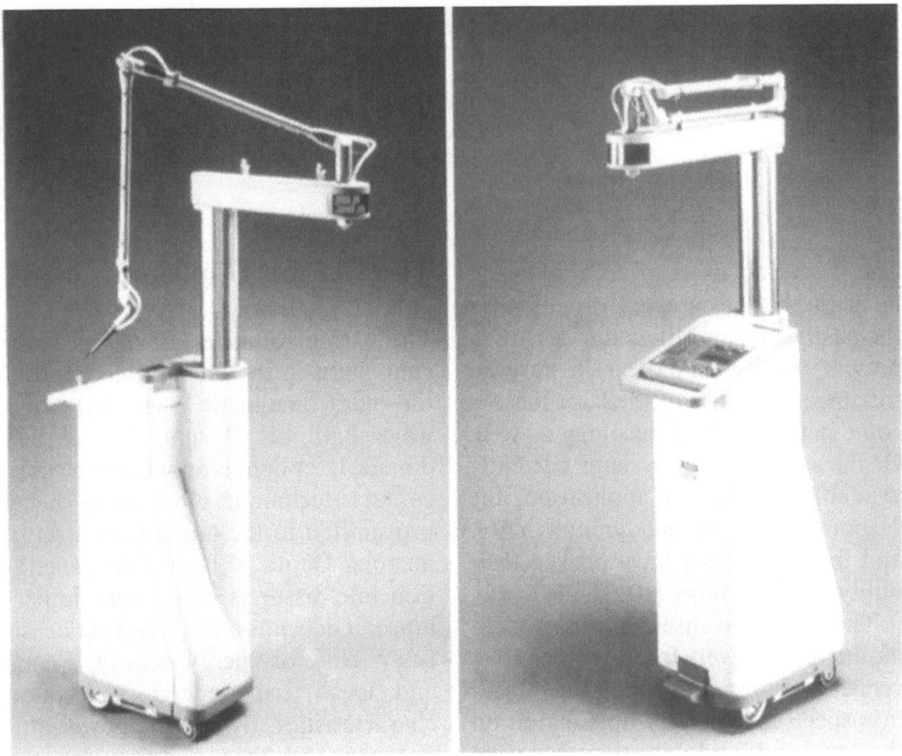

a b

Fig. 1 a, b. Sharplan 1050 carbon dioxide laser

a precisely defined energy spectrum, surface evaporation, vaporization, or shaving off of the finest tissue layers is possible, without thermal injury to the adjacent tissue. Focused application resulting in a very precise cutting effect is employed mainly in microsurgery, laparoscopy, and for cutting parenchymatous or tumor tissue.

Clinical handling is relatively simple: The laser beam is transmitted to the operation area over an articulating arm and focused on a target spot with a system of lenses. As the carbon dioxide laser light is in the far infrared invisible band, visual control is achieved with a helium-neon guiding beam, which marks the carbon dioxide focal spot. Sharp preparation is achieved by focused application, hemostasis by defocused use.

A major advantage is the minimal coagulation damage beneath carbon dioxide laser excision, determined by the low depth (fractions of a millimeter) of carbon dioxide radiation penetration into tissue, as well as by good diffusion of heat. The thermal effect is less marked, and coagulation, not cutting of tissue is achieved.

Newer generations of carbon dioxide lasers delivering in the so-called superpulsed-wave mode (Fig. 3), using short and ultrashort pulses discharged in extremely short busts at great power ranges, cause even less damage to surrounding tissue. Whereas the continuous-wave laser delivery is steady, high peak powers are emitted intermittently at precise but rapid repetition rates in

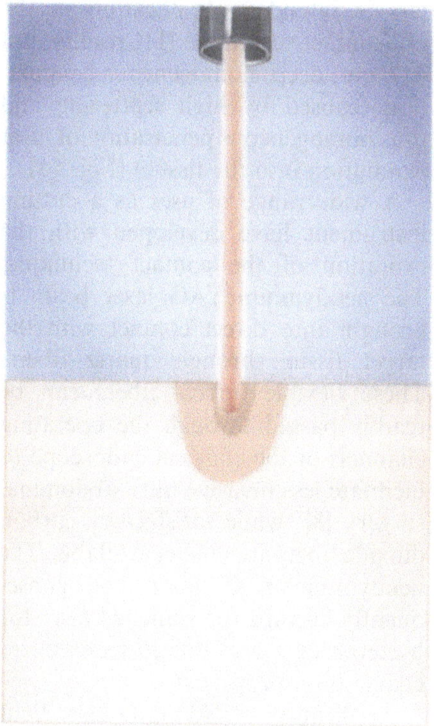

Fig. 2. Tissue effect of the $CO_2$ laser

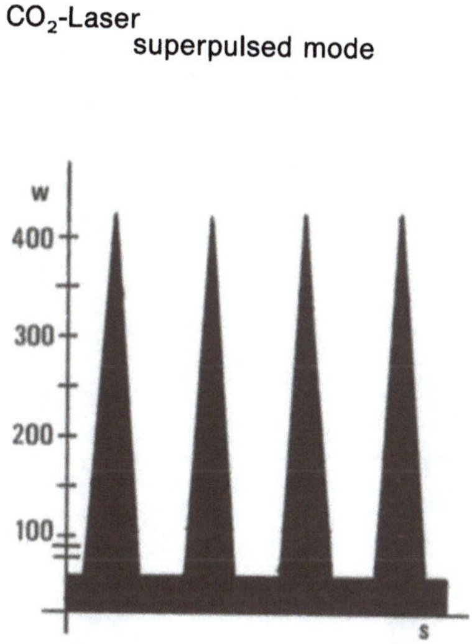


Fig. 3. Superpulsed wave mode

the superpulsed mode. Baggish and coworkers [1,2] were able to generate power peaks of a ten times higher energy density than with continuous-wave delivery by varying pulse duration, power repetition rates, and mean power density. The superpulsed mode permits cutting at a much slower velocity, yet with greater precision and less thermal injury. In the intervals between the power peaks sufficient cooling is achieved. New modes of laser delivery are currently being investigated. Histologic examination has revealed necrosis to be 2.5 times smaller than with the continuous-wave technique [1].

The carbon dioxide laser, valued for its ability to incise tissue accurately while leaving a minimal thickness of coagulation damage, is used largely in gynecologic fertility microsurgery on the open abdomen, as operative tissue traumatization is minimal. It is also employed for treatment of benign and premalignant diseases of the vulva, palliation of breast cancer recurrences, and, in some cases, for endoscopic procedures. The absorption characteristics of the carbon dioxide laser, a function of its wavelength, namely the high, almost complete absorption by water, also determine the hemostasis, albeit reduced, that can be achieved by carbon dioxide laser application. Yet it will not seal vessels more than 0.5 mm in diameter and electrosurgical coagulation will often be necessary.

## The Neodymium: YAG Laser

Unlike the carbon dioxide laser, the neodymium: YAG laser (Fig. 4) is a solid-state laser emitting light in the near infrared region at a wavelength of 1.06 or 1.32 μm. These wavelengths cause completely different tissue effects. Due to low absorption of water

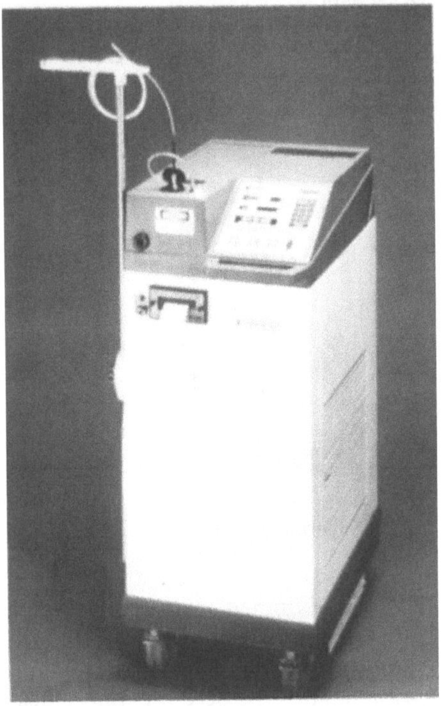

**Fig. 4.** Sharplan 3000 Nd:YAG laser

and high scattering in tissue, the YAG laser is essentially a coagulation instrument, with the capacity to seal vessels up to several millimeters in diameter [19], whereas the major application of the carbon dioxide laser is as a cutting device [13]. Only at high energy levels is effective cutting by vaporization possible, though tedious. Fine preparation, as with the carbon dioxide laser, is not possible.

Thermal damage to the adjacent tissue, however, is more marked than with a carbon dioxide beam (several millimeters), and the vaporization area is larger. The neodymium:YAG should be preferred to the carbon dioxide laser as an ablative instrument if a good coagulation effect associated with deep tissue penetration of several millimeters is intended. Application in the noncontact technique [11] results in a 4–6 mm deep, homogeneous coagulation (caused by high scattering and thus homogeneous penetration of laser irradiation into the tissue) (Fig. 5a).

A wide range of uses as a cutting instrument have developed with the evolution of the contact technique. The neodymium:YAG laser beam is brought into direct contact with the target tissue through quartz fibers. These flexible optical fibers can be readily passed through the operating channels of the classical endoscope to facilitate less invasive intra-abdominal surgery [8], while satisfactory carbon dioxide fibers are not yet available. The neodymium:YAG laser has consequently become the standard laser for hysteroscopy, also due to its excellent coagulation properties.

Application of Nd:YAG laser light with the contact technique results in immediate carbonization of the tissue at the tip of the laser beam transmission system. Due to complete absorption of the Nd:YAG light, flash vaporization with a low depth of irradiation penetration is achieved. Danger of secondary sequelae of the thermally damaged tissue beyond the vaporization crater is reduced. In handling, contact preparation resembles electro- or dermocautery rather than $CO_2$ laser application. Fine cutting movements in slight contact with the tissue surface are performed; otherwise, the fiber penetrates too deeply into the tissue and adheres laterally. No pressure should be exerted on the tissue. Experience has shown that for fine preparation, e.g., in tubal surgery, swinging like a pendulum in slight contact with the tissue is the best technique.

Endoscopic application of Nd:YAG light via bare fibers frequently led to

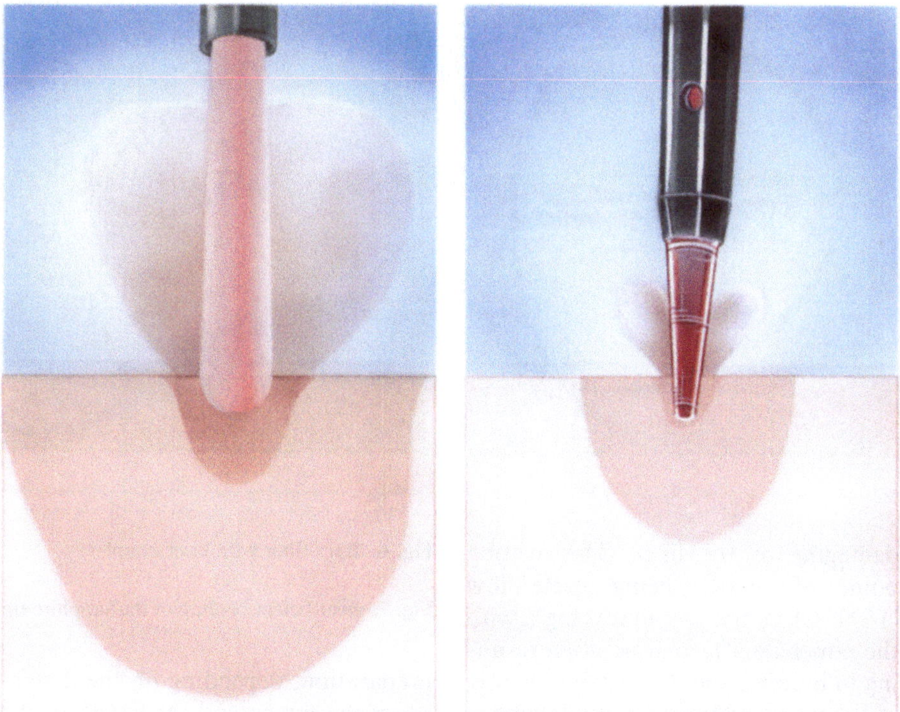

a b

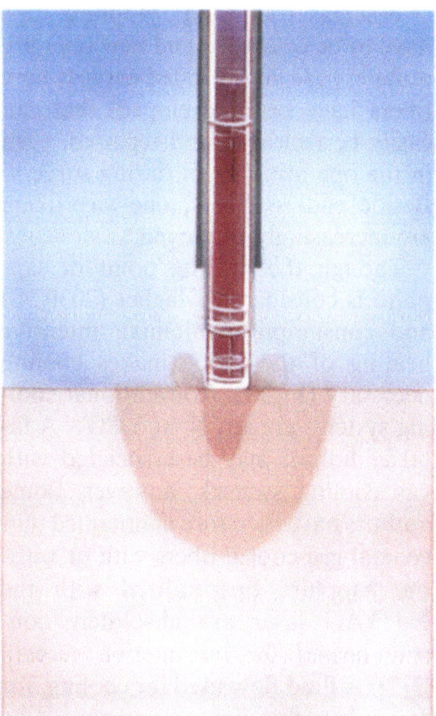

c

**Fig. 5a–c.** Tissue effects of the Nd:YAG laser: **a** Noncontact technique; **b** contact technique, sapphire tip; **c** contact technique, bare fiber

6
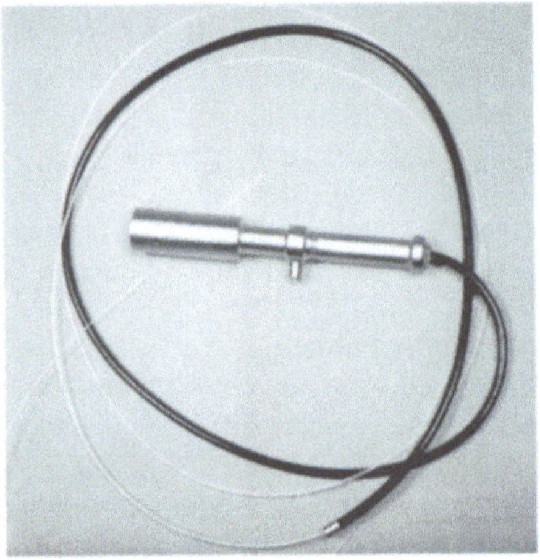

7

**Fig. 6.** Bare fiber with laser coupler

**Fig. 7.** Insufficient cooling of the sapphire tip

damaging of the fiber. The melting point of quartz being quite low (1600 °C), consumption was high, with the noncontact technique often resulting in burning and subsequent melting of the tip, and thus in thermal damage to surrounding tissue. The newly developed so-called intelligent surgical lasers with automatic energy regulation will solve this problem in the future.

Preparation with bare fibers was quite difficult, due to the extreme flexibility of the fibers. Therefore, an artificial sapphire contact probe was connected by a metallic holder to the distal end of the laser light guide. For further details on sapphire tip preparation see the literature [22, 23].

In the hands of experienced users, however, bare quartz fibers (300–600 μm in diameter) present certain advantages over sapphire tips. The handling of bare fibers was improved by fixing them in a stand and leaving only a few millimeters of mechanical load exposed (Fig. 6). This application system was especially designed for contact preparation. It can be employed for laparoscopic interventions or external preparation, depending on the diameter of the lumen and the length of the applicator.

Whereas damaged sapphire tips have to be discarded and new ones are quite expensive, so-called endless bare fibers have been developed, that can easily be recleaved and repaired, even in the operating room during surgery. Beside endless fibers, one-way fibers are increasingly employed.

Though the melting point of sapphire is considerably higher (2030 °C) and consumption minimal, intensive heating of the probe makes cooling necessary (Fig. 7). Conventional cooling systems are air, N, and $CO_2$. A fatal embolism may be associated with gas cooling systems, however. Some authors have therefore maintained that coaxial gas-cooled fibers with or without sapphire tips utilized with the Nd:YAG laser are absolutely contraindicated for intrauterine surgery [3, 7]. A fluid flow used for cooling, for example with saline, was suggested by

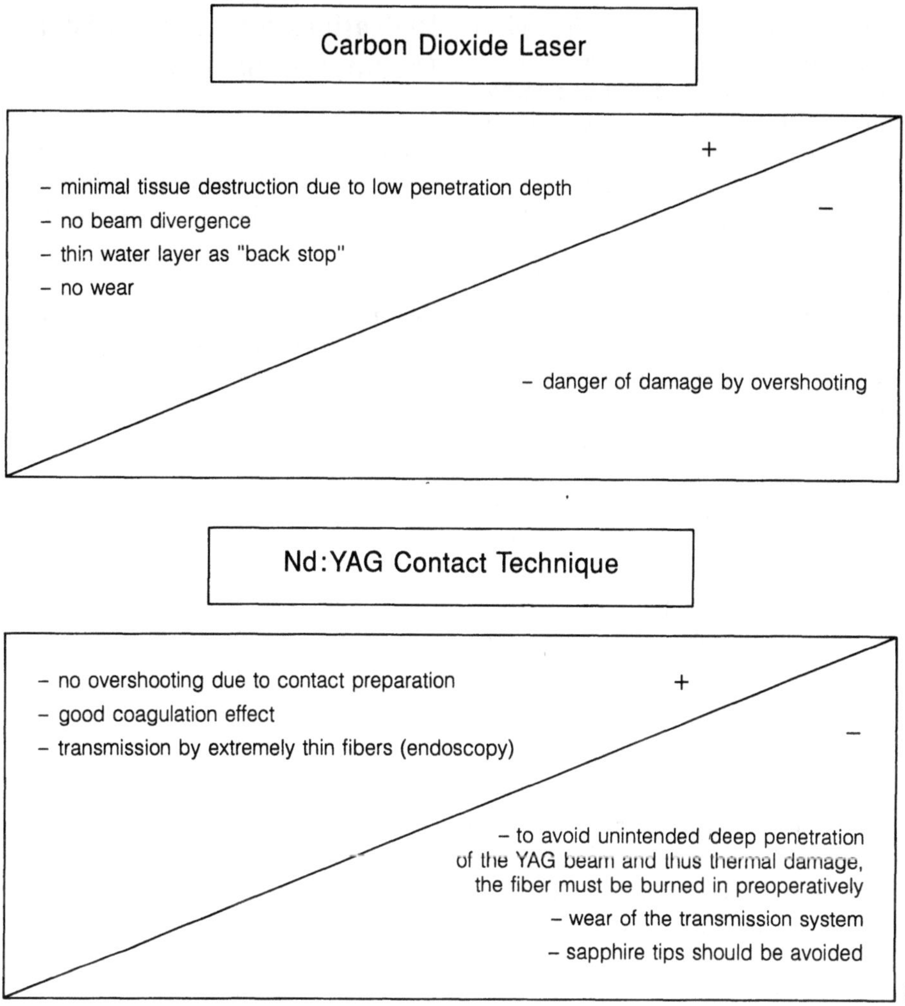


**Fig. 8. a** Advantages and disadvantages of the carbon dioxide laser; **b** advantages and disadvantages of the Nd:YAG laser

others [18]. However, our experience has shown that bare fibers are well suited for intra-abdominal surgery. No cooling at all of the bare fibers has to be performed, as in endoscopic application the fluid in the uterine cavity is sufficient to dissipate the heat from the tip of the fiber, and the thermal capacity, as well as the conductivity, of bare fibers is low anyway [21].

The tissue effects of sapphire tip and bare fiber application are also different (Fig. 5b, c). Bare fibers, even in a used condition, have a slender intensity emission pattern; the axis of energy divergence is low, resulting in the characteristic deep penetration of Nd:YAG laser energy into tissue (several millimeters). Deep coagulation is desirable for indications such as endometrial ablation. The greater divergence of Nd:YAG beams emitted through sapphire tips accounts for a different radiation geometry and different tissue effects of sapphire tips, also dependent on the form of the tip. Application of a sapphire tip reduces depth of penetration of Nd:YAG laser energy into tissue, but even at low power outputs, power density is high enough for sufficiently rapid vaporization and cutting. The vaporization area is larger than with bare fiber application, whereas lateral and deep coagulation necrosis is reduced. Hemostasis is good in most tissues. Sapphire tip preparation requires more time for cutting but is easier to learn and perform than bare fiber preparation.

Our experience has shown that cutting with the bare fiber results in a greater penetration depth (400-μm bare fiber versus sapphire tip: penetration 1.5 times deeper) and, in the hands of an experienced user, in smaller lateral thermal damage areas (0.5 times) [21].

Burning in of the tip of the laser light transmission system is essential to achieve good cutting effects and reduce thermal damage. A carbonization layer at the distal end of the sapphire tip or bare fiber absorbs laser energy, producing high temperatures and preventing penetration of laser light into tissue. New fibers and tips, but also used tips after sterilization, ought to be burned in, usually by applying a power output of 20–30 W/s on a drop of blood or a shred of tissue.

Figure 8 summarizes the advantages and disadvantages of carbon dioxide and Nd:YAG contact laser application.

## Indications for Laser-assisted Preparation in Gynecology

On the basis of the wavelength-specific tissue effects of lasers three main applications may be distinguished (Table 1).

**Table 1.** Spectrum of applicability of the carbon dioxide laser in surgery

| | | |
|---|---|---|
| Surface tissue vaporization | | |
| | Endoscopic tissue preparation | |
| | | Endoscopic tissue vaporization or coagulation |

**Table 2.** Indications for laser-assisted gynecologic surgery

*Nonendoscopic application*

1. Microsurgical tubal reconstruction
   - proximal tubal pathology
   - segmental occlusion
   - reversal of sterilization
2. Nonmalignant diseases of portio, vulva, vagina
3. Recurrences of malignoma

*Endoscopic application*

Laparoscopy

1. Distal tubal pathology
   - salpingolysis/ovariolysis
   - fimbriolysis (/fimbrioplasty)
   - salpingostomy
2. Adhesiolysis
3. Salpingotomy (tubal pregnancy)
4. Endometriosis
5. Ovarian cysts
   - cystectomy
   - fenestration
   - punction (PCOD)

Hysteroscopy

1. Intrauterine septa
2. Intrauterine synechiae
3. Submucous myoma
4. Endometrial ablation

By fine dosage of the laser beam an excellent *cutting effect* associated with minimal tissue traumatization and minimal thermal damage may be achieved. By variation in focusing and dosage *surface vaporization* is possible, and finally, high dosage may lead to thermal destruction or *carbonization* of tissue. Indications for laser application in gynecology are derived from these three main tissue effects. Table 2 summarizes laser application in gynecology.

## Choice of the Laser Wavelength Best Suited for Various Gynecologic Applications

The various technical applications and tissue effects described above have made the carbon dioxide laser the laser of first choice in gynecology. A Nd:YAG laser already available in other departments of a unit may, however, be alternatively employed for certain laparoscopic interventions. For hysteroscopy the Nd:YAG laser is the instrument of choice. Figure 9 shows the applicability of gynecologic lasers in other surgical disciplines.

## Optional and Obligatory Indications for the Nd:YAG Laser

With the development of bare fiber contact preparation, the Nd:YAG laser can be employed in operative laparoscopy. The tissue effects are comparable to those of the carbon dioxide beam. Noncontact preparation must be avoided, however, since divergent noncontact Nd:YAG beams penetrate very deeply and may thus cause tissue destruction or induce adhesions. A further optional indication is Nd:YAG coagulation of condylomata of the lower female genital tract. Dermatolo-

**Fig. 9.** Applicability of gynecologic lasers in other surgical disciplines

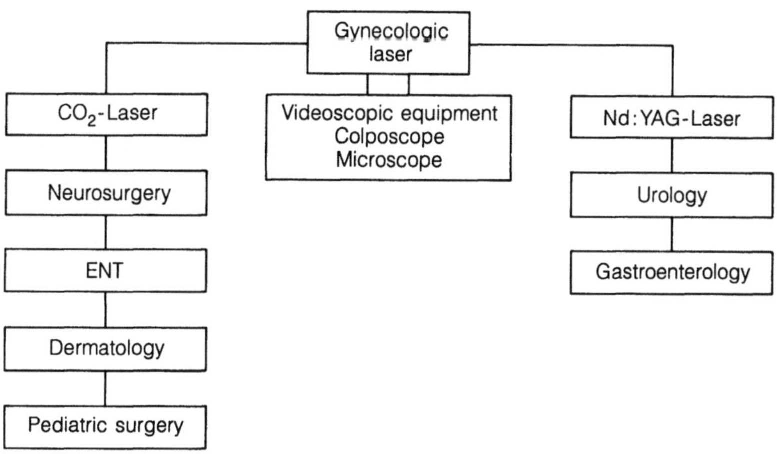

gists now advocate obligatory application of the Nd:YAG light, especially for therapy of condylomata in HJV-positive patients, since less smoke is generated, and fewer viral particles are thus dispersed in the air. An obligatory indication is operative hysteroscopy. The carbon dioxide laser may not be employed here. In some, yet rare, cases of local recurrences, e.g., of vulvar malignoma, with very solid tumor platelets the Nd:YAG laser has to be utilized.

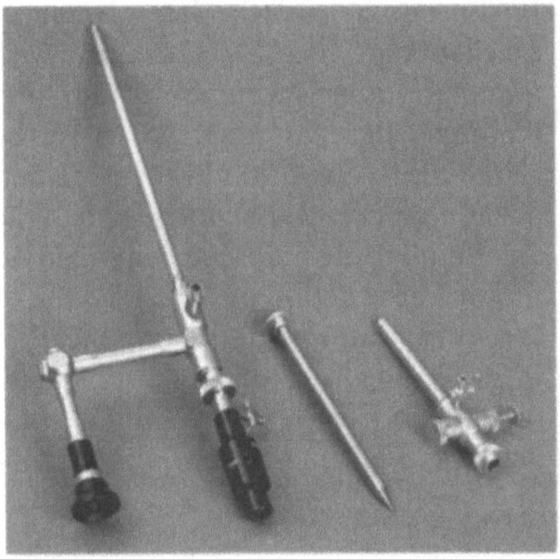

10

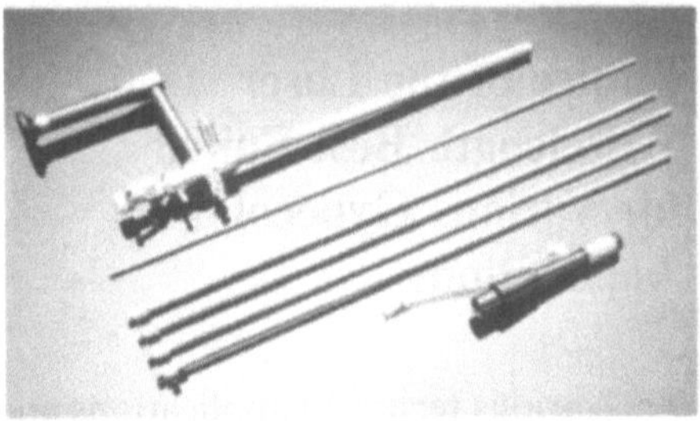

11

**Fig. 10.** Carbon dioxide laser laparoscope

**Fig. 11.** Carbon dioxide laser waveguides

## Technical Background

### Operative Carbon Dioxide Laser Laparoscopy

The carbon dioxide laser light was first applied laparoscopically via the operating channel (single-puncture laparoscope) by Bruhat and associates in 1975 [6]. In 1981, Tadir [20] introduced the second/double-puncture laparoscope, which became the clinical standard. Application of laser energy over a second or third puncture/incision with concomitant observation of the manipulations over the first incision offers a better panoramic perspective of the operation site. Experience in carbon dioxide laparoscopy has revealed certain technical deficiencies, however, especially of the coupling mechanism. Direct transmission or use of a transmission system are the alternatives for carbon dioxide light beamed via the operating channel of the laparoscope or a second incision to the operation site.

An established technique and the easiest to handle is $CO_2$ delivery through the articulated arm of the laser unit attached directly to the operating channel of the laparoscope. The articulating arm is a hollow metal tube with mirrors within movable joints to bounce the beam out of the end of the arm. The use of an additional laser endoscope coupled with a joystick for beam alignment has proved superfluous, being an additional weight hampering fine preparation movements. A Galilei lens system (focal lengths 100–315 mm) can be coupled in for focusing and defocusing the $CO_2$ beam at approximately 2 cm distal to the end of the laparoscope, so that varying the laparoscope–target distance is no longer necessary (Fig. 10).

For laparoscopic use of $CO_2$ laser light rigid, small, hollow waveguides were developed, which can be readily attached to the articulating arm [5]

(Fig. 11). They are slender, needle-like metal tubes, similar to fibers in appearance but not flexible. These dielectric and nonconducting waveguides have a high index of refraction and a low divergence of 3°, and they can transmit relatively high power densities, as loss of power is minimal. High power densities, and thus rapid excision (endometriosis, adhesions), are achieved with lower initial power outputs than with free beam transmission over lenses. The impact site of the laser beam transmitted via waveguide lies immediately beyond its tip, so it must be held in close proximity to the tissue for preparation [12] in order to avoid overshooting. Depth of incision, dependent on power density and duration of application, can be controlled by the surgeon and is connected with his skill and experience, as well as with his ability to visualize the operation field. In order to minimize inadvertent residual tissue damage, an optical backstop may be employed to decrease the focal length of the laser beam (approximately 300 mm), which is focused for a considerable distance behind the focal point. The waveguides have spot sizes of 0.8–1.2 mm, whereas available coupling lenses produce spot sizes of from 0.5 to over 2.0 mm through the laparoscope.

Flexible fibers for $CO_2$ laser light endoscopic application are currently being developed by different companies, but all still have problems that impede their use (toxicity, shielding, energy losses). Some reports [4, 9, 14], however, suggest that adequate fibers will soon be available.

It is essential to perform a beam-alignment check before each laparoscopic surgical procedure with the $CO_2$ laser. Radiation must pass freely through the working channel without touching the inner walls. Distal emission must be "midline". Only thus can damage by overshooting be avoided.

Readjustment of the reflexion transmission system is necessary if transport and/or intraoperative vibration has led to displacement, since a defocused beam renders preparation impossible. Furthermore, problems can develop in the attachment piece between the laser reflexion arm and the laser laparoscope. The reflexion system within the linkage adaptor can also become displaced.

New videoscopic developments have also led to further improvements in endoscopic surgery. Small cameras have been designed which allow the surgeon, the operating team, resident physicians in training, and the patient's family to observe the laser laparoscopic procedure. The surgeon's position is more convenient, as bending over during the whole procedure is no longer necessary. The video material also serves as a record.

## Laparoscopic Nd:YAG Contact Preparation

Various systems were tested for Nd:YAG contact laser laparoscopy. The single-puncture laparoscope was used for both first and second puncture. The modified Albarrán lever was too rigid to enable fine preparation movements, however. Newly developed rigid laparoscopic handpieces for "endless" bare fibers with almost no leakage of the $CO_2$ pneumoperitoneum are now available.

## Choice of Initial Laser Power Output

The user of laser energy must choose the initial laser power output according to the intended field of application. In case of the carbon dioxide laser, which has a relatively limited intrinsic depth of penetration, short pulses (sp mode) of high peak power density and rapid preparation ensure the cutting effect desired and minimal thermal damage to the adjacent tissue. Table 3 shows the minimal initial power output for various applications.

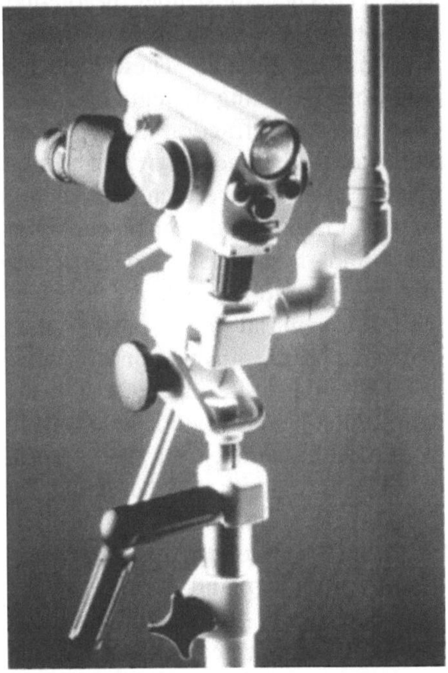

**Fig. 12.** Carbon dioxide laser colposcope

## Conclusion

Though laser preparation techniques have become well established for various indications in gynecology, the potential range of applications has barely been explored. Continuous scientific research is under way to enlarge the existing spectrum of specific laser wavelengths and transmission modes, to develop new fibers and new video techniques. Therefore, a final evaluation of laser techniques is not yet possible. Advantages of laser application, such as precise preparation associated with minimal tissue traumatization, as well as the possibilities offered by the new video techniques must, at least presently, still be weighed against the high costs of laser equipment. Moreover, the laser can be employed only for indications where its wavelength-specific tissue effects are desirable.

We can conclude that laser techniques are by no means an operative concept. Surgical lasers are merely, albeit almost perfect, preparation instruments, optional or obligatory, within the overall surgical concept. Experience with the various applications is of great relevance. Laser endoscopy should be the domain of the endoscopist, laser colposcopy that of the colposcopist (Fig. 12), and laser tumor palliation that of the oncologist (Table 4). It is evident that laser-assisted preparation can optimize surgical techniques, but only if the surgeon first acquires considerable knowledge of the specific laser-technical background and versatility in handling the lasers.

**Table 3.** Choice of power output

| Indication | Minimal initial power output |
|---|---|
| Laser colposcopy | 20–30 W/s |
| Laser laparoscopic or microsurgical preparation | 40–50 W/s |
| Tumor surgery | at least 50–60 W/s |

**Table 4.** Laser in a surgical perspective

| | | | |
|---|---|---|---|
| Laser | | | |
| supplements | → the operative technical spectrum | | |
| does not replace | → other operative technical equipment | | |
| | and | | |
| | → expertise of the specialist | | |
| | laser colposcope | → | colposcopist |
| | laser endoscope | → | endoscopist |
| | laser palliation | → | oncologist |

## References


1. Badawy S, Baggish MS (1986) Pulsed $CO_2$ laser versus conventional microsurgical anastomosis in gynecology. Fertil Steril 46:21–27
2. Badawy S, El Bakry MM, Baggish MS (1987) Comparative study of continuous and pulsed $CO_2$ laser on tissue healing and fertility outcome in tubal anastomosis. Fertil Steril 47:843–847
3. Baggish MS, Daniell JF (1989) Catastrophic injury secondary to the use of coaxial gas-cooled fibers and artificial sapphire tips for intrauterine surgery: a report of five cases. Lasers Surg Med 9:581–584
4. Baggish MS, El Bakry MM (1986) A flexible $CO_2$ laser fiber for operative laparoscopy. Fertil Steril 46:16–20
5. Baggish MS, Sze E, Badawy S, Choe J (1988) Carbon dioxide laser laparoscopy by means of a 3.0 mm diameter rigid wave guide. Fertil Steril 50:419–424
6. Bruhat M, Mage G, Manhes M (1975) Use of the $CO_2$ laser via laparoscopy. In: Kaplan I (ed) Proceedings of the 3rd International Society for Laser Surgery, Tel Aviv, p 275
7. ECRI (1989) Fatal gas embolism associated with intrauterine laser surgery. Health Devices 18:325–326
8. Frank F (1986) Biophysical basis and technical prerequisites for the endoscopic and surgical use of the Nd:YAG laser. Laser Med Surg 3:124–132
9. Fuller T, Nadkarni VJ (1985) Carbon dioxide laser fiber optics. Biolaser News 4:1–4
10. Keckstein J, Finger A, Steiner R (1988) Laser application in contact and noncontact procedures: sapphire tips in comparison to "bare-fiber", argon laser in comparison to Nd:YAG laser. Laser Med Surg 4:158–162
11. Keiditsch F, Hofstetter A, Zimmermann I, Stern J, Frank F, Babaryka I (1985) Histological investigation to substantiate the therapy of bladder tumors with the neodymium:YAG-laser. Laser Med Chir 1: 19–23
12. Marlow JL (1989) Laser equipment. In: Sanfilippo JS, Levine RL (eds) Operative gynecologic endoscopy. Springer, Berlin Heidelberg New York Tokyo, pp 201–210
13. Reid R, Elfont EA, Zirkin RM, Fuller TA (1985) Application of $CO_2$ lasers in obstetrics and gynecology. Am J Obstet Gynecol 152:261–271
14. Rontal M, Fuller TA, Rontal E, Jakob HJ (1985) Flexible nontoxic fiberoptic delivery system for the carbon dioxide laser. Ann Otol Rhinol Laryngol 94:357–360
15. Rouse SB, Minielly RW (1985) Carbon dioxide laser for combination excisional-vaporization conization. Am J Obstet Gynecol 153:343–344
16. Rubinstein E (1985) Carbon dioxide laser surgery. Am J Obstet Gynecol 153:345–346
17. Sasako M, Iwasaki M, Konishi T, Maruyama Y, Wada T (1982) Clinical application of the Nd:YAG laser endoscopy. Laser Med Surg 2:137–147
18. Schroeder TM, Puolakkainen PA, Hahl J, Ramo J (1989) Fatal air embolism as a complication of laser-induced hyperthermia. Lasers Surg Med 9:183–185
19. Stein BS, Kendall AR (1984) Lasers in urology. I. Laser physics and safety. II. Laser therapy. Urology 23:405–416

20. Tadir Y, Kaplan I, Zuckerman Z, Edelstein T, Ovadia J (1984) New instrumentation and technique for laparoscopic carbon dioxide laser operations: a preliminary report. Obstet Gynecol 63:582–585
21. Wallwiener D (1989) Fertilitäts-chirurgische Therapiekonzepte. Lasertechnische Neuentwicklungen. Habilitationsschrift, Ruprecht-Karls-Universität, Heidelberg, pp 40–41
22. Wallwiener D, Stolz W, Pollmann D, Spiller M, Bastert G (1988) Neodym YAG in-touch Laser (Kontakttechnik mit Saphirspitzen) in der operativen Laparoskopie. Laser Med Surg 4:172–179
23. Wallwiener D, Pollmann D, Stolz W, Kappler M, Bastert G, Krampe C (1989) Die Nd:YAG-Kontakttechnik mit „nackten" Glasfasern – lasertechnischer Background und Gewebeeffekte im Vergleich zur Präparation mit Saphirschneidespitzen. Laser Med Surg 5:31–35

# Lasers in Modern Medicine

A. Hofstetter [1]

It is not an exaggeration to state that no single high-tech system has recently opened such profoundly new possibilities in medicine as the laser. Our own 18 years of using laser technology experimentally and clinically have led to a number of standard procedures and revealed new aspects for future investigation.

## Laser Application

The versatility of laser applications is astonishing. The product of energy density and application time leads to a variety of interactions with tissue, some of which have become standard procedures in therapy; others are being used experimentally in therapeutic or diagnostic approaches. Photon energy, expressed in the wavelength of a laser, also plays an important role in these interactions (Fig. 1).

All possible laser-tissue interactions are of relevance in medicine, from low-power tissue irradiation, producing biological effects by energy densities well below the threshold for thermal interaction, over thermal interactions, to high-power effects.

[1] Klinikum Großhadern, Klinik und Poliklinik für Urologie, Ludwig-Maximilians-Universität, Marchioninistr. 15, W-8000 München 70, FRG.

### Photodynamic Therapy

The most promising use of the laser is in photodynamic therapy (Fig. 2). The fairly old principle of activating a photosensitizing agent by light has been developed into a therapeutic concept using high-power light sources such as the argon-ion Laser or the argon pumped-dye laser.

Integral photodynamic therapy is a concept adapted to the multifocal nature of superficial cancer. While patients are being treated worldwide with hematoporphyrin derivatives (HPD), photosensitizing agents with less toxic side effects have been sought and are currently being tested.

The field of low-power irradiation has also given rise to investigations that have deepened our understanding of the elementary laser-tissue interactions, such as mitosis synchronization and biochemical stimulation. The combination of both might enhance chemotherapeutic efficiency. Our group has tested a similar concept of adding phototoxic processes to cytostatics, which may possibly increase the efficacy of adjuvant chemotherapy while reducing the side effects.

### Thermotherapy

Clinical data on thermal interactions caused by higher power densities in

Lasers in Gynecology
Edited by G. Bastert and D. Wallwiener

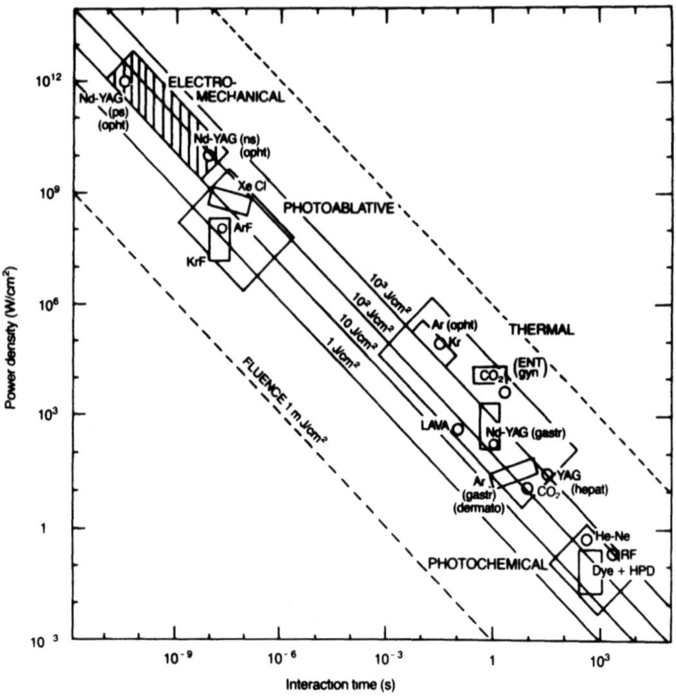


**Fig. 1.** Energy density-time diagram as expression of laser-tissue interaction

seconds or minutes are abundant. The conversion of electromagnetic radiation into heat increases the tissue temperature and leads to protein denaturation at temperatures over 60 °C and to vaporization and ablation at temperatures over 100 °C.

Our group experimented with tumor destruction from 1972 to 1976 [4]. To date, we have treated more than 1500 patients with superficial and infiltrating bladder carcinoma using the Nd:YAG laser. With this mode of treatment the local recurrence rate has dropped remarkably compared with that following transurethral resection. This is presumably connected with the total, contact-free destruction of the tumor-bearing area of the vesical wall by the simultaneous sealing of blood and lymphatic vessels. Apart from this, the absence of bleeding provides an excellent view during coagulation and aids therapy when multifocal tumor growth is present. Randomized studies on laser treatment of bladder carcinoma by Beisland, Meier et al., and Le Guillou et al. [1, 5, 7] verify the remarkably low recurrence rate after Nd:YAG laser therapy. Conservative, organ-saving therapy of superficial urothelial cancer of the upper urinary tract using the Nd:YAG laser is very promising; clinical application, however, should be restricted to specialized centers until enough clinical data are available to prove its superiority to other treatments.

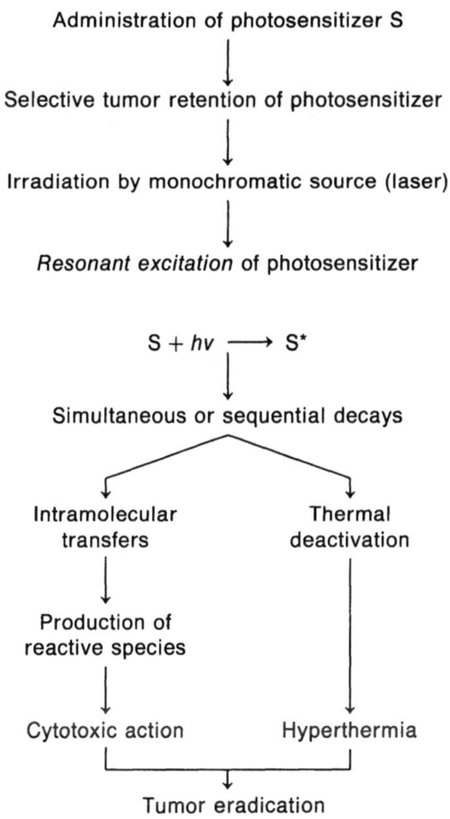


**Fig. 2.** Physical principles of photodynamics

Clinical experience with Nd:YAG laser irradiation of localized prostatic carcinoma was first reported by our group [3] in 1979, then by Beisland and Sander [2]. This technique is promising as an alternative to radical prostatectomy for apparently localized prostatic cancer in combination with pelvic lymphadenectomy. The latest, very interesting Nd:YAG laser use is the interstitial thermocoagulation in BPH (benign prostate hyperplasia).

Tumors of the external genitalia are easily accessible to Nd:YAG laser treatment. The results of Nd:YAG laser treatment of condylomata acuminata are unsurpassed so far; in our experience, the only relapses have occurred in the fossa navicularis, where irradiation is technically difficult [4]. The treatment of superficial penile cancer shows that the afflicted organ can generally be spared [4], and recurrence is seen in only 5%–10% of cases [6, 9].

The treatment of urethral strictures with lasers is still controversial. The $Ar^{++}$ laser urethrotomy as first performed by Rothauge [8] has not brought the desired benefit. The $CO_2$-laser, probably best suited for cutting, has not been adapted to small-caliber flexible application systems. The reports on pulsed, frequency-doubled Nd:YAG lasers are few, but nevertheless surprisingly good.

Thermal interactions are also used in other fields of medicine which have relevance to urology, such as laser-assisted microvascular anastomosis or vasostomy.

## Further Effects

Antiphlogistic and analgetic effects of laser therapy have been demonstrated in patients with interstitial or radiogenic cystitis. These results were confirmed by Shanberg et al. [10].

## Laser-induced Shock Waves

Very high power laser densities with micro- or nanosecond pulses have been utilized by our group for lithotripsy since 1978. Laser-induced shock waves for stone destruction are a result of plasma induction. Lasers supply sufficient energy to induce a plasma in a controlled fashion, so that lithotripsy is possible in any part of the urinary system, with minimal side effects to the adjacent tissue (Fig. 3).

**Fig. 3.** Different laser systems for lithotripsy

Short u. v. laser pulse (10 ns) focused on tissue — $I \sim 10^8$ W/cm$^2$

↓

Strong absorption in the u. v. (GeV) (proteins; amides; peptides) — Absorption depth $\sim 1$ µm

↓

Promotion to repulsive excited state

↓

Photodissociation

↓

Desorption

↓

No necrosis

**Fig. 4.** Physical principles of laser photoablation

Three different systems, one using a Q-switched Nd:YAG, the second a flash lamp pumped-dye laser, and the third an alexandrite laser, have been developed by our group for clinical use. The Q-switched solid-state alexandrite laser and the flash lamp pumped-dye laser with a spectroscopic stone recognition system have proved to be very effective tools in lithotripsy.

These methods constitute further steps forward, not only in endoscopic urology but also in general surgery, with regard to the endoscopic destruction of bile, pancreatic, and salivary calculi.

## Laser-induced Fluorescence

Fluorescence induced by micro- or nanosecond laser pulses makes possible a wide range of spectroscopic analysis, not only for urology. The differentiation of urinary calculi from tissue is already possible and has been incorporated in an optical feedback regulation for dye laser lithotripsy as mentioned above. Tissue analysis, especially tumor identification with tumor-selective photosensitizers, is within reach.

## Photoablation

Photoablation, one of the most recent physical principles that has been adapted to medicine, requires strong absorption of ultraviolet radiation with energy densities of several MW/cm$^2$ (Fig. 4). The ejection of molecules out of the irradiated structure produces an extremely precise, nonthermal cutting effect. This method could be very promising for spectroscopic tissue identification and laser angioplasty if it were possible to ensure clinical practicability.

## Summary

Different kinds of laser-tissue interactions have been tested as to their feasibility in surgery. Some applications, especially the thermal interactions of tumor destruction, have become standard procedures; others are becoming indispensable in difficult cases, such as laser-induced shock wave lithotripsy as a treatment for those calculi that elude extracorporeal shock wave lithotripsy.

Further applications, especially tumor diagnosis by laser spectroscopy or holography, are still in their infancy, and the scope of future developments cannot be estimated at present; however, it is a fact that interest in lasers has brought us a big step forward in medical care, and the success we have had justifies our continuing research.

## References

1. Beisland HO (1985) Neodymium: laser irradiation of urinary bladder tumors. Urology 25:559–563
2. Beisland HO, Sander S (1986) Experience with treatment of localized prostatic carcinoma using the Nd:YAG laser. Eur Urol 12 [Suppl 1]:37
3. Böwering R, Hofstetter A, Keiditsch E, Frank F (1977) Irradiation of prostatic carcinoma by Nd:YAG laser. Optics Photonics Appl Med 211:16
4. Hofstetter A, Frank F (1979) Der Nd :YAG Laser in der Urologie. Editiones Roche, Basel
5. Le Guillou M, Ferriere JM, Piechaud T, Gaston R, Brucher P (1988) Application du laser en urologie.
6. Malloy T (1986) Neodymium-YAG laser in transitional cell cancer of the bladder with emphasis on outpatient potential. Eur Urol 12 [Suppl 1]:25–27
7. Meier U, Hofstetter A, Pflüger H (1985) Effects of intravesical instillation of mitomycin after endoscopic treatment with TUR or laser on recurrence rate of bladder tumors. Twentieth Congr Intern Soc Urology, June 23–28, Vienna
8. Rothauge CF (1980) Ureteroscopic recanalization of urethra stenosis using argon laser. Urology 16:158
9. Rothenberger KH (1986) Value of the Nd:YAG laser in the therapy of penile carcinoma. Eur Urol 12 [Suppl 1]:34
10. Shanberg AM, Baghdassarian R, Tansey LA (1983) Treatment of interstitial cystitis with the Nd:YAG laser. J Urol 184: 885

# Future Applications of Lasers in Gynecology and Reproduction

Y. Tadir [1,2], B. Fisch, B. J. Tromberg, J. Ovadia, and M. W. Berns

## Introduction

When attempting to predict future developments, one should bear in mind the ancient Talmudic proverb: "Since the temple was destroyed, prophecy has been taken from prophets and given to fools and children" [3]. However, in an era when embryos are frozen in "banks" and lasers are "fired" into the pelvis to reopen blocked fallopian tubes, it appears that the present can actually be regarded as "the future", and further clinical developments can be extrapolated from recent advances in the basic sciences.

After its introduction, conventional surgery underwent no fundamental changes until the middle of the twentieth century. At that time, scissors and scalpel were still the main surgical instruments in use. Recent advances in endoscopic technology that enable detailed visualization of almost every cavity in the human body and the adaptation of various lasers with suitable delivery systems offer an optimal surgical/therapeutic combination. The developments have paralleled rapid progress in other innovative imaging technologies such as ultrasonography (US), computerized tomography (CT), and magnetic resonance imaging (MRI) – all enabling early, precise, and noninvasive detection of pathological lesions. The combination of these diagnostic and therapeutic modalities demands considerable flexibility in adapting new treatment protocols for daily clinical purposes. The introduction of in vitro fertilization (IVF) as a routine clinical service has similarly provoked the development of several other therapeutic modalities, generally defined as artificial reproductive technologies (ART). As a result, the clinician, as well as the patient, has to face a situation where different protocols may be offered to treat the same pathological condition, and, therefore, an individualized approach is sometimes mandatory [32]. The availability of these new technologies, in turn, underscores the importance of a well-informed approach to decision-making in gynecology and reproductive medicine.

## Laser-tissue Interaction

The effect of the laser beam on tissue is primarily thermal, since the absorbed photons can be converted into heat. Thermodynamic calculations indicate that the radiant energy density re-

[1] Beckman Laser Institute and Medical Clinic, 1002 Health Sciences Road East, Irvine, CA-92715, USA.
[2] Sackler School of Medicine, Beilinson Medical Center, Tel-Aviv University, Petah-Tiqva, 49100, Israel.

Lasers in Gynecology
Edited by G. Bastert and D. Wallwiener

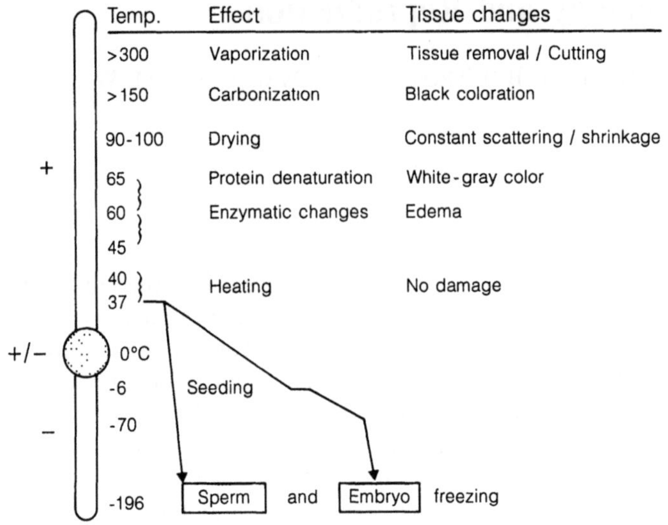


**Fig. 1.** Range of temperatures used in assisted reproductive technologies (ART) and their related effects

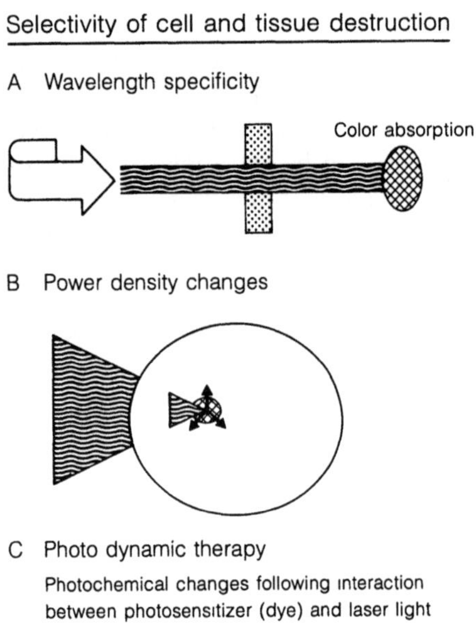


**Fig. 2.** Selectivity of tissue, cell, and subcellular organelles can be achieved in several ways: *A*, color specificity of target; *B*, spot diameter and power density changes; *C*, photodynamic therapy

quired to raise the temperature of living tissue from 37 °C to 150 °C is roughly 488 J/cm$^3$ and approximately 3200 J/cm$^3$ is required to vaporize a given volume element of tissue [2]. Of course, these values are dependent on the wavelength of light, the tissue absorption coefficient (at the incident wavelength), and the temporal characteristics of the laser (i.e., whether a pulsed- or continuous-wave laser is used). Although the heating effect may be harmful if not carefully monitored, it can be beneficial to several aspects of reproductive surgery and ART when properly applied (Fig. 1).

In fact, selective cell destruction is possible with careful choice of laser parameters which permit targeting of pigmented, structurally defined tissue types (Fig. 2). Similar effects can be achieved via photochemical mechanisms discussed below.

The carbon dioxide ($CO_2$) laser, operating at 10.6 μm in wavelength, is the

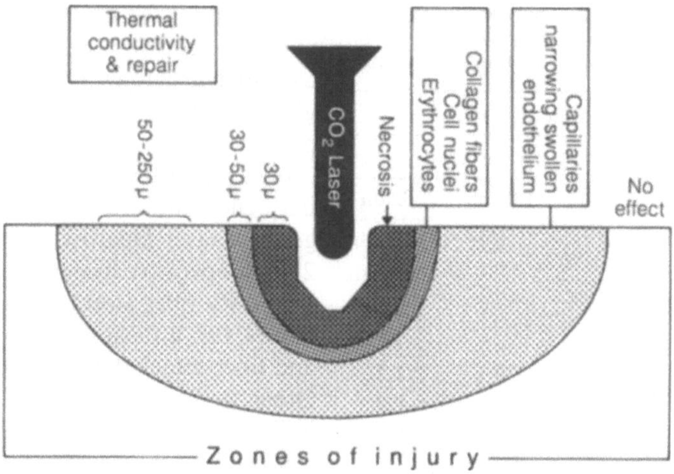


**Fig. 3.** Electron-microscopic-based study on skin and mucous membrane injury following $CO_2$ laser application. (From Ben-Bassat et al. [7] with permission)

first and most commonly used laser in gynecology [6,21]. The 10.6 µm laser light is strongly absorbed by water, thus allowing vaporization and precise incision of tissues with minimal thermal damage to adjacent cells (Fig. 3). This is especially relevant in reconstructive surgery, where tissue preservation is essential. Several other lasers have become available for investigational and clinical purposes. These include the neodymium: yttrium-aluminum-garnet [Nd:YAG] laser which operates at the fundamental wavelengths of 1.06 µm [25] and 1.32 µm [18]; the frequency-doubled Nd:YAG operating at 532 nm (typically a 1.06-µm Nd:YAG used in conjunction with a potassium-titanyl-phosphate (KTP) frequency-doubling crystal [15]); the argon-ion laser, with primary lines at 488 nm and 514 nm [24]; and the 2.1-µm holmium laser [22]. Each of these lasers has a unique effect on tissue which depends on a number of factors, including the "color" of the tissue, its water content, the power density, wavelength, and pulse structure of the laser beam.

Transmission of laser beams through fluids – e.g., distending, flushing, and contrast media – is of crucial importance when endoscopically or fluoroscopically monitored invasive procedures are performed. Transmission of the Nd:YAG wavelengths 1.06 µm and 1.32 µm through various fluids has been previously investigated [34] (Table 1). For most of the water-soluble solutions studied, the absorption coefficient (α) at both wavelengths is similar to that of water. There is almost no absorption of 1.32-µm light by Lipiodol, and the absorption by Urografin and Endografin is about half that of water. These data confirm that it is possible to perform invasive procedures by delivering laser light through liquid media guided by X-ray imaging techniques. The precise choice of media depends upon the specific medical indication and the absorption coefficient of the solution. Similar studies

**Table 1.** Absorption coefficient (α) of Nd:YAG lasers operating at wavelengths of 1.06 μm and 1.32 μm through various fluids

Absorption coefficient ($\alpha\,cm^{-1}$)

| Material | Wavelength | |
|---|---|---|
| | 1.06μm | 1.32μm |
| Water | 0.1 | ≧ 1.2 |
| Lactated Ringer | 0.1 | ≧ 1.2 |
| 5% Dextrose in water | 0.1 | ≧ 1.2 |
| 10% Dextrose in water | 0.1 | ≧ 1.2 |
| 0.9% Sodium chloride | 0.1 | ≧ 1.2 |
| Haemaccel | 0.1 | ≧ 1.2 |
| Urografin 60% | ≦ 0.1 | ≧ 0.6 |
| Endografin 70% | ≦ 0.1 | ≧ 0.6 |
| Lipiodol | ≦ 0.1 | < 0.01 |

with other lasers are required to broaden the spectrum of indications and develop suitable delivery systems.

## Delivery Systems

There is no doubt that future surgery will increasingly depend upon endoscopic equipment, video monitoring, and documentation. Small lesions, diagnosed at an early stage by specific blood markers, will accurately be located and treated endoscopically with optical fibers, flexible scissors, and biopsy attachments. The combination of laser light with endoscopic equipment is ideal, especially whenever the effect on tissue can be predetermined, adjusted by color and wavelength, transmitted via fluid, or targeted with drugs. The development of new delivery systems is a product of collaboration between surgeons, engineers, physicists, chemists, and other technical staff. Many new accessories have appeared on the market during the past decade and many others will surely be developed in the near future.

Several different laser endoscopes and accessories are currently being used in gynecology. These include: single- and double-puncture $CO_2$ laser laparoscopes [10, 36]; rigid $CO_2$ laser probes [33]; fiber guides for Nd:YAG, frequency-doubled Nd:YAG, and argon-ion laser laparoscopes [15, 18, 24, 25]; an operative hysteroscope with an Nd:YAG fiber steerable delivery system [4, 19]; a flexible fiberoscope [12]; and fiber-attachable sapphire tips [13]. Other accessories include: a flexible hollow wave guide for $CO_2$ lasers [5], a rigid hollow wave guide for $CO_2$ lasers, which holds a miniature zinc selenide lens (the lens allows for focusing, defocusing, and creating an "optical backstop" within short distances beyond the focal plane [35]); and contact (nonsapphire) scalped fibers for the Nd:YAG that convert the coagulating laser effect into a cutting device. These fibers, 600–800 μm in diameter, with conical or hemispherical tips, convert roughly half of the YAG energy into heat. The remaining light energy is distributed over the tip surface in a diffuse fashion. As a result, the tip immediately heats to tissue cutting temperatures and the coagulation side effect occurs at a significantly reduced depth compared with that of the noncontact YAG. The fiber does not cauterize tissue unless it is actually in contact, an effect which is especially important for laparoscopic use (Y. Tadir, unpublished data).

The original $CO_2$ laser laparoscopes [10, 36] were further modified by coupling to an articulating arm. These devices employ a joystick control for manual adjustment of the laser beam and an interchangeable lens housing to match the beam focal length to the length of the laparoscope. Some laser laparoscopes contain a continuous variable defocusing (CVD) lens that enlarges the beam spot diameter, thereby changing power density, temperature, and, consequently, surgical options (Fig. 1). Increased sophistication in laser technology, including the availability of tunable dye, diode, and excimer lasers, promises to expand the scope of this fascinating field.

## Laser Therapy with Complementary Medical Treatment

Laser effects can be augmented, in selected cases, by administering specific drugs before exposure to laser energy. The improved efficacy results either from hormonal suppression of the lesion's size or from photochemical cell disruption. Photochemical effects can be initiated by combining photosensitizing compounds with laser excitation in a process known as *photodynamic therapy*, or PDT [41].

PDT has been successfully applied to the selective destruction of malignant tumors [17] and hormone-dependent tissue [30]. Treatment is initiated by intravenous administration of a photosensitizing drug, and the therapeutic activity is triggered by laser irradiation. When laser light of sufficient energy and appropriate wavelength interacts with the photosensitizer, a cytotoxic species, singlet molecular oxygen, is generated. Singlet oxygen, in turn, rapidly reacts with nearby cells by vigorously oxidizing molecules which participate in cell structure and function. This selective "photodestruction" of crucial organelles in tumor cells and vasculature ultimately causes tumor necrosis [11]. It has been proposed that tumor vasculature may be the primary site of photodynamic activity. Based on this assumption, we evaluated the potential applicability of PDT to vascularized, hormone-dependent, endometrial tissue which has some similarity to neoplasm. This feature may be exploited during treatment of several pathological conditions including menorrhagia, endometriosis, and endometrial cancer. By increasing the concentration of photosensitizer in "target" tissues, significant therapeutic improvements can be achieved [31]. Several experimental studies suggest that endometrial ablation [31] and treatment of endometriosis [27] can be achieved using these principles.

Currently, non-PDT laser-induced tissue ablation is routinely performed in order to treat menorrhagia [19] and endometriosis [14, 23, 28]. It is our estimation that photodynamic treatment of these conditions may ultimately prove to be superior to conventional therapy. This is because PDT exhibits excellent "per photon" destructive capabilities (i.e., significantly lower laser powers are required with PDT) and greater tissue selectivity, since the combination of light, photosensitizer, and hormone can be manipulated in order to elicit an optimal response. Moreover, simple outpatient treatment modalities for selective ablation of endometrial tissue may be developed.

The combination of danazol or gonadotropin-releasing hormone (GnRH)-analogue administration with endometrial ablation has been suggested by several authors. Both medications suppress endometrial proliferation prior to hysteroscopic Nd:YAG application. This promotes laser beam penetration into inner layers and therefore enhances fibrosis. Based on the same concept, the administration of these drugs prior to laser photovaporization of endometriosis is a promising treatment. Similarly, the size-reduzing effect of the GnRH agonists on endometriotic implants [16] and uterine fibroids may well serve, in combination with laser ablation, to allow a definitive surgical treatment for otherwise endoscopically inoperable conditions.

## From Pelvic Organ Laser Surgery to Cellular Laser "Microsurgery"

Delivery systems in endoscopic and "conventional" laser microsurgery facilitate treatment of a variety of macroscopic lesions simply by changing the laser beam wavelength and spot size (in the range of 300–1000 µm). In contrast, laser microscope technology generally employs a tunable dye laser microbeam (wavelength ranges from 217 to 800 nm) with spot diameters of 1–5 µm, thus permitting selective alteration of single living cells and organelles [8]. Laser microscopes can also be used to perform optical micromanipulation or "optical trapping" (see below) [1]. Comparative parameters for the application of lasers to reconstructive pelvic surgery and single cell surgery are summarized in Fig. 4.

Clearly, the rapid proliferation of infertility diagnostics and therapeutics requires development of a variety of new technologies. Optical micromanipulation procedures may prove to be particularly useful in single cell interventions. As a result, we have exposed human sperm, oocytes of various species, and other living cells to laser microbeams in order to evaluate the efficacy of this technique. Some of these results are reported below.

### Micromanipulation of Human Sperm with Laser Optical Trap

Flagellar movement is a fundamental expression of the viability of sperm and is essential for its reproductive capabilities. In order to develop a microma-

**Fig. 4.** Laser parameters used in pelvic reconstructive surgery and cell microsurgery.*, Typical wavelengths used in our studies; **, peak powers, pulsed mode (power varies according to wavelength and application); ***, power range in the optical trap

| | Reconstructive pelvic surgery and laparoscopy | Cell microsurgery and optical trap |
|---|---|---|
| Spot size Ø Micrometer | 300-1000 µm | 1-5 µm |
| Wavelength Nanometer | $CO_2$ 10,600 nm<br>Nd:YAG 1,064 nm<br>Frequency-doubled Nd:YAG (KTP) 532 nm<br>Argon 515 nm | 1064 nm<br>532 nm<br>366 nm<br>355 nm<br>266 nm * |
| Power Watt & Milliwatt | 5-30 W Continuous wave | $10^2$-$10^6$ W** @ 532 nm<br>Pulse width 14 ns<br>10-1000 mW *** |

nipulation device for testing sperm force or assisting sperm penetration, we measured the influence of radiation pressure forces (an optical trap) on living human sperm [37].

A single-beam gradient-force optical trap consists of a Gaussian energy profile laser beam focused (by a microscope objective) to a spot smaller than the particle being trapped. It is essential that the laser light is not absorbed by the sample. Since cells absorb minimally in the near infrared, a continuous-wave Nd:YAG laser operating at 1.06 μm was used to construct the trap. The trap confines and centers the cell just below the focal plane of the laser beam.

Sperm was placed on a motorized X-Y stage that allows positioning in the optical trap. Real-time viewing of the sperm in the trap was performed using a video camera colinear with the path of the trapping beam. The recorded video images were analyzed using a computer-assisted image processing system. This system was used to measure specific characteristics in sperm movement before and after exposure to the optical trap. By assuming that the trap (laser) power is proportional to the force generated by the sperm, measurements of the relative force generated by single spermatozoa were performed [38]. Sperm swimming patterns were classified as straight and zig-zag motile. Linear velocities were subdivided into three groups: (a) slow (≦20 μm/s), (b) medium (21–40 μm/s) and (c) fast (≧41 μm/s) motile. The power measurements were compared with the linear velocity and the motility pattern. There was a correlation between velocity and power for the entire sperm population. Zig zag motile sperm required significantly higher mean trapping powers than straight motile sperm with similar mean linear velocities. Trap power requirements at various time intervals following ejaculation were also determined. These results indicate that, in addition to its power as a micromanipulation device, the optical trap may contribute to better understanding of sperm physiology and locomotion.

## Laser Micromanipulation of Oocytes

Mammalian fertilization is an inefficient process. Only one sperm of the millions initially deposited in the female reproductive tract penetrates the egg. In vitro fertilization may provide a means of overcoming infertility in oligospermic men. However, the ratio of sperm to eggs needed for successful fertilization is several thousand to one. By drilling a small hole through the zona pellucida of the mouse oocytes with acidified Tyrode solution [20] it was possible to achieve IVF with only 1% of the sperm concentration that would otherwise be required. Similar aptitude has been demonstrated in human oocytes following mechanical partial zona dissection [26]. In order to assess the potential of laser microbeams for controlled, "noncontact" laser zona drilling (LZD), 420 oocytes (hamster, mouse, and human oocytes) were irradiated under varying wavelength (532 nm, 366 nm, 355 nm, and 266 nm), power (5–9 mJ, 14 ns pulse duration), and spot size (1–5 μm diameter spots) conditions [39]. Light and scanning electron microscopy confirmed the laser effects. Well-shaped, controlled-diameter craters in the zona pellucida were obtained. The possible pathophysiological effects and applications of LZD are currently being assessed.

### Intracellular Application of Laser Beams

One of the key questions in cell biology centers on the forces involved in the movement of chromosomes during mitosis. Various models have been proposed to explain how forces are exerted on chromosomes, but they have not been adequately tested due to the difficulty involved in the measurement of forces within a cell. An optical trap provides a new approach to the measurement of intracellular forces without physically touching the cell with a transducer.

Berns et al. [9] used an optical trap similar to that described earlier to alter in vitro the movement of chromosomes in the prometaphase and metaphase of *Potorus tridactylis* kidney (Pt K 2) cells. Microdissection of selected nucleolar genetic regions of mitotic chromosomes and cytoplasmic organelles is also possible [8]. A combination of the single-beam gradient optical trap with a pulsed UV laser microbeam was used to induce fusion of one pair of cells [42].

Polyspermic fertilization (incidence rate approximately 4%–5%) is a potential complication encountered during human IVF. Rawlins et al. [29] reported identification of the supernumerary male pronucleus in three tripronuclear (3 PN) human zygotes, removal of one male pronucleus, and syngamy without subsequent cleavage of the resultant diploidized zygotes. We were able to damage extra pronculei using a subcellular laser microsurgery technique.

Nine 3 PN human oocytes were exposed to 5 mJ (10-ns pulses) of 532 nm and 355 nm laser light focused through a 100 X objective. The beam was directed towards one of the extra pronuclei. The high magnification and the marked changes in power density within short distances (see Fig. 2) allowed nontraumatic penetration through the zona pellucida and selective destruction of the pronucleus. Two of the nine fertilized ova exposed to the laser light were cleaved to the 6–8 cell stage and further processed for chromosomal analysis. Although technical difficulties limited our ability to determine whether selective inactivation of the extra *male* pronucleus was accomplished, our experiments demonstrate that laser-induced pronuclear inactivation is efficacious.

## Monitoring and Documentation

During the past 5 years there has been an increasing tendency to use videocameras [28] and computerized image processors [40] in research as well as in clinical facilities. Initially, surgeons viewed the operative field in endoscopic procedures through the eyepiece. Currently, however, operations are performed using a video monitor for guidance. This approach offers several advantages: (a) magnification of the target organ and the pathological tissue; (b) good depth perception; (c) stable position and less fatigue to the surgeon; (d) team work and better assistance; (e) documentation via recorded video files.

The increased market demand has stimulated the industry to further develop high-resolution light-weight endocameras and bright light sources. There is no doubt that future surgeons will perform most, if not all endoscopic operations with endocameras. One may even expect some legal demand for video tapes, at least for special procedures.

## Envoi

The fields of assisted reproduction and reconstructive gynecological surgery have developed rapidly in the past decade. Many questions are being raised and potential solutions are being suggested as a result of exposure to new technologies, some of which reside on investigative frontiers.

Current techniques of pre-embryo biopsy and DNA amplification have made it possible to determine precise chromosome-specific repeat sequences in a single cell. This combination enables preimplantation diagnosis of certain genetically transmitted diseases and will, undoubtedly, become a clinical service. This is the beginning of a new era in which geneticists and scientists will be able to add treatment to a field that was limited to counseling, and the term "invasive procedure" will also refer to human embryos at the preimplantation stage. Since many ethical, legal, and religious questions will be provoked by the practical availability of such novel "invasive procedures," we will be faced with the prodigious task of establishing the proper use of these technologies.

As mentioned in the introduction, "prophecy" has no place in a scientific manuscript. However, it seems as if the biblical quotation: "and there was light," is particularly relevant to the practice of contemporary laser biomedicine.

*Acknowledgment.* The authors express their thanks to WH. Wright, MSc., for his invaluable contribution in setting up the optical trap facility and his assistance in the design of these experiments.

## References

1. Ashkin A, Dziedzic JM (1987) Optical trapping and manipulation of viruses and bacteria. Science 235:1517
2. Ashamead AC (1987) Technological trends in $CO_2$ medical laser. In: Proc. International Society of Optical Engineering, (SPIE) 737, pp 67–74
3. Baba Batra 12-B (5OO AC) Talmud
4. Baggish MS (1988) New laser hysteroscope for Neodymium-YAG endometrial ablation. Lasers Surg Med 8:99–103
5. Baggish MS, El-Bakri MM (1986) A flexible $CO_2$ laser fiber for operative laparoscopy. Fertil Steril 46:16–20
6. Bellina JH (1974) Gynecology and the laser. Contemp Obstet Gynecol 4:24–29
7. Ben-Bassat M, Ben-Bassat M, Kaplan I (1976) A study of the ultrastructural features of the cut margin of skin and mucous membrane excised by carbon dioxide laser. J Surg Res 21:77–84
8. Berns MW, Edwards JJ, Strahs K et al (1981) Laser microsurgery in cell and development biology. Science 213:
9. Berns MW, Wright WH, Tromberg BJ et al (1989) Use of a laser-induced optical force trap to study chromosome movement on the mitotic spindle. Proc Natl Acad Sci 86:4539–4543
10. Bruhat M, Mage C, Manhes M (1979) Use of the $CO_2$ laser via laparoscopy. In: Kaplan I (ed) Laser Surgery III, Proceedings of the Third International Society for Laser Surgery. Tel-Aviv, Ot-Paz, pp 274–276
11. Chaudhuri K, Keck RW, Selman SH (1987) Morphologic changes of tumor microvasculature following hematoporphyrin derivative sensibilized photodynamic therapy. Photochem Photobiol 46:823–827
12. Cornier E (1986) J Gynecol Obstet Biol Reprod 15:661–664
13. Daikuzono N, Joffe SN (1985) Artificial sapphire probe for contact photocoagulation and tissue vaporization with the Nd:YAG laser. Med Instrum 19:173–178
14. Daniell JF (1984) Laparoscopic salpingostomy utilizing the $CO_2$ laser. Fertil Steril 42:815–823
15. Daniell JF, Miller W, Tosh R (1986) Initial evaluation of the use of potassium-titanylphosphate (KTP/532) laser in gynecologic laparoscopy. Fertil Steril 46:373–377

16. Donnez J, Lemaire M, Karman Y et al (1987) $CO_2$ laser laparoscopy in infertile women with endometriosis and women with adnexal adhesions. Fertil Steril 48: 390–393
17. Dougherty TJ (1987) Photosensitizers: therapy and detection of malignant tumors. Photochem Photobiol 45:879–889
18. Frank F, Beck OJ, Hessel et al (1987) Comparative investigations of the effect of the Neodymium:YAG laser at 1.06 microns and 1.32 microns on tissue. Lasers Surg Med 7:546–551
19. Goldrath M, Fuller T, Segal S (1981) Laser photovaporization of endometrium for the treatment of menorrhagia. Am J Obstet Gynecol 140:14–19
20. Gordon JW, Talansky BE (1986) Assisted fertilization by zona drilling: a mouse model for correction of oligospermia. J Exp Zool 239:347–354
21. Kaplan I, Goldman J (1973) The treatment of erosion of the uterine cervix by means of $CO_2$ laser. Obstet Gynecol 4:795–796
22. Kaplan I, Giler S, Sagie D et al (1987) The clinical potential of the Holium laser. Lasers Surg Med 7:207–209
23. Kelly RW, Roberts DK (1983) $CO_2$ laser laparoscopy: a potential alternative to danazol in the treatment of stage I and II endometriosis. J Reprod Med 28:638–640
24. Key WR, Dixon J (1983) Photocoagulation of endometriosis by the argon laser through the laparoscope. Obstet Gynecol 62:363–383
25. Lomano JM (1985) ND:YAG Laser applications in gynecology. In: Joffe SN, Ogurs Y (eds) Advances of ND:YAG Laser Surgery. Springer, Berlin Heidelberg New York, pp 201–207
26. Malter EH, Cohen J (1989) Partial zona dissection of the human oocyte: a nontraumatic method using micromanipulation to assist zona pellucida penetration. Fertil Steril 51:139–148
27. Manyak MJ, Nelson LM, Solomon D (1989) Photodynamic therapy of rabbit endometrium transplants: a model for treatment of endometriosis. Fertil Steril 52:140–145
28. Nezhat C (1986) Videolaseroscopy: A new modality for the treatment of diseases of the reproductive organs. Colposc Gynecol Laser Surg 2:221–224
29. Rawlins RG, Binor Z, Radwanska E et al (1988) Microsurgical enucleation of ripronuclear human zygotes. Fertil Steril 50:266–272
30. Schneider D, Schellhas HF, Wessler TA et al (1988) Hematoporphyrin derivative uptake in uteri of estrogen treated ovariectomized rats. Colposc Gynecol Laser Surg 4:67–71
31. Schneider D, Schellhas HF, Wessler TA et al (1988) Endometrial ablation by DHE phoatoradiation therapy in estrogen treated ovariectomized rats. Colposc Gynecol Laser Surg 4:73–77
32. Tadir Y, Fisch B, Pinkas H et al (1989) Ovadia individualized approach towards selection of ovum pick-up method for in vitro fertilization. J Gynecol Surg 5:133–143
33. Tadir Y, Kaplan I, Zukerman Z et al (1984) New instrumentation and technique for laparoscopic carbon dioxide laser operations: a preliminary report. Obstet Gynecol 63:582–585
34. Tadir Y, Karni Z, Fisch B et al (1988) $CO_2$ laser probe for operative laparoscopy. Colposc Gynecol Laser Surg 4:105–109
35. Tadir Y, Kami Z, Fisch B et al (1989) Tissue effects of a new laparoscopic carbon dioxide laser probe. Fertil Steril 51:1046–1049
36. Tadir Y, Ovadia J, Zukerman Z et al (1981) Laparoscopic application of $CO_2$ laser. In: Atsumi K, Nimsakul N (eds) Laser Tokyo 81. Tokyo, Inter Group Corp, pp 13–27
37. Tadir Y, Wright WH, Vafa O et al (1989) Micromanipulation of sperm by a laser generated optical trap. Fertil Steril 52:870–873
38. Tadir Y, Wright WH, Vafa O et al (1990) Force generated by human sperm correlated to velocity and determined using a laser trap. Fertil Steril 53:944–947
39. Tadir Y, Wright WH, Vafa O et al (1991) Micromanipulation of gametes using laser microbeams. Human reproduction 6:1011–1016
40. Walter RJ, Berns MW, Arditti J et al (1987) Computer-enhanced Video microscopy; Digitally processed microscope images can be produced in real time. Environ Exp Biol 27:177–184
41. Weishaupt KR, Gomer CJ, Dougherty TJ (1976) Identification of singlet oxygen as the cytotoxic agent in photoactivation of a murine tumor. Cancer Res 36:2326–2329
42. Wiegand-Steubing R, Cheng S, Wright WH et al (1991) Laser induced cell fusion in combination with optical tweezers: The laser cell fusion trap. Cytometry 12:505–510

# A Change in Fertility Surgery?

# A Change in Fertility Surgery?

D. Wallwiener [1], S. Rimbach, and G. Bastert

## Introduction

The expression "gynecological fertility surgery" is inseparable from operation techniques which are carried out endoscopically (laparoscopy/hysteroscopy) or by conventional microsurgery on the open abdomen. The aim of such surgical interventions is the re-establishment of reproductive functions in sterile or infertile women with minimal tissue traumatization.

The patients have a mechanical inhibition in either oocyte transport (pick-up mechanism, tubal transport), conception (tubal occlusion), embryo transport (blind loop syndrome of the tubal mucosa, extrauterine pregnancies), and/or nidation (uterine adhesions, submucous myoma, uterine septa).

Since the end of the past century, numerous and diverse attempts have been made to develop successful surgical methods to re-open occluded tubes in order to correct such mechanical sterility factors and thus make pregnancy possible [18].

The success rates remained dissatisfactory for decades. A decisive step in the right direction was, however, the introduction of microsurgery into gynecological fertility surgery. Gomel [7], Winston [21], and Scheidel and Hepp [20] were the pioneers in this field and furnished proof of the value of microsurgical tubal reconstruction with excellent clinical results.

Microsurgery enables improved hemostasis due to the exact localization of the smallest hemorrhagic sites. Furthermore, tissue trauma is minimized compared with that which can develop using forceps or clamps. Under ideal magnification, preparation adapted to the anatomic circumstances and a conservation of tubal mucosa is possible. Thus, the maintenance or re-establishment of the functional integrity of the microanatomical structures of the tubo-ovarian unit can be realized under the operating microscope.

Three reconstructive surgical methods can be utilized for the correction of postinflammatory or endometriotic tubal disorders:

- Adhesiolysis, including salpingolysis, ovariolysis, and fimbrioplasty as simple surgical or combined interventions
- The revision of partial or complete distal tubal occlusions
- The resection of segments of tubal occlusions with subsequent anastomosis

Another domain for gynecological microsurgery is re-anastomosis after artificial interruption of tubal continuity.

[1] Universitäts-Frauenklinik, Abt. für Geburtshilfe und Gynäkologie, Voßstr. 9, W-6900 Heidelberg, FRG.

Lasers in Gynecology
Edited by G. Bastert and D. Wallwiener

Excellent surgical results with success rates in excess of 70% have been reported [6, 11, 17, 22].

The introduction of microsurgery into gynecology has enabled optimum organ-preserving treatment of tubal pregnancies. Tissue-conserving organ-preserving microsurgical techniques are extremely important because of the increasing incidence of extrauterine pregnancies due to the increase in chlamydia infections as well as sterility after conventional, i.e., macrosurgical, techniques (70% of all cases, with a 30% [10] risk of extrauterine pregnancy recurrence). Tubal segment resection with unilateral or bilateral re-anastomosis is possible, as is salpingotomy with removal of the pregnancy and retention of tubal continuity.

A change in the operating technique has become evident in gynecological fertility surgery during the past few years; endoscopic techniques have become increasingly significant. Comparison of the current therapeutic concepts of leading groups with both conventional microsurgical techniques on the open abdomen and endoscopic know-how at their disposal reveals that the endoscopic procedure tends to dominate [1, 16, 19].

Furthermore, surgical hysteroscopical intervention is replacing metroplastic surgery of intrauterine disturbances such as septa, submucous myoma, and severe bleeding after permanent anticoagulation (extirpation of the endometrium) [2]. The trend towards surgical endoscopy has also been promoted by the development of new technical methods such as surgical videoscopy.

## The Trend Towards Surgical Endoscopy

Conventional microsurgery is still highly significant for gynecological fertility surgery. Good microsurgical refertilization results are sometimes hindered, however, by dissatisfactory microsurgical revision of postinflammatory tubal alterations. The results of surgical reconstruction of distal tubal occlusions are especially unfavorable [11].

The question then arises as to whether operative endoscopic methods lead to comparable or better results. Consideration of this question allows one to differentiate between reparable and irreparable tubal damage to avoid surgical intervention in the latter group, a challenge which has recently developed. This is relatively obvious, as endoscopic surgical techniques are far more reasonable than techniques necessitating a laparotomy; they not only involve less operative trauma; they are also less of a burden to the patient and thus not unreasonable even for diagnostic purposes.

The endoscopic operation technique is utilizable not only for the revision of postinflammatory distal tubal occlusions, but also for the treatment of adhesions, endometriosis, ovarian cysts, and tubal pregnancies, as well as for all pathological intrauterine factors [16].

In medicine one often encounters the phenomenon that the euphoric reaction of the users of new methods and techniques is completely incomprehensible to and condemned by those who are not yet able to assess the value of the new development. In fertility and sterility medicine the interest in microsurgery was replaced by euphoria for in-vitro fertilization with embryo transfer and intratubal gamete transfer. Experience has shown that each method has its particular indications, but also its limits. Even under ideal conditions pregnancy can be attained in only some 30% of cases within a collective, through embryo transfer or intratubal gamete transfer [8].

The following constellation is evident when one reviews the success rates of microsurgical tubal correction after postinflammatory distal tubal affection. A live birth rate of approximately 30% is attained after adhesiolyses and fimbrioplasty; this is accompanied by a 5% tubal pregnancy rate. The results after salpingostomy are notably worse: a pregnancy rate of 30% including a tubal pregnancy rate of 10% [3, 5, 8, 21, 22].

At this point the possibility of operative laparoscopy is worthy of consideration. A change in the technical procedure, in favor of surgical endoscopic intervention, is in progress in organ-preserving treatment of tubal pregnancy, comparable to that which has already been noted in the treatment of postinflammatory tubal affection with adhesions and distal tubal occlusions. This is the second major field for the utilization of endoscopic techniques, the endoscopic surgical possibilities having led to a continuous expansion in the set of indications for surgical laparoscopy [12–14, 16].

A survey of the results obtained by the leading groups discloses a set of indications for surgical endoscopy or microsurgery by abdominal section in fertility surgery (Table 1). The conventional microsurgical method under the operation microscope is, however, retained for surgical interventions which are difficult due to extensive adhesions or because of the danger of intestinal damage; the same applies to tubocornual anastomosis, and recently, with a few exceptions, to refertilization.

**Table 1.** Indications for surgical endoscopy/microsurgery

I. Operative laparoscopy
- adhesiolysis, salpingolysis, ovariolysis
- salpingostomy
- fimbrioplasty
- salpingotomy (extrauterine pregnancy)
- endometriosis
- ovarian cysts
- polycystic ovary disease

II. Surgical hysteroscopy
- uterine septa
- synechiae
- submucous myomas

III. Microsurgery
- resection of occluded tubal segments
- indications in I. above, when endoscopic surgery is not possible

But can this euphoria change the entire spectrum of fertility surgery? Is it even possible to submit fertility surgery as a whole to DeCherney's provocative thesis: "An idea becomes reality"?

> The future is boundless, but the new generation of endoscopic surgeons must gain experience and develop new techniques and instruments. The challenge is present: should, in those cases in which reconstructive sterility surgery cannot be performed endoscopically, a laparotomy ensue at all, or should the patient participate in an IVF program? Is this the deathblow to laparotomy for reconstructive surgery in the minor pelvis? Is this not a concept which has now become reality? [4]

If this question cannot be answered at present (perhaps it will never be answered completely), we must at least try to come to a satisfactory consensus regarding the large field of fertility surgery. As the experience of the past years has shown, however, DeCherney's thesis is a little exaggerated, the truth lying somewhere in between. Conventional microsurgery, laser endoscopic operations, and in-vitro fer-

**Table 2.** Conventional electrosurgical techniques in operative laparoscopy

| | Mode | Technical risks | Disadvantages |
|---|---|---|---|
| High-frequency technique | Mono- and bipolar high-frequency current | Transmission by tissue compartments and electrolytes | Difficult to restrict tissue destruction and sequestration with extensive tissue damage and danger of rebleeding |
| Endocoagulation technique | Weak current in separate circuit | None | Secondary sharp tissue preparation after coagulation necessary; thermally damaged zone equal to diameter of endocoagulator (minimum of 5 mm) |

tilization will all be retained for their particular set of indications, and they are thus of equal importance in the entire concept of fertility surgery.

On the basis of our experience, we think that the three main pillars of fertility surgery, namely operative laparoscopy, hysteroscopy, and conventional microsurgery, will continue to exist, each with its own spectrum of indications. The concept of fertility surgery based on these three pillars must (by all means) be integrated into the general concept of reproductive medicine, which is schematically presented in Table 1.

Though this general concept of reproductive medicine has been unanimously accepted, the choice of the surgical procedure for tubo-ovarian and/or uterine pathologies is still very controversial, with endoscopic, microsurgical, or even macrosurgical procedures being discussed. The next few years will show which of these will be accepted by the majority of fertility surgery centers. The surgical spectrum, furthermore, will have to be differentiated, optimizing and establishing new preparation techniques such as laser-assisted preparation, modified high-frequency electrosurgery, or adapted microsurgical techniques, a combination of vasoconstrictive and sharp preparation.

Thus far, electrosurgical, especially high-frequency, and endocoagulation techniques have dominated in operative laparoscopy. Both have been adequately described in the literature [12–14]. Table 2 shows the advantages and disadvantages of conventional electrosurgical techniques.

Diverse tests have been undertaken to modify or extend the operation procedure in gynecological fertility surgery [16]. Table 3 gives an overview of the current conventional preparation techniques.

All preparation techniques, however, either are subject to technical problems or involve a certain danger for the patient. Endoscopic application of unipolar high-frequency electrodes for tissue preparation has been replaced by the safer endocoagulation technique in most German-speaking countries [15]; forensic requirements have also had an influence. Extensive experimental studies on animals in whom secondary tissue sequestration was evident after unipolar high-frequency preparation have also strengthened this view [9].

In contrast, the endocoagulation technique is harmless for the patient because the current flows in a separate

**Table 3.** Conventional surgical procedures

| | |
|---|---|
| I. Thermal | |
| a) direct preparation | unipolar HF-electrode |
| b) indirect preparation (electrical hemostasis, sharp preparation) | 1. unipolar HF-coagulator<br>2. endocoagulation<br>3. bipolar coagulation |
| II. Nonthermal | |
| indirect preparation; sharp preparation subsequent to: | |
| a) endocoagulation | |
| b) application of vasoconstrictive substances (vasopressin) | |

HF, High-frequency

circuit within the instrument. The tissue destruction is extensive, however, as the point coagulator already has a diameter of 5 mm, resulting in tissue damage of at least 5 mm.

The endocoagulation technique is often replaced by endoscopic vasopressin application to attain a local ischemia and thus reduce the thermally damaged zone. Administration of this substance necessitates high-risk anesthesiological monitoring, however, as numerous cardiovascular risks are involved. Vasopressin application is furthermore contraindicated per se in cardiac risks; it is viewed extremely critically in most centers, as there have been reports of complications with a fatal outcome.

Two problems are evident from the above: The first is the need for the development of new preparation techniques. The second results from the fact that conventional preparation techniques, as mentioned in the introduction, are not basically satisfactory in all points. The disadvantages of conventional preparation techniques (high-frequency surgery and endocoagulation; endoscopic ligation and the application of vasoconstrictive substances) are contrasted by minimalized tissue lesions during laser preparation.

# Appendix: A Change in Fertility Surgery

The change in fertility surgery from microsurgery per laparotomy towards endoscopic applications, generally termed "minimal invasive surgery", must, however, be viewed critically. Despite the major advantages of laparoscopic or hysteroscopic procedures such as reduced traumatization of the abdominal wall, minimal adhesion induction, and short convales-

**Table 4.** Advantages

Minimized
- traumatization
- induction of adhesions
- convalescence

**Table 5.** Operative laparoscopy risks

| | |
|---|---|
| Adhesions: | organ lesions |
| Ectopic pregnancy: | residuals in case of organ preserving procedures |
| Ovarian Cysts/ Endometriomas: | risk of converting ovarian neoplasia or borderline tumors in case of organ preserving surgery → Ia → Ic (?) ← |

**Table 6.** Disadvantages

| |
|---|
| 1. Risk of suboptimal tissue handling (instruments vs. hand of the microsurgeon) |
| 2. Risk of suboptimal utilization of reconstructive possibilities (generalized adhesions) |
| 3. Maximal high tech equipment |

cence, certain risks and disadvantages are also inherent as it is the case with all new techniques in medicine, especially if minimal invasiveness is attempted. Also, the results do not yet compare favourably with those of established microsurgical procedures (Tables 4–6).

In any case the central aim of the surgeon must be a wellperformed operation to the benefit of the patient. An endoscopic reconstructive procedure must not be attempted by all means.

## References


1. Bruhat M, Mage G, Manhes M (1975) Use of the $CO_2$ laser via laparoscope. III. In: Kaplan I (ed) Proceedings of the 3rd International Society for Laser Surgery meeting, Tel Aviv, p 275
2. Cornier E (1982) La fibroscopie en gynecologie: la fibro-hysteroscopie et la fibrotuboscopie. Nouv Presse Med 11:2841–2843
3. Daniell JF, Diamond MP, McLaughlin DS, Martin DC, Feste J, Surrey MW, Friedman S, Vaughn WK (1986) Clinical results of terminal salpingotomy with the use of $CO_2$-laser: report of the intraabdominal laser study group. Fertil Steril 45:175–178
4. DeCherney AH (1987) Eine Idee wird Wirklichkeit. In: Medizin, Gesundheit, Politik – operative Pelviskopie in Klinik und Praxis. Mettler L, Weichert-von Hassel M (eds) Deutscher Ärzte-Verlag, Köln, pp 177–182
5. Diamond MP, Daniell JF, Martin DC, Feste J, Vaughn DS, McLaughlin DS (1984) Tubal patency and pelvic adhesions at early second-look laparoscopy following intra-abdominal use of the carbon dioxide laser: initial report of the intra-abdominal laser study group. Fertil Steril 42:717–723
6. Frantzen C (1981) Experimentelle Untersuchungen zur Physiologie der Eileitersegmente bei Kaninchen und deren Bedeutung für die mikrochirurgische Refertilisierung beim Menschen. Habilitationsschrift, Düsseldorf
7. Gomel V (1977) Tubal reanastomosis by microsurgery. Fertil Steril 1:59–65
8. Karbowski B, Schneider HPG (1986) Stellenwert mikrochirurgischer Operationen der tubarbedingten Sterilität-Abgrenzung zur extrakorporalen Befruchtung. Geburtshilfe Frauenheilkd 46:825–828
9. Riedel HH, Cordts-Kleinwort G, Semm K (1983) Tierexperimentelle, morphologische und endokrinologische Untersuchungen nach Anwendung verschiedener Koagulationstechniken. Zentralbl Gynäkol 105:665–670
10. Scheidel P (1982) Mikrochirurgie in der Gynäkologie – tierexperimentelle und klinische Untersuchungen zum Einsatz mikrochirurgischer Techniken in der rekonstruktiven Tubenchirurgie. Habilitationsschrift, Homburg/Saar
11. Schlösser HW, Frantzen C, Mansour N, Verhoeven HC (1983) Sterilisation-Refertilisierung. Geburtshilfe Frauenheilkd 43:213–217
12. Semm K (1976) Endocoagulation: a new field of endoscopic surgery. J Reprod Med 16:194–203
13. Semm K (1977) Pelviskopische Chirurgie in der Gynäkologie. Geburtshilfe Frauenheilkd 37:909–920
14. Semm K (1977) Die Mikrochirurgie in der Gynäkologie. Geburtshilfe Frauenheilkd 37:93–102
15. Semm K (1984) Operationslehre für endoskopische Abdominal-Chirurgie: operative Pelviskopie – operative Laparoskopie. Schattauer, Stuttgart
16. Semm K, Mettler L (1980) Technical progress in pelvic surgery via operative laparoscopy. Am J Obstet Gynecol 138:121–127
17. Silber SJ, Cohen R (1980) Microsurgical reversal of female sterilization: the role of tubal length. Fertil Steril 33:598–601
18. Swolin K (1967) 50 Fertilitätsoperationen. Acta Obstet Gynecol Scand 46:234–267

19. Tadir Y, Kaplan I, Zuckerman Z, Edelstein T, Ovadia J (1984) New instrumentation and technique for laparoscopic carbon dioxide laser operations: a preliminary report. Obstet Gynecol 63:582–585
20. Wiedeman R, Scheidel P, Hepp H (1990) Zur differenzierten Indikationsstellung der operativen Techniken in der Reproduktionsmedizin – Mikrochirurgie, IVF und ET, GIFT und TET. In: Scheidel P, Hepp H, DeCherney AH (eds) Operative Techniken der Reproduktionsmedizin. Urban & Schwarzenberg, München, pp 31–45
21. Winston RML (1980) Microsurgery of the fallopian tube: from fantasy to reality. Fertil Steril 34:521–525
22. Winston RML (1980) Reversal of tubal sterilization. Clin Obstet Gynecol 23: 1261–1266

# The Treatment of Infertile Patients with Tuboperitoneal Disease. Microsurgery Versus Endoscopy

M. K. Hohl [1]

Historically, surgical reconstruction of structural tubal and ovarian damage has been notoriously unsuccessful and frustrating. Inadequate surgical techniques and gross instruments and suture materials often led to greater postoperative iatrogenic damage, and the most extensive and complicated operations were followed by the most complicated and extensive postoperative tuboperitoneal adhesions [7]. The average pregnancy rate reported was 15.6% [12], surprisingly enough with no difference between surgical correction of proximal or distant tubal occlusion.

Postoperative conception rates were significantly enhanced by the introduction of microsurgical techniques in the mid 1960s. Microsurgical techniques aim to minimize trauma and injury to the delicate tissues surrounding and within the fallopian tube. The main principles are:

1. A minimal handling of tissues, preferably with atraumatic instruments
2. Prevention of tissue dehydration by means of careful irrigation with warm isotonic fluid
3. Meticulous preparation and unequivocal removal of all diseased tissues with the aid of an operation microscope
4. Precise hemostasis
5. Use of extremely fine, nonreactive suture material
6. Reconstruction of normal anatomy and tubo-ovarian relationship
7. Avoiding residual areas of raw surfaces and subsequent adhesions by careful reperitonization with or without the use of free peritoneal grafts

Early reports on the successful lysis of adhesions using the laparoscope [9] were followed in the mid 1980s by a wider application of minimal invasive surgery with and without the laser for the treatment of distant tubal occlusion [2, 4, 5] and even reanastomosis [6].

Nevertheless, structural changes in the fallopian tubes that commonly arise from infectious processes from previous operations remain incorrectable by surgical means. In addition, adhesion formation can occur after reproductive surgery despite meticulous technique and the use of adjuvant therapy. The nature, location, and extent of pelvic disease are important variables in predicting the chances of a favorable outcome with reconstructive surgery.

Since the 1978 report of the first human birth resulting from in vitro fertilization and embryo transfer (IVF-ET), the reproductive surgeon has been faced with new perspectives. Refine-

[1] Kantonsspital, Frauenklinik, 5404 Baden, Switzerland.

Lasers in Gynecology
Edited by G. Bastert and D. Wallwiener

ments in the technique of IVF-ET resulted in improved pregnancy rates. In the light of these achievements it has become necessary to reassess what constitutes the appropriate mode of therapy for women with anatomic abnormalities of the pelvis.

The choice between reconstructive tuboperitoneal surgery using the microscope or laparoscope and IVF is not always easily made. Although the different approaches try to achieve the same objective, namely the establishment of at least one pregnancy resulting in a live baby, their success is influenced by different variables. These variables must be assessed indvidually by the patient and physician before a decision is made as to which therapy is pursued. In many instances (e.g., reconstructive tubal surgery) therapy can be supplemented by IVF if the initial approach is not successful. A less common pattern of therapy is to perform tubal reconstructive surgery after IVF attempts have failed. Several issues must be addressed before a decision is made:

1. The extent of tubal and/or peritubal pathology
2. The cause of tubal disease
3. The presence of other infertility factors besides tubal disease that might strengthen the indication for IVF
4. The skill and experience of the physician in performing reconstructive tubal surgery and the viable pregnancy rate in his/her personal experience
5. The skill and efficiency of the IVF program as far as viable pregnancy rate is concerned
6. The age of the patient
7. The physical, emotional, and financial resources of the couple, which might influence their ability to undergo repeated cycles of IVF
8. The presence of risk factors that would make an abdominal surgical procedure hazardous or technically impossible
9. The time that elapses between surgery and establishment of pregnancy thereafter
10. The need for only one surgical procedure to facilitate several future conceptions versus the requirement of additional IVF treatment cycles in order to establish more than one pregnancy over time
11. The ethical acceptability of IVF for the couples concerned

Success rates which can be achieved by different surgical procedures determine, to a great extent, the choice between surgery and IVF. The comparison of success rates reported by different authors is not easy, however. Reasons for this include: heterogeneity of patient material, surgeons, and mode of reporting data. When reporting our own data we will adhere to the following principles:

1. All operations were performed by one surgeon (MKH).
2. Patients lost to follow-up are reported as failures.
3. Only patients with bilateral disease or disease on the single remaining tube were considered.
4. Cumulative pregnancy rates were calculated using the life table method.

## Reversal of Tubal Sterilization

Microsurgical tubal anastomosis for the reversal of tubal sterilization has proven over the years to be highly suc-

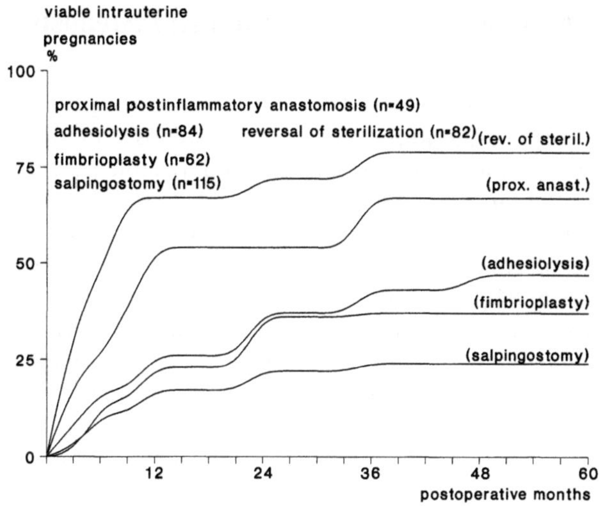


**Fig. 1.** Cumulative probability of viable intrauterine pregnancy after microsurgery

cessful. In our own series comprising 82 patients the cumulative probability of achieving a viable intrauterine pregnancy was 68% after 1 year, 78% after 2 years, and 82% after 3 years (Fig. 1). Three factors determined the outcome:

1. Length of the tube at the end of the operation (Table 1)
2. The patient's age (Table 2)
3. Ancillary fertility factors such as male subfertility and ovulatory dysfunction [3]

In patients under the age of 40 with at least one tube longer than 6 cm, a very high success rate of 90% was achieved. On the other hand, first attempts to reverse tubal sterilization by laparoscopy have been disappointing [6]. Microsurgical anastomosis should therefore be recommended to patients who have undergone previous sterilization procedures. In cases where the fimbriae have been excised, where the residual tubes are short (< 4 cm), or where important additional infertility factors are present IVF-ET should be considered.

**Table 1.** Reversal of tubal sterilization: influence of tubal length

| Length | Patients (*n*) | Viable pregnancies (%) |
|---|---|---|
| < 4 cm | 1 | 0 |
| 4–6 cm | 29 | 18 (62) |
| > 6 cm | 52 | 41 (79) |
| Total | 82 | 59 (72) |

**Table 2.** Reversal of tubal sterilization: influence of patient's age

| Age (years) | Patients (*n*) | Viable pregnancies (%) |
|---|---|---|
| Up to 30 | 15 | 14 (93) |
| 31–35 | 35 | 24 (69) |
| 36–40 | 28 | 20 (71) |
| Over 40 | 4 | 1 (25) |
| Total | 82 | 59 (72) |

## Adnexal Adhesions

The highest success rate achieved in microsurgical lysis of adhesions occurs when the previous inflammatory event did not involve the fimbriae. Associated fimbrial damage results in a marked fall in the rate of conception. In our unselected series of 84 patients (including patients with abnormal fimbriae) the viable pregnancy rate was 24% after 1 year, 42% after 2 years, 48% after 3 years, 51% after 4 years, and 57% after 5 years (Fig. 1). Microsurgical salpingolysis in patients with grossly normal fimbriae resulted in a success rate of 55% within the first year and 74% after 3 years [11]. Four percent of our patients experienced an ectopic pregnancy, which is similar to the 3.5% reported in a recent series [8]. Recently, similar results have been published for the surgical treatment of adhesions via laser laparoscopy [2,4,5]. For most patients with mild and moderate degrees of adhesions surgical lysis via laser laparoscopy is the treatment of choice because it is a less invasive procedure. However, even advocates of laser laparoscopy admit that there is a definite place for microsurgery, especially in patients with dense and fibrous adhesions between the fimbriae and the ovary [5] or between the ovary and the lateral pelvic wall. These favorable results support a primary surgical therapy for patients with adhesive disease not associated with intrinsic damage to the structure of the tube.

## Fimbrial Stenosis

After microsurgical correction of fimbrial stenosis in 62 patients we obtained a viable pregnancy rate of 25% after 1 year, 31% after 2 years, and 46% after 3 years (Fig. 1). Eleven percent of patients experienced an ectopic pregnancy. Good results have also been reported with $CO_2$ laser laparoscopy [4]. Comparison of results is difficult, however, due to the heterogeneity of patients. Hence, IVF should be considered for younger patients only when surgery has failed after 2 or even 3 years. In women over 35 years of age either IVF or surgical reconstruction could be recommended as initial therapy.

## Distal Tubal Obstruction

Distal tubal occlusion is accompanied by the partial destruction of the endosalpinx with deciliation and loss of steroid receptors. In some instances this is even associated with more or less intense fibrosis of the muscular wall. Because of these irreversible changes, restoration of tubal patency in conjugation with adhesiolysis results in a lower overall success rate. The average reported pregnancy rate in microsurgical series is 21% [10]. In our own unselected series (115 patients including cases with bad prognostic features) we observed viable pregnancies in 15% of cases after 1 year, 21% after 2 years, and 23% after 3 years. The potential for a favorable outcome is influenced by the extent of associated adhesions, amount of endosalpingeal damage, and thickness of the tubal wall. Boer-Meisel et al. [1] reported a 59% term pregnancy rate when they observed a macroscopically normal endosalpinx, thin tubal wall, and few adhesions. However, only 25% of patients fell into this favorable prognostic category. The pregnancy rate fell to 3% in the 34% of patients who had dense peritubal adhesions and a thickened tubal wall. Laparoscopic salpingostomy is more complicated than the microsurgical procedure, and its use is still controversial. Future experience will have to show which cases are best treated by microsurgery and which by laparoscopy. At the present time, $CO_2$ laser salpingostomy seems to be a good alternative for patients with large thin-walled hydrosalpinx not densely adherent to the ovary. Despite the relatively low mean success rate, careful patient selection for salpingostomy may provide an excellent therapeutic response. Patients over 35 years or those belonging to the unfavorable prognostic group may be more appropriately treated by IVF.

## Proximal Postinflammatory Tubal Occlusion

Many diverse disease processes lead to postinflammatory proximal tubal occlusion. The preferred surgical correction of intramural and proximal obstruction is resection and tubocornual anastomosis. Microsurgical cornual anastomosis preserves the tubal blood supply and integrity of the uterine musculature much better than an implantation procedure. In our own series of 49 patients with pure proximal tubal occlusion we observed a viable pregnancy rate of 48% after 1, 59% after 2, and 71% after 3 years (Fig. 1). However, the extent and etiology of the obstructive disease process determines the outcome to a large extent. Negative prognostic factors include presence of extensive nodular disease (salpingitis isthmica nodosa), chronic inflammation, and bipolar disease (proximal and distal occlusion). In these instances our own experience and that of others has consistently shown a viable pregnancy rate of less than 10%. The performance of a tubocornual anastomosis requires considerable expertise because of the relative inaccessability of the tissues being dissected and the small diameter in the intramural fallopian tube. Hence, reconstructive microsurgery remains the treatment of choice in selected patients with cornual occlusion, and IVF is reserved for those patients with negative prognostic criteria.

## Conclusion

When we keep in mind the aim of all our endeavors, i.e., to overcome an individual patient's tuboperitoneal infertility, the different therapeutic options (microsurgery, endoscopic surgery, IVF-ET) should not be looked at as competitive methods. Each of them has its place in a multimodal strategy to overcome tuboperitoneal infertility.

Microsurgery remains the treatment of choice for most patients desiring a reversal of sterilization. The same is true for patients with proximal postinflammatory tubal obstruction, provided that negative prognostic criteria are taken into consideration.

Reconstructive surgery (microsurgery or operative laparoscopy) provides favorable conception rates in select cases of adhesive disease and distal tubal obstruction.

Surgery requires a single period of days (operative laparoscopy) or 2–4 weeks (microsurgery) for recuperation. The cumulative time commitment necessary for participation in an IVF program, particularly if the patient undergoes multiple treatment cycles or if she is required to travel long distances to receive treatment, may be more disruptive to her life-style and emotionally more consuming. The expense of surgery may be less than that of repeated attempts at IVF.

Nevertheless, the conception rate per menstrual cycle is generally higher with IVF-ET than that achieved by surgery. This is of special importance for patients over 35 years of age, since fecundity declines markedly after this time. Ultrasonically guided follicular aspiration during IVF is less invasive and less hazardous than laparoscopy or laparotomy. The chance of an ectopic pregnancy is lower through IVF-ET than with reproductive surgery.

In conclusion, each patient should be provided with a realistic appraisal of her potential for a successful pregnancy based on the statistics of the individual surgeon or IVF program under consideration. Only then can she make an informed decision regarding her care.

## References


1. Boer-Meisel ME, Te Velde ER, Habbema JDF, Karduen JWPF (1986) Predicting the pregnancy outcome in patients treated for hydrosalpinx: a prospective study. Fertil Steril 45:23
2. Bruhat MA, Wattiez A, Mage G, Pouly JL, Canis M (1989) $CO_2$ Laser laparoscopy. Clin Obstet Gynaecol 3:487
3. DeCherney AH, Mezer HC, Naftolin F (1983) Analysis of failure of microsurgical anastomosis after mid-segment, non-coagulation tubal ligation. Fertil Steril 39:618
4. Donnez J, Nisolle M (1989) $CO_2$ Laser laparoscopic surgery. Clin Obstet Gynaecol 3:525
5. Donnez J, Nisolle M, Casanas-Roux F (1989) $CO_2$ Laser laparoscopy in tubal infertility. In: Donnez J (ed) Laser operative laparoscopy and hysteroscopy. Nauwelaerts, Leuven, pp 161
6. Gauwerky, Wallwiener D, Human Reprod
7. Grant A (1971) Infertility surgery of the oviduct. Fertil Steril 22:496
8. Lavy G, Diamond MP, De Cherney AH (1987) Ectopic pregnancy: its relationship to tubal reconstructive surgery. Fertil Steril 47:543
9. Madelenat P, Palmer R (1973) Etude critique des libérations per-coelioscopiques des adhérences péri-annexielles. J Gynecol Obstet Biol Reprod (Paris) 8:347
10. Maran R, Quagliarello J (1988) Distal tubal occlusion: microsurgery versus in vitro fertilization – a review. Int J Fertil 33:107
11. Schoysman R (1984) Tubal microsurgery versus in vitro fertilization. Acta Eur Fertil 15:5
12. Siegler AM, Hellman LM (1956) Tubal plastic surgery. Report of a survey. Fertil Steril 7:170

# Microsurgical Laser Tubal Reconstruction

# Proximal Tubal Pathology, Segmental Occlusion, Refertilization

D. WALLWIENER [1], D. POLLMANN, S. RIMBACH, W. STOLZ, and G. BASTERT

Though to date there is a clear trend in fertility surgery away from open microsurgery and toward pelviscopic procedures, for certain indications an open approach is still to be preferred.

For repair of proximal tubal pathology, microsurgical resection of the occluded segment with subsequent anastomosis is still the treatment of choice (see Hohl, this volume). Postinflammatory or iatrogenic (poststerilization) segmental occlusions are still the province of open reconstructive surgery.

Due to its wavelength-specific tissue effects and excellent cutting properties, the carbon dioxide laser has been successfully employed for tubal preparation. By exact focusing, finest preparation of tubal tissue can be performed under microscopic view. Compared with HF electrosurgery, the zones of thermal damage are less marked [10]. Concomitant laser coagulation of finest blood vessels reduces parenchymatous hemorrhage. Additional local application of vasopressin for preventive hemostasis is useful. Since minimal thermal damage is the ultimate goal, however, laser sealing of bleeding vessels is limited. Neither in the continuous-wave nor in the superpulsed mode is the carbon dioxide laser able to coagulate vessels larger than 0.5 mm in diameter. Bleedings from mesosalpingeal vessels or vessels of the ramus tubarius must be stopped by electrocoagulation.

The carbon dioxide laser light is applied either by manipulating the handpiece directly under the microscope, or by coupling the laser to the microscope and carrying out preparation with a joystick.

After extensive animal experiments had been done, the carbon dioxide superpulsed laser was applied for microsurgical reversal of sterilization, since it had proved to be an excellent instrument for preparation of tubal tissue and caused minimal thermal effects on subjacent and adjacent tissue and no tissue loss. Additional concomitant coagulation resulted in reduced damage to the mucosa. All tubal layers can be transected by one clean laser cut, without a change of instruments being necessary. It is, furthermore, also entirely unproblematic to guide a discoid resection through all tubal layers until an intact tubal lumen is attained. Bipolar high-frequency coagulation is necessary in laser surgery as in preparation with the microelectrode, when transecting the vascular anastomosis between the tubal and ovarian branches.

The question often arises as to how far the microsurgical preparation method utilized influences the result of tubal anastomosis. The success of surgical intervention depends first and foremost on the severity of the tubal

[1] Universitäts-Frauenklinik, Abt. für Geburtshilfe und Gynäkologie, Voßstr. 9, W-6900 Heidelberg, FRG.

Lasers in Gynecology
Edited by G. Bastert and D. Wallwiener

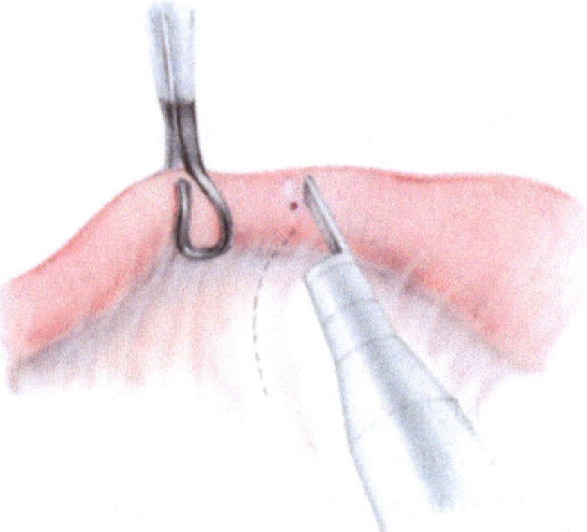

a

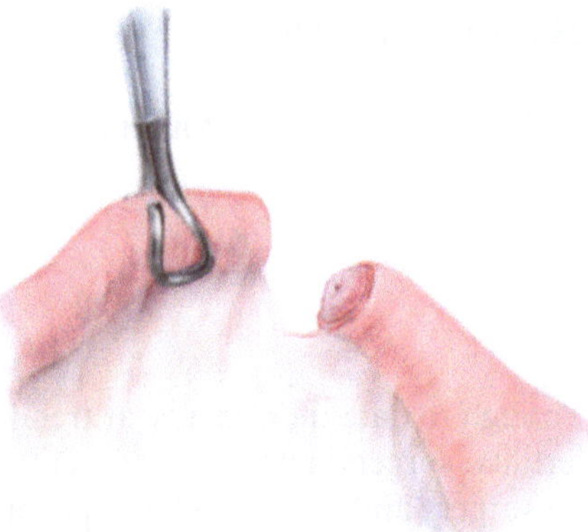

b

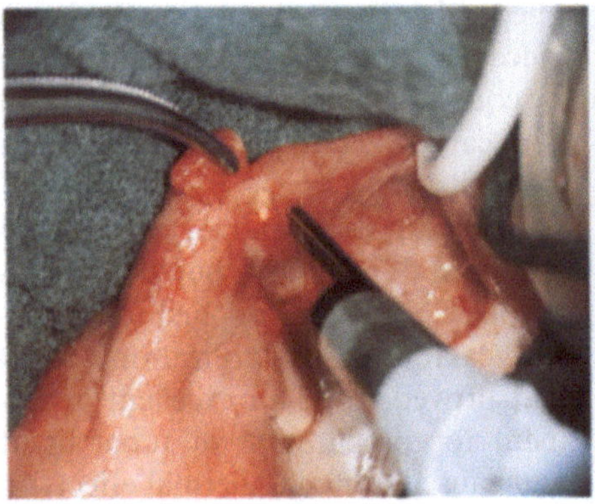

c

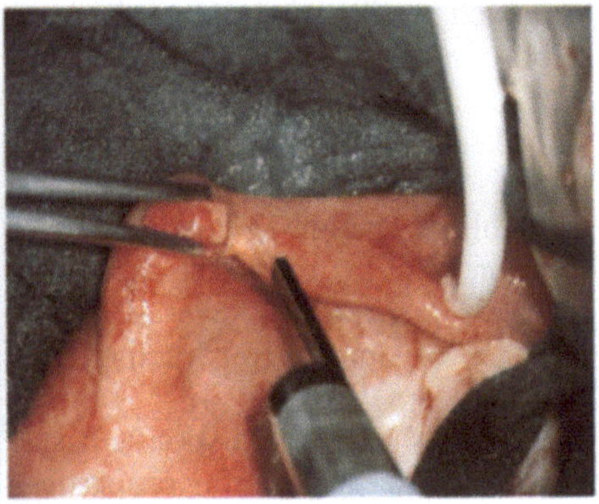

d

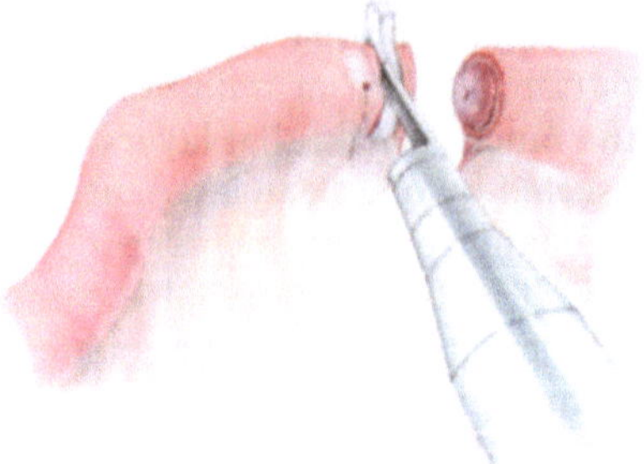

e

**Fig. 1 a–g**. Reversal of sterilization with the carbon dioxide superpulsed laser. **a, b** Laser excision of tubal segmental occlusion; **c–e** resection of the isthmic tubal segment of a patient by means of $CO_2$ laser subsequent to occlusion with Hulka clips; **f** anastomosis of the muscularis subsequent to laser preparation of the tubal stumps; **g** isthmic end-to-end-anastomosis

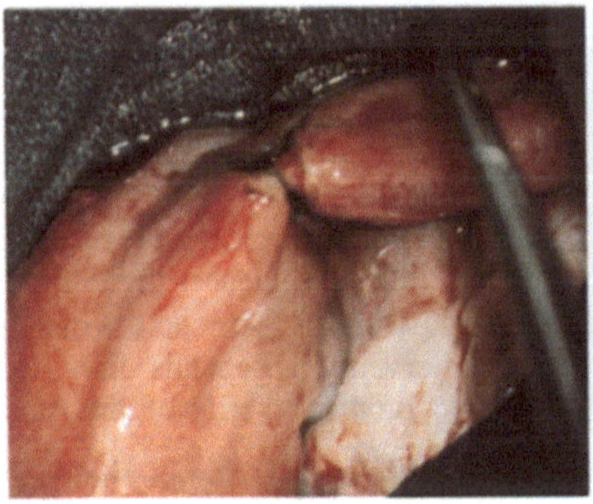

f

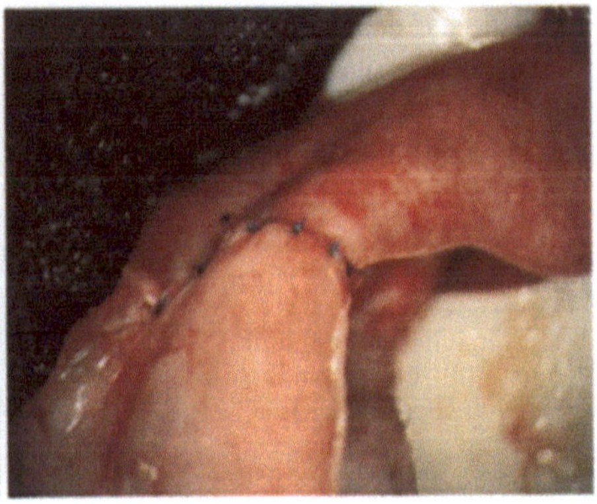

g

damage [9, 11, 14, 15, 18]; i.e., the success of refertilization or the outcome of an anastomosis is dependent upon the residual tubal length subsequent to end-to-end anastomosis [11].

Further reflexion leads one to question which microsurgical technique ensures the best preservation of intact tubal tissue during preparation. A pilot study thus aimed to clarify whether or not microsurgical preparation for resection of occluded tubal segments using the $CO_2$ superpulsed laser was superior to conventional microsurgical preparation using monopolar high-frequency electrodes.

The patients were strictly selected to exclude as many influencing factors as possible. Only patients with residual tubal segments sufficing for an end-to-end anastomosis, in the sense of an isthmo-isthmic or isthmo-ampullary anastomosis, were included. If the microelectrode is replaced by a $CO_2$ superpulsed laser, one can dispense with sharp preparation using scissors or a scalpel.

Microsurgical isthmo-isthmic end-to-end anastomosis is carried out after microsurgical resection of the occluded tubal segments. It is generally unnecessary to insert a polyethylene splint into the tubal lumen to ensure exact approximation after laser preparation, due to the clean cut made through all tissue layers. Microsurgical suture of the tubal muscles is carried out with 8-0 strength nylon or polyamide, the peritonealization of the anastomosis with a 6-0 strength prolene serosa suture. To illustrate this procedure the individual surgical steps of an isthmo-isthmic end-to-end anastomosis in refertilization, subsequent to sterilization with Hulka clips, are shown in Fig. 1.

Eight of 20 patients have had a safe delivery subsequent to microsurgical preparation by means of the $CO_2$ superpulsed laser within the framework of microsurgical refertilization; a further two patients are presently pregnant.

## Conclusion

The development of the $CO_2$ superpulsed mode provides microsurgery with a preparation technique resulting in substantially less tissue damage than when the conventional $CO_2$ laser is used [19]. The question thus arose as to whether or not this technique, especially as far as microsurgical resection of tubal occlusions and the minimized tissue destruction using the $CO_2$ laser technique are concerned, would result in a higher tubal permeability and pregnancy rate due to the improved functional integrity of the tubal tissue.

The surgical procedure is obviously strongly marked by the utilization of the $CO_2$ laser, which requires greater experience than most other instruments. Pilot operations on animals, which were not included in the evaluation, were carried out for adaptation purposes. Handling errors were avoided during the test series, although it must be admitted that the handling was substantially easier with increasing experience, and certainly not entirely without influence on the operative results.

The greatest difficulty in comparing various laser studies lies in the diversity of the technical differentiation of the instruments. In 1983, using a weaker $CO_2$ laser with a relatively large spot-size diameter and a basic laser energy density of 700 $W/cm^2$, Choe attained poorer experimental surgical results as far as postoperative adhesion formation and the success of tubal preparation were concerned [5].

In a study undertaken by Baggish [2], an average output density of 4000–5000 W/cm$^2$ and peak power values of 100000–450000 W/cm were attained. Bellina carried out adhesiolysis using a laser energy ranging from 800 to 4000 W/cm$^2$, emphasizing that the adhesion insertion sites were prophylactically relasered with 300–600 W/cm$^2$ laser energy. He carried out distal tubal laser surgery using the same energy range, performing the eversion, however, with a laser energy of 100–300 W/cm$^2$. Bellina [3] did not undertake mucosa preparation using the conventional laser.

Fayez et al. [10] also reported poor tubal laser surgical results, describing in particular mucosa coagulation, as well as postoperative fibrosis and relatively extensive tissue necrosis. However, they had also utilized a 900 W/cm$^2$ power density with an adjustment of 20–25 W/s and a focal spot diameter of 1.8–2 mm.

The statements made concerning the damaged tissue zone are just as diverse. Daniell [6] reported a maximum laser penetration depth of 1 mm when appropriately focused. In 1985 he described a damaged zone of less than 500 µm and a necrotic area under 1000 µm. On the other hand, Baggish [1] demonstrated that necrotic zones of 0.2 mm (17-W, superpulsed laser) and 0.5 mm (20-W, continuous-wave mode) can be attained using the newest type of $CO_2$ laser instruments.

In conclusion, we may say that tubal preparation by means of the $CO_2$ laser is more elegant than that using the microelectrode. It enables a highly precise, quick and easy transection through all layers. Negative side effects on pregnancy and/or the nidation index have been precluded on the basis of animal experiments.

The clinical results attained within the framework of this pilot study cannot be seen as significant, but merely as preliminary trend values, owing to the small number of cases and the presently still rather short follow-up.

Table 1 sums up the advantages and disadvantages of carbon dioxide laser application in reconstructive microsurgery. Handling of the laser in the proximal tubal part, especially for intramural preparation, is difficult. The microscope must be placed carefully to ensure an optimal view, the laser must be vertically pointed at the target tissue surface, and the preparation area must lie exactly in the focal point. Intensive training is therefore necessary.

A review of the literature shows no significant differences in terms of subsequent pregnancies between groups of patients treated with the carbon dioxide laser and the microelectrode.

**Table 1.** The carbon dioxide laser in microsurgical tubal reconstruction

Advantages
- Elegant handling
- Time-effective preparation, due to concomitant coagulation
- Reduction of parenchymous bleeding
- Minimal thermal damage to adjacent tissue
- No mucosal agglutination
- Postoperative course without excessive formation of dense fibrotic tissue or highly vascular adhesions

Disadvantages
- No coagulation of vessels more than 0.5 mm in diameter
- Handling in the intramural tubal part impaired by anatomy
- Reduced to insufficient cutting effect, if tissue covered by a fluid (blood, irrigation fluid) layer
- No statistically relevant improvement in fertility

Mage [13] and Tulandi [16, 17] found comparable pregnancy rates after salpingostomy, and also no differences in adhesion formation. Tulandi [16] and Daniell [7], however, also observed shorter intervals to conception after laser therapy. After tubal repair and adhesiolysis, Choe [5] and Bellina [4] attained more pregnancies and less adhesions in the laser groups.

Diamond [8] reported on higher pregnancy rates after laser tubal surgery, anastomosis, and salpingolysis, but no difference in adhesion formation.

Kelly and Roberts [12] used the same technique that we employed and their overall term pregnancy rate was 70%.

## References


1. Baggish MS (1981) Complications associated with carbon dioxide laser surgery in gynecology. Am J Obstet Gynecol 139: 568–574
2. Baggish MS (1986) Comparison of electronically superpulsed and continuous-wave $CO_2$ laser on the rat uterine horn. Fertil Steril 45:120–127
3. Bellina JH (1981) Reconstructive microsurgery of the fallopian tube with the carbon dioxide laser – procedures and preliminary results. Reproduction 5:1–17
4. Bellina JH (1984) Analysis of electronically pulsed versus quasi-continuous-wave carbon dioxide lasers in an animal model. Am J Obstet Gynecol 150:934–940
5. Choe JK, Dawood MY, Bardawil WA, Andrews AH (1984) Clinical and histological evaluation of laser reanastomosis of the uterine tube. Fertil Steril 41:754–760
6. Daniell JF, Brown DH (1981) Carbon dioxide laser laparoscopy: initial experience in experimental animals and humans. Obstet Gynecol 59:761–764
7. Daniell JF, Diamond MP, McLaughlin DS, Martin DC, Feste J, Surrey MW, Friedman S, Vaughn WK (1986) Clinical results of terminal salpingotomy with the use of $CO_2$ laser: report of the intraab dominal laser study group. Fertil Steril 45:175–178
8. Diamond MP, Daniell JF, Martin DC, Feste J, Vaughn DS, McLaughlin DS (1984) Tubal patency and pelvic adhesions at early second-look laparoscopy following intra-abdominal use of the carbon dioxide laser: initial report of the intraabdominal laser study group. Fertil Steril 42:717–723
9. Döring GK (1970) Über die Häufigkeit der verschiedenen Sterilitätsursachen bei der Frau und die Erfolgsaussichten ihrer Behandlung. Geburtshilfe Freuenheilkd 30:302–307
10. Fayez JA, McComb JS, Harper MA (1983) Comparison of tubal surgery with the $CO_2$ laser and the unipolar microelectrode. Fertil Steril 40:476–480
11. Frantzen C, Schlösser H, Beck L (1979) Die mikrochirurgische Refertilisierung nach Tubensterilisation. Dtsch Ärztebl 42:2715–2719
12. Kelly RW, Roberts DK (1983) Experience with the $CO_2$ laser in gynecologic microsurgery. Am J Obstet Gynecol 146:585
13. Mage G, Bruhat MA (1983) Pregnancy following salpingostomy: comparison between $CO_2$ laser and electrosurgery procedures. Fertil Steril 40:472–475
14. Rock J (1982) Tubal anastomosis following unipolar cautery. Fertil Steril 37:613–618
15. Silber SJ (1980) Microsurgical reversal of female sterilization: the role of tubal length. Fertil Steril 33:598–601
16. Tulandi T (1987) Adhesion reformation after reproductive surgery with and without the carbon dioxide laser. Fertil Steril 47:704–706
17. Tulandi T, Vilos GA (1985) A comparison between laser surgery and electrosurgery for bilateral hydrosalpinx: a 2-year follow-up. Fertil Steril 44:846–848
18. Wallwiener D, Morawski A, Damböck R, Bastert G (1988) Histological differences in the tissue effect as well as the healing process after the uterotomy performed on rats with the use of $CO_2$ laser (Sharplan 1060) – superpulsed wave versus continuous wave. In: Waidelich W, Waidelich R (eds) LASER Optoelectronics in medicine. Springer, Berlin Heidelberg New York Tokyo
19. Wheeless CR (1977) Problems with tubal reconstruction following laparoscopic sterilization using the electrocoagulation and resection technique. Fertil Steril 28:723–727

# Laparoscopic Laser Application

# Laser Laparoscopy

D. WALLWIENER [1], D. POLLMANN, S. RIMBACH, W. STOLZ, and G. BASTERT

The laser of first choice in operative laparoscopy is still the carbon dioxide laser. As discussed in the chapter entitled "Overview", single-puncture and second-puncture application of the free carbon dioxide laser light have proved efficacious.

In contrast to free beam transmission, preparation via waveguides is currently being investigated. The carbon dioxide beam is transmitted over a nonflexible, thin tube, into the peritoneal cavity. Whether preparation via such a transmission system will replace free application is still controversial.

The entire equipment for operative laparoscopy must be available and in working order for a laser laparoscopic procedure. The instrument cart should hold the complete instrumentarium, from at least one hemostatic source, e.g., a bipolar generator, to the finest small atraumatic grasping tongs and the endoloop, an excellent instrument for ablative procedures or lysis and resection of highly vascularized adhesions. Some authors [2] also emphasize the use of laparoscopic hooks, especially for fimbrioplasty, and of ancillary probes with a backstop to prevent overshooting into the parietal peritoneum. The laser endoscopist with little experience should make use of these probes for security reasons. Pooled fluid in the cul-de-sac of Douglas may also serve as a backstop and thus prevent lesions, mainly during distal tubal procedures. We have dispensed with such backstops, since in our study collectives second-look laparoscopies, e.g., after carbon dioxide endometrial vaporization, have not shown increased adhesion formation in the vaporized areas of the peritoneal cavity.

The main requirement for operative laser laparoscopy, as for any kind of pelviscopic surgery, is excellent vision, which can be obtained only with optimal optical or videoscopic equipment. It is also almost mandatory that the insufflator producing the pneumoperitoneum contain a permanent-flow aspiration system with an integrated smoke evacuation device. Only when these requirements are met can the advantages of endoscopic laser preparation be used optimally.

## Indications

Reconstructive procedures of the tuboovarian functional unit are the main indications for laser laparoscopy, performed with the carbon dioxide laser, or alternatively, with the Nd:YAG

[1] Universitäts-Frauenklinik, Abt. für Geburtshilfe und Gynäkologie, Voßstr. 9, W-6900 Heidelberg, FRG.

Lasers in Gynecology
Edited by G. Bastert and D. Wallwiener

contact laser. Minute preparation movements, causing only minimal thermal damage, and a certain concomitant hemostasis are essential for successful restoration of tubal patency and function. Laser preparation is thus well suited for distal tubal surgery. It is an elegant procedure, careful preparation through various tissue layers is possible, and neither a change of instruments, e.g., for sharp preparation or coagulation, nor constant lavage for removal of blood and other debris are necessary. The main indications for laser application in gynecologic laparoscopy are

1. Distal tubal pathology
2. Adhesiolysis
3. Tubal pregnancy
4. Endometriosis
5. Ovarian cysts

# Distal Tubal Pathology: Salpingolysis/Ovariolysis, Fimbriolysis/Fimbrioplasty, Salpingostomy

D. WALLWIENER [1], J. GAUWERKY, W. STOLZ, and G. BASTERT

## Salpingolysis/Ovariolysis

Adhesions in the areas of the fallopian tube and ovary are grasped by means of small atraumatic forceps; then a certain traction is applied. The carbon dioxide laser, sharply focused and used in a power range of 20–35 W/s in a continuous-wave mode, is applied for lysis and resection of adhesions in the region of tubes and ovaries. Exposure times should be kept short. Adhesions that are thought to contain blood vessels should be coagulated before division. Ancillary probes with a backstop are useful.

Salpingolysis reestablishes the normal intra-anatomic relationships of the pelvic organs, the fallopian tubes regaining their mobility with respect to the cortical surface of the ovary. Only thus are successful procedures for distal tubal pathology possible (Fig. 1).

## Fimbriolysis/Fimbrioplasty

Fibrinous postinfectious adhesions between the tubal fimbriae and the distal endosalpingeal walls – mostly thin, filmy, and relatively avascular – respond well to blunt dissection. After being grasped with an atraumatic probe, they can usually be freed by gentle traction and dissected, no coagulation being necessary. The carbon dioxide beam is set at about 20 W/s, focused. To minimize re-formation of adhesions, removal is desirable, as far as possible without damage to adjacent organs. According to Donnez [2], blanching of the serosa is usually sufficient for adequate eversion of the fimbriae and to avoid agglutination.

## Salpingostomy

Distal closure of the fallopian tube is diagnosed in the infertile woman by hysterosalpingography, diagnostic laparoscopy, or is simply suspected from the patient's history of previous pelvic infection. After lysis of associated adhesions a laparoscopic laser procedure can be performed to open the distal tube. In cases of extensive, dense, fibrous adhesions, in which it is not possible to free the tube sufficiently for terminal salpingostomy, or in cases of thick-walled sactosalpinges, reconstruction should be abandoned in favor of IVF-ET. Experience and training of the laparoscopic surgeon are major factors in this decision, however.

In cases of thin-walled hydrosalpinges with healthy-appearing fimbri-

[1] Universitäts-Frauenklinik, Abt. für Geburtshilfe und Gynäkologie, Voßstr. 9, W-6900 Heidelberg, FRG.

Lasers in Gynecology
Edited by G. Bastert and D. Wallwiener

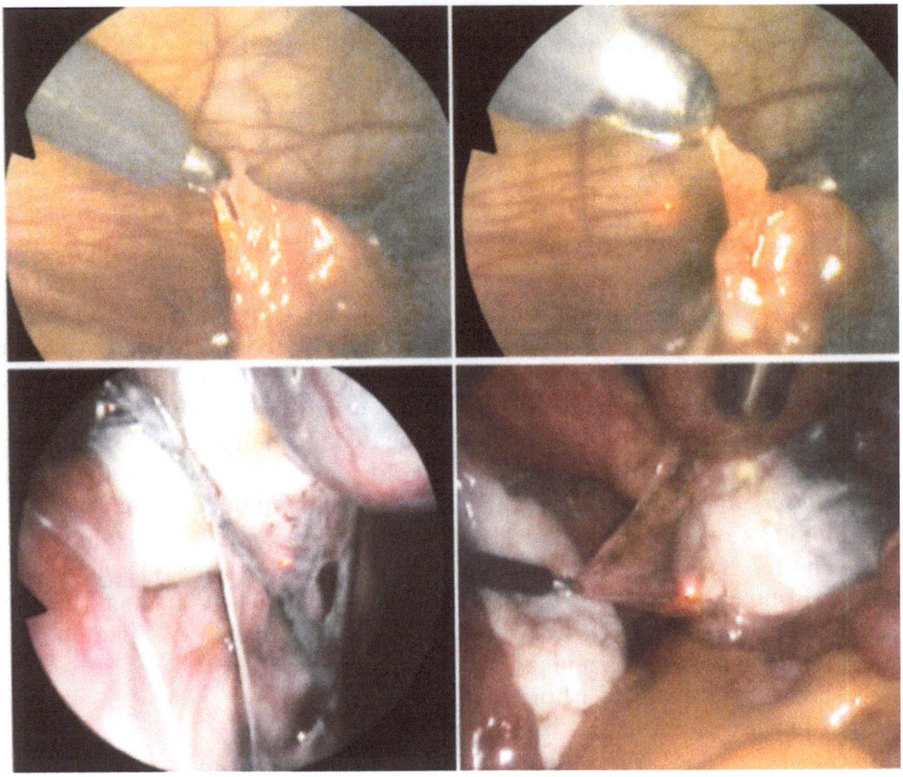

a

b

**Fig. 1.** **a** Agglutination of fimbriae (fimbriolysis); **b** carbon dioxide laser, cw, 30 W/s, ovariolysis

ae folded into the closed ampulla, a laser salpingostomy is performed; a thick-walled hydrosalpinx with no ampullary folds should not be treated by salpingostomy.

The tube is grasped with two forceps and manipulated into clear vision. The carbon dioxide laser light, beam in focus, is applied at an angle of 90°. The initial power is set at 20–35 W/s, continuous-wave mode. At the proposed site, incision in the tube with the laser light in focus is started centrally and conducted in a radial fashion along the previously devascularized area, linear incisions being made from anterior to posterior. Since the number of fimbriae kept intact after the procedure correlates directly with subsequent pregnancy, a procedure causing minimal thermal damage is preferred. Blue dye is used to keep the tube distended. Then the opening is enlarged sufficiently to allow eversion of the mucosa. To maintain patency, it is necessary to establish a certain eversion of the tube. Sealed-in fimbriae tend to evert with the blue dye instilled continuously. However, additional artificial eversion is desirable to improve chances of subsequent pregnancy. This can be performed by serosal coagulation with the defocused carbon dioxide laser, producing the so-called flowering effect (Fig. 2).

Evaluation of results in terms of subsequent pregnancy after laparoscopic treatment of distal tubal pathology is rather difficult (see chapter by Hohl). Of 40 women in whom a laser laparoscopic salpingostomy had been performed at our department, 12 became pregnant (follow-up 1 year). However, since these collectives are relatively small compared with microsurgically treated groups (see chapter by Hohl) and follow-up is rather short, our laparoscopic results are only preliminary and should be regarded as trendsetters. Table 1 compares our results with those after laparoscopic (closed) and microsurgical (open) salpingostomy performed by Leventhal [6]. The overall pregnancy rates are comparable: 28% in the microsurgery group, 30% in the laparoscopic group [6], and 30% in our collective of laser laparoscopic salpingostomies. Abortion rates in all collectives were also comparable. The differences in tubal pregnancy between the two laparoscopic groups are significant however, (8% versus 33%), as are the differences in term pregnancy (77% versus 50%).

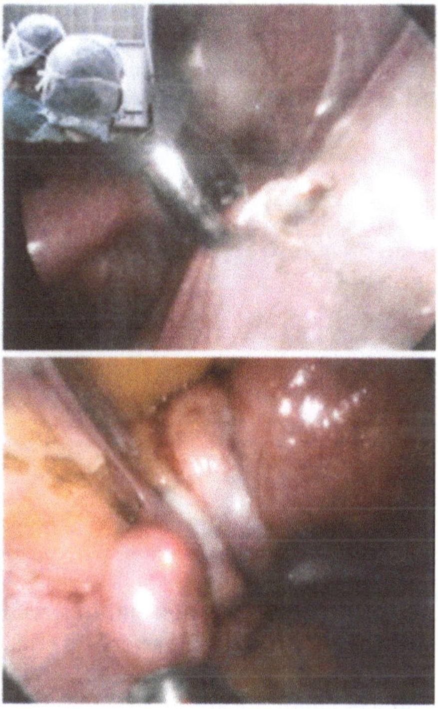

a

b

**Fig. 2a, b.** Salpingostomy. **a** preparation by means of the $CO_2$ laser, cw, 30 W/s; **b** radial incisions

## Laser Laparoscopy by Means of the Nd:YAG Contact Laser ("Bare Fiber")

A pilot study was started at our department to test the applicability of the Nd:YAG contact laser bare-fiber tech-

**Table 1.** Comparison of results with various methods of salpingostomy

| Result | Leventhal [6] Microsurgery $n = 25$ $n$ (%) | | Laparoscopy $n = 43$ $n$ (%) | | Heidelberg[a] Laser laparoscopy $n = 40$ $n$ (%) | |
|---|---|---|---|---|---|---|
| Pregnancy | 7 | (28) | 13 | (30) | 12/40 | (30) |
| Overall | 1/7 | (14) | 2/13 | (15) | 2/12 | (17) |
| Abortion | 2/7 | (28) | 1/13 | (8) | 4/12 | (33) |
| Tubal pregnancy | 2/7 | (58) | 10/13 | (77) | 6/12 | (50) |
| Baby take home | 4/25 | (16) | 10/43 | (23) | 12/40 | (30) |

[a] Follow-up 1 year, selected

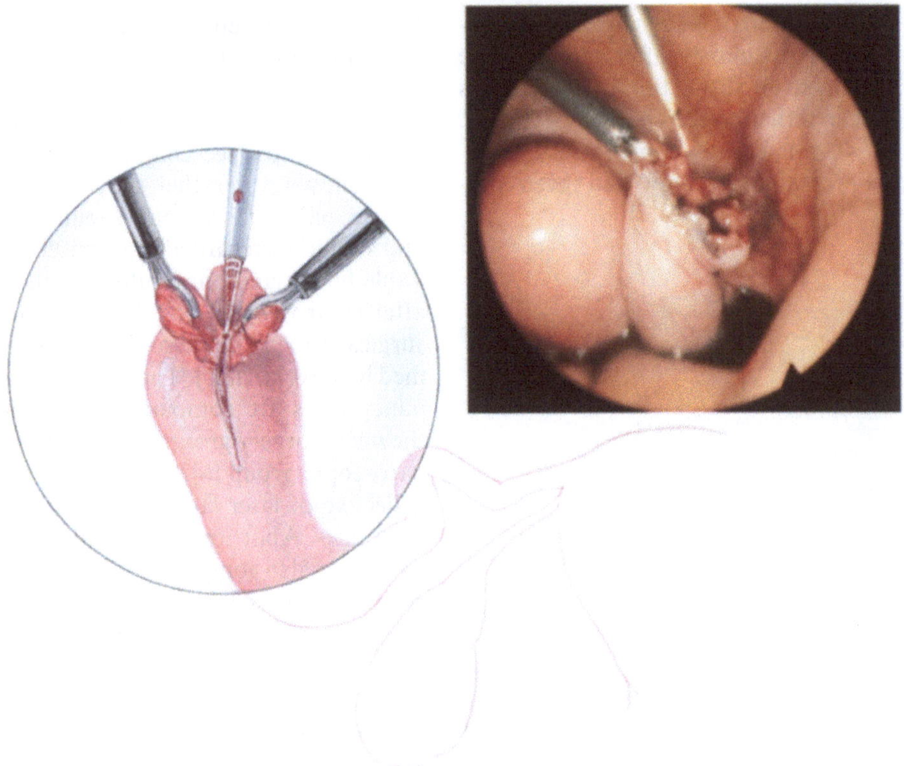

a

b

**Fig. 3a, b.** Laser laparoscopic salpingostomy with the "bare fiber" Nd:YAG contact technique. **a** sactosalpinx; **b** salpingostomy

nique for laparoscopic removal of distal tubal pathology, since in many centers a carbon dioxide laser might not be available.

Terminal, mobile sactosalpinges, or sactosalpinges movable due to salpingolysis were grasped by means of atraumatic forceps so that good vision into the cicatricial navel of the former fimbria was ascertained. The sactosalpinx was then opened with the Nd:YAG contact laser under direct laparoscopic view. Fimbrioplasty is performed in a similar manner. In the majority of cases eversion sutures were unnecessary due to the eversion effect attained by laser preparation, which induces slight, locally restricted, superficial tissue exsiccation (Fig. 3).

Based on experimental-technical development of the contact technique (for details see "Overview"), and following successful animal experiments, the Nd:YAG bare fiber contact technique was applied and proved efficacious. Noncontact preparation is not advisable, even for endometrial coagulation, since the considerable divergence of the Nd:YAG beam and its deep penetration, even through water layers, might result in tissue sequestration, and thus in induction of adhesions. Therefore, the carbon dioxide

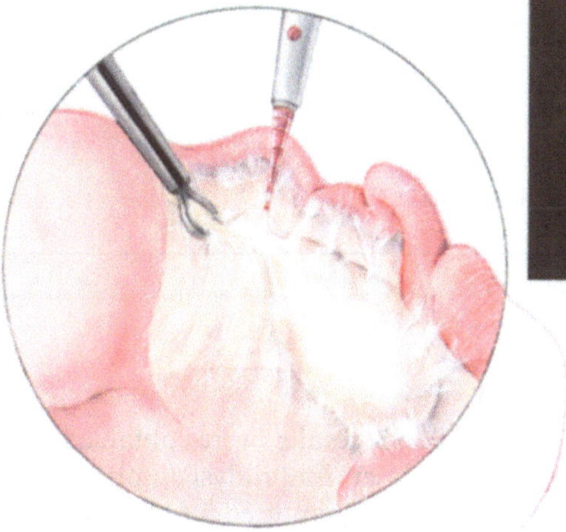

a

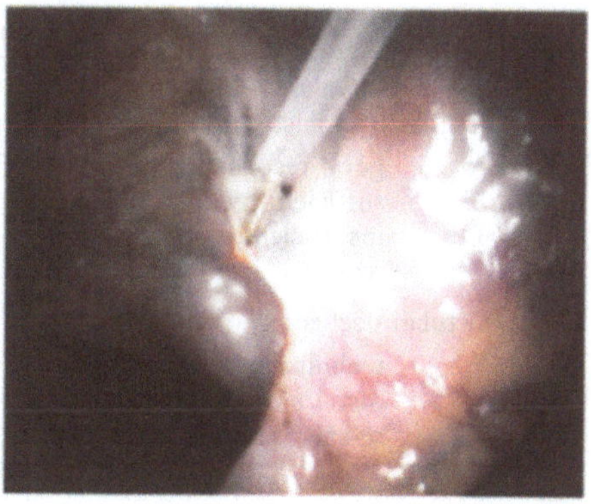

b

**Fig. 4a, b.** Salpingolysis by means of the Nd:YAG contact technique

laser is still the laser of first choice in reconstructive tubal surgery.

The results of this pilot study are shown in Table 2. In 30 patients with tubal patency, adhesions were treated with the Nd:YAG contact technique (salpingolysis, ovariolysis, fimbriolysis). A salpingostomy was performed in 15 patients with bilateral sactosalpinges (Fig. 4).

Second-look laparoscopy revealed technically good results (or at least an intact ovum pick-up mechanism) in 25 patients with distal tubal pathology and initially patent tubes and in 12 patients after treatment of bilateral sactosalpinges.

Adhesiolysis was not successful in three patients; in these cases, however, the laparoscopic approach had been critical from the outset, since the patients had undergone numerous operations and the extensive adhesions present were not completely amenable to laparoscopy. Still, the patients had wished to avoid laparotomy.

In two patients adhesions were more extensive after laparoscopic adhesiolysis. In four patients a second-look laparoscopy could not be performed (I: group 3* and 4*, II: group 2* and 4* in Table 2). They had undergone laparotomies for ectopic pregnancies in other units. They were evaluated retrospectively on the basis of operation reports.

In accordance with other authors [4], our preliminary results show the applicability of the Nd:YAG contact laser for laparoscopy. If all requirements described in "Overview" are met, adhesion induction by this technique is excluded.

**Table 2.** Laser laparoscopy with Nd:YAG contact laser: second-look laparoscopy/laparotomy*

| | | |
|---|---|---|
| I. Distal tubal changes with patent tube ($n = 30$; fimbrioplasty, salpingolysis, ovariolysis, adhesiolysis) | | |
| Group 1: | good reconstruction/no significant readhesions | 12 |
| Group 2: | sufficient reconstruction/intact ovum pick-up | 13 |
| Group 3*: | insufficient reconstruction | 3 |
| Group 4*: | worse | 2 |
| II. Distal tubal occlusion – sactosalpinx ($n = 15$) | | |
| Group 1: | technically good results | 8 |
| Group 2: | distal ampulla, periampullar adhesions | 4 |
| Group 3: | re-sactosalpinx | 3 |

## Discussion

We have found the laparoscopic approach to distal tubal pathology to be as effective in terms of patency and subsequent pregnancy as conventional microsurgical procedures. However, low morbidity, short hospital stay, cost-effectiveness, and shorter operating times are arguments for a laparoscopic approach whenever possible.

Lysis of pelvic postinflammatory, and often also postoperative adhesions is possible with the carbon dioxide laser, which, compared with cautery, allows a more precise dissection with minimal thermal damage to adjacent and subjacent tissue. Thus, preparation of adhesions near the ureter or on the bowels can be performed with the laser. Optimization of the endoscopic equipment for laser application, associated with the cutting precision of the $CO_2$ laser, allow surgery even in sites unapproachable by electrocautery.

Dense, vascular, fibrous adhesions, especially in the fimbrial and ovarian areas, are sometimes not amenable to an open approach. This also depends on the experience and training of the laparoscopic surgeon, however.

Good results were obtained after $CO_2$ laser fimbriolysis/fimbrioplasty; Donnez [2] reported a 61% pregnancy rate and stated that laser laparoscopic fimbrioplasty must be considered the treatment of choice for a phymotic ostium.

Laparoscopic laser salpingostomy seems to be a viable alternative – with increased experience, even the treatment of choice for infertility patients with the diagnosis of distal tubal occlusion (hydrosalpinx). Poor results of conventional microsurgery are reported by most authors [5]: 20%–30% term pregnancies. In a series of 20 patients, Donnez et al. [3] achieved a 6-week postoperative patency rate of 80%. Follow-up of their collective is too short to allow final evaluation in terms of pregnancy, however. Daniell et al. [1] reported a live-birth rate of 31% and a tubal pregnancy rate of 16% in a series of 200 laser laparoscopic salpingostomies.

## References

1. Daniell JF, Diamond MP, McLaughlin DS, Martin DC (1986) Clinical results of terminal salpingostomy with the use of the $CO_2$ laser: report of the Intra-abdominal Laser Study Group. Fertil Steril 45:175–178
2. Donnez J (1989) Instrumentation and operational instructions. In: Donnez J (ed) Laser operative laparoscopy and hysteroscopy. Nauwelaerts, Leuven, pp 15–22
3. Donnez J, Nisolle M, Casanas-Roux F (1989) $CO_2$ laser laparoscopy in tubal infertility. In: Donnez J (ed) Laser operative laparoscopy and hysteroscopy. Nauwelaerts, Leuven, pp 161–179
4. Keckstein J, Finger A, Steiner R (1988) Laser application in contact and noncontact procedures: sapphire tips in comparison to "bare-fiber", argon laser in comparison to Nd:YAG laser. Laser Med Surg 4:158–162
5. Kelly RW, Roberts DK (1983) Experience with the $CO_2$ laser in gynecologic microsurgery. Am J Obstet Gynecol 146:585
6. Leventhal JM (1989) Tubal reconstructive surgery. In: Sanfilippo JS, Levine RL (eds) Operative gynecologic endoscopy. Springer, Berlin Heidelberg New York, pp 140–147

# $CO_2$ Laser Laparoscopy

M. A. BRUHAT[1], A. WATTIEZ, G. MAGE, J. L. POULY, and M. CANIS

Lasers have been used via laparoscopy for 10 years now. The first prototype was developed by the Polyclinique team Clermont-Ferrand, France, in 1979 [4]. Since then techniques have evolved considerably, and the indications have widened to the extent that many laparoscopic surgeons now have a decided preference for the laser.

There are a number of different lasers available for endoscopy. The $CO_2$ laser is by far the most commonly used. It has the advantage of combining three functions: incision, by which it provides a real scalpel effect without opening the abdomen; vaporization, enabling pathological surfaces to be destroyed; and coagulation of the tissues, making it possible to "shape" the surfaces in question. Each of these functions is governed by physical parameters which are, for the most part, under the operator's control. A knowledge of laser theory and training in the use of these lasers is therefore absolutely essential.

[1] Polyclinique de l'Hôtel-Dieu, Service de Gynécologie-Obstétrique Médecine de la Reproduction, 13, boulevard Charles de Gaulle, 63003 Clermont-Ferrand, France.

## The $CO_2$ Laser

In practice, the $CO_2$ laser is the most suitable for gynecological applications [4]. It cuts well, the peripheral coagulation effect is low but nevertheless sufficient to achieve hemostasis of vessels less than 0.5 mm, and it can be used for surface coagulation. It is thus excellent for tubal surgery but can also be used for tissue vaporization – in endometriosis, for example. There are a number of problems involved with its use via laparoscopy, but most of these should be resolved with time.

1. The $CO_2$ laser cannot be transmitted by optical fibers, but has to travel via an articulated optical arm, in a heavy and awkward apparatus. This system cannot give as wide a range of angles for the laser beam as a fiberoptic system can. Recently, however, flexible fibers capable of transmitting the $CO_2$ laser have been put on the market [2] (J. Colles, personal communication).
2. The laser handpiece needs to be leakproof and must accommodate the $CO_2$ supply required for the pneumoperitoneum, which also keeps the lens free of carbonized debris.
3. Part of the energy is absorbed by the $CO_2$ of the pneumoperitoneum [7].
4. The impact of the laser shots produces smoke, which has to be evacu-

Lasers in Gynecology
Edited by G. Bastert and D. Wallwiener

ated. There is, in fact, an automatic pneumoperitoneum ventilation system which goes into operation as soon as the laser is started up, but it is extremely expensive.

5. The red sighting laser (helium-neon laser) is sometimes difficult to see because of the light source of the laparoscope; this is particularly true where defocused shots are concerned.

### The Route of Entry

The ideal route of entry into the body for the $CO_2$ laser was the subject of debate during early trials. The suprapubic operative route appeared best for many reasons but did not seem to be safe enough because the operator had no direct visual control over the laser as compared with the coaxial transumbilical route. Comparative tests demonstrated that handling via the suprapubic route was decidedly superior and that the risk of shots going astray had been grossly overestimated. The advent of endoscopic cameras today tips the balance back in favor of the coaxial route because it makes the endoscope-laser assembly easier to maneuver. A further advantage of this route is that it frees an introduction route and one hand for the operator.

### Lasers and Safety

The laser is a light beam of very high energy; its use requires proper training and the strict observance of basic safety rules. The literature contains very few reports of accidents due to use of the laser. Most of these can be explained by the use of an unsuitable wavelength for the desired result and by the surgeon's inexperience.

As the laser completely destroys the tissues, no pathological tests can be carried out when it is used for vaporization. This raises a very real possibility of making mistakes due to incorrect assessment of the severity and extent of the lesion being treated, e.g., vaporization of a malignant tumor [6].

The main types of damage reported in endoscopy are lesions due to breaching of a large vessel and perforation of hollow organs by burs (mainly the urinary and digestive tracts). The main causes of these accidents are stray shots and overshots, or excessively high power. All such accidents can be avoided if extreme diligence is exercised at all times when the laser is used.

## Indications for Laser Laparoscopy

At present, the use of the laser is indicated essentially in three pathological conditions: adhesions, endometriosis, and tubal obstruction.

### Adhesions

The $CO_2$ laser is capable of simulating the scalpel with its high precision, and can thus be used to incise adhesions. It has the additional advantages of causing little peripheral tissue damage and enabling filmy adhesions on the surface of the ovary to be vaporized. Finally, in the case of very tight, type-C adhesions, incisions can be carried out with simultaneous coagulation of small vessels; however, there is a certain danger with this last application, due to the creation of an artificial cleavage plane which may result in damage to neighboring structures.

**Table 1.** Results of adhesiolysis via laparoscopy alone

| Reference | Number of cases | Intrauterine pregnancy (%) | Ectopic pregnancy (%) |
|---|---|---|---|
| Madelenat and Palmer [18] | 144 | 42 (29) | 11 (7.6) |
| Mintz | 65 | 24 (37) | 7 (11.0) |
| Gomel | 92 | 57 (62) | 5 (5.4) |
| Audebert | 50 | 15 (30) | 0 |
| Abeille | 19 | 7 (37) | 3 (16.0) |
| Mettler et al. [23] | 44 | 13 (29) | |
| Fayez [11] | 50 | 30 (60) | 2 (4.0) |
| Henry-Suchez | 38 | 15 (39) | 1 (3.0) |
| Bruhat et al. [4] | 93 | 48 (51) | 7 (5.0) |

The technique is similar to that of conventional laparoscopic surgery: the adhesion is put under tension and vaporized as close as possible to the least vulnerable organ. That is to say, for a tubo-ovarian adhesion the section should be as close as possible to the ovary, whereas with a tubointestinal adhesion, it should be as close as possible to the tube. Stray shots or damage to the organ behind the adhesion can be avoided either by creating a physiological "backstop" by flooding the cavity with irrigation fluid or by stretching the adhesion over the laparoscopic forceps and cutting it with laser shots.

Thus far, no studies have definitively settled whether laparoscopic laser adhesiolysis or laparoscopic scissor adhesiolysis is the most effective. The laser has certain points in its favor, however: (a) simultaneous hemostasis, (b) high precision, and (c) the ability to free very tight adhesions. The disadvantages are the risk of stray shots and the risk of damage to fragile tissues when a tight adhesion is being freed. The results are at least equal to those with conventional laparoscopic adhesiolysis, that is, up to 60% positive results for fertility (Table 1). How efficient adhesiolysis is in reducing pain is more controversial, since the exact relationship of adhesions to abdominal pain is still unresolved.

## Endometriosis

Endometriosis is a pathology which gives rise to peritoneal implants, adhesions that are often dense, and ovarian cysts. The laser can be used to treat all of these by vaporizing the peritoneal implants, cutting the adhesions free, and vaporizing the endometriotic cyst walls when it is not possible to find a cleavage plane for them. Laser vaporization of the peritoneal implants is carried out using a spot diameter of 0.5–2 mm and between 10 and 20 W of power in continuous mode. If the implants are close to the ureter or to large retroperitoneal vessels, these can be protected by infiltrating normal saline solution. When vaporization is completed a thorough peritoneal lavage is necessary to eliminate all carbonized debris which might otherwise provoke postoperative adhesions or be taken for a recurrence of endometriosis at a later checkup [10].

Our team usually treats ovarian endometriomas by intraperitoneal cystectomy, i.e., complete removal of the cyst wall, rather than by puncturing,

**Table 2.** Treatment of pain

| Level of pain | Number of patients (%) | | Improvement (%) | Cure (%) | Success (%) |
|---|---|---|---|---|---|
| Minimal | 20 | 45 | 35 | 80 | 20 |
| Mild | 9 | 55.6 | 22.2 | 77.8 | 22.2 |
| Moderate | 31 | 29 | 48.4 | 77.4 | 22.6 |
| Severe | 31 | 29 | 42 | 71 | 29 |
| Total | 91 | 35.2 | 40.7 | 75.8 | 24.2 |

**Table 3.** Results of laser laparoscopy for infertility (Clermont-Ferrand series)

| | Stage (AFS, 1985) | | | | | |
|---|---|---|---|---|---|---|
| | I (%) | II (%) | III (%) | IV (%) | IV < 70 (%) | Total (%) |
| No. | 28 | 16 | 33 | 33 | 20 | 110 |
| IVF | 0 | 1 | 4 | 4 | 2 | 9 |
| Lost | 4 | 1 | 4 | 5 | 1 | 14 |
| No. | 24 | 14 | 25 | 24 | 17 | 87 |
| IUP | 12 (50.0) | 4 (28.5) | 10 (40.0) | 9 (37.5) | 9 (52.9) | 35 (40.2) |
| EP | 2 (8.3) | 1 (7.1) | 2 (8.0) | 0 (0.0) | 0 (0.0) | 5 (5.7) |

IVF, In vitro fertilization; IUP, intrauterine pregnancy; EP, ectopic pregnancy

opening, and destroying the cyst wall with the $CO_2$ laser. The cystectomy technique has the advantage of providing a sample for the pathologist to check on the nature and differentiation of the disease. Some authors recommend that the bottom of the ovary be vaporized after removal of the cyst wall to eliminate deep-seated implants of endometriosis. The $CO_2$ laser is, however, extremely useful for dealing with endometriotic adhesions that are very dense and thick, which is often the case.

In a recent study carried out in our department, we operated on 91 patients using laser laparoscopy. All 91 complained of chronic pain; 20 had stage I endometriosis, nine stage II, 31 stage III, and 31 stage IV. Ten patients were lost to follow-up; the remainder were followed up for over 9 months. Whatever the stage, 70% of the patients showed a distinct functional improvement. The stage involved was equally unimportant as regards the failure rates (Table 2). One hundred and ten patients were treated for infertility, of which 14 were lost to follow-up and nine were included in an in vitro fertilization program immediately after laparoscopy because they each had a very advanced stage of endometriosis

**Table 4.** Results reported in the literature of laparoscopic laser treatment for endrometriosis (all stages together)

| Reference | Number of pregnancies | Number of patients (%) |
|---|---|---|
| Feste [13] | 140 | 82 (59) |
| Daniell [6] | 48 | 34 (54) |
| Adamson 1986 | 65 | 35 (54) |
| Bowman 1986 | 35 | 18 (51) |
| Davis [9] | 64 | 37 (58) |
| Martin [22] | 115 | 54 (47) |
| Nezhat et al. [24] | 102 | 62 (61) |
| Paulsen and Asmar [27] | 282 | 150 (53) |
| Total | 851 | 472 (55) |

and very extensive adhesions. The results after treatment show that the uncorrected rate of intrauterine pregnancies was 40.3%. The fertility results are given in Table 3, and it should be noted that 26 of these 35 intrauterine pregnancies occurred within 8 months after the operation. Table 4 gives the results according to the literature for laparoscopic laser treatment of endometriosis.

## Tuboplasty

Distal tuboplasty via laparoscopy is not a new idea. The technique was abandoned, but then it received renewed interest with the appearance of the $CO_2$ laser [8,12,14].

The first phase in operating is adhesiolysis. This is of prime importance to create the best conditions possible for the tuboplasty, as the adhesiolysis has to restore normal tubo-ovarian mobility. Next, a methylene blue test should be carried out, when the tension induced in the blocked tube helps in locating the scar of the former ostium. Then a focused shot at high-power density is used to make three or four radial incisions on the tube, enabling it to be opened. Once this is achieved, the new tube fringes are exposed using two forceps, and the incisions are completed. After this, the serous surface of these new fringes is subjected to defocused shots of low-power density which coagulate the peritoneal surface. The consequent retraction of the tube everts the fringes; this eversion is stable with time, and sutures to the tube fringes can be avoided.

Only distal obstruction of the tube can be considered a good indication. Prior assessment by hysterosalpingography followed by laparoscopy is essential before starting treatment. It has been demonstrated without a doubt that the prognosis depends essentially on the tube status (Table 5) and the degree of the adhesions (AFS; Table 6). It would seem at present that results giving approximately 29% intrauterine pregnancies for all stages taken together can be expected [19]. When the stage is taken into account, stage I gives good results with 45%, stage II 22%, stage III 7.7%, and stage IV 0%; the rate of ectopic pregnancies rises in inverse parallel (Tables 7 and 8). These rates are comparable to those obtained with microsurgery to the tube (Table 9). It is not possible to make any definite statement as to relative efficacy when comparing our series with the literature, because the success rates given

**Table 5.** Distal tube scoring system (Mage et al. [20])

| Tube permeability | Ampulla mucosa | Tube wall |
|---|---|---|
| | Normal 0 | Normal 0 |
| Phimosis 2 | Folds diminished 5 | Thin 5 |
| Hydrosalpinx 5 | No folds or alveolar salpingitis 10 | Thickened or sclerotic 10 |

Stage I: 2–5; stage II: 6–10; stage III: 11–15; stage IV: > 15

**Table 6.** Adhesion scoring system (AFS 8: Buttram [5])

| Organ | Type | Amount of surface covered by adhesions | | |
|---|---|---|---|---|
| | | 1/3 | 2/3 | 3/3 |
| Ovaries | Filmy | 1 | 2 | 4 |
| | Thick | 4 | 8 | 16 |
| Distal tube | Filmy | 1 | 2 | 4 |
| | Thick | 4 | 8 | 16 |

None: 0; mild: 1–10; moderate: 11–16; severe: > 16

**Table 7.** Clermont-Ferrand results with laparoscopic salpingoneostomy according to the tubal stage

| | Patients n % | IUP n % | EP n % |
|---|---|---|---|
| Stage I | 22 (32.35) | 10 (45.5) | 1 (4.5) |
| Stage II | 27 (39.70) | 6 (22.2) | 4 (14.8) |
| Stage III | 13 (19.11) | 1 (7.7) | 0 (0) |
| Stage IV | 6 (8.80) | 0 (0) | 0 (0) |
| Total | 68 | 17 25 | 5 7.4 |

**Table 8.** Clermont-Ferrand results with laparoscopic salpingoneostomy according to the tubal stage and the adhesion stage (Mage et al. [21])

| | Patients n % | IUP n % | EP n % |
|---|---|---|---|
| Stage I, II, with adhesions O, L, M | 42 (61.8) | 15 (35.7) | 4 (9.5) |
| Stage III, IV or severe adhesions | 26 (38.2) | 2 (7.7) | 1 (3.8) |
| Total | 68 (100.0) | 17 (25) | 5 (7.4) |

**Table 9.** Clermont-Ferrand results with microsurgery for distal lesions

| Stages | Number of cases | Intra-uterine pregnancies n % | Ectopic pregnancies n % |
|---|---|---|---|
| Tube score | | | |
| Stage I | 12 | 7 (58.3) | 1 (8.3) |
| Stage II | 30 | 11 (36.6) | 3 (10.0) |
| Stage III | 21 | 2 (9.5) | 3 (14.2) |
| Stage IV | 13 | 0 (0.0) | 0 (0.0) |
| Total | 76 | 20 (26.3) | 7 (9.2) |
| Adhesion score | | | |
| Stage O | 18 | 7 (38.8) | 3 (16.6) |
| Stage L | 25 | 8 (32.0) | 0 (0.0) |
| Stage M | 15 | 4 (26.6) | 1 (6.6) |
| Stage S | 18 | 1 (5.5) | 3 (16.6) |
| Total | 76 | 20 (26.3) | 7 (9.2) |

in the various articles published range from 26% [23] over 13% [16] to 0% [12]. Not all these series involved the $CO_2$ laser, so no conclusions can be drawn. Nevertheless, the progress made with the $CO_2$ laser and laparoscopic techniques in general give reason to hope that distal tubal obstruction will, in the future, be treated entirely via laparoscopy using the laser.

## Conclusions

The use of the laser via laparoscopy originated in our department in 1979. After a slow beginning it is now being widely developed because it overcomes certain barriers in operative laparoscopy. Laser technology is also undergoing rapid changes, and even more impressive progress can be confidently expected. This being said, the laser is not totally indispensable, and a large number of techniques can be applied perfectly satisfactorily using conventional instrumentation.

## References


1. Baggish M (ed) (1985) Basic and advanced lasers in gynecology. Appleton-Century-Crofts, East Norwalk, CN
2. Baggish M, Elbakry MM (1986) A flexible $CO_2$ laser fiber for operative laparoscopy. Fertil Steril 46:16–20
3. Beytout M, Mage G, Marquet C, Pouly JL, Bruhat MA (1983) Etude expérimentale de l'influence de puissance sur la vitesse de section et des lésions tissulaires en microchirurgie tubaire avec le laser $CO_2$. Laser Medical 81/82. Masson, Paris: ESI publications
4. Bruhat MA, Mage G, Manhes H (1979) Use of the $CO_2$ laser via laparoscopy. In: Kaplan I (ed) Laser surgery, vol III. Proc

3rd International Society for Laser Surgery. International Society for Laser Surgery, Tel Aviv
5. Buttram VC (1985) Evolution of the revised American Fertility Society classification of endometriosis. Fertil Steril 43:347
6. Daniell JF (1985) Basic and advanced laser surgery in gynecology, pp 345–356
7. Daniell JF, Brown D (1982) $CO_2$ laser laparoscopy: initial experience in experimental animals and humans. Obstet Gynecol 59:761
8. Daniell JF, Herbert CM (1984) Laparoscopic salpingostomy utilizing the carbon dioxide laser. Fertil Steril 41:558
9. Davis GD (1986) Management of endometriosis and its associated adhesions with the $CO_2$ laser laparoscope. Obstet Gynecol 68:422
10. Donnez J (1987) $CO_2$ laser laparoscopy in infertile women with endometriosis and women with adnexal adhesions. Fertil Steril 48:390
11. Fayez JA (1983) An assessment of the role of operative laparoscopy in tuboplasty. Fertil Steril 39:476
12. Fayez JA, Collazo L (1986) Comparison of laparotomy in the treatment of moderate and severe stages of endometriosis. AFS meeting abstract 317
13. Feste JR (1985) Endoscopic laser surgery in gynecology. In: Reproductive surgery. Postgraduate course syllabus. American Fertility Society, Chicago, pp 51–69
14. Gordts S, Boeckx W, Brossens I (1984) Microsurgery of endometriosis in infertile patients. Fertil Steril 42:520
15. Keye WR, Dixon J (1983) Photocoagulation of endometriosis with the argon laser through the laparoscope. Obstet Gynecol 62:383
16. Lauritzen JG, Pagel JD, Vangsted P, Starup J (1983) Result of repeated tuboplasties. Fertil Steril 40:472
17. Lomano JM (1984) Photocoagulation of early pelvic endometriosis with the Nd:YAG laser through the laparoscope. Lasers Surg Med 3:328
18. Madelenat P, Palmer R (1973) Etude critique des libérations percoelioscopiques des adhérences péri-annexielles. J Gynecol Obstet Biol Reprod (Paris) 8:347
19. Mage G, Bruhat MA (1983) Pregnancy following salpingostomy: comparison between carbon dioxide laser and electrosurgery procedures. Fertil Steril 40:472
20. Mage G, Pouly JL, Bouquet de Joliniere J et al. (1986) A preoperative classification to predict the intrauterine and ectopic pregnancy rates after distal tubal microsurgery. Fertil Steril 46:807
21. Mage G, Bruhat MA, Bouquet J et al. (1987) Score d'opérabilité tubaire. 33rd Assises Françaises de Gynécologie, Poitiers
22. Martin DC (1986) Operative laparoscopy with the carbon dioxide laser for the treatment of endometriosis associated with infertility. J Reprod Med 31:1089
23. Mettler L, Giesel H, Semm K (1979) Treatment of female infertility due to tubal obstruction by operative laparoscopy. Fertil Steril 32:384
24. Nezhat C, Crowgey SR, Garrisson CP (1986) Surgical treatment of endometriosis via laser laparoscopy. Fertil Steril 45:778–783
25. Olive DL, Martin DC (1987) Treatment of endometriosis-associated infertility with $CO_2$ laser laparoscopy: the use of one- and two-parameter exponential models. Fertil Steril 48:18
26. Palmer R (1980) La coelioscopie dans le diagnostic et le traitement des adhérences pelviennes. Contraception Fertil Sex 7:797
27. Paulsen JD, Asmar P (1986) Analysis of the first 150 pregnancies after laser laparoscopy. Presented at the ASCCP/GLS meeting, Boston
28. Pouly JL, Bruhat MA, Mage G, Manhes H (1982) Utilisation du laser $CO_2$ par coelioscopie: comparaison entre un branchement coaxial et un branchement sur trocard suspubien. Acta Med Romana 20:257–260
29. Tadir Y, Kaplan I, Zuckerman Z, Edelsein T, Ovadia J (1984) New instrumentation and technique for laparoscopic $CO_2$ laser operations; a preliminary report. Obstet Gynecol 63:582

# $CO_2$ Laser Laparoscopy in Infertile Women with Adnexal Adhesions and in Women with Tubal Occlusions

J. DONNEZ [1], M. NISOLLE, and F. CASANAS-ROUX

Several procedures have been carried out with the $CO_2$ laser laparoscope [1–4, 7, 8, 10–12]. The most frequent include: (a) Vaporization of peritoneal and ovarian endometriosis; (b) pelvic adhesiolysis, salpingolysis, and ovariolysis; (c) uterosacral ligament ablation; (d) fimbrioplasty and salpingostomy; (e) salpingotomy in case of tubal pregnancy; (f) vaporization of small uterine fibroids.

So far, more than 1550 laparoscopic operations with the $CO_2$ laser have been carried out in our department. The most frequent indication is endometriosis [7, 8] (Table 1); the tissue can be vaporized by means of $CO_2$ laser laparoscopy with a good postoperative pregnancy rate [7, 8, 11, 12]. Peritubal adhesions ($n$ = 372) and/or tubal occlusion ($n$ = 203) (occlusion degree I–IV according to the classification of Donnez and Casanas) [9] are important factors in infertility.

The aim of this report is to evaluate the role of $CO_2$ laser laparoscopy in the management of pelvic adhesions and/or tubal occlusion.

## Materials and Methods

From January 1983 until December 1988, 1550 patients were treated by means of $CO_2$ laser laparoscopy. The operative laser laparoscopes were developed by Wolf (Richard Wolf, Knittlingen, FRG) and Storz (Karl Storz, Tuttlingen, FRG). Accessories for second and third punctures were developed in our department (Fig. 1).

*Salpingolysis* is usually performed by applying traction to adhesions with a suprapubic atraumatic grasping forceps and another probe (smooth manipulating probe, hook, probe with its backstop). Adhesions of degree I (filmy and avascular) and II (dense and vascular, but not very thick) are easily va-

**Table 1.** Indications for $CO_2$ laser laparoscopy in a series of 1550 infertile women

| Indication | No. of patients |
|---|---|
| – Laser uterine nerves ablation | 143 |
| – Tuboplasty | 203 |
| Fimbrioplasty | 162 |
| Salpingoneostomy | 41 |
| – Endometriosis | 714 |
| Mild (peritoneal only) | 343 |
| Moderate (ovarian cyst) | |
| Severe (ovarian cyst) | 371 |
| – Pelvic adhesions | 372 |
| Salpingolysis | 168 |
| Salpingo-ovariolysis | 204 |
| – Tubal pregnancy | 18 |

[1] Infertility Research Unit, Dept. of Gynecology, Catholic University of Louvain, Cliniques Universitaires St. Luc, Avenue Hippocrate 10, 1200 Brussels, Belgium.

Lasers in Gynecology
Edited by G. Bastert and D. Wallwiener

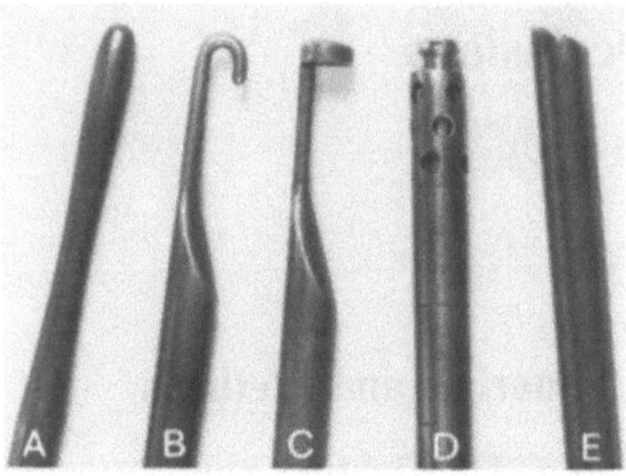


**Fig. 1.** Operating instruments for second and third punctures: *A*, Atraumatic probe; *B*, hook for fimbrioplasty; *C*, probe with backstop for use in vaporizing adhesions near blood vessels; *D*, smoke suction and rinsing tube; *E*, double-channel probe for rinsing and suction

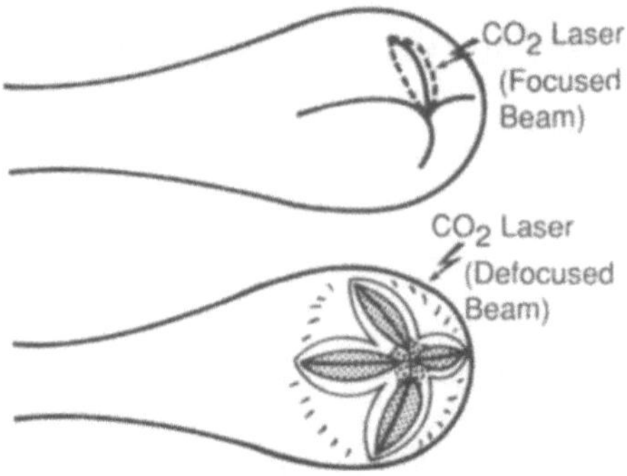


**Fig. 2.** $CO_2$ laser salpingostomy. Two linear incisions are made with *focused beam*; when the incision is enlarged, *defocused beam* is used to evert the serosal aspect of the incised edge

porized with the operative laser laparoscope. The probe with its backstop can be used to simplify the procedure. When this probe is used, the adhesion is placed across the "firing" platform and the laser is fired to vaporize the band. The use of the probe with a backstop eliminates the risk of inadvertent injury to intraperitoneal structures. With a power output of 25 W, adhesions can be both coagulated and incised. Short exposure times are adequate to vaporize the adhesions around the fallopian tubes and ovaries and will prevent the laser beam from penetrating more than 100–200 μm. In the hands of more experienced laparoscopists the continuous mode can be used.

*Salpingostomy* can be performed with the $CO_2$ laser and is indicated in cases of thin-walled hydrosalpinx where both proximal tubal patency and the presence of ampullar folds have been confirmed by hysterosalpingography. In our department, in cases of thick-walled hydrosalpinx without ampullary folds, a salpingectomy is carried out. Two grasping forceps are introduced for traction and manipulation at the ampullary-fimbrial segment. The blocked tube is held so that the focused laser beam can be aligned at a 90° angle to the dimple (Fig. 2). The laser is set on continuous mode (25 W), and two linear incisions are made, cutting from anterior to posterior along blood vessels. As soon as the lumen is entered, the tube collapses, and continuous dye injection keeps it distended. Only then is the incision enlarged. At this point, the probes and grasping forceps gently hold the incision edges, and a reduced-power, defocused beam is used to evert the serosal aspect of the incised edge.

In our department, *fimbrioplasty* is also carried out during laparoscopy. If fimbrial adhesions are found when the blue dye begins to spill through the open tube, these adhesions between fimbrial folds are carefully grasped by means of the probe with a hook passed through a third-puncture trocar and cut in a bloodless fashion with the finely focused $CO_2$ beam set at about 20 W

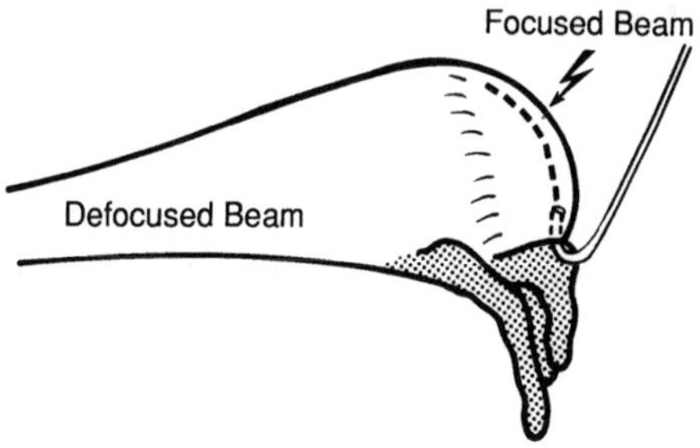


**Fig. 3.** Fimbrioplasty: the adhesions are carefully cut by means of the hook

**Table 2.** Viable pregnancy rates in series of patients for a follow-up period of 18 months

| | Patients *n* | Pregnancies *n* | % |
|---|---|---|---|
| Adhesiolysis | | | |
| Degree I | 102 | 65 | 64 |
| Degree II | 84 | 43 | 51 |
| Total | 186 | 108 | 58 |
| Fimbrioplasty | 100 | 61 | 61 |
| Salpingoneostomy | 25 | 5 | 20 |

(Fig. 3). It is usually sufficient just to cause blanching of the serosa to obtain the desired effect; this allows adequate eversion of the mucosa and avoids recurrence of adhesions. At the end of the procedure, the peritoneal cavity is irrigated with Ringer's solution to remove carbonized particles. Thereafter, following the removal of fluid from the cul-de-sac, 100–200 ml of 32% dextran 70 is instilled into the peritoneum.

## Results

Results are shown in Table 2. The pregnancy rate was evaluated for a follow-up of 18 months. In our series of *adhesiolysis*, pregnancy rates of 64% and 51% respectively were obtained after lysis of degree-I (filmy and avascular) and degree-II (dense and vascular but not very thick) adhesions.

After *fimbrioplasty*, the postoperative pregnancy rate is 61% in a first series of 100 patients. This rate was similar to that obtained in a series of 142 fimbrioplasties carried out by microsurgery [9].

In our department, hysterosalpingography was systematically performed 3 months after laparoscopic *salpingostomy*. Recovery of a normal tubal pattern was found (Fig. 4) when postoperative patency was achieved. A postoperative patency rate of 80% was achieved, with an intrauterine pregnancy rate of 20% in a first series of 25 patients.

In our series of 1550 infertile women operated on with the $CO_2$ laser laparoscope, no case required laparotomy for bleeding and no bladder or bowel injury was reported. During the operation, some difficulties arose as decreased visibility of the HeNe beam or back strain associated with operating directly through the laparoscope were encountered. Indeed, the high intensity of the scope's light source for photographic purposes obscures the visibility of the HeNe beam. Video monitoring systems can be helpful to avoid the back strain of direct laparoscopic visualization.

The operating time varied from 30 to 60 min. All patients were released in good condition the following day. Typical postlaparoscopy pain, such as pelvic discomfort and signs of diaphragmatic irritation, was the most common postoperative complaint. None of the patients experienced any postoperative infection. No patient sustained urinary bladder damage. In

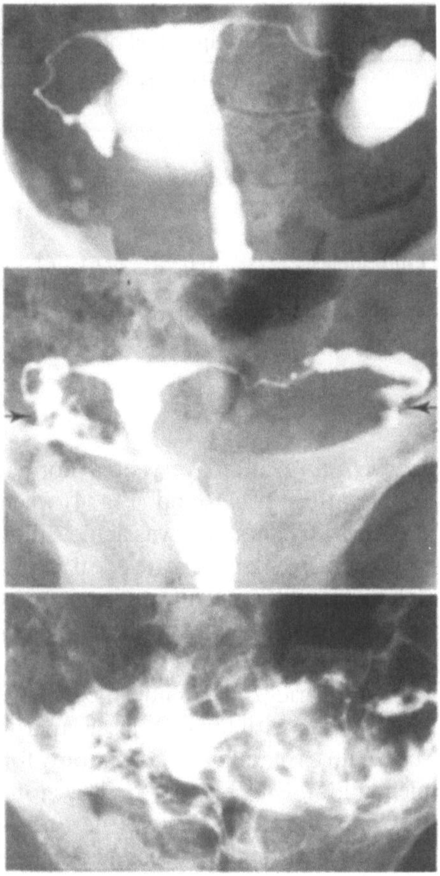

a

b

**Fig. 4a, b.** Hysterosalpingography: **a** Hydrosalpinx; **b** 3 months after laparoscopic salpingostomy, bilateral tubal patency is proved

28 patients requiring salpingo-ovariolysis, incision of vascular adhesions was followed by intraoperative bleeding that required electrocautery. No patient required laparotomy for control of bleeding. In our series no complication such as bladder injury, bowel injury, or ureteral injury occurred.

# Discussion

Conservative microsurgical procedures for pelvic adhesions and tubal occlusion are efficacious [5, 6, 9]. Pregnancy rates depend on preexisting lesions of the tubal mucosa [9]. In this study, the surgical treatment via laser laparoscopy was evaluated.

In cases of laser adhesiolysis, the pregnancy rate (58%) was similar to that obtained after microsurgical salpingolysis by laparotomy (56%) [9]. In many patients, postoperative or postinfectious adhesions are amenable to vaporization by laser laparoscopy. Compared with the standard technique using cautery and the laparoscope scissors or blunt dissection, there is probably no difference in the outcome when the adhesions are small and avascular. With more vascular adhesions or particularly thick tubo-ovarian adhesions, however, the $CO_2$ laser allows more precise destruction of the adhesions with minimal injury to the adjacent normal tissue. The precision attributed to the $CO_2$ laser allows the experienced operator to vaporize adhesions near the ureter or on the bowel. The magnification attained with the laparoscope and the precision of the $CO_2$ laser beam allow precise removal of diseased tissue in many sites often unapproachable by electrocautery.

However, laser procedures cannot replace microsurgical procedures for certain indications. Indeed, laser procedures permit incision of numerous adhesions by laparoscopy, but adhesiolysis of very dense and fibrous adhesions (degree III) between the fimbria and the ovary must be carried out by microsurgical procedures. In our series of 372 infertile women with pelvic adhesions operated on with the $CO_2$ laser laparoscope, no case required laparotomy for bleeding and no bladder or bowel injury was reported. Having investigated clinical gynecologic surgical applications with great care, we are able to conclude that, for selected patients, $CO_2$ laser laparoscopy will become the preferred procedure for resection of postoperative or postinfectious adhesions. Very dense adhesions are probably best treated by microsurgical techniques, however.

Results of fimbrioplasty carried out by laparotomy and laparoscopy are similar. Vaporizing fimbrial adhesions is easy with the $CO_2$ laser and the operating time is not greater than 20 min. The high pregnancy rate obtained in our study confirms that laser laparoscopic fimbrioplasty must be considered the procedure of choice in cases of phimotic ostium.

The results of microsurgical salpingostomy for hydrosalpinx are poor. Indeed, a 20%–30% term pregnancy rate is generally reported [11]. $CO_2$ laser laparoscopic salpingostomy seemed like a reasonable alternative in selected patients who desired pregnancy but did not wish to undergo a major surgical procedure for hydrosalpinx. In an initial series of 20 patients, a 6-week postoperative patency rate of 80% was achieved [4]. Three intrauterine pregnancies, one ectopic pregnancy, and one spontaneous abortion were obtained. The rate obtained in our series is similar to those obtained with laparotomy and microsurgery [9]. Additional follow-up of these patients is necessary before any conclusions concerning the efficacy of this new operative procedure can be made, but laser salpingostomy will probably prove to be the ideal procedure for management of large hydrosalpinx.

In conclusion, by reducing the hospitalization time and the morbidity rate, $CO_2$ laser laparoscopy offers advantages over other operative techniques for adhesions and tubal occlusion.

## References


1. Baggish MS, El Bakry MM (1986) A flexible $CO_2$ laser fiber for operative laparoscopy. Fertil Steril 46:16
2. Bruhat M, Mage G, Manhes M (1979) Use of the $CO_2$ laser in laparoscopy. In: Kaplan I (ed) Proceedings of the Third International Symposium on Laser Surgery. Sept 24–28, 1979. O. T. Paz, Tel-Aviv, pp 235–238
3. Daniell JF, Brown DH (1982) Carbon dioxide laser laparoscopy: initial experience in experimental animals and humans. Obstet Gynecol 59:761
4. Daniell JF, Feste JR (1985) Laser laparoscopy. In: Keye WR Jr (ed) Laser surgery in gynecology and obstetrics. G. K. Hall, Boston, pp 147–163
5. Donnez J (1984) La trompe de Fallope: histophysiologie normale et pathologique. Nauwelaerts, Leuven, p 105
6. Donnez J (1985) Microchirurgie et fécondation in vitro: méthodes concurrentes ou complémentaires? Rev Med Brux 6: 627
7. Donnez J (1987a) $CO_2$ laser laparoscopy in infertile women with endometriosis and women with adnexal adhesions. Fertil Steril 48:390
8. Donnez J (1987b) $CO_2$ laser in intraepithelial neoplasia and in infertility. Nauwelaerts, Leuven
9. Donnez J, Casanas-Roux F (1986) Prognostic factors of fimbrial microsurgery. Fertil Steril 46:200
10. Donnez J, Fernandez C, Willems T, Casanas-Roux F (1987) Experimental ovarian $CO_2$ laser surgery. In: Donnez J (ed) $CO_2$ laser in intraepithelial neoplasia and in infertility. Nauwelaerts, Leuven, p 25
11. Kelly RW, Roberts DK (1983) Experience with the $CO_2$ laser in gynecologic microsurgery. Am J Obstet Gynecol 146:585
12. Nezhat C, Crowgey SR, Garrison CP (1986) Surgical treatment of endometriosis via laser laparoscopy. Fertil Steril 45:778

# Adhesiolysis

D. WALLWIENER[1], J. GAUWERKY, W. STOLZ, and G. BASTERT

The application of laser techniques has enlarged the spectrum of adhesions amenable to laparoscopic therapy. Thus besides salpingo- and ovariolysis, as described in previous chapters, dissection of adhesions between bowels and uterus or the tubo-ovarian functional unit can be performed laparoscopically with the carbon dioxide laser (Fig. 1).

Additionally, we have performed laparoscopic lysis of adhesions not involving, or after removal of, inner genital organs in a small collective of patients. The carbon dioxide and – to a lesser degree – the Nd:YAG contact laser were employed. Adhesions between intestinal convolutions and peritoneum parietale were lysed in 45 patients who had presented at our department with considerable pain after gynecologic surgery or appendectomy. Pain relief and improved digestion were achieved in 42 cases; three patients required an open procedure due to persistent symptoms. Elegant preparation, reduction of operating time (since no change between coagulation and cutting instruments was necessary), and minimal risk of damage to adjacent structures are only some of the advantages of the laser laparoscopic approach. Whether laser application plays any role in prevention of re-adhesions is still unclear. It is certain, however, that induction of re-adhesions is correlated with the volume of thermally damaged tissue. The logical conclusion would thus be that the minimal thermal destruction caused by the laser should optimize success rates of adhesiolysis.

[1] Universitäts-Frauenklinik, Abt. für Geburtshilfe und Gynäkologie, Voßstr. 9, W-6900 Heidelberg, FRG.

**Fig. 1 a, b.** Lysis of adhesions in the vicinity of the bowels ($CO_2$ laser, cw, 20 W/s)

a 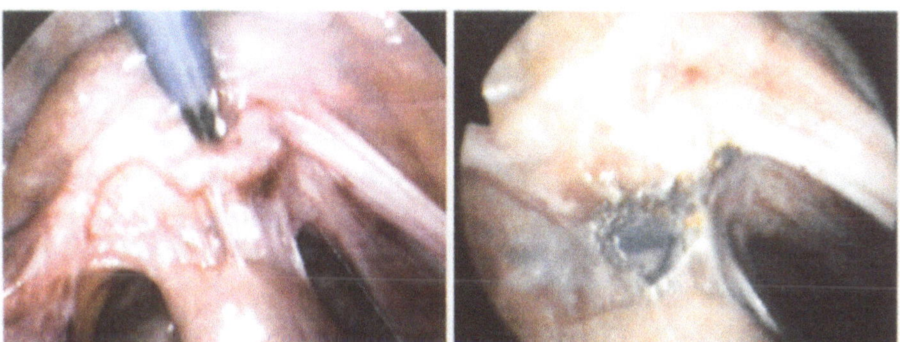 b

Lasers in Gynecology
Edited by G. Bastert and D. Wallwiener
© Springer-Verlag Berlin Heidelberg 1992

# Application of the Nd:YAG Contact Laser in Operative Pelviscopy

H. MECKE [1] and I. FREYS

Pelviscopy was, at its beginning, a purely diagnostic procedure. It was only through the introduction of safe methods of hemostasis, such as endocoagulation, coagulation with high-frequency current, endoligature, and endosuture [8], that operative pelviscopy became possible. The necrotic zone produced by endocoagulation or bipolar high-frequency current is determined by the size of the pelviscopic instrument and is at the very least 0.5 cm in diameter. In the case of high-frequency current, the size of the necrotic zone is in most cases appreciably greater. When employed for procedures on the adnexa, where an atraumatic technique is mandatory in order to maintain the fertility of women in their reproductive years, these methods seem to produce a disproportionate amount of trauma to the tissue. Furthermore, when used in the direct proximity to the bowels bipolar high-frequency current may be problematic, and endocoagulation [7], owing to the largeness of the instrument, may be impossible. In these particular cases the application of laser systems broadens the range of endoscopic operations which may be performed.

[1] Klinikum der Christian-Albrechts-Universität, Abt. Frauenheilkunde und Michaelis-Hebammenschule, Michaelisstr. 16, W-2300 Kiel, FRG.

Three types of lasers may be used pelviscopically: the $CO_2$ laser, the Nd:YAG as a no-touch and contact laser, and the argon laser. The YAG contact laser unites the advantage of the YAG laser, i.e., its ease of use via its flexible quartzfiber light cable, with that of the $CO_2$ laser, i.e., its sharply defined narrow necrotic zone.

The YAG contact laser has a sapphire contact tip which is fixed via a screw mechanism to the end of its light cable. Cone-shaped tips ar particularly suited for cutting, half-moon-shaped for vaporization, and cyclindrical for coagulation of tissue. Power densities high enough to produce a cutting or vaporization effect are found only on-contact at the immediate tip of the sapphire. The laser beam emitted at the sapphire tip is 0.4 cm in diameter [10].

When cutting is done with the cone-shaped tip the necrotic zone directly parallel to the incision margins measures less than 1 mm in diameter. Small vessels on both sides of the incision margins are coagulated, thus generally ensuring a virtually blood-free operative field. Where large vessels are injured hemostasis must be achieved employing the conventional pelviscopic techniques of endocoagulation, coagulation with bipolar high-frequency current, or application of the endoloop.

The light cable with a sapphire tip is used in combination with a laser probe which fits into the 5-mm-in-diameter

Lasers in Gynecology
Edited by G. Bastert and D. Wallwiener

1
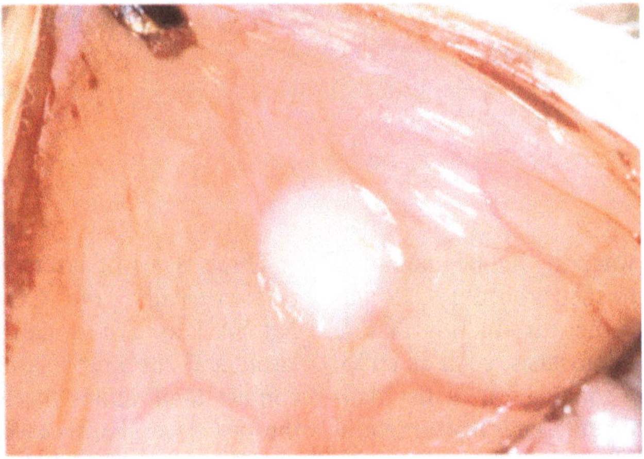

2
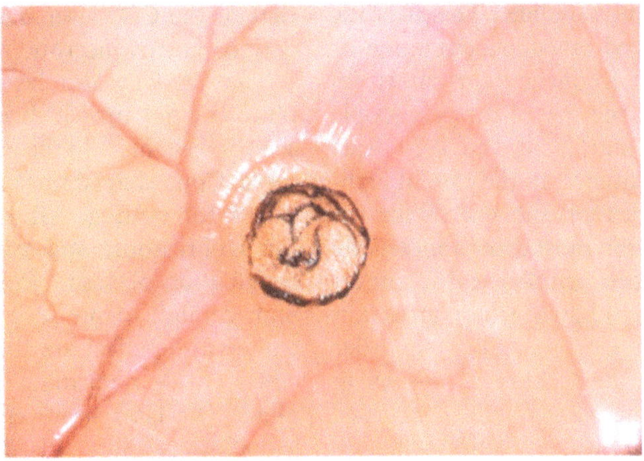

**Fig. 1.** Area of rat peritoneum endocoagulated at 120 °C for 20 s

**Fig. 2.** Area of rat peritoneum vaporized with the Nd:YAG contact laser

trocar sheath. Evacuation of smoke is not required, as only a small amount of smoke is produced. After a short warm-up period at 18–20 W with the tip immersed either in blood or in tissue, the cutting or vaporization effect may be produced with 10–15 W. Cutting or vaporization with the sapphire tip requires tissue contact, as otherwise the tip melts and is thus damaged.

Varying opinions have been reported on the frequency of postoperative adhesions following use of the YAG contact laser. In animal experiments, Wallwiener et al. [10] found no significant difference in the recurrence of adhesions after initial adhesiolysis between the $CO_2$ and the YAG contact laser.

In our own investigations using groups containing 20 rats each, after coagulation of equal-sized areas of peritoneal tissue of the anterior abdominal wall, we found that both bipolar high-frequency current and en-

3

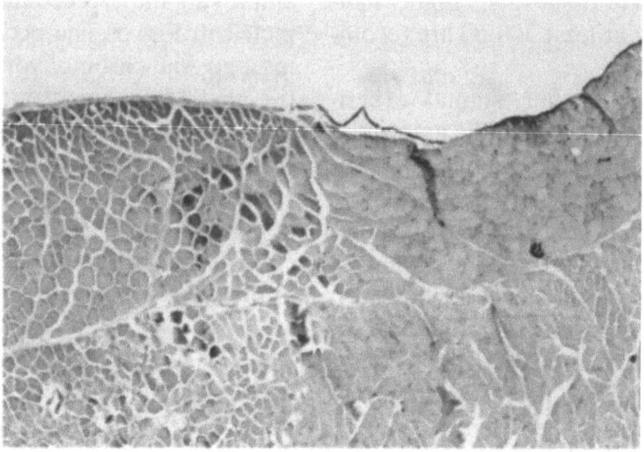

4

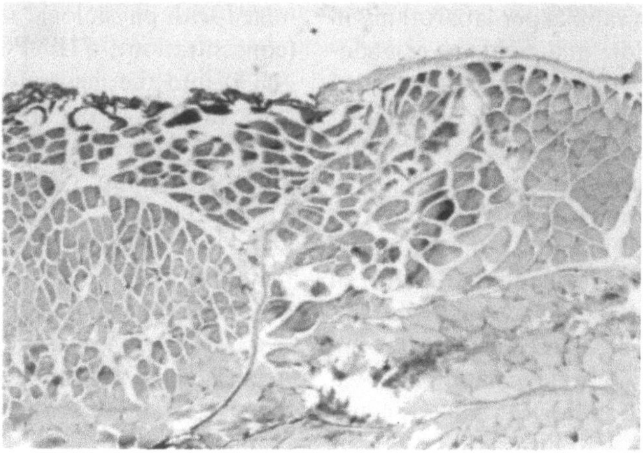

docoagulation at 120 °C produced significantly fewer adhesions as compared with vaporization of equal-sized areas of rat peritoneum using the YAG contact laser (Figs. 1,2; H. Mecke, unpublished data) or the $CO_2$ laser [3].

Coagulation and vaporization are two forms of thermal tissue distruction (Figs. 3,4). In order to coagulate biological tissue via either endocoagulation, bipolar high-frequency current, or the defocused laser beam, temperatures of at least 58 °C are recommend ed. Carbonization and charring of biological tissue requires temperatures of

**Fig. 3.** Rat abdominal wall after endocoagulation at 120 °C for 20 s. *Right*, wall is intact; *left*, thermally destroyed tissue with peritoneal and subperitoneal tissue thickening. Muscle fibers also show thickening. Note that distance between muscle fibers is enlarged. Modified Goldner

**Fig. 4.** Rat abdominal wall after vaporization with the Nd:YAG contact laser. *Right*, intact abdominal wall; *left*, vaporized area is cut. Peritoneum and superficial muscle layer are vaporized. The zone of carbonization superficially covers a layer histologically similar to that of coagulation. Modified Goldner at least 150 °C. For vaporization and evaporation of tissue much higher temperatures of at least 300 °C are recommended [1].

As shown in our animal experiments, vaporization and the associated carbonization of the remaining superficial tissue layers seem to produce more adhesions than coagulation of equal-sized peritoneal areas. The extent of postoperative adhesions is, however, determined not only by the type of thermic lesion, either coagulation or vaporization, but also by the amount of tissue damaged [9].

Adhesions of identical intensity were produced in rabbits per laparotomy in the left middle and right lower abdomen. These adhesions were then lysed per pelviscopy using the YAG contact laser, endocoagulation, or coagulation with bipolar high-frequency current and sharp transection. We found no significant difference in the frequency of the recurrence of adhesions between the various methods employed [4]. When the laser is used the greater potential for adhesion formation after vaporization of biological tissue is counterbalanced by the reduced tissue destruction, i.e., the narrow necrotic zone.

Clinically, the YAG contact laser is routinely used for operative pelviscopy in the following procedures:

1. Ovariolysis, salpingolysis, and salpingostomy in cases of sterility due to complete tubal occlusion
2. Pelviscopic treatment of tubal pregnancy
3. Adhesiolysis of the middle and lower abdomen in cases of chronic middle and lower abdominal complaints

*Ad 1:* The technique of salpingostomy employing the YAG contact laser is depicted in Fig. 5. The occluded tube is inflated via chromohydropertubation (i.e. methylene blue solution); then in the area of the former fimbrial end the tube is opened hemostatically using the YAG contact laser. The tubal mucosa is everted and fixed using a 4-0 PDS endosuture with an intracorporeal knotting technique.

*Ad 2:* The pleviscopic treatment of tubal pregnancy is shown in Fig. 6. After instillation of 10–20 ml of the vasoconstrictive agent POR 8 (Sandoz) mixed with physiologic saline solution (concentration: 5 IE POR 8/100 ml NaCl) into the mesosalpinx, a hemostatic incision of the tube directly over the area of the tubal pregnancy is produced using the YAG contact laser. The narrow necrotic zone on both sides of the wound margins guarantees that the tube is only minimally traumatized. This is a great advantage, especially in cases of early ectopic pregnancy when there is relatively small distortion of the tube and therefore a proportionately thicker tubal wall.

The emerging trophoblastic tissue is removed using a spoon forceps, 11 mm in diameter; this is followed by irrigation of the tubal lumen and careful inspection for any remaining trophoblastic tissue. The wound margins are adapted with a 4-0 PDS endosuture using the intracorporeal knotting technique.

*Ad 3*: A further indication for the use of the YAG contact laser is middle and lower abdominal adhesiolysis in cases of chronic abdominal complaints. Due to the narrow necrotic zone produced lateral to the wound margins, pelviscopic cutting may be performed in di-

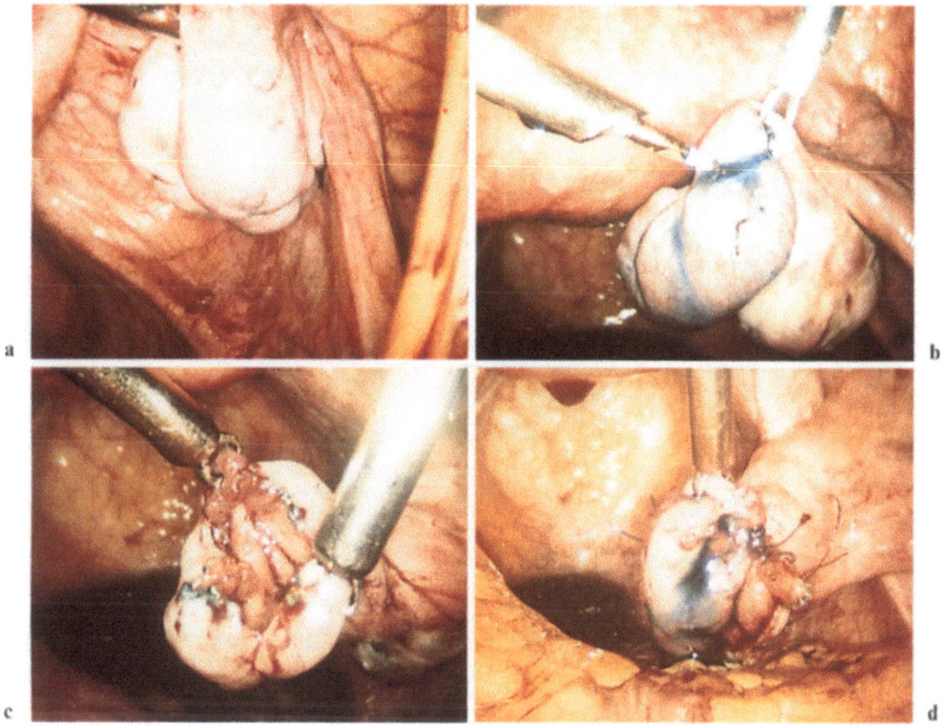

**Fig. 5. a** Terminally occluded right tube. **b** Opening of tube with the Nd:YAG contact laser. **c** Eversion of the fimbrial end. **d** Everted tubal mucosa is fixed using three 4-0 PDS endosutures [8]. At chromohydropertubation the tube was patent

**Table 1.** Division of adhesions into degrees of severity

| | |
|---|---|
| Grade I: | Filmy avascular adhesions |
| Grade II: | Thick and vascularized adhesions |
| Grade III: | Broad thick and vascularized adhesions and/or dense adhesions between organs |

rect proximity to the bowel (Fig. 7). The hemostatic effect of the laser ensures that the operative field remains clear. In cases of omental adhesions containing larger vessels, hemostasis is best performed using the endoloop or endocoagulation.

As regards both salpingostomy and the treatment of tubal pregnancies, we have not yet been able to follow up a sufficient number of patients to confirm the theoretical advantage of the laser, i.e., the minimal tissue destruction compared with the conventional pelviscopic techniques.

In 50 patients with chronic lower and middle abdominal pain where adhesions were the only pathological finding, complete adhesiolysis employing the YAG contact laser was performed. The adhesions were subdivided into three grades (Table 1). A follow-up procedure 6 months later was performed in 25 patients. Table 2 shows the degree of complaints 6 months following the complete pelvis-

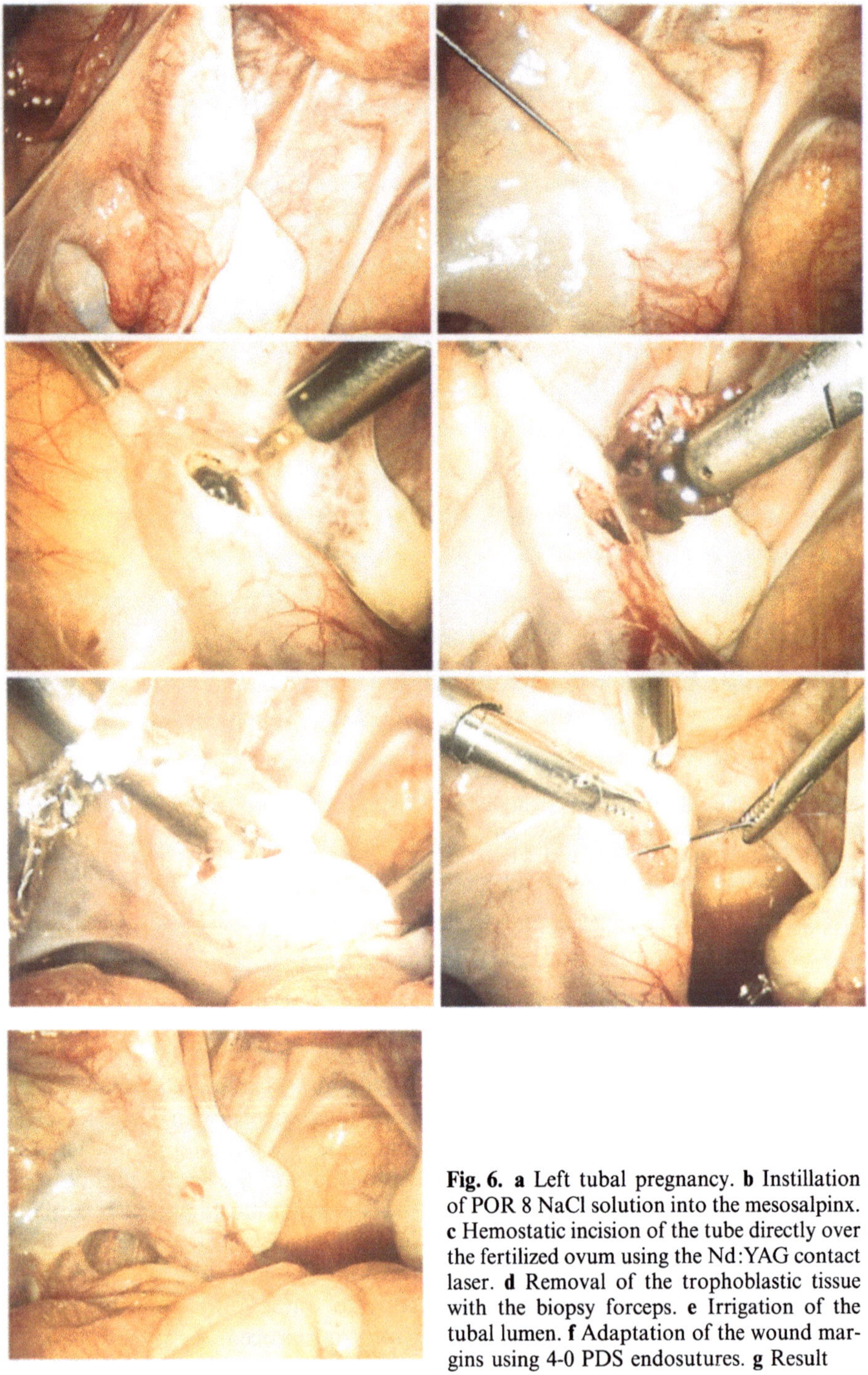

a b c d e f g

**Fig. 6.** **a** Left tubal pregnancy. **b** Instillation of POR 8 NaCl solution into the mesosalpinx. **c** Hemostatic incision of the tube directly over the fertilized ovum using the Nd:YAG contact laser. **d** Removal of the trophoblastic tissue with the biopsy forceps. **e** Irrigation of the tubal lumen. **f** Adaptation of the wound margins using 4-0 PDS endosutures. **g** Result

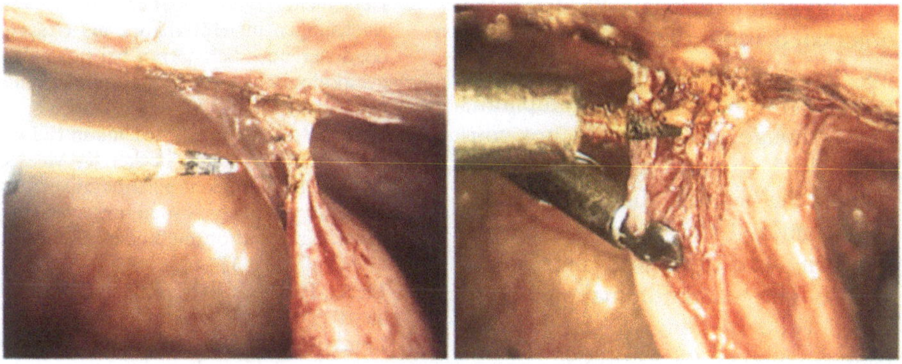

a b

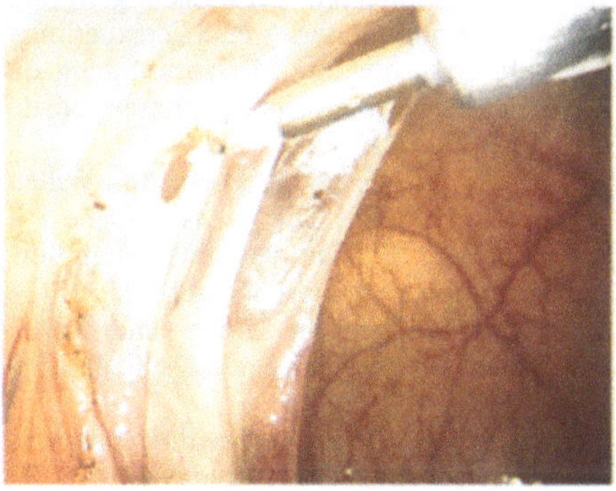

c

**Fig. 7. a, b** Dissection of the intestine from the anterior abdominal wall using the Nd:YAG contact laser. **c** Dissection of intestine from the lateral abdominal wall with the YAG contact laser

**Table 2.** Complaints 6 months following complete adhesiolysis using the YAG contact laser according to the degree of severity of the adhesions (from the Dept. of Obstetrics and Gynecology, University of Kiel, 1989)

| No. of patients | Grade of adhesions | Complaints | | | | |
|---|---|---|---|---|---|---|
| | | None left | Improved | Improved for a short period | Unchanged | Worsened |
| 0 | I | 0 | 0 | 0 | 0 | 0 |
| 10 | II | 5 | 3 | 0 | 1 | 1 |
| 15 | III | 3 | 5 | 4 | 3 | 0 |
| ($n = 25$) | | 8 (32%) | 8 (32%) | 4 (16%) | 4 (16%) | 1 (4%) |

copic adhesiolysis using the YAG contact laser.

Thirty-two percent of these 25 follow-up patients were free of complaints, and another 32% had at least noticed an improvement in their symptoms. Six months after complete pelviscopic adhesiolysis using conventional techniques we found that 38% of 39 patients were complaint free and 21% [5] had some improvement. No significant difference is demonstrated between the two groups.

The significance of adhesions in the etiology of chronic abdominal pain is controversial. On the one hand, adhesions are the only diagnosed pathological finding in 26%–86% of all cases of chronic abdominal pain [2,6]; on the other hand, intra-abdominal adhesions are found in up to 34% of operated patients who have no such complaints [6]. Although adhesions do not routinely lead to intra-abdominal complaints, the fact that they can produce lower and mid abdominal pain means they cannot be ruled out. Therefore, in cases of chronic abdominal pain associated with adhesions, we aim for complete pelviscopic adhesiolysis.

The laser, however, does not replace the conventional pelviscopic techniques. Adhesions which may be lysed without bleeding, either bluntly or sharply using a scissors, do not call for the application of a laser. Hemorrhage from larger vessels is very difficult to control with the laser. Endoligature, endosuture, and endocoagulation remain the methods of choice for hemostasis here. The advantages of the laser are that it allows for a practically hemostatic incision and produces narrow parallel necrotic zones which cause less traumatic destruction of the adjacent tissue. It is therefore our opinion that for the aforementioned indications the laser is a sensible and valuable addition to the traditional pelviscopic instrumentarium.

## References


1. Berlien HP, Müller GJ (1989) Angewandte Lasermedizin. 'Ecomed, Landsberg
2. Herschlein HJ, Lechner W (1974) Verwachsungen im Unterbauch als Ursache rezidivierender Schmerzen bei normalem gynäkologischen Tastbefund. Geburtshilfe Frauenheilk 34:303–306
3. Mecke H, Riedel HH, Möhlmann L, Lehmann-Willenbrock E, Becker K, Mees C (1988) Die Häufigkeit postoperativer Verwachsungen nach Anwendung verschiedener Koagulationsmethoden. Zentralbl Gynakol 110:433–442
4. Mecke H, Riedel H-H, Schloe U, Nekarda P (1988) Pelviskopische Verwachsungslösung mit verschiedenen Techniken unter besonderer Berücksichtigung des YAG-Kontaktlasers – eine tierexperimentelle Untersuchung. 4. Jahrestagung der Deutschen Gesellschaft für Lasermedizin, Wien, 5–8 Oct 1988
5. Mecke H, Semm K, Lehmann-Willenbrock E (1988) Die pelviskopische Adhäsiolyse. Erfolge bei der Behandlung chronischer Unterbauchschmerzen bei Verwachsungen im Unter- und Mittelbauch. Geburtshilfe Frauenheilk 48:155–159
6. Rapkin AJ (1986) Adhesions and pelvic pain: a retrospective study. Obstet Gynecol 68:13–15
7. Semm K (1983) Physical and biological considerations militating against the use of endoscopically applied high-frequency current in the abdomen. Endoscopy 15:282–288
8. Semm K (1984) Operationslehre für endoskopische Abdominal-Chirurgie. Schattauer, Stuttgart
9. Stangel JL, Nisbet JD, Settles H (1984) Formation and prevention of postoperative abdominal adhesions. J Reprod Med 29:143–161
10. Wallwiener D, Morawski A, Planterer G, Bastert G (1982) Ist die Adhäsiolyse mittels Neodym-YAG-Laser eine Alternative zur $CO_2$-Laser-Adhäsiolyse? Laser Med Surg 3:142–147

# Operative Treatment of Tubal Pregnancy

D. WALLWIENER [1], D. POLLMANN, J. GAUWERKY, C. SOHN, and G. BASTERT

## The Change in the Manifestation of Tubal Pregnancies

Tubal pregnancies are presently still a challenge for surgeons. The difficulties they are confronted with are different, however, from those encountered in past decades or even at the end of the past century. This is due partially to the change in the manifestation of tubal pregnancies and partially to the increasing significance of the sterility problems usually associated with tubal pregnancies.

Until well toward the end of the first half of this century, the acute and potentially fatal threat to the patient as a result of a tubal pregnancy and thus ablation as a life-saving measure were of central importance. The mortality, which was still well in excess of two thirds of the afflicted women, was reduced only gradually as of the beginning of this century with the introduction of ablative surgical techniques [45, 66].

Modern diagnostic procedures, such as a control of the $\beta$-HCG levels and transvaginal sonography, have lead to a complete change in the manifestation of tubal pregnancies [32, 39]. A planned, instead of an emergency operation, can be performed on the majority of the patients due to an early diagnosis of the extrauterine pregnancy. In approximately a third of the cases one is even confronted with a unruptured tubal pregnancy in which neither the tubal wall has ruptured nor has there been loss of blood into the abdominal cavity [42].

A further important aspect of extrauterine pregnancies is their socioeconomic significance, an issue which is gradually becoming the central point of interest. The incidence of extrauterine pregnancies has tripled during the course of the past decade [30], but no definite explanation for the etiologic pathogenesis has been found. Against this background, the full significance of the sterility problems associated with tubal pregnancies becomes evident.

One must reckon with secondary tubal sterility – even when the contralateral tube is intact – after radical surgical treatment of tubal pregnancies in up to 70% of all cases, and in 90% of afflicted first pregnancies. The earliest possible diagnosis and the choice of the specific operation method for the individual patient thus present major clinical and scientific challenges [73].

[1] Universitäts-Frauenklinik, Abt. für Geburtshilfe und Gynäkologie, Voßstr. 9, W-6900 Heidelberg, FRG.

Lasers in Gynecology
Edited by G. Bastert and D. Wallwiener

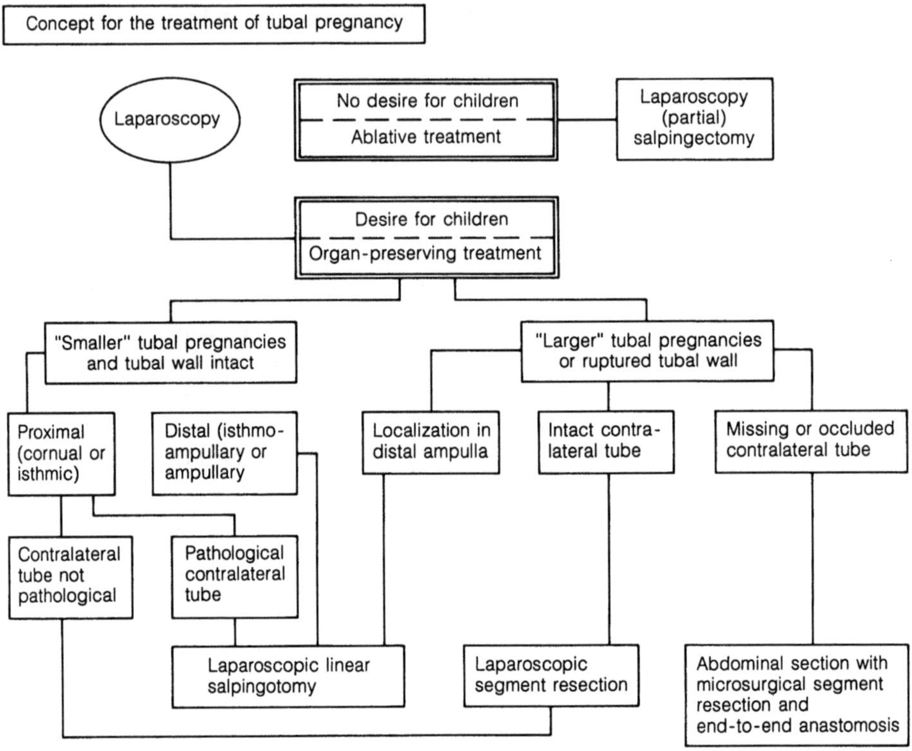


**Fig. 1.** Surgical concept of the operation methods utilized for tubal pregnancy

# The Operation Techniques

Figure 1 gives an overview of the concept of operative treatment of tubal pregnancy.

## Ablation

Radical surgical intervention, such as a total or partial salpingectomy or an adnexectomy, was first described over 100 years ago [66].

## Organ-preserving Microsurgical Methods

Numerous microsurgical and/or laparoscopic operation techniques are available for the conservative (or better said organ-preserving) treatment of tubal pregnancies. As a matter of principle one must differentiate between techniques which maintain continuity and those which interrupt continuity.

### Indirect Organ-preserving (Continuity-interrupting) Techniques

*Simple Segment Resection*
During simple segment resection the tubal segment which carries the gestational product is removed without a primary end-to-end anastomosis of the tubal stumps ensuing (Fig. 2). The mesosalpinx is preserved during resection (Fig. 3). The tubal stumps are submerged by peritonization to enable a

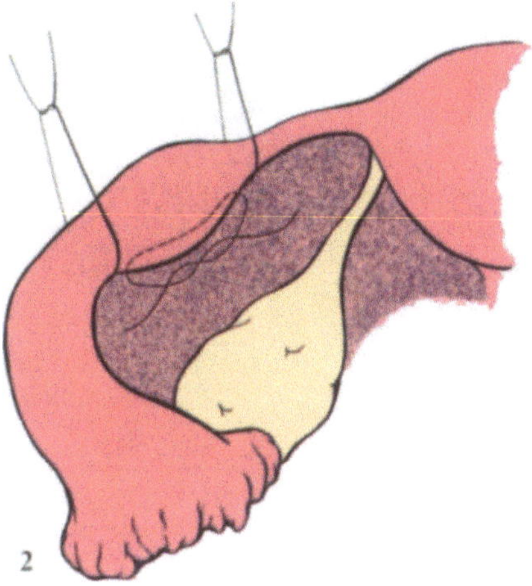

2

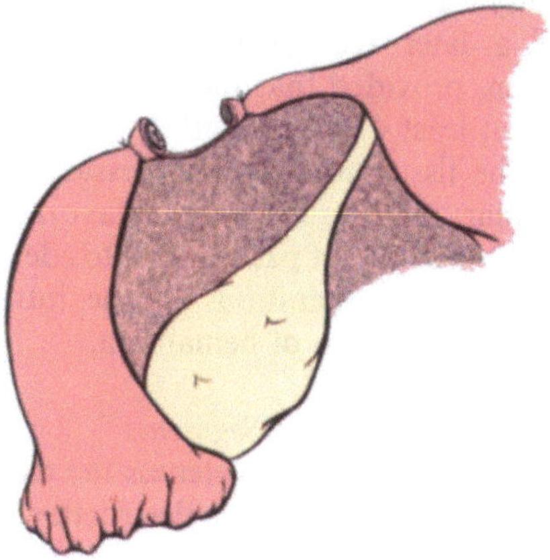

4

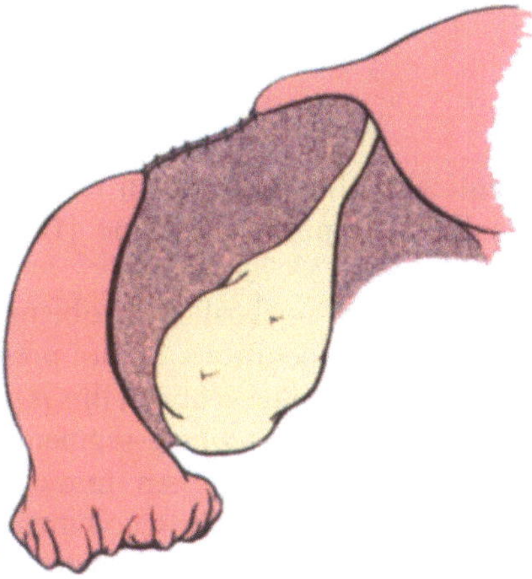

3

**Fig. 2.** Isthmic tubal pregnancy: microsurgical segment resection

**Fig. 3.** Isthmic tubal pregnancy (after microsurgical segment resection)

**Fig. 4.** Isthmic tubal pregnancy (after peritonization and submersion of the tubal stumps subsequent to microsurgical segment resection)

future tubo-tubal end-to-end anastomosis [10, 11, 29, 31, 37, 44, 46, 49, 69]. During segment resection the tube is first ligated on the proximal and peripheral sides of the tubal pregnancy (vicryl 4-0 to 6-0).

As the tubal stumps might have to be anastomosed at a later date, the atraumatic handling of the tube, where possible with microsurgical instruments, is imperative. Microsurgical precautions, such as the utilization of an operating microscope are superfluous, however. Either a sharp resection is then performed or a microelectrode is used. Thereafter, the tubal stumps are submerged by peritonization using absorbable thread (vicryl 6-0) (Fig. 4).

The advantage of segment resection lies in the fact that – at least hypothetically – the damaged tubal segment has been excised and a recurrence is thus no longer possible. Furthermore, there is no risk of trophoblast residues being left behind; the infiltration of trophoblasts into the muscles and/or mesosalpinx can never be fully precluded [2, 9, 15, 34].

Thus segment resection is advisable for those patients in whom tubal continuity is not imperative. The possibility of a later reconstruction is always given, provided that the tubal stumps are at least 4 cm in length.

The tissue conditions necessary for primary anastomosis are unfavorable in the majority of patients due to edematous tissue alterations or the intramural formation of hematoma.

*The Coagulation of Segments*
For the sake of completeness the technically extremely simple destruction of tubal pregnancies by electrocoagulation ought to be mentioned [28, 41–43, 54]. Such segment coagulation is possible only in extremely small tubal pregnancies. Furthermore, a histological safeguard is not present. Extensive electrocoagulation is necessary to ensure that the trophoblast cells have been entirely destroyed.

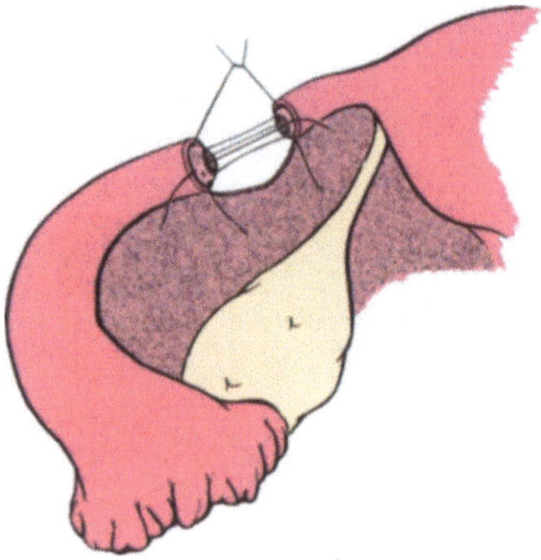

**Fig. 5.** Isthmic tubal pregnancy: end-to-end anastomosis

## Direct Organ-preserving (Continuity-maintaining) Techniques

*Segment Resection with Simultaneous End-to-end Anastomosis*
Primary end-to-end anastomosis directly after segment resection is reserved for those cases in which the contralateral tube is missing or severely damaged [60, 61]. The basic prerequisite is, however, that conditions are favorable for a first microsurgical intervention. The anatomical conditions must be appropriate and the damage caused by the tubal pregnancy must not be too extensive; i.e., a parietal infraction, an intramural hematoma, or a hematoma in the mesosalpinx must not be present. Furthermore, one must bear in mind that an entire microsurgical team must be available for such a complex operation. The technique employed in end-to-end anastomosis is identical to the well-known microsurgical anastomosis technique [18–22, 25–27, 31, 51, 52, 70] (Fig. 5).

A further unclarified problem is whether two operations are not preferable to a single intervention; this is particularly important where segment resection is concerned. An immediate reconstruction is to be recommended only in those rare cases in which an extremely early diagnosis of the tubal pregnancy has been made and in which it is located in the isthmus of the tube or in the isthmoampullary section; there would then be no great difference in lumina after excision of the tubal pregnancy. Above all else, the tissue should not be too edematous and no hematoma should be present in the strata; an infectious reaction must also be precluded. The residual tubal length must be at least 4–5 cm.

In the majority of cases the tissue conditions are most unfavorable for an anastomosis, however; hence, two interventions are usually advocated. During the first operation only the segment resection is carried out; an anastomosis can then be performed in irritation-free tissue at a later date, should it be impossible for an intrauterine pregnancy to develop via the contralateral tube or should an extrauterine pregnancy reoccur. Due to the small number of cases no definitive answer to this important question has yet been found. This is also true of the question as to whether or not a pathological tube on the contralateral side is an indication for a simultaneous reconstructive intervention using either a macro- or microsurgical technique. In our opinion, the same arguments are valid as for anastomosis in one or two interventions.

*Linear Salpingotomy*

The significance of continuity-maintaining treatment by means of a linear salpingotomy increases to the same extent as the number of tubal pregnancies diagnosed in the early stages increases. The tubal wall is incised above the gestational product and the tubal pregnancy is removed carefully. Salpingotomy is ideal for these early cases in which the tubal wall has not yet ruptured, especially if the tubal pregnancy is located in an isthmoampullary or ampullary position.

After atraumatic fixation of the tube an antimesenteric or a mesenteric linear salpingotomy is performed, depending upon the anatomical situation [3,4]. If atraumatic precautions are maintained and the trophoblasts are thoroughly removed, a linear salpingotomy is a very easy surgical technique, excluding those few cases in which the bed of the tubal pregnancy bleeds heavily; microsurgical instruments and an operating microscope are not necessary. As far as the salpingotomy preparation technique is concerned, we refer to the comparison of techniques at the end of this paper.

The toilet of the peritoneal area is performed after the bleeding from the bed of the wound of the tubal pregnancy has been stilled. Some teams of surgeons close the salpingotomy wound with a single interrupted suture or with a continuous seromuscular suture. An increasing number of teams, however, make do without a suture after a salpingotomy if the wound is only 2–3 cm in length [19,20]. In surgical experiments on the tubes of minipigs we were able to demonstrate that complete healing of a linear salpingotomy wound was possible in almost all cases without dehiscence or fistulization resulting [70].

*Transampullary "Expression" of the Tubal Pregnancy ("Milk-out")*

In some cases tubal pregnancies may be "expressed" through the fimbriated end of the tube. (Fig. 6). A possible problem is to be found in the total removal of the trophoblast tissue or in intra-ampullary control of bleeding in a diffuse hemorrhage. An alternative method to be advocated is a peripheral transampullary linear salpingotomy (Fig. 7); this enables a complete exploration as well as a more exact control of the blood flow. In experiments recurrences were found to be more frequent after "expression" of the tubal abortion than after distal ampullary salpingotomy [3]. According to Timonen, the pregnancy rate is also higher after invasive treatment than after a simple "milk-out" [67]. We did not lay a suture after a fimbria-near or ampullary linear salpingotomy; this was also the case after treatment of isthmic or isthmoampullary tubal pregnancies.

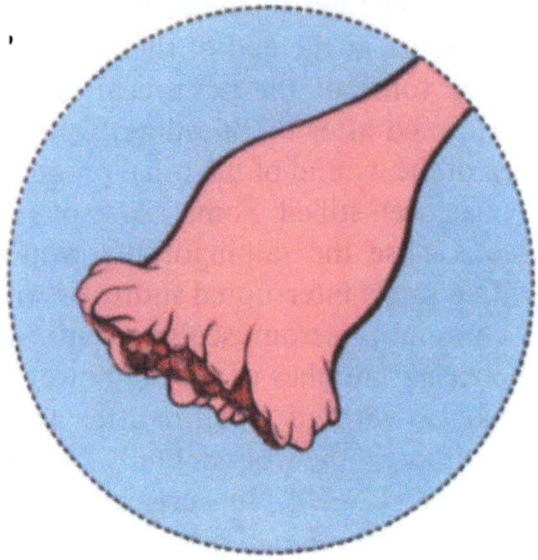

6

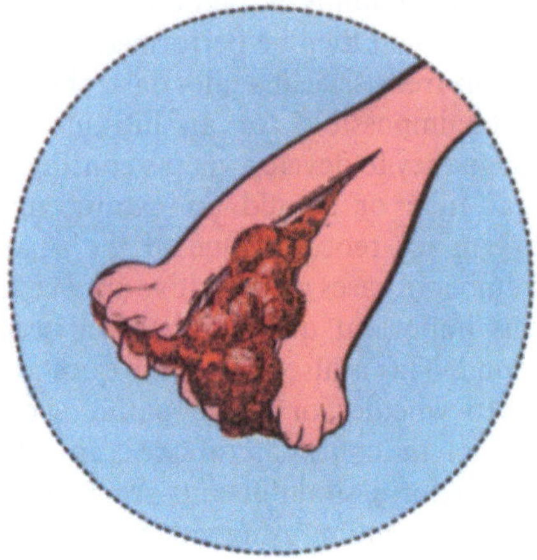

7

**Fig. 6.** Tubal abortion in statu nascendi

**Fig. 7.** Ampullary tubal pregnancy: linear ampullary salpingotomy

## Organ-preserving Laparoscopic Methods

Due to the optimization of surgical laparoscopic methods the treatment of tubal pregnancies by abdominal section has increasingly lost in significance and is reserved merely for a particular range of indications. Both ablative and the above-described indirect or direct tube-preserving methods can be laparoscopically performed.

Presently, however, the question remains as to the extent to which a primary re-anastomosis after segment resection can or ought to be laparoscopically performed. Gauwerky and Wallwiener have reported about laparoscopic end-to-end anastomization within the framework of refertilization; how far these techniques are transferable remains to be established in further studies [19–21, 70–72].

For over a decade now, various teams have described the modification of individual surgical techniques for laparoscopic purposes. Bruhat, de Cherney, and Semm reported on their initial series in which tubal pregnancies were removed by laparoscopic linear salpingotomy [5–7, 12–14, 58]. The first endosopic segment resections were performed by Gomel.

### Operative Technique

The laparoscopic techniques differ depending upon the choice of an ablative or a direct or indirect organ-preserving preparation technique. Three auxiliary punctures are always made suprasymphyseally in the cranial region of the pubic hair, the first puncture in the median line, the other two laterally of the epigastric artery. In all cases the tube is exposed with atraumatic fixation forceps through the lateral punctures. Scissors, bipolar forceps, an endocoagulator, or the various laser applicators are inserted through the middle auxiliary puncture, i.e. via the working channel of the laparoscope. When utilizing endoloops as developed by Roeder, the applicator for the endoligature is inserted intraperitoneally through the middle puncture.

The tissue is extracted through the laparoscopic working channel with an instrument of appropriate diameter to remove the tubal pregnancy and the excised parts of the tube or the adnexa from the intraperitoneal area; if larger pieces of tissue are present the median puncture is redilatated to allow for a 10-mm instead of a 5-mm trocar. After careful resection – the extraction of the tubal pregnancy from the interperitoneal area should be as total as possible – a toilet of the intraperitoneal area is carried out, all coagulated blood being removed in the process.

A thorough documentation of the condition of the contralateral tube, including a chromopertubation, is imperative. Important details concerning the various laparoscopic preparation techniques (partial salpingectomy, segment resection, linear salpingotomy) are found below.

## (Partial) Salpingectomy

A salpingectomy or partial salpingectomy is performed like an adnexectomy following the three-loop technique according to Semm and utilizing a endoloop. The endoloops are inserted intraperitoneally one after another through the applicator and placed over the tissue to be removed. After the prepared loops have been drawn together they are tightly knotted over the tubal segment or the stalk of the adnexa; the superfluous catgut is cut off with a pair of endoscissors and extracted. Three loops are applied for safety's sake. Thereafter, the tissue is removed using endoscissors [55–58].

## Segment Resection

Laparoscopic segment resection is carried out similarly to abdominal segment resection. It is an alternative to a salpingotomy, especially in proximally located tubal pregnancies, as the risk of secondary tubal stenosis and/or the retention of trophoblast residues must be minimized in such cases (Fig. 8). Laparoscopic segment resection was first described by Shapiro and Adler [59], as well as by Gomel and his team [23, 24].

We prefer segment resection by laser preparation or segment resection using tubal clips as introduced by Gauwerky to electrosurgical preparation, as this allows for maximum protection of the still-intact tissue. With the clip technique, commercial Filshie clips are applied proximal and distal to the tubal pregnancy using the applicator; the

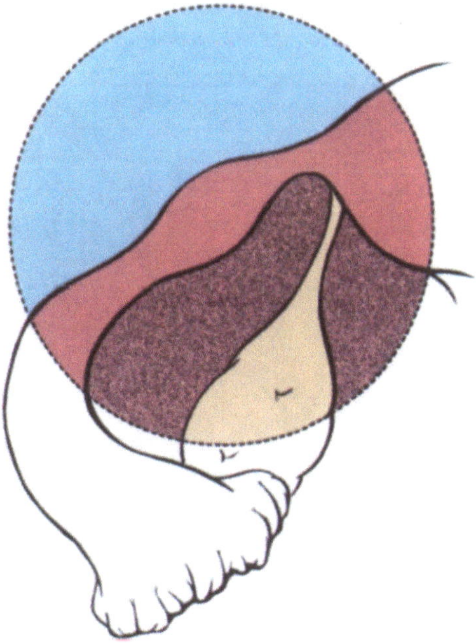

**Fig. 8.** Isthmic tubal pregnancy: laparoscopic view

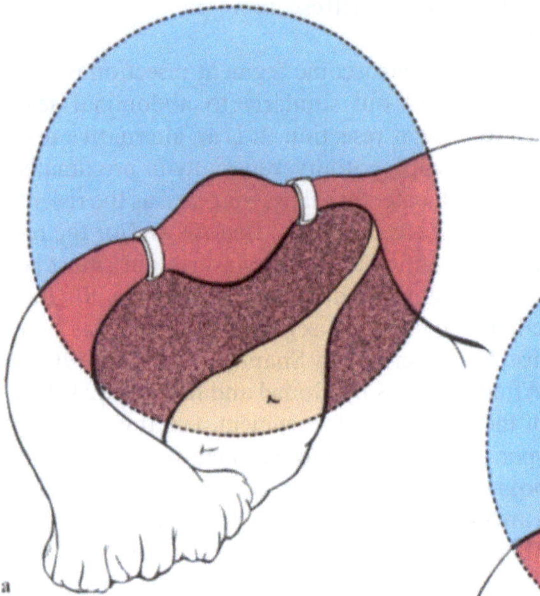

a

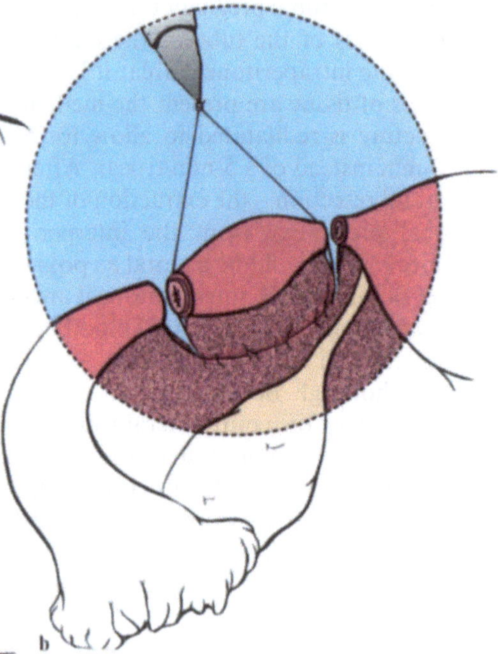

b

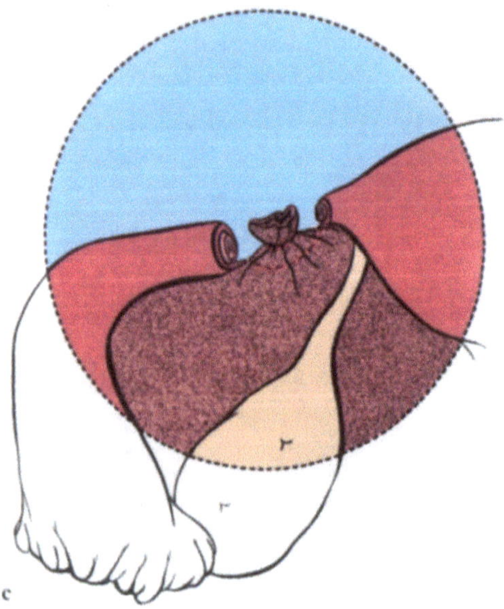

c

**Fig. 9a–c.** Isthmic tubal pregnancy: **a** laparoscopic isthmic segment resection using endoclips; **b** laparoscopic segment resection with endoloop; **c** after laparoscopic segment resection with endoloops tubal segment between the clips is then removed [20].

To avoid hemorrhaging from the mesosalpinx the tubal segment from the mesosalpinx can also be detached after it has been made entirely blood free by bipolar coagulation or the placement of an endoloop (Fig. 9).

## Linear Salpingotomy

A laparoscopic salpingotomy is preferably performed with a surgical laser, as the atraumatic technique allows for the greatest tissue protection [1,8,16,17, 70–73] (Fig. 10). Tissue infiltration with vasopressin derivatives is not necessary prior to laser preparation (Fig. 11).

A bipolar coagulation set and suction rinsing equipment must be available in all cases of laparoscopic surgical intervention; bleeding can occur in every laparoscopic operation. The blood can flow from one of the smaller arteries in the wound bed of the tubal pregnancy, or the hemorrhage can be in the form of abundant spurting blood from, for example, the tubal branch of the uterine artery; this cannot be coagulated with the laser, and bipolar coag-

**Fig. 10 a**. Isthmic tubal pregnancy: linear salpingotomy using the $CO_2$ laser and **b** extirpation of tubal pregnancy

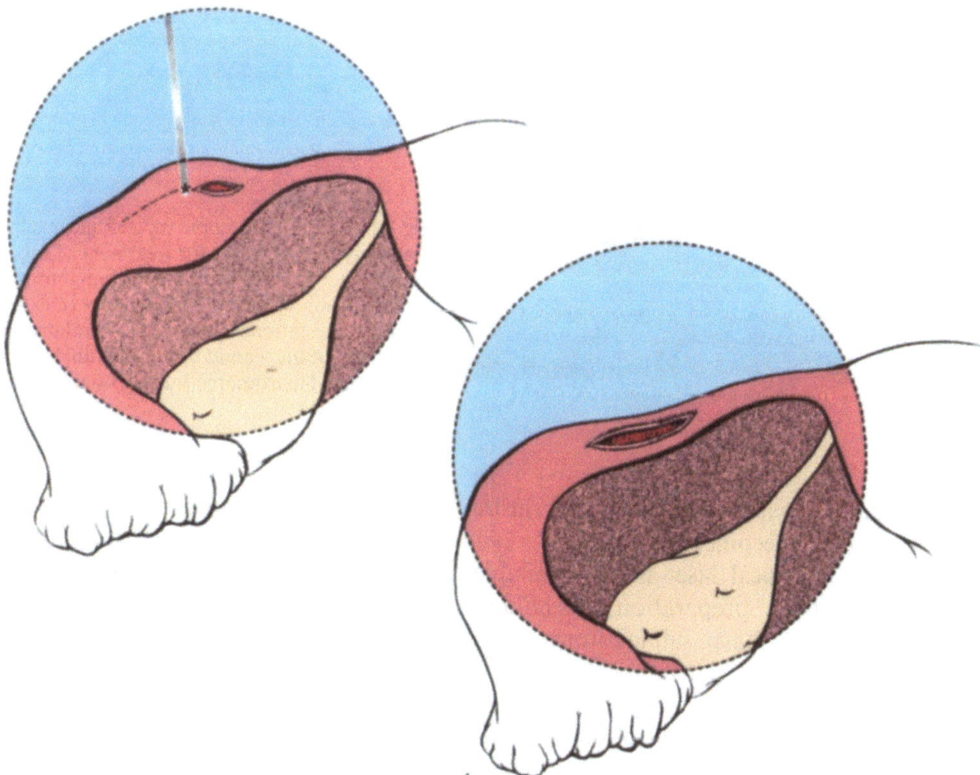

a

b

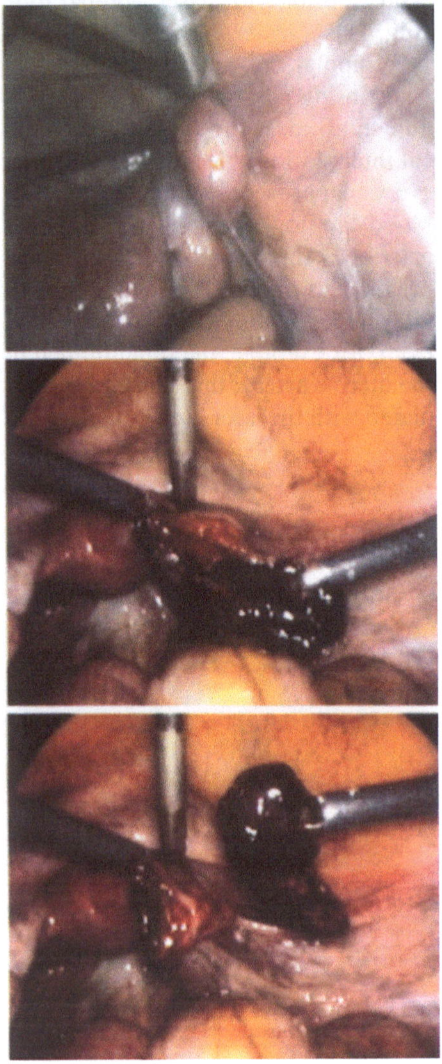

a b c

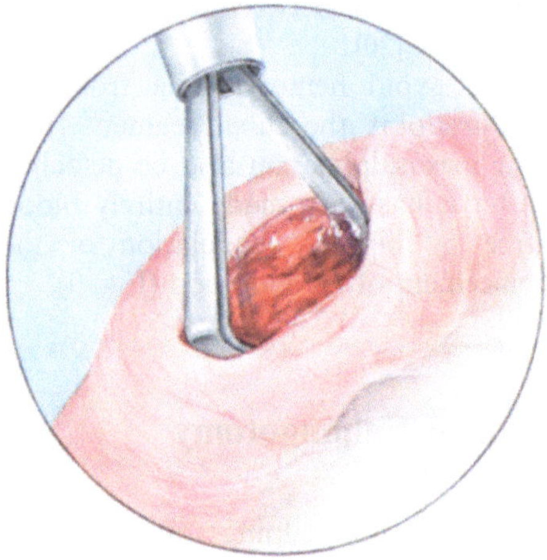

d

**Fig. 11 a–d.** Laparoscopic view of linear salpingotomy by $CO_2$ laser for treatment of tubal pregnancy: **a** linear isthmic salpingotomy; **b** preparation of gestational product; **c** extirpation of gestational product; **d** intraluminal inspection of the wound bed of the tubal pregnancy after laparoscopic salpingotomy

ulation with simultaneous rinsing therefore becomes necessary.

If a surgical laser is not available, alternative preparation techniques such as a sharp salpingotomy using scissors or a laparoscopic scalpel can be employed after vasopressin application. A vasopressin solution, which must be highly diluted for anesthesiological reasons, is applied (por-8, a few milliliters of a solution in which a por-8 ampulla with 2.5 IU has been diluted to 25 ml; 10 ml thereof are drawn off). We do not utilize the endocoagulation technique, as it implies the coagulation of a relatively wide tenia.

A laparoscopic salpingotomy necessitates an extremely careful, and if possible blunt excision of the tubal pregnancy, as the risk of trophoblast residues is at highest using this technique. As already mentioned above, suture of the salpingotomy wound can be foregone in the majority of cases. Meticulous hemostasis and toilet of the intra-abdominal area are, as after every laparoscopic operation, imperative.

## Laser Operating Techniques

To date, electrosurgical techniques such as the high-frequency and endocoagulation techniques have predominated in operative laparoscopy [54, 55]. Numerous and diverse attempts have been made to modify and supplement the preparation techniques used in endoscopic gynecological fertility surgery [58].

However, all preparation techniques either are confronted with operative technical problems or harbor risks for the patients [50, 54–58]. The endocoagulation technique, for example, is not dangerous for the patient, as the current flows in a separate electric circuit within the instrument. However, the tissue destruction is extensive, as a point coagulator already has a diameter of 5 mm. Instead of the endocoagulation technique thermal tissue destruction by means of endoscopic vasopressin application is utilized to obtain local ischemia and thus reduce this area. The administration of this substance necessitates anesthesiological high-risk control, however, as cardiovascular problems are to be feared. This is the reason for the increased significance and popularity of the laser preparation techniques.

## Laser Laparoscopic Salpingotomy as Organ-preserving Treatment of Tubal Pregnancies

Due to their technical modification both the $CO_2$ laser and the Nd:YAG contact preparation techniques are ideal for the organ-preserving treatment of tubal pregnancies. The tubal pregnancies are removed by means of a laser laparoscopic linear salpingotomy, and the organs are preserved [38] (Fig. 9).

With the help of these endoscopic operating techniques a salpingotomy and the removal of the tubal pregnancy in its early stages (i.e., prior to rupture of the wall and a maximum size of 4–5 cm) can be performed laparoscopically. If a laparoscopic sharp salpingotomy after electrocoagulation of the tubal tissue to be cut is planned, suture of the salpingotomy wound is necessary as the wound edges are rigid and do not adapt spontaneously due to the electrocoagulation. The majority of authors also consider a local vasopressin injection into the tubal wall necessary prior to the intervention to ensure sufficient hemostasis [54–58]. In animal experiments we were able to demonstrate that a precise preparation of the tubal tissue with simultaneous hemostatic effect and, furthermore, without the application of vasopressin is possible when operating with a laser.

Twenty-five tubal pregnancies were removed by laser laparoscopic salpingotomy within the framework of our clinical pilot study. Only two of the 12 patients who have already undergone a follow-up examination (hysterosalpingography, relaparoscopy, spontaneous pregnancy with only one tube) were found to have an occluded tube. Furthermore, a wound toilet was not necessary in most cases due to the negligible tubal tissue destruction [30].

## Technical Surgical Procedure and Summarizing Evaluation

The number of surgical interventions due to extrauterine pregnancies with associated maintenance of fertility has increased to the same extent as the number of emergency operations of tubal pregnancies has decreased. As early as in 1895, Broschownik summarized the problems associated with the conservative treatment of tubal pregnancies in one sentence: "It must be our aim to remain conservative, yet reduce the risks...!" [51].

The fundamental aim of the conservative operating techniques is thus to preserve the tube whenever this is justifiable, desired by the patient, and medically possible. An early diagnosis of the tubal pregnancy and a selection of the patients are thus the main criteria for organ-preserving intervention. An organ-preserving operation is possible in almost 95% of cases [28,41,42].

### Indications for an Organ-preserving Procedure

In patients with a tubal pregnancy who would like to have children an organ-preserving operation is always indicated. An organ-preserving procedure is also to be recommended for patients who are of child-bearing age, even if they do not wish to have children in the future. The advantages of a continuity-maintaining procedure and the associated risks must be carefully considered in such patients, however.

By now it has been established that the fertility rate is higher after conservative treatment of a tubal pregnancy than after an ablative procedure. This is also true for patients with an intact contralateral tube. Twenty percent of patients are confronted with fertility disturbances prior to a tubal pregnancy; this rises to 50% after a tubal pregnancy.

### Risks

The risks associated with organ-preserving treatment must always be discussed with the patient. The complaints typically associated with tube-preserving operations, in which the bed of the gestational product is not entirely excised, rebleeding occurs, and tubal pregnancies in the form of trophoblast residues persist, are the most significant. These risks become reality in 5% of all cases [41,46]. A routine postoperative control of the serum $\beta$-HCG level up to its detection limit is thus indispensable.

### Contraindications for an Organ-preserving Procedure

In an inquiry conducted by Gauwerky [19] the advanced age of a patient, a possible multiparity, an intact contralateral tube, and an unfavorable anatomical position were the most common factors quoted by various surgical teams as contraindications for organ-preserving treatment of a tubal pregnancy. An intact contralateral tube is by no means an indication for an ablative procedure, however. Under such conditions one must merely decide between a continuity-maintaining and a continuity-interrupting intervention.

## Results

Table 1 gives an overview of the results of organ-preserving and ablative treatment of tubal pregnancies as described in the literature. As far as objective evidence of the success of conservative operations of tubal pregnancies is concerned, only those cases can be evaluated in which only one tube is still intact or the contralateral tube is damaged in the sense of a total occlusion and in which an intrauterine pregnancy has developed after an organ-preserving and continuity-maintaining operation. Unfortunately, such information is difficult to obtain in the literature and, furthermore, such cases are extremely rare.

Either way, one can deduce from the available figures that the intrauterine pregnancy rate subsequent to tube-preserving intervention and with only one remaining fallopian tube lies between 56% and 60% [35,36,41,43,53], and that the chances of success are therefore much higher than for in-vitro fertilization.

The average tubal pregnancy recurrence rate is 22%. Particularly worthy of mention is the fact that the 36% recurrence rate after isthmic localization of the previous tubal pregnancy is significantly higher than that after ampullary localization, where it is only 18%. Whether or not this is due to the

**Table 1.** Overview of the results found in the literature

| | Organ-preserving treatment | | | | Ablative treatment | |
|---|---|---|---|---|---|---|
| | contralateral normal tube | | contralateral pathological tube | | contralateral nonpathological tube | |
| | X̄ | Reference | X̄ | Reference | X̄ | Reference |
| Intrauterine pregnancy rate | 64% | 64% Pouly [47]<br>75% Bruhat [5–7]<br>83% Langer [37]<br>50% Bukovsky [10]<br>60% Gauwerky [19,20]<br>Wallwiener [73]<br>50% de Cherney [12–14] | 53% | 46% Pouly [47]<br>54% Neeser [41,42]<br>60% Oelsner [43] | 42% | Schenker [53] |
| Recurrence rate | 18% | 22% Pouly [47]<br>12% Bruhat [5–7]<br>29% Henry-Suchet [26]<br>Kucera [35]<br>Stromme [62,63] | 22% | 29% Pouly [47]<br>16% Neeser [41,42]<br>22% Oelsner [43] | salpingotomy<br>16 | Schenker [53]<br>salpingo-oophorectomy<br>5 |
| Persistence rate after laparoscopic salpingotomy | | 5% Bruhat [5–7] | | | | |

stenosis subsequent to linear salpingotomy of the distinctly narrower isthmic lumen compared with the wider ampullary lumen is at present still unclear.

## The Surgical-technical Procedure

The choice of surgical method, particularly of the conservative intervention method depends, as mentioned, upon

- The current or future desire for children
- The age of the patient
- The condition of the contralateral tube
- The localization of the tubal pregnancy

When considering which of the technical operation methods ought to be used, the requirement to minimize the surgical trauma, which is guaranteed by the endoscopic method, is of fundamental importance. A precondition for laparoscopy is that the patient has a stable circulatory system. Evidence of intraperitoneal blood is, on the other hand, not a contraindication for a diagnostic and/or therapeutic laparoscopy. This is also the reason why a puncture of the pouch of Douglas has lost its significance as far as planning the technical method of operation is concerned.

After an operative laparoscopic intervention by means of a linear salpingotomy the risk of secondary hemorrhages and/or trophoblast residues remaining [19, 20] exists; this method should thus be reserved for patients who would like to have children. Those patients who have no desire for further children can be offered the greatest level of security by utilizing the ablative method, i.e., a partial or total salpingotomy. In patients with a desire for children and an occluded or missing contralateral tube either a linear salpingotomy or a segment resection by means of an abdominal section with one-sided end-to-end anastomosis is the preferred method.

As a higher recurrence rate is to be expected after linear salpingotomy when the tubal pregnancy is located in the proximal rather than in the ampullary section, it is worth considering a segment resection in such cases. Segment resection is without doubt a safer method than linear salpingotomy when the contralateral tube is intact, due to the relatively high hemorrhage rate and danger of trophoblast residues remaining when the latter method is employed.

## References


1. Beck OJ, Frank F, Kreiditsch E, Wondrazek F (1985) Klinische und experimentelle Untersuchungen zur Erweiterung der Nd:YAG-Laseranwendung in der Neurochirurgie. Laser Med Chir 1:13–18
2. Bobrow M, Bell HG (1962) Ectopic pregnancy. Obstet Gynecol 20:500
3. Brosens IA, Boeckx W, Gardts S, Vasquez G (1980) Funktionserhaltende Operation bei Ovarialendometriose, Tubenschwangerschaft und Tubenokklusion. Gynäkologe 13:153
4. Brosens IA, Gardts S, Boeckx W (1983) Tubal pregnancy: Salpingostomy versus Salpingotomy. Fertil Steril 39:384
5. Bruhat MA, Mage G, Manhes M (1979) Use of the $CO_2$-Laser via laparoscopy laser surgery. III In: Kaplan I (ed) Proceedings of the 3rd International Society for laser surgery. O. T. Paz, Tel Aviv, p 275

6. Bruhat MA, Manhes H, Mage G, Pouly JL (1980) Treatment of ectopic pregnancy by means of laparoscopy. Fertil Steril 33:411–414
7. Bruhat MA, Mage G, Cani M, Pouly JL, Manhes H, Wattiez A (1988) $CO_2$-laser laparoscopy. A ten-year experience. Lasers Med Surg 4:149–151
8. Brunestaud JM, Mosquet L, Houcke M, Scopelliti JA, Rance FA, Cortot A, Paris JC (1985) Villous adenomas of the rectum – results of endoscopy treatment with argon and Nd:YAG lasers. Gastroenterology 89:832–837
9. Budowik M, Johnson TRP, Genadry R, Parmley TH, Woodruff JG (1980) The histopathology of the developing tubal ectopic pregnancy. Fertil Steril 34:169–171
10. Bukovsky I, Langer R, Hermann A, Caspi E (1979) Conservative surgery of tubal pregnancy. Obstet Gynecol 53:706–711
11. Caffier B (1942) Die konservative Operation des schwangeren Eileiters (ein Beitrag zur Sterilitätsverhütung). Zentralbl Gynakol 66:119–124
12. De Cherney A, Kase N (1979) The conservative therapy of unruptured ectopic pregnancy. Obstet Gynecol 54:451–454
13. De Cherney A, Romero R, Naftolin F (1981) Surgical management of unruptured ectopic pregnancy. Fertil Steril 35:21–24
14. De Cherney A, Maheaux R, Naftolin F (1982) Salpingostomy for ectopic pregnancy in the sole patient oviduct: reductive outcome. Fertil Steril 37:619–622
15. Dietl J, Buchholz F (1986) Zur Histologie der Tubargravidität. Geburtshilfe Frauenheilkd 46:829–830
16. Emslander HP, Prauer HJ, Muntaenu J, Heinl KW, Hinke K, Sebening H, Daum D (1985) Palliative endobronchiale Tumorverkleinerung durch Laserbehandlung: Behandlungsmodus – Sofortergebnisse – Langzeitergebnisse. Laser Med Chir 1:28–34
17. Frank F (1986) Biophysical basis and technical prerequestions for the endoscopical use of the Nd:YAG-laser. Lasers Med Surg 3:124–132
18. Franklin EW, Zeidermann AM, Lämmle P (1973) Tubal ectopic pregnancy: etiology and obstetric and gynecologic sequelae. Am J Obstet Gynecol 117:220–225
19. Gauwerky JFH, Kubli F (1987) Die operative Behandlung der Extrauteringravidität. Fertilität 3:125–132
20. Gauwerky JFH (1989) Eine einfache Technik zur endoskopischen Segmentresektion der Tubargravidität. Fertilität 5:65–67
21. Gauwerky JFH, Wallwiener D, Bastert G (1990) Die endoskopische Refertilisierung – operative Techniken, Indikationen, erste Erfahrungen. Tagung der Arbeitsgemeinschaft für gynäkologische und geburtshilfliche Endoskopie, Berlin 26.–28.1.1990
22. Giana M, Dolfin GC, Siliquini PN (1978) Trattamento chirurgico conservativo in 51 casi di gravidanza tubarica. Minerva Clin 30:99–104
23. Gomel V (1978) Laparoscopy in diagnosis and treatment of ectopic gestation. Ginedips 2:85–88
24. Gomel V (1983) Microsurgery in female infertility. Little Brown, Boston
25. Hallat JG (1986) Tubal conservation in ectopic pregnancy: a study of 200 cases. Am J Obstet Gynecol 154:1216–1221
26. Henry-Suchet J, Tesquier V, Loffredo Y (1979) Chirurgie conservatrice de la grosesse extrauterine. In: Brosens (ed) Oviducté et Fertilité. Masson, Paris
27. Hepp H, Scheidel P (1982) Operatives Vorgehen bei der Extrauteringravidität. In: Künzel W, Rauskolb R (Hrsg) Giessener gynäkologische Fortbildung. Thieme, Stuttgart
28. Hirsch HA, Neeser E (1985) Uncommon indications for laparoscopy. J Reprod Med 9:651
29. Jaervinen PA, Nummi S, Pietala K (1972) Conservative operative treatment of tubal pregnancy with postoperative daily hydrotubations. Acta Obstet Gynecol Scand 51:169–170
30. Johns DA, Hardie RP (1986) Management of unruptured tubal pregnancy with laparoscopic carbon dioxide laser. Fertil Steril 56:703–705
31. Jones OH (1966) Ectopic pregnancy: an analysis of 100 conservative cases. Br J Clin Pract 20:377–384
32. Kadar N, De Vore G, Romero R (1981) Discriminatory HCG zone: its use in the sonographic evaluation for ectopic pregnancy. Obstet Gynecol 58:156
33. Keiditsch F, Hofstetter A, Zimmermann I, Stern J, Frank F, Babaryka I (1985) Histological investigation to substantiate the therapy of bladder tumors with the Nd:YAG-laser. Laser Med Chir 1:19–23
34. Klass A (1984) Laser in der Gastroenterologie. Mich Med 83:284–291

35. Kucera E, Macku F, Novak J, Andrasova V (1969) Fertility after operation of tubal pregnancy. Int J Fertil 14:127–134
36. Künzig HJ, Nittner G, Seitz E (1983) Tubargravidität: aktuelle Aspekte in Diagnostik und Therapie. Geburtshilfe Frauenheilkd 43:658–663
37. Langer R, Bukovsky I, Hermann A, Sherman D, Sadovsky G, Caspi E (1982) Conservative surgery for tubal pregnancy. Fertil Steril 38:427–430
38. Lavy G, Diamond M, De Cherney (1987) Ectopic pregnancy: its relationship to tubal reconstructive surgery. Fertil Steril 47:543–556
39. McBain JC, Evans JH, Pepperell RJ, Smith MA, Brown JB (1980) An unexpected high rate of tubal pregnancy following the induction of ovulation with human pituitary and chorionic gonadotropin. Br J Obstet Gynaecol 87:5
40. Moore KC, Steger A, Hira N (1986) The operative care of patients for Nd:YAG-laser surgery. In: Oguro Y, Atsumi K, Joffe SN (eds) Nd:YAG-laser in medicine and surgery: fundamental and clinical aspects. Professional Postgraduate Services, pp 124–127
41. Neeser E, Hirsch H (1985) Vier Tubargraviditäten bei derselben Patientin: Vor- und Nachteile der tubenerhaltenden Operation. Gebursthilfe Frauenheilkd 45:706
42. Neeser E, Hirsch H (1987) Diagnostische und therapeutische Eingriffe bei Extrauteringravidität. Geburtshilfe Frauenheilkd 47:149
43. Oelsner G, Rabinovich O, Morard J, Mashiach S, Serr DM (1987) Reproductive outcome after microsurgical treatment of tubal pregnancy in women with a single fallopian tube. J Reprod Med 31:149
44. Palmer R (1972) Resultats et indications de la chirurgie conservatirice au cours de la grosesse extrauterine. CR Fr Gynecol 62:317–319
45. Parry JS, Lea HC (1976) Extrauterine pregnancy. Fertilität 87:125–132
46. Polmann L, Wichsel F (1960) Fertility after conservative surgery of tubal pregnancy. Acta Obstet Gynecol Scand 39:143–152
47. Pouly JL, Manhes H, Magl G, Canis M, Bruhat MA (1986) Conservative laparoscopic treatment of 321 ectopic pregnancies. Fertil Steril 46:1093–1097
48. Reid R, Elfont EA, Zirkin RM, Fuller TA (1985) Application of $CO_2$-lasers in obstetrics and gynecology. Am J Obstet Gynecol 152:261–271
49. Reist A (1961) In: Burger K. Über konservative Operationen der Extrauteringravidität. Geburtshilfe Frauenheilkd 21: 633
50. Riedel HH, Cordts-Kleinwort G, Semm K (1983) Tierexperimentelle, morphologische und endokrinologische Untersuchungen nach Anwendung verschiedener Koagulationstechniken. Zentralbl Gynakol 105:665–670
51. Scheidel P (1982) Mikrochirurgie in der Gynäkologie – Tierexperimentelle und klinische Untersuchungen zum Einsatz mikrochirurgischer Techniken in der rekonstruktiven Tubenchirurgie. Habilitationsschrift, Homburg/Saar
52. Scheidel P, Hepp H (1985) Organerhaltende Chirurgie der Tubargravidität. Geburtshilfe Frauenheilkd 45:691–701
53. Schenker JM, Eyal E, Polishuk WZ (1972) Fertility after tubal pregnancy. Surg Gynecol Obstet 135:74–76
54. Semm K (1976) Endocoagulation: a new field of endoscopic surgery. J Reprod Med 16:194–203
55. Semm K (1977) Pelviskopische Chirurgie in der Gynäkologie. Geburtshilfe Frauenheilkd 37:909–920
56. Semm K (1977) Die Mikrochirurgie in der Gynäkologie. Geburtshilfe Frauenheilkd 37:93–102
57. Semm K, Mettler L (1980) Technical progress in pelvic surgery via operative laparoscopy. Am J Obstet Gynecol 138:121–127
58. Semm K (1984) Operationslehre für endoskopische Abdominalchirurgie: operative Pelviskopie – operative Laparoskopie. Schattauer, Stuttgart
59. Shapiro HI, Adler DH (1973) Excision of an ectopic pregnancy through the laparoscope. Am J Obstet Gynecol 117:1093–1097
60. Stangel JJ, Reyniak JV, Stone MC (1976) Conservative surgical management of tubal pregnancy. Obstet Gynecol 48:241–244
61. Stangel JJ, Gomel V (1980) Techniques in conservative surgery for tubal gestation. Clin Obstet Gynecol 23:1221-1228
62. Stromme WB (1953) Salpingotomy for tubal pregnancy. Obstet Gynecol 1: 472

63. Stromme WB (1973) Conservative surgery for ectopic pregnancy. Obstet Gynecol 41:215–223
64. Tadir Y, Kaplan I, Zuckerman Z, Edelstein T, Ovadia J (1984) New instrumentation and technique for laparoscopic carbon dioxide laser operations: a preliminary report. Obstet Gynecol 63:582–585
65. Tadir Y, Kaplan I, Zuckerman Z, Ovadia J (1986) Actual effective $CO_2$-laser-power on tissue in endoscopic surgery. Fertil Steril 45:492–495
66. Tait RL (1884) Pathology and treatment of extrauterine pregnancy. Br Med J 2:317–323
67. Timonen S, Niemenen U (1967) Tubal pregnancy, choice of operative method of treatment. Acta Obstet Gynecol Scand 46:327–339
68. Totani R, Karasave T, Suziki Y (1986) Thermal influence on neighbouring tissues during the contact Nd:YAG-laser conisation. In: Oguro A, Atsumi K, Joffe SN (eds) Nd:AG-laser in medicine and surgery: fundamental and clinical aspects. Professional Postgraduate Services, pp 578–582
69. Valle JA, Lifchez AS (1983) Reproductive outcome following conservative surgery for tubal pregnancy in women with a single fallopian tube. Fertil Steril 39:316–320
70. Wallwiener D, Stolz W, Pollmann D, Morawski A, Spiller M, Bastert G (1989) Nd:YAG-In-Touch-Laser (Kontakttechnik mit Saphirspitzen) – Eine tierexperimentell-chirurgische Studie an Minischweinen. Lasers Med Surg 4:172–179
71. Wallwiener D, Morawski A, Pollmann D, Stolz W, Bastert G (1989) Der Nd:YAG In-Touch-Laser in der operativen Laparoskopie – Erste klinische Ergebnisse. Lasers Med Surg 5:36–41
72. Wallwiener D, Balde MD, Stolz W, Plantener G, Bastert G (1989) Application du laser Neodym YAG dans l'adhesiolyse. Rev Fr Gynceol Obstet 84/11:721–723
73. Wallwiener D (1989) Fertilitätschirurgische Therapiekonzepte – Lasertechnische Therapiekonzepte. Habilitationschrift, Universität Heidelberg
74. Wood JW, Innes JW (1985) Tumor ablation by endoscopic Nd:YAG-laser. Am J Gastroenterol 80:715–718

# Laparoscopic Treatment of Ectopic Pregnancy With and Without Laser

J. L. Pouly [1], C. Chapron, A. Wattiez, G. Mage, M. Canis, H. Manhes, and M. A. Bruhat

With the development of the rapid HCG assay and echography, the diagnosis of ectopic pregnancy (EP) has become easier and can be made earlier. At the same time, great improvements have been made in operative laparoscopy. These two points logically led to the laparoscopic treatment of ectopic pregnancy; we treated the first such cases in 1974 [1]. Following some initial success, the instrumentarium and technique have been progressively modified to obtain the current procedure, which is safe and effective. Moreover, it soon became clear that this surgical technique was followed by a low rate of secondary infertility. Later, in 1981, Dubuisson et al. [5] described laparoscopic salpingectomy for the treatment of ectopic pregnancies. Recently, laser treatment [13] has also been described, but it is of little value.

## Conservative Laparoscopic Treatment

### Instrumentarium

Conservative laparoscopy for EP requires a special instrumentarium, consisting minimally of:

- One 5-mm suprapubic trocar
- One 7-mm suprapubic trocar
- One atraumatic forceps
- One Triton (The Triton is a special tool devised for treating EP pregnancy. It has three functions: (a) strong aspiration through a 7-mm channel, (b) serum pressurized injection, and (c) a retractable monopolar electrode for electrosection [18].)
- One spinal needle (gauge 19)
- One 5-IU ornithine vasopressin ampule (POR 8, Sandoz Laboratories, Basel, Switzerland)

### Installation

The laparoscopy is performed with the patient under general anesthesia. Two suprapubic trocars are placed, the 5-mm one to introduce the forceps and the 7-mm one for the Triton. They must be set at each end of an imaginary Pfannenstiel incision, 4–5 cm from the middle line. Generally, it is best to introduce the Triton as far distal as possible to the EP.

### Operative Procedure

First, the hemoperitoneum is aspirated with the Triton. Then the EP is exposed to evaluate the possibility of treating it by means of laparoscopy.

[1] Polyclinique Gynécologie, Obstétrique Médecine de la Reproduction, 13, Boulevard Charles de Gaulle, 63003 Clermont-Ferrand, France.

Lasers in Gynecology
Edited by G. Bastert and D. Wallwiener

Next, preventive hemostasis is performed. A 5-IU ampule of ornithine vasopressin (POR 8) is diluted in 20–30 ml saline. The spinal needle is introduced into the abdomen under laparoscopic control. Then 10–20 ml of the solution are injected into the mesosalpinx on each side of the EP. Before injection a slight aspiration must insure that the end of the needle is not in a vessel. This injection can be performed just along the EP, but it is often easier to administer a double injection along the lumbo-ovarian artery and along the initial part of the tubal artery. This injection induces a strong vasoconstriction. The tube becomes white and the bleeding stops rapidly.

The salpingotomy is performed with the monopolar electrode of the Triton. It is done on the antimesenteric side and on the proximal part of the hematosalpinx. This incision does not induce any bleeding. Clots and the trophoblast begin to drip out of the tube through the salpingotomy.

Then the Triton is pushed into the tube through the salpingotomy. The aspiration is started. In most cases, clots and the trophoblast can be removed from the tube and the abdomen. Sometimes aspiration is difficult, however. If the EP is large, extraction from the tube may be easy, but not from the abdomen; the EP must be fragmented in the abdomen in order for it to be aspirated. In rare cases the aspiration is not powerful enough to pull the trophoblast out of the tube, and it must be removed by grasping it with a forceps. It seems that POR 8 induces a cleavage between the trophoblast and the tube that facilitates the removal of the EP. In all cases, the operation must be completed with a check that no trophoblast remains in the tube. An atraumatic forceps is inserted through the salpingotomy and then gently opened. Trophoblast appears as a very white tissue.

The peritoneum and the tube are washed by repeated injection and aspiration of saline. The Triton makes it an easy procedure. It is often necessary to push the bowel with the forceps to facilitate the washing. Moreover, the patient must be moved from Trendelenburg's position to permit the bloody irrigation fluid to flow from the upper part of the abdomen into the cul-de-sac and then to be completely aspirated.

A drain is placed in the abdomen to detect any acute early secondary bleeding. It is removed after 12–24 h. It is introduced into the abdomen through a suprapubic trocar.

The postoperative hospitalization is 48 h. The decrease in HCG must be monitored; this point will be discussed below. No medication is prescribed, except for cases with associated adhesiolysis; then doxycycline and corticoids are prescribed for 15 days.

## Laparoscopic Salpingectomy for Ectopic Pregnancy

In the initial publication by Dubuisson et al. [5], salpingectomy is performed with thermocoagulation and a scissors. The tube is grasped with a forceps and pulled. The thermocoagulator forceps is placed on the proximal part of the tube and the thermal coagulation is applied. Then the coagulated tissue is cut with a scissors. The same procedure is applied to the distal part of the mesosalpinx and continued until the tube is completely separated from the mesosalpinx.

Then the tube is removed from the abdomen through a suprapubic trocar if the tube is not too enlarged, or through a culdotomy. Sometimes it is necessary to aspirate the hematosalpinx from the tube to facilitate the extraction of the latter from the abdomen.

Laparoscopic salpingectomy can also be performed using other techniques [30]. In place of the thermocoagulation it is possible to use instillation of vasopressin and bipolar coagulation or laparoscopic sutures (Roder loop).

In all cases the procedure must be completed by cleaning the abdomen. A drain must be also placed in the abdomen to detect acute secondary hemorrhage. The postoperative course is the same as that after conservative treatment, but monitoring of the HCG decrease is not necessary.

## Laser Laparoscopic Treatment of Ectopic Pregnancy

Two different laser procedures are reported. In the first, the $CO_2$ laser is used in place of electrosection to perform the salpingotomy. The remaining operative procedures are the same [13].

In the second method, the EP is treated by coagulation with a YAG laser. The procedure is easy, but there are some disadvantages: the YAG laser is a dangerous machine; as it has a wide lateral thermal spread (4–6 mm), a significant portion of the tube can be damaged.

In fact, laser is of little interest for the treatment of ectopic pregnancy; a salpingotomy can easily be performed by means of a monopolar current, which is far less expensive than laser. Also, coagulation of the tube with a YAG laser is a radical treatment that can also be accomplished with cheaper techniques.

Other methods have been proposed recently, such as the injection of various drugs into the mesosalpinx, e.g., methotrexate or prostaglandins [6, 14, 16, 20, 26].

## Comments on the Operative Procedures

The first laparoscopic treatment of ectopic pregnancy was conservative. The current technique, described in this paper, is the result of 10 years' development.

At the beginning we tried to treat most cases of ampullar EP by aspiration through the fimbria. This method resulted in many failures due to incomplete removal of the trophoblast. Therefore, we decided to perform a salpingotomy in all cases where there was a hematosalpinx, even if the aspiration through the fimbria appeared to be successful. Salpingotomy can be avoided only in cases of purely fimbrial pregnancies.

On the other hand, the problem of hemostasis was very difficult at the beginning. In EP the bleeeding comes mainly from the implantation area, rather than from the salpingotomy. Even during a laparotomy, hemostasis is sometimes difficult to complete. In early cases the hemostasis was achieved by compression of the tube with an atraumatic forceps. But in 20% of the cases it was necessary to perform a laparotomy just to complete the hemostasis. Vasopressin instillation has been used for 8 years [18]. It induces a strong vasoconstriction that stops all bleeding from the small vessels. The pharmacological action lasts 2–4 h, which is sufficient to permit effective physiological hemostasis. The instillation of vasopressin is very efficient and safe. Among more than 300 cases treated, only two problems have occurred. A secondary hemorrhage occurred 4 h after the laparoscopy in a woman who was treated with antivitamin K drugs. In another case, pulmonary edema occurred 4 h after the laparoscopy in a very anemic woman (Hb = 4 g/l); her cardiac insufficiency was related to probable coronary vasoconstriction associated with severe hypoxia.

It appears that vasopressin should be used preventively because it has some advantages. The surgeon can operate under better conditions, and vasopressin induces a better cleavage of the trophoblast from the tube.

The effects of vasoconstriction on tubal function are difficult to evaluate. Nevertheless, a comparison of post-EP fertility rates shows no difference when vasopressin is used.

## Failures

Among the first 500 ectopic pregnancies treated by laparoscopy, 22 failures occurred (4.4%). They were due to trophoblast remaining in the tube. This failure rate appears very high, but it must be regarded as the hazard of a new technique. Moreover, the same failures have been reported in 1%–2% of cases treated conservatively during a laparotomy.

An analysis of the failures shows three reasons: First, most occurred under a new surgeon not experienced in operative laparoscopy. Second, the technique of aspiration through the fimbria caused 30% of these failures. Third, some cases cannot be treated safely during a laparoscopy and must be considered contraindications: hematosalpinx larger than 6 cm, shock, excessive acute bleeding, interstitial EP, EP associated with extensive pelvic adhesions, and EP in very obese women. Therefore, these failures can be prevented by:

- Respecting contraindications
- Always performing a salpingotomy
- Always operating with a trained laparoscopist

The clinical features of these failures were very remarkable. In most cases the patient felt well for 10–15 days. Pelvic pain returned after this period but remained weak, and excessive acute bleeding occurred from the 20th to the 30th postoperative day. Therefore, we aim for early detection by monitoring the relative postoperative HCG decrease (% of remaining HCG based on the HCG rate at the time of laparoscopy).

From this study we can conclude that early failure detection is possible on the second or third postoperative day [23]. A survey diagram indicated the risk according to the decreased HCG rate and the postoperative day. Thus, prolonged monitoring of HCG until it becomes undetectable can be avoided in cases where the initial drop is more than 90% during the second or third postoperative day. The best way to treat this failure is still unclear, but we have had seven successful experiences with methotrexate [26].

## Fertility

Fertility following an EP is problematic. EP has been claimed to be a main cause of secondary infertility [9], but it is obvious that many patients have a

**Table 1.** Fertility results according to patient history

| History | Number | IUP | | EP | | Infertile | |
|---|---|---|---|---|---|---|---|
| | | *n* | % | *n* | % | *n* | % |
| Total | 223 | 149 | 66.8 | 27 | 12 | 74 | 33.2 |
| Infertility | 102 | 102 | 38 | 18 | 17.6 | 64 | 74 |
| Tubal | 62 | 16 | 25.8 | 11 | 17.8 | 46 | 74.2 |
| Other | 40 | 22 | 55 | 7 | 17.5 | 18 | 45 |
| Solitary tube | 47 | 11 | 23.5 | 10 | 21.3 | 36 | 76.5 |
| One EP | 24 | 5 | 20.8 | 11 | 45.8 | 18 | 75 |
| Two or more EPS | 10 | 1 | 10 | 4 | 40 | 9 | 90 |
| Salpingitis | 29 | 11 | 37.9 | 7 | 24 | 18 | 62.1 |
| No history | 101 | 91 | 90.1 | 5 | 5 | 7 | 7 |

tubal pathology prior to the EP. In any case, the efficacy of the surgical procedure into preventing infertility must be discussed. During a laparotomy the question of radical or conservative treatment remains controversial. It has been widely agreed but never proved that conservative treatment results in a lower infertility rate [27]. The current consensus is that any new treatment for ectopic pregnancy must guarantee high post-EP fertility.

## After Conservative Laparoscopic Treatment

Of the first 500 cases that we treated conservatively with laparoscopy, 223 can be evaluated with regard to postoperative fertility; 168 were excluded because they did not want any more children and 70 were lost to follow-up during the first postoperative years. There were 39 recurrences. Of the 223, 149 (66.8%) have achieved an intrauterine pregnancy (IUP), six of them after a second EP. Twenty-seven (12.1%) have had one or more recurrence of EP (39 EPs). Of these patients, only 18 have both tubes and the recurrences were equally distributed. A total of 74 (33.2%) patients remained infertile after the first EP or after one or more recurrences.

A patient's medical history influences her post-EP fertility. The results of this series are presented in Table 1. Among the 102 patients previously suffering from infertility the IUP rate was 37.2%. But of the 62 patients who have been operated on for tubal infertility by microsurgical or laparoscopic procedures such as neosalpingostomy or adhesiolysis, only 25.8% have achieved an IUP; the recurrence rate was 17.8% and the sterility rate was 74.2%. Where the previous infertility was not tubal, the results are much better. The rates of IUP, recurrence, and sterility are 55%, 17.5%, and 42.5% respectively. Among the 47 patients with a solitary tube the IUP rate is also very poor 23.4% and the recurrence rate reached 21.3%. The prognosis for recurrence of EP is also poor; among 24 patients that presented with two EPs, only 20.8% obtained an IUP, but 45.8% a third EP, and after three EPs only 10% became pregnant. The occurrence of an EP as the first pregnancy after a laparoscopically investigated salpingitis has a high probability, among 29 patients the IUP rate was 37.9% and the recurrence rate 24%.

**Table 2.** Comparison of post-EP fertility according to treatment procedures

| Reference | Year | Number | IUP | | EP | |
|---|---|---|---|---|---|---|
| | | | *n* | % | *n* | % |
| Surgical treatment | | | | | | |
| Jarvinen [12] | 1972 | 43 | 22 | 51.2 | 4 | 9.3 |
| Stromme [29] | 1973 | 37 | 20 | 54.1 | 7 | 18.9 |
| Decherney and Kase [3] | 1979 | 48 | 19 | 39 | 4 | 8 |
| Giana [7] | 1979 | 51 | 17 | 33 | 4 | 7.8 |
| Henri-Suchet [10] | 1979 | 52 | 22 | 42 | 10 | 19 |
| Sherman et al. [28] | 1982 | 47 | 39 | 83 | 7 | 15 |
| Paavonen et al. [21] | 1985 | 116 | 68 | 59 | 11 | 10 |
| Langer et al. [15] | 1987 | 73 | 56 | 76 | 9 | 12 |
| Thorburn [31] | 1987 | 112 | 62 | 55 | 23 | 20 |
| Tuomivaara and Kauppila [32] | 1988 | 323 | 265 | 82 | 42 | 13 |
| Microsurgical treatment | | | | | | |
| Decherney et al. [4] | 1980 | 9 | 5 | 55 | 0 | 0 |
| Janecek [11] | 1980 | 10 | 6 | 60 | 2 | 20 |
| Oelsner et al. [19] | 1987 | 51 | 29 | 57 | 13 | 25 |
| Laparoscopic treatment | | | | | | |
| Decherney and Diamond [2] | 1987 | 69 | 36 | 52 | 7 | 10 |
| Reich et al. [24] | 1988 | 38 | 19 | 50 | 11 | 29 |
| Current series | 1989 | 223 | 149 | 66 | 27 | 12 |

On the other hand, among patients who have no previous history of infertility, no EP, and no salpingitis, the IUP rate is a very high 90% and the recurrence rate a very low 5%.

### After Laparoscopic Salpingectomy

No important series has been published regarding post-EP fertility after laparoscopic salpingectomy. In our experience with 39 patients, there were 18 IUPs (46.2%) and no recurrence of EP. These rates cannot be compared with those after conservative treatment because most of our cases were selected and the series is small.

### After Laser Treatment

As laser methods are new and not yet widespread, no significant series have been published. Nevertheless, it is probable that the results after $CO_2$ laser salpingotomy will not differ from those we report here.

### Comparison with Other Procedures

Any comparison of post-EP fertility is difficult and doubtful. Improved methods of diagnosis have led to earlier treatment of less damaged tubes, and have therefore enhanced the post-EP fertility. The development of microsurgery makes it easier for the surgeon to prevent postoperative adhesions. Finally, most series consist mainly of patients with single tubes; the numbers are very small and the results are favorable, and they do not reflect reality.

Nevertheless, in Table 2 we present a comparison of large series reported in the literature over the past 20 years. It clearly shows that conservative laparoscopic treatment is at least as effective as conservative surgical treatment and probably does more to prevent infertility. There are three reasons for these good results. First, during a laparoscopy there is no drying of the peritoneum and postoperative adhesions are therefore less frequent. This point was largely demonstrated by Ryan et al. [25], who showed that adhesions are more frequent after drying of the peritoneum than after an isolated hemoperitoneum. Moreover with laparoscopy the chance of postoperative adhesion formation is significantly lower than with laparotomy; the surgeon cannot grasp the tubes with a forceps, and therefore less direct traumatic damage is caused. Second, hemostasis with vasopressin is less aggressive than hemostasis obtained with sutures or electrocoagulation. This is true for the involved tube, but is is even more important for the contralateral one.

Third, animal studies by MacComb and Gomel [17] and by Gordji et al. [8] have demonstrated that the spontaneous healing of the tube is better after salpingotomy than after sutures. The latter group demonstrated that fewer tubes are occluded and the former that the architecture of the mucosal folds is less disturbed.

We performed 50 second-look laparoscopies. In 60% of cases the salpingotomy scar was not visible. In the other cases, only the mucosal and the serosal layer were scarred but not the muscle, and the tube appeared to be herniated. It seems that this does not affect post-EP fertility. In only two cases did we find a tubal fistula, after treatment of heavily ruptured tubes.

Thus, the advantages of laparoscopic treatment lie in better prevention of damages to the opposite tube and less aggressive treatment of the involved tube.

However, after conservative laparoscopic treatment of EP the recurrence rate is high. There are two explanations for this: There is a correlation in all series between a high rate of IUP and a high recurrence rate, which indicates that prevention of sterility carries a high risk of recurrence if all patients receive conservative treatment. This series was started 10 years ago, before the development and clinical application of in vitro fertilization (IVF). Therefore, until 1984, all patients who still wanted to become pregnant were treated conservatively, without regard to their history or their relative risk for recurrence [22]. Since 1984, we have tried to reduce the recurrence rate without causing a significant reduction in the IUP rate. Patients with a high risk of recurrence are treated by laparoscopic salpingectomy, in some cases associated with a controlateral sterilization, and are then referred to the IVF unit. This rapidly reduced the recurrence rate from 18% to 12% but did not affect the general fertility results.

## Indications and Contraindications for Laparoscopic Treatment of Ectopic Pregnancy

### Contraindications

Laparoscopic treatment is contraindicated in all cases where the procedure may be dangerous for the patient; it must be kept in mind than even in developed countries some patients still die from ectopic pregnancies. Therefore, safety must remain an important point with this treatment, as with all surgical laparoscopic procedures.

Three factors make laparotomy preferable: a lack of material or experience, an inoperable EP, and poor local conditions. Inexperience is certainly the main problem. The hemostasis of a massive acute hemorrhage during a laparoscopy can be easy with vasopressin or electrocoagulation, but if it is not achieved rapidly it can induce shock. On the other hand, it is dangerous to aspirate a 4-cm tubal pregnancy through an aspiration channel 4 mm in diameter, as it takes a long time and is often incomplete. Therefore, we think that any new laparoscopic surgeon should first be provided with the appropriate instrumentarium and then treat only small, unruptured EPs. Later, with more experience, he or she will be able to treat most cases.

Even for a trained laparoscopist, some EPs must be considered inoperable through the laparoscope, e.g., ruptured hematosalpinx with massive acute hemorrhage and shock, or interstitial pregnancy. Poor local conditions are given where laparoscopic exposition of the tube is difficult, or it is impossible to achieve a safe treatment. This is the case with extended pelvic adhesions or in very obese patients. Nevertheless, a trained laparoscopist can treat at least 95% of EPs with laparoscopy.

### The Choice Between Radical and Conservative Treatment

Basically, conservative treatment is always preferable, but sometimes it is not possible and radical treatment is necessary for technical reasons. For instance, hematosalpinx larger than 6 cm and isthmic, ruptured EP require a salpingectomy. In addition, salpingectomy must be preferred if the relative risk of recurrence is too high. We consider a probability of one EP to two IUPs justification for removing the tube. This is the case with patients previously treated for tubal infertility, after a second EP, or after salpingitis with tubal sequellae. The most difficult question in such a situation is whether to perform a controlateral sterilization, as EP often recurs in the controlateral tube. No definitive rules can be given for these cases; the decision depends on the tubal status, on the wish of the patient, and on her accessibility to an IVF program.

Of course, for patients who do not want any more children, salpingectomy is often performed but it can be replaced by conservative treatment with sterilization using clips.

## Conclusion

With good laparoscopic experience and the proper instrumentarium, 95% of ectopic pregnancies can be treated during a laparoscopy. The choice between conservative or radical treatment is based on the patient's history and the status of the EP. Laparoscopic treatment has several advantages: reduction in cost, reduction in the patient's discomfort, and an increase in post-EP fertility.

## References


1. Bruhat MA, Manhes H, Mage G, Pouly JL (1980) Treatment of ectopic pregnancy by means of laparoscopy. Fertil Steril 33:411
2. Decherney AH, Diamond MP (1987) Laparoscopic salpingostomy for ectopic pregnancy. Obstet Gynecol 70:948
3. Decherney AH, Kase N (1979) The conservative surgical management of unruptured pregnancy. Obstet Gynecol 54:451
4. Decherney AH, Polan ML, Kort H, Kase N (1980) Microsurgical technique in the management of tubal ectopic pregnancy. Fertil Steril 34:324
5. Dubuisson JB, Aubriot F, Cardone V (1987) Laparoscopic salpingectomy for tubal pregnancy. Fertil Steril 47:225
6. Feichtinger W, Kemeter P (1987) Conservative treatment of ectopic pregnancy by transvaginal aspiration under sonographic control and methotrexate injection. Lancet 1:381
7. Giana M (1979) Tratamento chirurgico conservativo in 51 caza di gravidenza tubarica. Minerva Ginecol 30:99
8. Gordji M, Henri-Suchet J, Pigeaud F et al. (1981) Etude comparée des salpingotomies avec et sans suture. Recherche microchirurgicale expérimentale sur la trompe de lapine. J Gynecol Obstet Biol Reprod (Paris) 10:765
9. Grant A (1962) The effect of ectopic pregnancy on fertility. Clin Obstet Gynecol 5:861
10. Henri-Suchet J (1979) Chirurgie conservatrice de la grossesse extra-utérine. In: Oviducte et fertilite. Masson, Paris, p 393
11. Janecek J (1979) Résultats de la chirurgie reconstructrice dans les grossesses extra-utérine non rompues. Rev Med Suisse Romande 99:603
12. Jarvinen PA (1972) Conservative operative treatment of tubal pregnancy with postoperative daily hydrotubation. Acta Obstet Gynecol Scand 51:169
13. Johns DA, Hardie R (1986) Management of unruptured ectopic pregnancy with laparoscopic dioxide laser. Fertil Steril 46:703
14. Kojima E, Morita M, Abe Y (1988) Local injection of methotrexate for the treatment of unruptured tubal pregnancy. AAGL meeting, Dallas, November
15. Langer R, Bukovsky I, Sherman A, Ron-El R, Lifshitz Y, Caspi E (1987) Fertility following conservative surgery for tubal pregnancy. Acta Obstet Gynecol Scand 66:649–652
16. Lindblom B, Kallfelt B, Hahlin M, Hamberger C (1987) Local prostaglandin $F_{2a}$ injection for termination of ectopic pregnancy. Lancet 1:776–777
17. MacComb P, Gomel V (1984) Linear ampullary salpingotomy heals better by secondary versus primary closure. Fertil Steril 41:45
18. Manhes H, Mage G, Pouly JL, Ropert JF, Bruhat MA (1983) Traitement coelioscopique de la grossesse extra-utérine non rompue: ameliorations techniques. Nouv Presse Med 12:1431
19. Oelsner G, Morad J, Mashiach S, Serr DM (1987) Reproductive performance following conservative microsurgical management of tubal pregnancy. Br J Obstet Gynaecol 94:1078–1083
20. Ory SJ, Villanueva AL, Scand PK, Tamura RK (1986) Conservative treatment of ectopic pregnancy with methotrexate. Am J Obstet Gynecol 154:1299–1306
21. Paavonen J, Varjonen-Toivonen M, Komulainen M, Heinonen PK (1985) Diagnosis and management of tubal pregnancy: effect on fertility outcome. Int J Gynaecol Obstet 23:129–133
22. Pouly JL, Manhes H, Mage G, Canis M, Bruhat MA (1986) Conservative laparoscopic treatment of 321 ectopic pregnancies. Fertil Steril 46:1093
23. Pouly JL, Gachon M, Gaillard G, Mage G, Bruhat MA (1987) La décroissance de l'HCG après traitement coelioscopique conservateur de grossesse extra-utérine. J Gynecol Obstet Biol Reprod (Paris) 16:195
24. Reich H, Johns DA, Decaprio J, Mc Glynn F, Reich E (1988) Laparoscopic treatment of 109 consecutive ectopic pregnancies. J Reprod Med 33:885
25. Ryan GB, Grobety J, Majno G (1971) Postoperative peritoneal adhesions. Am J Pathol 65:117
26. Sauer MV, Gorrill MI, Rodi IA et al. (1987) Nonsurgical management of unruptured ectopic pregnancy: an extended trial. Fertil Steril 48:752–755
27. Schenker JG, Evron S (1983) New concepts in the surgical management of ectopic pregnancy and the consequent postoperative results. Fertil Steril 40:709

28. Sherman D, Langer R, Sadovsky G, Bukovsky I, Caspi E (1982) Improved fertility following ectopic pregnancy. Fertil Steril 37:497
29. Stromme WB (1973) Conservative surgery for ectopic pregnancy. Obstet Gynecol 41:251
30. Tarasconi JC (1981) Endoscopic salpingectomy. J Reprod Med 26:541
31. Thorburn JO (1987) Ectopic pregnancy. Clinical features, risk factors and fertility prognosis. Thesis, Goteborg
32. Tuomivaara L, Kauppila A (1988) Radical or conservative surgery for ectopic pregnancy? A follow-up study of fertility of 323 patients. Fertil Steril 50:580

# Argon Laser for Laparoscopic Treatment of Unruptured Ectopic Pregnancy

J. KECKSTEIN [1]

## Introduction

Linear salpingotomy during laparotomy or laparoscopy has become a standard surgical technique for the treatment of unruptured ectopic pregnancy. As the tube shows little change in the early stage of gestation, very delicate and precise instruments are required for preserving the tube. Scalpels or microelectrodes used in microsurgery are not well adapted for laparoscopic surgery. Various authors have reported on their experience with electrocautery [3, 13]. Others have introduced the $CO_2$ laser or the Nd:YAG contact laser because of their ability to produce precise incisions [2, 3, 6, 8]. Due to the possible danger of burns using monopolar instruments and the limited effect of the $CO_2$ laser on bleeding, we used the argon laser for linear salpingotomy on a selected group of 60 patients with unruptured ectopic pregnancy. The aim of this study was to demonstrate the value of this technique by evaluation of tubal status and formation of adhesions postoperatively.

[1] Universitäts-Frauenklinik, Abt. für Geburtshilfe und Gynäkologie, Prittwitzstr. 43, W-7900 Ulm, FRG.

## Material and Methods

From October 1987 to January 1990, 88 patients with unruptured ectopic pregnancy in the ampullary portion of the tube were treated at the University Hospital of Ulm. Of those, 60 patients were treated with linear laparoscopic salpingotomy using the argon laser (Model 20 Surgical Laser System, HGM, Salt Lake City, USA). The remaining 28 patients were treated by $CO_2$ laser ($n = 3$) or Nd:YAG contact laser ($n = 2$) or by salpingectomy ($n = 23$) if preservation of fertility was no longer desired or in case of recurring ectopic pregnancy.

The laser fiber was passed into the abdomen through a special steering device for laser fibers (Wolf instruments, Knittlingen, FRG) [7], which allows deflection of the fiber tip during the procedure. For the delivery of the argon laser light a 300-µm fiber was used either in the noncontact or in the contact technique. After aspiration of the hemoperitoneum, 2.5 IU Ornipressin (POR-8, Sandoz Laboratories, Basel, Switzerland), diluted in 20 ml saline was injected into the mesosalpinx. Usually a 5- to 20-mm incision was made with the bare fiber (diameter 0.3 mm) on the antimesenteric side of the tube, by directly touching the tissue. The procedure was performed with a power of 3–6 W. Small vessel hemorrhage was controlled using the

Lasers in Gynecology
Edited by G. Bastert and D. Wallwiener

laser with the noncontact technique. The fiber was situated 1–2 mm from the bleeding site. After the initial laser incision and rinsing, the trophoblast and blood clots were evacuated from the tube.

In patients with ovarian adhesions, adhesiolysis was performed using the same technique. The instruments were removed but an intraperitoneal drainage, introduced into the pouch of Douglas, was left in place for 4–12 h.

The patients were discharged between 4 h and 3 days after the operation. Beta-subunit human chorionic gonadotropin (HCG) titers were determined in all patients before and after the operation until levels of less than 5 mU/ml were reached.

The effect of this laser application on the healing process and postoperative formation of adhesions was evaluated by hysterosalpingography or second-look laparoscopy.

## Results

All 60 patients had an unruptured tubal pregnancy located in the distal part of the tube. Linear salpingotomy was performed in all patients with the argon laser. There was no significant bleeding. Hemorrhage from wound margins or the interior wall of the tube was controlled with the argon laser by the noncontact technique. Tubal ovarian adhesions of varying degree were found in 28 patients and all were lysed with the argon laser. No intraoperative complications occurred and no additional instruments for cutting or coagulation were necessary.

Persistently high $\beta$-HCG levels and suspicious clinical signs called for a relaparoscopy in two women. In one patient small implants of persisting trophoblastic tissue near the salpingotomy site were detected and endoscopically removed. In the other patient trophoblastic tissue was washed out by the rinsing procedure and became secondarily implanted in the omentum majus; this was also removed through the laparoscope. In all other patients the serum $\beta$-HCG levels fell rapidly and were negative after a mean of 15.7 days (range: 17–39 days).

Postoperatively, two patients were examined by hysterosalpingography and 22 patients by second-look laparoscopy for tube patency.

In all patients the involved tube proved to be patent. In one patient the tube failed to close, and a fistula was seen on second-look laparoscopy. The ectopic pregnancy in this patient was the largest in the series, and the salpingotomy required extension to 3.5 cm for evacuation of the trophoblast. In this patient the fistula was welded by argon laser in the noncontact technique during second-look laparoscopy. Six weeks later hysterosalpingography showed a patent tube with no fistula. The patient conceived 3 months later.

The status of the tube was also evaluated with respect to postoperative adhesions on the site of salpingotomy and on the fimbriae. Four patients had filmy adhesions covering the tube near the incision site, which, however, did not impair tubal function. There were no adhesions covering the distal part of the tube in any of these patients.

## Discussion

Operative laparoscopy has become a new method of treatment for ectopic pregnancy. Compared with laparotomy this technique offers the advantages of lower operative risk, reduced postoperative morbidity, and reduced postoperative formation of adhesions [5].

The use of electrocautery to treat extrauterine pregnancies has been described by several authors [2, 3, 13, 16]. With this technique an incision into the antimesenteric border of the ampullary segment of the tube is made. With the laparoscopic scissors attached to the unipolar cutting current electrocoagulator, or with a special cutting and aspiration instrument, "TRITON", the tube is incised. Semm [15] described a similar technique using thermocoagulation followed by incision with scissors [11]. The strong coagulation effect of electrocautery instruments, which may endanger adjacent tissue [1, 12], and the awkwardness of these instruments called for more delicate techniques, however. The $CO_2$ laser has therefore gained popularity as a safe and precise cutting instrument in ectopic pregnancy [2, 6].

The pitfall of this technique is the poor coagulative property of the $CO_2$ laser, which is further diminished or altogether nonexistent in blood- or saline-covered operative fields. Any bleeding therefore had to be controlled by bipolar or monopolar instruments. The use of the $CO_2$ laser may also be hampered by large amounts of laser plume produced during application, and by the rigid structures of its delivery system.

In contrast, the argon laser passes through flexible fibers, allowing convenient endoscopic manipulation [7] with less laser plume production in comparison with the $CO_2$ laser. The bare fiber can be used either in contact or in noncontact techniques [4]. Incision and coagulation can be performed with one and the same instrument, in contrast to the techniques discussed above. The coagulation zone produced by the argon laser is small but sufficient for local hemostasis [9]. With this technique invisible tissue damage to adjacent healthy tissue is reduced and can be better controlled in comparison to other coagulation techniques such as monopolar coagulation [14]. With the argon laser, no other instruments were necessary for hemostasis, in contrast to the experience of Johns [6] using the $CO_2$ laser. Since it has been shown that linear salpingotomy is associated with better healing by secondary intention rather than by primary closure [2, 10], the salpingotomy wound was left open in our studies. Despite the coagulation effect produced by the argon laser [9], wound healing was excellent as demonstrated by our second-look laparoscopy series.

The tubal patency rate of 100% was evidenced by hysterosalpingography or second-look laparoscopy. However, in one case the salpingotomy wound was large and failed to close. The ectopic pregnancy in this patient was the largest in the series and the salpingotomy required extension to 3.5 cm for evacuation of the trophoblast. For the repair of the fistula successful laparoscopic sealing with the argon laser was performed (data submitted for publication).

Mild postoperative peritubal adhesions were seen in four patients on the site of salpingotomy, and these may have been caused by tubal bleeding in the postoperative period. We did not see any instrument-related postoperative complication, e.g., tubal occlusion, after linear salpingotomy by argon laser.

This study demonstrates that the application of the argon laser in both the contact and the noncontact technique for conservative management of unruptured tubal pregnancy is advantageous compared with conventional laparoscopic operation techniques. The surgical procedure is greatly simplified, thanks to the flexibility of the laser fibers and to its combined coagulation and incision properties.

## References


1. Bahig CS (1968) Electrosurgical burn injuries and their prevention. JAMA 204:1025
2. Cartwright P, Herbert C, Maxson W (1986) Operative laparoscopy for the management of tubal pregnancy. J Reprod Med 31:589–591
3. DeCherney A, Diamond M (1987) Laparoscopic salpingostomy for ectopic pregnancy. Obstet Gynecol 70:948–950
4. Diamond M, DeCherney A, Lake Polan M (1986) Laparoscopic use of the argon laser in nonendometriotic reproductive pelvic surgery. J Reprod Med 31:1011–1013
5. Fayez JA (1983) An assessment of the role of operative laparoscopy in tuboplasty. Fertil Steril 39:476–479
6. Johns DA, Hardie RP (1986) Management of unruptured ectopic pregnancy with laparoscopic carbon dioxide laser. Fertil Steril 46: 703–705
7. Keckstein G (1989) A new steering device for laparoscopic application for laser fibers. Presented at the 2nd world congress of gynecologic endoscopy, June 5–8 1989 in Clermont-Ferrand, abstracted in the meeting proc., Bruhat M (ed.), p 147
8. Keckstein G, Wolf A, Steiner R (1987) The use of contact laser probe in gynecological endoscopy. In: Laser Optoelectronics in Medicine 1987. Proc. of 7th Congress International Society for Laser Surgery and Medicine, Munich. Waidelich W, Waidelich R (eds). Springer, Berlin Heidelberg New York Tokyo, pp 267–272
9. Keckstein G, Finger A, Steiner R (1988) Laser application in contact and noncontact procedures; sapphire tips in comparison to "bare-fiber", argon laser in comparison to Nd:YAG laser (in German). Laser Med Surg 4:158–162
10. McComb PF, Gomel V (1984) Linear ampullary salpingotomy heals better by secondary versus primary closure. Fertil Steril 41:45S
11. Mecke H, Semm K, Lehmann-Willebrock E (1989) Results of operative pelviscopy in 202 cases of ectopic pregnancy. Int J Fertil 34:93–100
12. Murphy AA (1987) Operative laparoscopy. Fertil Steril 47:1–18
13. Pouly J, Mahnes H, Mage G et al. (1986) Conservative laparoscopic treatment of 321 ectopic pregnancies. Fertil Steril 46:1093–1097
14. Riedel H (1986) Die Anwendung verschiedener Koagulationstechniken zur Eileitersterilisation und ihre Bedeutung für das Auftreten von ovariellen Ausfallserscheinungen. Habilitationsschrift, Universität Kiel
15. Semm K, Mettler L (1980) Technical progress in pelvic surgery via operative laparoscopy. Am J Obstet Gynecol 138:121–127
16. Vermesh M, Silva PD, Rosen GF et al. (1989) Management of unruptured ectopic gestation by linear salpingostomy: a prospective, randomized clinical trial of laparoscopy versus laparotomy. Obstet Gynecol 73:400–404

# Endocrine Treatment of Endometriosis

L. Kiesel [1], T. Rabe, and B. Runnebaum

## Introduction

Endometriosis is one of the most common benign gynecological diseases, first described a century ago. Endometriosis is defined as ectopic endometrial tissue, usually located in the pelvis. In spite of a large number of investigations, the pathogenesis of this disorder has remained unknown. Symptomatic disease is characterized generally by pain (dysmenorrhea, dyspareunia, and pain on vaginal examination) and occasionally by a palpable mass (e.g., ovarian cysts). It is assumed that endometriosis accounts for 10%–20% of all female infertility. Presently, various therapeutic approaches can be taken, once the disease is detected by laparoscopy or laparotomy. Although the surgical removal of the implants or cysts is often attempted, medical treatment is usually administered.

Various hormonal preparations have been used in the past in order to suppress ovarian function and relieve symptoms and to suppress the growth of endometriotic implants [26]. The rationale for this therapeutic approach is the observation that ovarian hormones are required for the establishment and presence of ectopic endometrial tissue [29].

## Treatment with Steroids

A wide range of sex steroids, including estrogens, progestins and androgens, have been used to treat endometriosis.

*Estrogens* were among the first to be used. Although patients improved clinically during therapy, the severity of side effects (i.e., thromboemboli, hypertension, endometrial hyperplasia) has eliminated this approach today [7].

In the 1950s the use of *testosterone* demonstrated some transient relief of symptoms. The androgenic side effects (i.e., hirsutism, acne, deepening of the voice) and the lack of marked regression of the endometrial tissue were the reasons that this mode of treatment had to be abandoned.

The *pseudopregnancy* regimen was introduced by Kistner [19]. The rationale for this treatment was based on the assumption that the improvement of endometriosis observed during pregnancy could be reproduced by high-dose oral contraceptives. High doses of estrogens and progestins are thought to transform endometrial tissue into decidua followed by necrosis. This regimen relieves symptoms of endometriosis, but side effects are common.

[1] Universität Heidelberg, Abt. für Geburtshilfe und Gynäkologie, Abt. für gynäkologische Endokrinologie und Fertilitätsstörung, Voßstr. 9, W-6900 Heidelberg, FRG.

Lasers in Gynecology
Edited by G. Bastert and D. Wallwiener

Two decades ago the synthetic drug *danazol* was introduced, and it has remained a well-accepted hormonal agent since [8]. The mechanism of this drug appears to be complex: it inhibits gonadotropin surge and is therefore often referred to as an "antigonadotropin"; it interacts with androgen and progesterone receptors and inhibits the action of steroidogenic enzymes [3]. Danazol is effective at a dosage of 200–800 mg/day. The adverse effects include weight gain, hirsutism, acne, deepening of the voice, and some undesirable alterations of lipoproteins and hepatic enzymes.

The administration of *progestins* has been an acceptable approach for several years. Oral (e.g. lynestrenol) or depot (e.g. medroxyprogesterone acetate) progestins are presently in use. The advantages are the low cost and generally well-tolerated side effects (i.e., breakthrough bleeding, weight gain, breast tenderness). In a few countries the progestin and partial antiprogestin *gestrinone* has been investigated. It is administered at a low dose (twice a week 2.5 mg) and has shown relief of symptoms.

## Treatment with GnRH Analogues

A new endocrine tool in the treatment of the disease has been established by the introduction of gonadotropin-releasing hormone (GnRH) analogues. They have been shown to induce a reversible hypogonadism due to a blockade of the pituitary-gonadal axis, allowing their use for the treatment of several sex-hormone-dependent disorders [27]. Initial studies have demonstrated the beneficial effect of GnRH agonists on endometriosis [9, 20, 21, 25], and their use in the therapy of this disease has been reviewed previously [2, 10]. This new type of hormonal agent has a novel mechanism of action [14, 15] which is summarized below.

### Mechanism of GnRH Analogue Action

Since the structural elucidation of GnRH, over 2000 GnRH analogues have been synthesized [4, 23]. Prolonged chronic treatment with the GnRH analogues will cause a down-regulation of GnRH receptors of the pituitary and a desensitization of the gonadotroph in regard to gonadotropin release. As a consequence of the suppression of gonadotroph function, a reversible castration will follow.

The lack of gonadotropin secretion, and especially the lack of biologically active LH and FSH, will result in a hypoestrogenic state due to inhibition of follicular maturation.

Native GnRH binds to its receptor and provokes postreceptor effects, resulting in LH and FSH release. Due to the short half-life and the small quantity of hypothalamic GnRH released, gonadotropin secretion is pulsatile [16]. The prolonged administration of a GnRH agonist will initially cause upregulation of GnRH receptor numbers as well as increased hormone release. Due to the higher resistance to enzyme degradation as well as higher binding affinity to the GnRH receptor, agonist action is 100–200 times more potent than that of native GnRH. During the second phase of chronic GnRH agonist treatment, desensitization of the pituitary occurs [11]. The mechanisms involved in this action are the down-regulation of GnRH receptors and the desensitization of postreceptor effects. As a result, the release of bioactive LH and FSH is markedly reduced. The decrease of receptor numbers is due to internalization of the receptor. The precise manner of receptor down-regulation and pituitary desensitization is still a matter of research.

## Clinical Efficacy of the GnRH Analogue Buserelin

In an international multicenter study the efficacy and safety of the GnRH analogue buserelin was investigated in patients ($n = 275$) suffering from endometriosis as demonstrated either by laparoscopy or laparotomy [17].

Patients with active sites of endometriosis diagnosed by laparoscopy or laparotomy were treated with 3 × 300 μg buserelin intranasally per day for 6 months. Following therapy, a second-look laparoscopy was requested to review endometriotic implants and adhesions. While receiving buserelin treatment, women were seen at least once a month at outpatient clinics so that compliance, tolerance, and efficacy, such as changes of symptoms and endocrine values, could be monitored.

First the *changes in AFS score* were analyzed according to The American Fertility Society Score [1]. If implants were considered alone (excluding adhesions), in 38% of patients a complete and in 40% a partial resolution of implants occurred. No change was seen in 16% and a progression of disease was observed in 6% of cases. When total AFS score 10 was compared before and after 6 months of intranasal buserelin therapy, a marked improvement of the score was noted. In 18% of the cases a complete and in 62% a partial resolution was seen. No change of total AFS score occurred in 9% of patients. The disease progressed in 11% of cases.

Looking at the number of women at the various AFS stages before and after therapy, a remarkable shift to lower stages was observed. Prior to treatment, stage-I endometriosis was demonstrated in 83 patients, stage-II in 73, stage-III in 28, and stage-IV in five. Following therapy, no endometriosis was seen in 34 patients, stage I of the disease in 96, stage II in 47, stage III in 11, and stage IV in one patient.

AFS stage for implants only (excluding adhesions) was even more favorable when compared before and after buserelin therapy (Fig. 1). After treatment (vs prior to treatment) 69 patients had no endometriotic implants. Stage I occurred in 57 (vs 63 prior to treatment), stage II was seen in 43 (vs 77) patients, stage III in 12 (vs 37) cases, and stage IV in no (vs 4) cases.

*Changes of symptoms* (dysmenorrhea, pelvic pain, and dyspareunia) were caused by buserelin treatment.

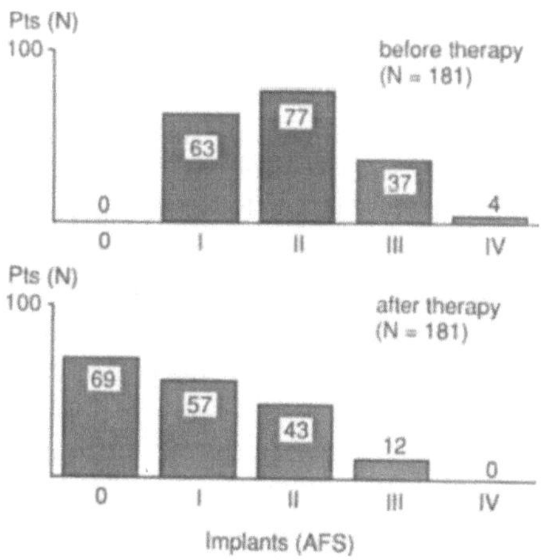


**Fig. 1.** Effect of buserelin treatment (6 months) on endometriotic implants. Staging of disease (implants only) was according to the American Fertility Society Score [1]

Dysmenorrhea occurred in 86% of patients (33% mild and 53% moderate-to-severe complaints). At the end of treatment only 2% of patients complained of mild dysmenorrhea.

52% of patients complained of pelvic pain prior to treatment, 30% of mild and 22% of moderate or severe pain. At the end of treatment, more than 75% were symptom free and only 3% complained of moderate-to-severe pelvic pain. Dyspareunia occurred in 44% of patients (29% mild and 15% moderate to severe). Following GnRH-analogue treatment, only 3% suffered from moderate or severe dyspareunia, 12% complained of mild dyspareunia, and the remainder were symptom free.

Buserelin treatment resulted in a *suppression of serum estradiol* levels. Estradiol levels were already suppressed in 83% of patients after 1 month of GnRH-analogue therapy. This figure rose to 91% during the further course of treatment, with a subgroup of 76% having markedly suppressed levels (estradiol less than or equal to the upper limit of the postmenopausal range).

The *bleeding pattern* of patients during hormone treatment was monitored. Prior to treatment all patients ($n$ = 272) had an active cycle and menstruated regularly. During buserelin therapy, the number of women without menses increased from 53% in month 1 to 91% in month 6 (Fig. 8). In addition, the intensity of bleeding decreased during treatment.

During the 6 months of intranasal buserelin therapy, a number of *side effects* occurred. The most common of these related to estrogen deficiency were hot flushes (81%), dry vagina (18%), decreased libido (13%), and sweating (8%). Other common side effects were headaches (30%), nausea (12%), mood swings (10%), and nervousness (10%).

To monitor the safety of the drug, a number of *laboratory parameters* were measured before and at the end of treatment and 6 months after the end of treatment. The values for hematology, clotting system, clinical chemistry, electrolytes, and liver metabolism were determined. No statistically significant abnormalities concerning mean values were seen during treatment.

The data from this multicenter trial demonstrate that GnRH analogues are effective in reducing symptoms and ectopic endometrial implants in women suffering from endometriosis. Major side effects were generally related to the hypoestrogenic state.

More recently, further trials have been performed to investigate the effects of different GnRH agonists [5, 9, 22, 28]. In addition, comparative studies (GnRH agonists versus danazol) have demonstrated that GnRH-agonist treatment can provide results similar to those obtained with conventional therapy [12, 13, 24]. These data support the findings of our study that GnRH agonists can reversibly suppress ovarian function and thereby inhibit the presence of endometriotic implants. The long-term effects of these agonists are being investigated at present with regard to metabolic changes [6, 18, 31].

## References

1. American Fertility Society (1985) Revised American Fertility Society classification of endometriosis. Fertil Steril 43:351–352
2. Barbieri RL (1988) New therapy for endometriosis. N Engl J Med 318:512–514

3. Barbieri RL, Ryan KJ (1981) Danazol: endocrine pharmacology and therapeutic applications. Am J Obstet Gynecol 141:453
4. Burgus R, Butcher M, Amoss M, Ling N, Monahan MW, Rivier J, Fellows R, Blackwell R, Vale W, Guillemin R (1972) Primary structure of the ovine hypothalamic luteinizing hormone releasing factor (LRF). Proc Natl Acad Sci USA 69:278
5. Cirkel U, Schweppe KW, Ochs H, Schneider HPG (1987) Metabolic effects and general side effects of endometriosis treatment with an LHRH agonist. Geburtshilfe Frauenheilkd 47:154–157
6. Dlugi AM, Rufo S, D' Amico JF, Seibel MM (1988) A comparison of the effects of buserelin versus danazol on plasma lipoproteins during treatment of pelvic endometriosis. Fertil Steril 49:913–915
7. Douglass CF, Weed JC (1957) Endometriosis treated with prolonged administration of diethylstilbestrol. Obstet Gynecol 13:744
8. Greenblatt RB, Dmowski WP, Mahesh VP et al. (1971) Clinical studies with an antigonadotropin danazol. Fertil Steril 22:102
9. Hardt W, Schmidt-Gollwitzer M, Schmidt-Gollwitzer K, Genz T, Nevinny-Stickel J (1986) Initial results in the treatment of endometriosis with the LH-RH analog buserelin. Geburtshilfe Frauenheilkd 46: 483–489
10. Healy DL (1986) LHRH analogues in endometriosis. Lancet 1016–1018
11. Helm K, Kiesel L, Rabe T, Runnebaum B (1988) Desensitization of pituitary cells by gonadotropin-releasing hormone in vitro. In: Runnebaum B, Rabe T, Kiesel L (eds) Female contraception. Berlin Heidelberg New York, Springer, pp 184–191
12. Henzl MR, Corson SL, Moghissi K, Buttram VC, Berquist C, Jacobson J (1988) Administration of nasal nafarelin as compared with oral danazol for endometriosis. N Engl J Med 318:485–490
13. Jelley RY (1987) Multicentre open comparative study of buserelin and danazol in the treatment of endometriosis. Br J Clin Pract 41:64–68
14. Kiesel L, Runnebaum B (1989) Intracellular aspects of gonadotropin-releasing hormone. In: Vickery B (ed) Proceedings of the International Symposium on GnRH Analogue in Reproduction and Cancer. MTP Press, pp 53–70
15. Kiesel L, Runnebaum B (1990) Mechanisms involved in the hormonal treatment of endometriosis. In: Chadha DR, Buttram VC jr (eds) Current concepts in endometriosis. Alan R. Liss, New York, pp 179–196
16. Kiesel L, Rabe T, Schleef J, Schweizer M, Klinga K, Runnebaum B (1986) Pulsatile secretory pattern of gonadotropins and steroids during the follicular and luteal phase of the menstrual cycle in women. In: Pulsatile GnRH 1985. Proceedings of the 3rd Ferring symposium on pulsatile GnRH. Noordwijk, September 11–13, 1985. pp 43–49 Haarlem, Ferring
17. Kiesel L, Thomas K, Tempone A, Trabant H, Widdra W, Runnebaum B (1989) Efficacy and safety of buserelin treatment in women with endometriosis – a multicentre open-label study. Gynecol Endocrinol 3 [Suppl 2], pp 5–19
18. Kiesel L, Braun S, Bertges K, Schaible A, Scharla S, Minne HW, Runnebaum B (1990) Bone mineral density in patients treated with intranasal buserelin for endometriosis: a randomized study for calcium supplementation. In: Brosens I, Jacobs HS, Runnebaum B (eds) LHRH analogues in gynaecology. Parthenon, Carnforth (Lancs), pp 179–188
19. Kistner R (1959) The treatment of endometriosis by inducing pseudopregnancy with ovarian hormones. A report of 58 cases. Fertil Steril 10:539
20. Lemay A, Quesnel G (1982) Potential new treatment of endometriosis: reversible inhibition of pituitary-ovarian function by chronic intranasal administration of a luteinizing hormone-releasing hormone (LH-RH) agonist. Fertil Steril 38:376–379
21. Lemay A, Maheux R, Faure N, Jean C, Fazekas ATA (1984) Reversible hypogonadism induced by a luteinizing hormone-releasing hormone (LHRH) agonist (buserelin) as a new therapeutic approach for endometriosis. Fertil Steril 41:863–871
22. Lemay A, Maheux R, Huot C, Blanchet J, Faure N (1988) Efficacy of intranasal or subcutaneous luteinizing hormone-releasing hormone agonist inhibition of ovarian function in the treatment of endometriosis. Am J Obstet Gynecol 158:233–236
23. Matsuo H, Baba Y, Nair RM, Arimura A, Schally AV (1971) Structure of the porcine LH- and FSH-releasing hormone. I. The

proposed amino acid sequence. Biochem Biophys Res Commun 43:1334
24. Matta WH, Shaw RW (1987) A comparative study between buserelin and danazol in the treatment of endometriosis. Br J Clin Pract 41:69–72
25. Meldrum DR, Chang RJ, Lu J, Vale W, Rivier J, Judd HL (1982) Medical oophorectomy using a long-acting GnRH agonist – a possible new approach to the treatment of endometriosis. J Clin Endocrinol Metab 54:1081–1083
26. Olive DL, Haney AF (1986) Endometriosis-associated infertility: a critical review of therapeutic approaches. Obstet Gynecol 41:538–555
27. Sandow J (1983) Clinical applications of LHRH and its analogues. Clin Endocrinol 18:571–592
28. Schriock E, Monroe SE, Henzl M, Jaffe RB (1985) Treatment of endometriosis with a potent agonist of gonadotropin-releasing hormone (nafarelin). Fertil Steril 44:583–588
29. Schweppe KW, Wynn RM (1984) Endocrine dependency of endometriosis: an ultrastructural study. Eur J Obstet Gynecol Reprod Biol 17:193–208
30. Shaw RW, Fraser HM, Boyle H (1983) Intranasal treatment with luteinizing hormone-releasing hormone agonist in women with endometriosis. Br Med J 287:1667–1669
31. Tummon IS, Ali A, Pepping ME, Radwanska E, Binor Z, Dmowski WP (1988) Bone mineral density in women with endometriosis before and during ovarian suppression with gonadotropin-releasing hormone agonists or danazol. Fertil Steril 49:792–796

# Laser Therapy of Endometriosis

D. WALLWIENER [1], D. POLLMANN, W. STOLZ, C. SOHN, and G. BASTERT

Endometriosis is a complex disease whose etiology, pathogenesis, and treatment are still being investigated [7]. It may be associated with dysmenorrhea and dyspareunia and is often a cause of infertility. Clinical management of the disease varies according to the patient's child-bearing desire and is controversial. It may consist of medical therapy with a variety of hormones, all designed to suppress the disease by decreasing size, quantity and vascularization. Subsequent or alternative surgical therapy aims at removal of the endometrial implants and repair of anatomic pathology. Often, combined treatment schemes are applied.

For operative laparoscopic management of endometriosis laser and electrosurgical techniques are alternatively applicable, since the goal of surgical therapy of thermal destruction of endometrial implants can be achieved with both above-mentioned methods. The laparoscopic approach presents the advantage of treatment at diagnosis and reduced adhesion induction.

Gentle, atraumatic resection or accurate vaporization of endometrial implants through the various tissue layers is possible with the carbon dioxide laser. By noncontact application, finest, precise, layer-specific thermal destruction in an exactly controlled vaporization process can be performed, even in the delicate areas of the tuboovarian functional unit, the urethra, or even the bowel serosa. Damage to adjacent tissues is minimal and the operating field is bloodless. The carbon dioxide laser has been successfully applied laparoscopically especially for treatment of endometriosis associated with infertility. Pregnancy rates after $CO_2$ vaporization of endometriomas of 73%/75%/66% (mild, moderate, and severe endometriosis) [3], 75%/63%/42%/50% (stage I–IV AFS) [6], and 61%/52%/40% (mild, moderate, and severe endometriosis) [2] have been reported in the literature.

In our experience, thermal destruction of endometrial implants with the Nd:YAG contact laser does not differ from that achieved by electrosurgical procedures. Additional coagulation zones caused by leaking currents are a slight disadvantage of electrosurgery, which, however, plays a role only for finest tissue structures. Application of the Nd:YAG laser energy in a noncontact technique should be avoided. Though intraoperatively thermal destruction of large tissue aspects is achieved within fractions of seconds and endometrial implants of every size can be rapidly vaporized with this tech-

[1] Universitäts-Frauenklinik, Abt. für Geburtshilfe und Gynäkologie, Voßstr. 9, W-6900 Heidelberg, FRG.

Lasers in Gynecology
Edited by G. Bastert and D. Wallwiener

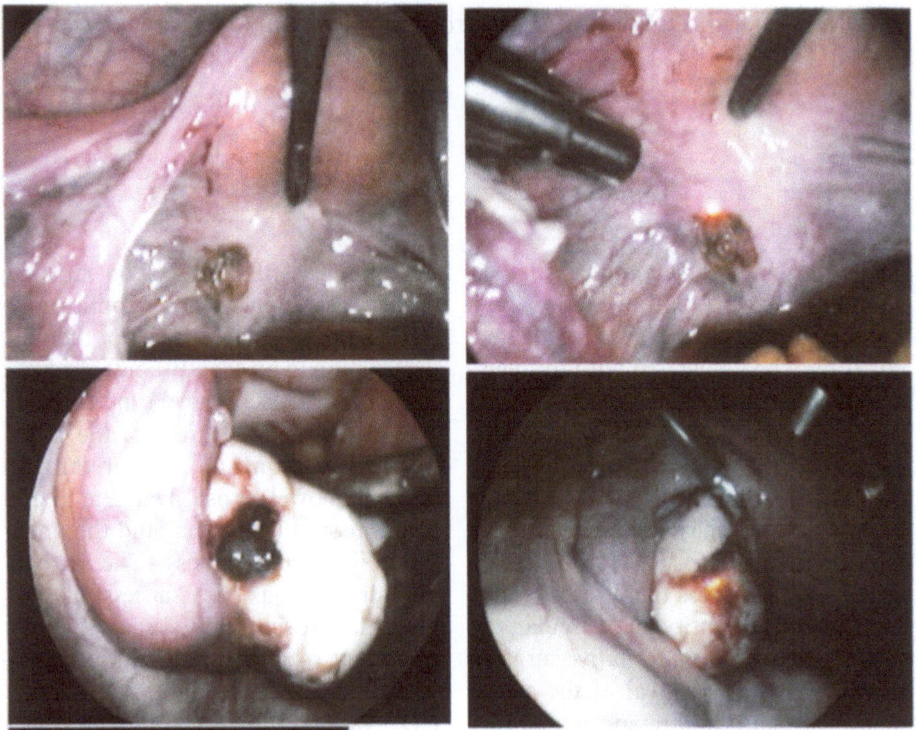

a

$b_I$

$b_{II}$

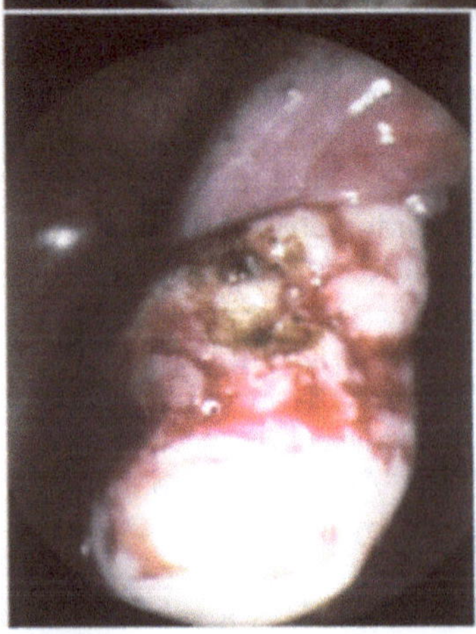

$b_{III}$

**Fig. 1 a–c.** Endometriotic implants – carbon dioxide laser vaporization. **a** Cul-de-sac of Douglas; **b** Ovary (**$b_I$** pre-operatively, **$b_{II}$** $CO_2$-Laser-beam in situ, **$b_{III}$** post-operatively); **c** Sacrouterine ligament

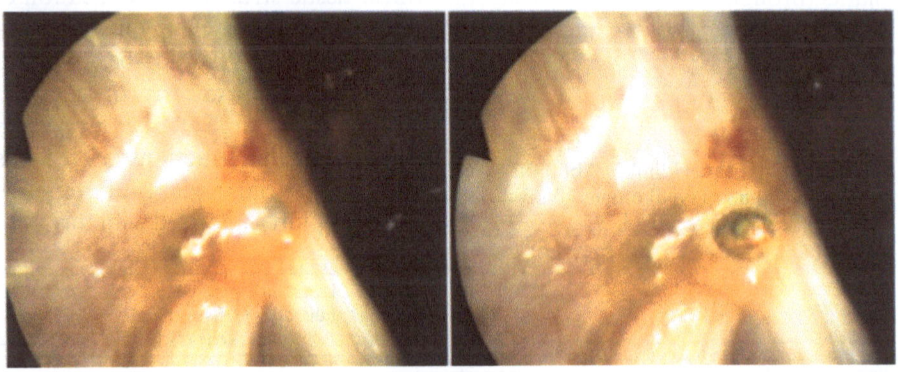

c

nique, the characteristic deep penetration of noncontact Nd:YAG light of up to 7 mm might lead to inadvertent tissue sequestration with unwanted and uncontrolled adhesion induction. Shirk [8] reported an overall conception rate of 67% after Nd:YAG laser treatment of patients in various stages of endometriosis. Corson et al. [1] achieved restoration of fertility in 39.7% (short follow-up) and found pain relief to be excellent after treating 84 patients with a Nd:YAG tissue-contact laser probe via laparoscopy.

An evaluation of operative-laparoscopic procedures for therapy of endometriosis is almost impossible, especially since a large variety of combination treatments of hormonal therapy (one- to multi-step) with second look laparoscopy and surgical removal of residual endometriomas are also being investigated. Especially problematic is the therapy of recurrent disease after various therapeutic approaches, recurrence rates being still relatively high. Rates between 2% and 47% are described by various investigators [4, 5].

In a collective of 50 patients with endometriosis recurrent after various therapies, a combination management of preoperative GnRH analog medication and $CO_2$ laser vaporization of residuals proved successful in our department. During follow-up of 6 months to 2 years, in 91% of the patients endometriosis did not recur Long-term follow-up is, however, necessary.

## References

1. Corson SL, Woodland M, Frishman G, Batzer FR, Gocial B, Maislin G (1989) Treatment of endometriosis with a Nd:YAG tissue-contact laser probe via laparoscopy. Int J Fertil 34(4):284–288
2. Donnez J, Nisolle M, Wayembergh M, Clerckx F, Casanas-Roux F (1989) $CO_2$ laser laparoscopy in peritoneal endometriosis and in ovarian endometrial cyst. In: Donnez J (ed) Laser operative laparoscopy and hysteroscopy. Nauwelaerts Printing, Leuven, 53–75
3. Feste JR (1985) Laser laparoscopy: a new modality. J Reprod Med 30:1089–1094
4. Green R (1966) Conservative surgical treatment of endometriosis. Clin Obstet Gynecol 9:293
5. Gordts S, Boeckx W, Brosens I (1984) Microsurgery of endometriosis in infertile patients. Fertil Steril 42:520–525
6. Nezahat C, Crowgui S, Carrison C (1986) Surgical treatment of endometriosis via laser laparoscopy. Fertil Steril 45:778–783
7. Sanfilippo JS (1989) Endometriosis. In: Sanfilippo JS, Levine RL (eds) Operative gynecologic endoscopy. Springer, New York Berlin Heidelberg, pp 236–250
8. Shirk GJ (1989) Use of the Nd:YAG laser for treatment of endometriosis. Am J Obstet Gynecol 160(6):1344–1348

# Endometriosis Associated with Infertility: Therapeutic Approaches

J. DONNEZ [1] and M. NISOLLE

The controversy surrounding the optimal treatment of women with endometriosis is not simply a battle between the endocrinologist [11] and the surgeon; it is also the dilemma of choosing between expectant management and any active treatment [13–15]. Since Sampson first described endometriosis as a unique entity in 1921 [16], the most common locations of this disease have remained the surfaces of the most dependent intra-abdominal structures in the upright woman. The lesions are diagnosed according to various classification systems [1–3, 12 ].

## Hormone Therapy

### Pseudopregnancy and Progestins

The amelioration of endometriosis that seemed to accompany pregnancy was the subject of much debate in the 1950s and fostered initial treatment with progestin therapy and oral contraceptives (OCs). The fact that endometriosis was infrequently encountered in the multiparous female but most often seen in nulliparous women was felt to indicate the protective effect of pregnancy.

[1] Infertility Research Unit, Dept. of Gynecology, Catholic University of Louvain, Cliniques Universitaires St. Luc, Avenue Hippocrate 10, 1200 Brussels, Belgium.

Combination Estrogen-Progestogen

The combination of an estrogen and a progestogen for therapy of endometriosis, the so-called pseudopregnancy regimen, has been utilized for many years. Pseudopregnancy is believed to produce an initial decidualization and growth of endometrial tissue which is followed in several months by atrophy. The efficacy of combination therapy is difficult to assess, despite the great number of years this treatment has been utilized.

Progesterone

In a similar manner, progestin therapy was fashioned on the production of decidualization and eventual atrophy of endometrium-like tissue. While injectable forms of progestogens are available, oral administration may be preferable for the patient who desires conception, because of rapid reversibility of the effect.

Medroxyprogesterone acetate, lynestrenol, and norethindrone acetate have been used orally [8]. The most common side effect is intermittent breakthrough bleeding. Other side effects include nausea, weight gain, edema, fluid retention, and depression. Littlc is known about thc cxact rccurrence rate of endometriosis after treatment with progestogens.

Lasers in Gynecology
Edited by G. Bastert and D. Wallwiener

## Androgens

The use of methyltestosterone to treat infertility associated with endometriosis is of mainly historical interest. The daily dose of 5–10 mg of methyltestosterone sublingually, although effective in relieving dysmenorrhea and dyspareunia, does not consistently inhibit ovulation. However, higher dosages of methyltestosterone can be flagrantly masculinizing.

## Danazol

The isoxasol derivative of 17α-ethinyl testosterone, danazol, was supposed to effect a state of pseudomenopause by creating an endocrine environment that inhibits the growth of endometrial tissue [4, 17]. A number of side effects have been attributed to danazol. Most are a result of the anabolic and androgenic effects of the drug and some of its metabolites, although progestational effects are occasionally seen, such as weight gain, muscle cramps, decreased breast size, flushing, oily skin, depression, edema, acne, hirsutism, and decreased libido.

Although the diminutional effect of danazol on active endometriotic tissue has been well documented and pregnancies have followed treatment [6], the efficacy of danazol was not confirmed in our studies. Indeed, in our series of 40 women with moderate and severe endometriosis treated for 6 months, danazol reduced endometriotic lesions in only 36% and 23% of the cases, respectively. Moreover, in more than 30% of the cases the diameter of ovarian endometriotic cysts was increased.

## Gonadotropin-releasing Hormone Agonist

In addition to pseudopregnancy and pseudomenopause, the availability of the GnRH agonist offers the most recent weapon in hormone therapy. This "medical oophorectomy" allows temporary suppression of ovarian function.

GnRH agonists initially stimulate the secretion of follicle-stimulating hormone (FSH) and luteinizing hormone (LH). After a single injection, gonadotropins remain elevated for 6–24 h (stimulation phase). Upon repeated daily administration, gonadotropin secretion declines progressively (suppression phase), and the production of gonadal steroids may reach castration-like low levels.

Gonadotropin inhibition is readily reversible, because synthesis of new receptors requires only 2–7 days. Buserelin nasal spray is rapidly absorbed by the nasal mucosa (2.5%–5% of the dose, depending on the formula of the nasal solution). Sustained release systems currently under clinical investigation can be administered by injection (implants or microcapsules). Although treatment effects castratory estrogen levels associated with regression of lesions, menopausal symptoms of vaginal thinning and bone loss are also induced [5].

The Buserelin nasal spray or sustained release formulas to suppress endometrial proliferation can be used to reduce the size of lesions prior to surgical intervention. In our study of 40 women with moderate and severe ovarian endometriosis, Buserelin spray was given intranasally for 6 months for this purpose. According to the AFS classification, the average decrease of the endometriosis score was 1.3 (from

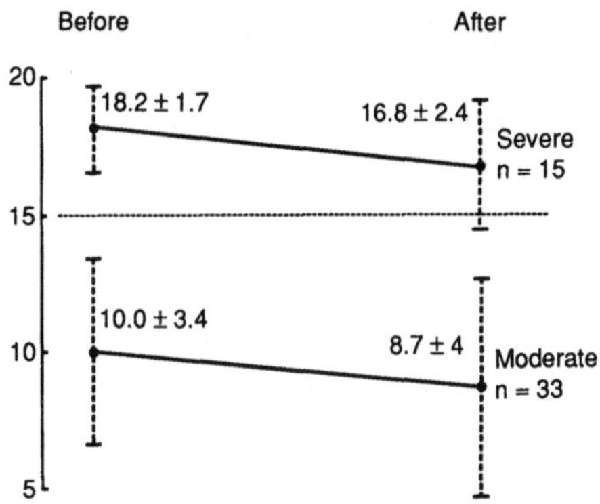


**Fig. 1.** Comparison of American Fertility Society (AFS) scores before and after 6 months of GnRH agonist therapy for severe and moderate endometriosis

10.0 ± 3.4 to 8.7 ± 4) in moderate endometriosis and 1.4 (from 18.2 ± 1.75 to 16.8 ± 2.4) in severe endometriosis (Fig. 1). Preoperative hormone therapy resulted in a reduction of pelvic vascularity and inflammation. Unfortunately, these parameters are not evaluated in any staging system.

## Microsurgery

Although the definition of conservative surgery has remained "the destruction of as much endometriosis as possible with preservation of childbearing function," the technical advances of in vitro fertilization-embryo transfer and embryo donation permit maintenance of a uterus to be the sole prerequisite for implantation and delivery of a viable infant.

Improved pregnancy rates have been reported with atraumatic microsurgical techniques that employ loupes and the microscope for magnification. Such procedures were used in 176 women whose cumulative pregnancy rates using life-table methods were 28.8% after 1 year and 40.5% after 2 years [9].

Similarly, the carbon dioxide laser has been proposed to be optimal for endometriotic resection [10]. Vaporization of focal implants, excision of endometriotic cysts, and lysis of adhesions were performed in 64 patients with mild or moderate endometriosis. Patients were followed up for a period of 1 year. Pregnancy rates of 57% and 50% respectively were obtained in patients with mild and moderate endometriosis.

## Combined Therapy: Hormones and Microsurgery

The coupling of preoperative hormone therapy with microsurgery was suggested by Donnez in 1978 [8]. Pregnancy rates after surgery were 53% among women with moderate endometriosis and 39% among those with severe endometriosis (Fig. 2). The type of hormone therapy did not influence the

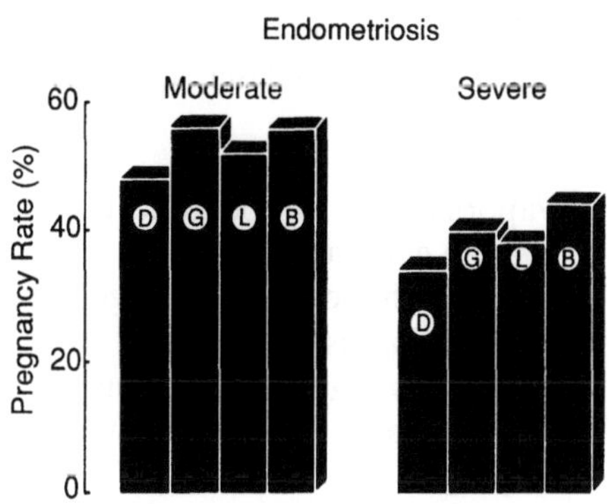


**Fig. 2.** Pregnancy rate after combined hormone therapy and microsurgery. Hormone preparations: *D*, danazol; *G*, gestrinone; *L*, lynestrenol; *B*, Buserelin

pregnancy rates in our series. Because of the high pregnancy rates found in our study, the coupling of preoperative hormone suppression with microsurgery was suggested to the primary treatment approach for infertile women with moderate and severe endometriosis [2, 3].

The actual combined therapy is the coupling of the GnRH agonist with microsurgery. Our preliminary results with this combination suggest that this therapeutic approach should be recommended for such patients. The improvement in pregnancy rates after combined therapy is probably the result of reducing the duration of the surgical procedure by facilitating it and of reducing the risk of postoperative adhesion by decreasing pelvic vascularity.

## $CO_2$ Laser Laparoscopy

The endoscopic use of the laser is not new in medicine; laser endoscopy has already been used in otolaryngology and gastroenterology and is currently being investigated for clinical use in orthopedics, urology, and gynecology.

The procedures most frequently carried out with the $CO_2$ laser laparoscope are:

1. Vaporization of peritoneal and ovarian endometriosis
2. Pelvic adhesiolysis, salpingolysis, and ovariolysis
3. Uterosacral ligament ablation
4. Fimbrioplasty and salpingostomy
5. Salpingostomy in case of tubal pregnancy

Less frequently, vaporization of small uterine fibroids is done.

### Peritoneal Endometriotic Implant

Vaporizing an endometriotic implant provokes the bubbling of old blood, followed by a curdy white material representing vaporization of the stromal layer. After the entire endometriotic lesion has been vaporized, retroperitoneal fat is encountered, and the appearance of "bubbling water" confirms the complete vaporization of the lesion. The absorption of the $CO_2$ laser by water prevents deeper penetration of the laser beam for a few seconds after the complete destruction of the implant.

### Ovarian Endometriosis

Ovarian endometriosis is treated if penetration into the ovary is no greater than 3 cm and cysts no larger than 3 cm in diameter are present. Any chocolate material is aspirated with the help of an aspirating cannula and the cyst is washed with irrigation fluid. The interior wall of the endometrioma is then vaporized until follicles are encountered or no further pigmented tissue is seen. The ovaries are left open (Fig. 3). At the end of the laser procedure, copious irrigation of the pelvic cavity is carried out to prevent deposits of carbon from the plume, which may be irritating to the peritoneum. At the end of the procedure, and following the removal of blood from the cul-de-sac, 100–200 ml of 32% dextran 70 can be instilled into the peritoneum.

In our series of 64 infertile patients treated by laser laparoscopy, who had been followed up for 1 year, 42 had mild endometriosis and 22 had moderate endometriosis, and pregnancy rates of 57% and 50% respectively were obtained (Fig. 4). No patient required la-

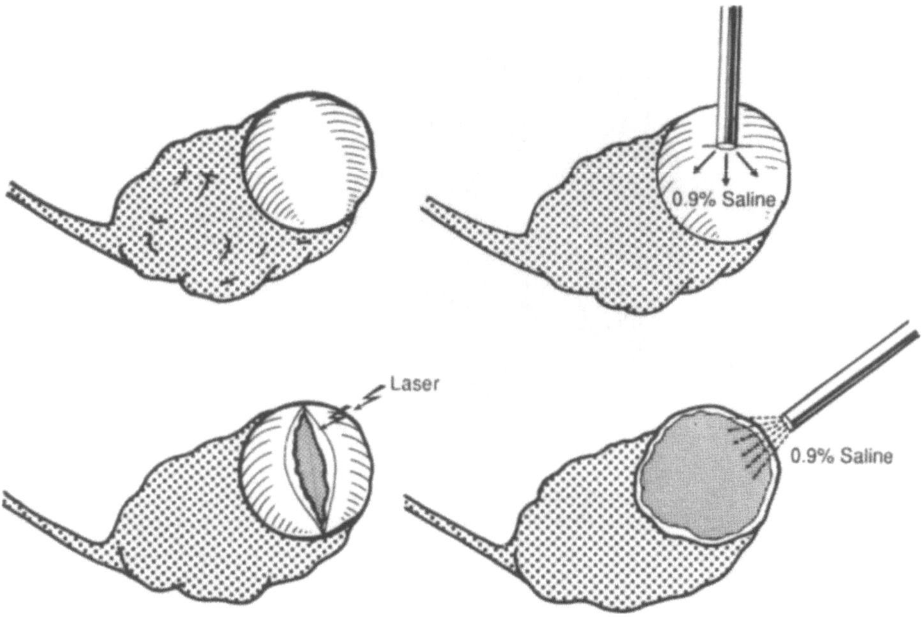


**Fig. 3.** Ovarian endometriosis: $CO_2$ laser laparoscopy

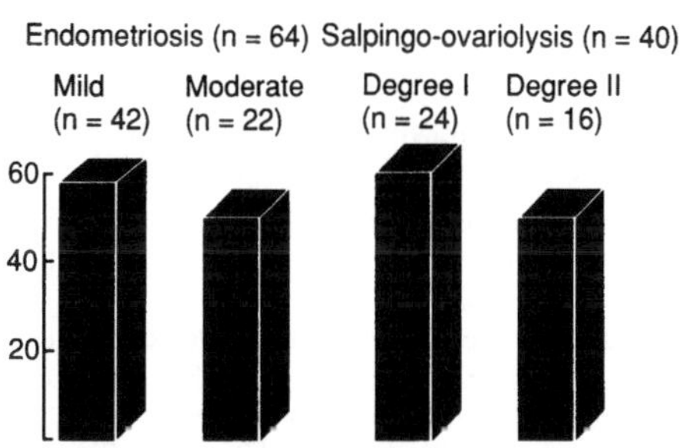


**Fig. 4.** Pregnancy rates after $CO_2$ laser laparoscopy

parotomy for control of bleeding. All patients were released the following day in good condition. The pregnancy rate obtained after laser vaporization of ovarian endometriomas is similar to that achieved with combined hormone therapy and microsurgery.

In cases of more severe endometriosis, the procedures we used (Fig. 5) consisted of:

1. Vaporization of endometriotic implants as much as possible and laser salpingo-ovariolysis
2. Hormone therapy (GnRH agonist such as Buserelin)
3. Laparotomy for microsurgical resection of endometriotic cysts

## Conclusion

Used endoscopically, the $CO_2$ laser offers some advantages over other operative techniques for endometriosis and adhesions. But to demonstrate the safety and efficacy of such treatment, a comparison of postoperative pregnancy rates in a prospective study with other forms of therapy (microsurgery, combined hormone therapy and surgery) must be carried out.

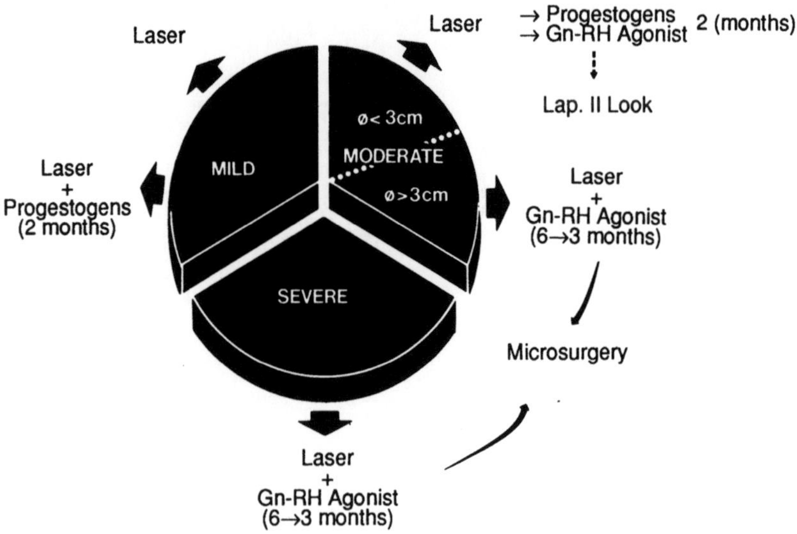


**Fig. 5.** Actual treatment in case of endometriosis-related infertility

In the hands of an experienced laparoscopist, the $CO_2$ laser appears safe and effective for vaporization of endometriotic lesions, uterosacral neurectomy, adhesiolysis, and salpingostomy. The judicious use of these techniques, combined with carefully planned further investigations by adequately trained and experienced laparoscopists and continuing improvements in the delivery systems, will soon demonstrate the true efficacy of the $CO_2$ laser laparoscope. If studies continue to show pregnancy rates equivalent to those among patients treated by laparotomy, $CO_2$ laser laparoscopy will become the preferred procedure for the management of pelvic endometriosis and its associated adhesions.

## References


1. Acosta AA, Buttram CJ jr, Besch PK (1973) A proposed classification of pelvic endometriosis. Obstet Gynecol 42:19
2. American Fertility Society (1979) Classification of endometriosis. Fertil Steril 32:633
3. American Fertility Society (1985) Revised American Fertility Society classification of endometriosis. Fertil Steril 43:351
4. Barbieri RL, Canick JA, Makris A (1977) Danazol inhibits steroidogenesis. Fertil Steril 28:809
5. De Vogelaer JP, Nagant de Deuxchaisnes C, Donnez J, Thomas K (1987) LH-RH analogues and bone loss. Lancet 4:1498
6. Dmowski WP, Cohen MR (1975) Treatment of endometriosis with an antigonadotropin, danazol: a laparoscopic and histologic evaluation. Obstet Gynecol 46:147
7. Donnez J (1987) $CO_2$ laser laparoscopy in infertile women with adhesions or endometriosis. Fertil Steril 48:390
8. Donnez J, Lemaire M, Karaman Y, Nisolle-Pochet M, Casanas-Roux F (1987) Combined (hormonal and microsurgical) therapy in infertile women with endometriosis. Fertil Steril 48:239
9. Gordts S, Boeckx W, Brosens I (1984) Microsurgery of endometriosis in infertile patients. Fertil Steril 42:520

10. Kelly RW, Roberts DK (1983) $CO_2$ laser laparoscopy: a potential alternative to danazol in the treatment of stage-I and stage-II endometriosis. J Reprod Med 28:638
11. Kistner RW (1966) Current status of the hormonal treatment of endometriosis. Clin Obstet Gynecol 9:271
12. Kistner RW, Siegler AM, Behrman SJ (1977) Suggested classification for endometriosis: relationship to infertility. Fertil Steril 28:1008
13. Olive DL, Haney AF (1987) Endometriosis-associated infertility. A critical review of therapeutic approaches. Obstet Gynecol Surv 41:538
14. Olive DL, Stohs GF, Metzger DA (1985) Expectant management and hydrotubations in the treatment of endometriosis associated with infertility. Fertil Steril 44:35
15. Portuondo JA, Echanojauregui AD, Herran C (1983) Early conception in patients with untreated mild endometriosis. Fertil Steril 39:22
16. Sampson JA (1921) Perforating hemorrhagic (chocolate) cysts of the ovary: their importance and especially their relation to pelvic adenomas of endometrial type. Arch Surg 3:245
17. Seibel MM, Berber MJ, Weinstein FG (1982) The effectiveness of danazol on subsequent fertility in minimal endometriosis. Fertil Steril 38:534

# Vaginal Endosonography: Its Value for Planning and Follow-up of Endometriosis Therapy

W. Stolz [1], C. Sohn, D. Wallwiener, E. Grischke, and G. Bastert

Sonography is an established method for evaluating the internal genitals as well as pathological findings in the lower abdomen. In particular, vaginal sonography offers a decisive advantage, being able to depict directly at the location of interest. In contrast to abdominal sonography, there is no need for a filled urinary bladder [1], a bothersome and time-consuming condition.

The pelvic wall is easily depicted. With the ultrasound probe introduced into the vagina the structures of the perineum present no hindrance as in a perineal scan [3]. With bimanual palpation tumors can be manipulated and intestinal obstructions can be removed. This method is as well established as digital palpation [5] (Table 1).

At our disposition we have a 240° panorama scan, 5 MHz, as well as a 90° probe, 5 MHz, with integrated pulsed Doppler. The 240° presentation (Fig. 1) allows a complete overview; the 90° probe, however, has the advantage of better imaging quality (Fig. 2). The Doppler allows the identification of veins (Fig. 3) whereas the arteries are depicted without the aid of doppler by their pulsation (Fig. 4).

Cysts are easy to interpret (Fig. 5), whereas conglomerates, consisting partly of densities and partly of cystic structures (Fig. 6), may present difficulties in interpretation [4]. It is absolutely necessary to exactly describe location, measurements, and structure of the findings. In the case of well palpable tumors the evaluation of the structure may result in the diagnosis of "ovarial endometriosis" (Fig. 7). Even in the case of an echo-free cyst (Fig. 8) we may discover a "chocolate cyst." The vagino-sonographically guided puncture (Fig. 9) offers new diagnostic and therapeutic possibilities [2]. The puncture can be done as an ambulatory service and has so far ($n = 152$) been without any complication. Defects at the pelvic wall (Fig. 10) are minimal.

All ultrasonically discovered findings of endometriosis ($n = 101$) are

**Table 1.** Possibilities for evaluating the lower abdomen and the internal genitals

| |
|---|
| Transabdominal access |
| – filled bladder |
| – interpretation of pelvic wall difficult |
| – problem of gas-filled intestine |
| Perineal scan |
| – filled bladder |
| – perineum to be overcome |
| – intestinal problems |
| Transvaginal access |
| – no filled bladder |
| – can be done directly at the place of interest |
| – easy evaluation of pelvic wall |
| – intestine to be removed can be discovered by palpation |

[1] Universitäts-Frauenklinik, Abt. für Geburtshilfe und Gynäkologie, Voßstr. 9, W-6900 Heidelberg, FRG.

Lasers in Gynecology
Edited by G. Bastert and D. Wallwiener

1

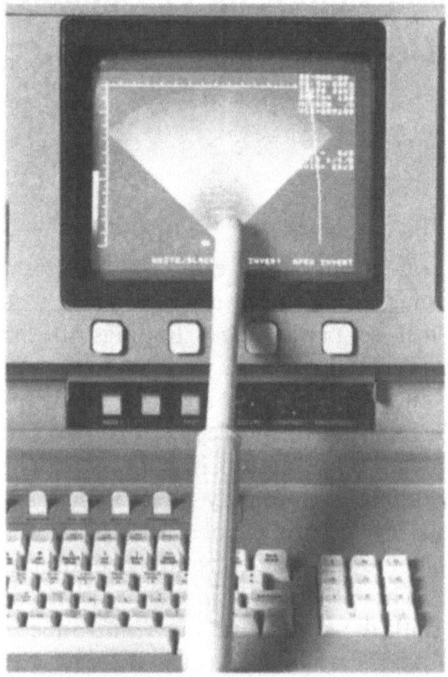

2

3

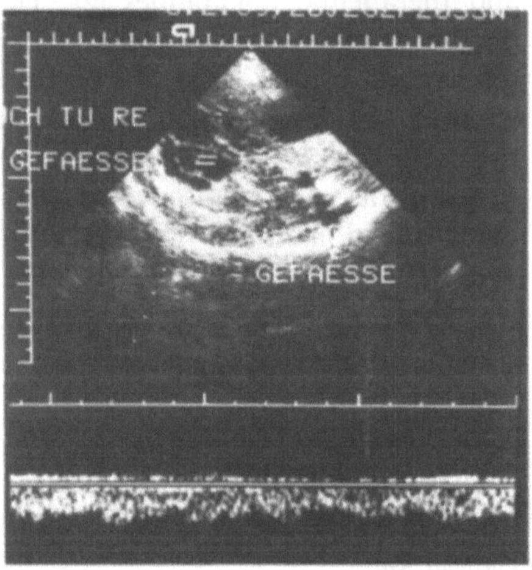


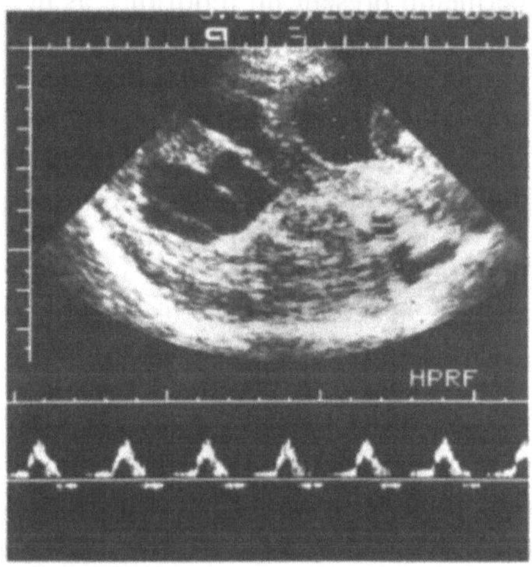


4

**Fig. 1.** 240° panorama scan offers good overview

**Fig. 2.** 90° probe offers valuable depiction of detail, high quality of picture, integrated Doppler

**Fig. 3.** Depiction of veins at the pelvic wall and venous flow

**Fig. 4.** Arteries at the pelvic wall and their flow

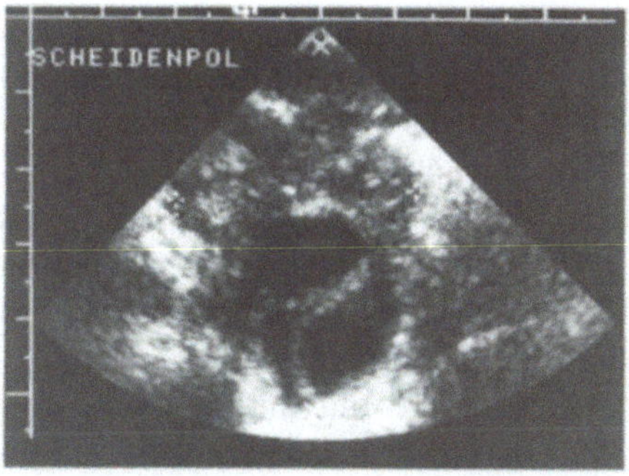


5

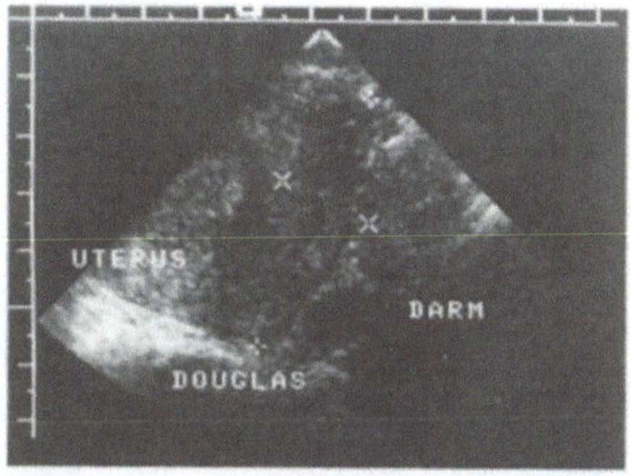


6

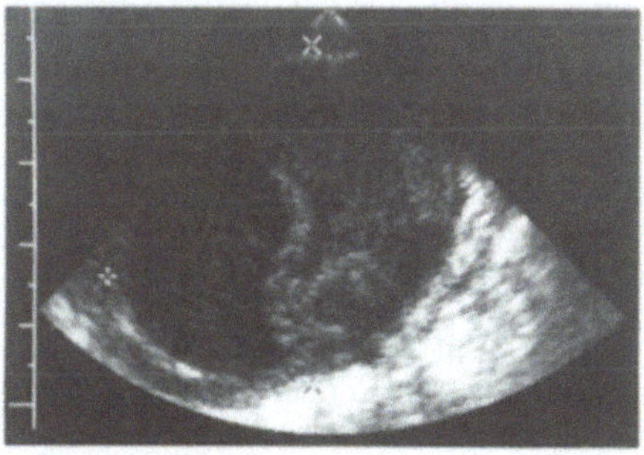

7

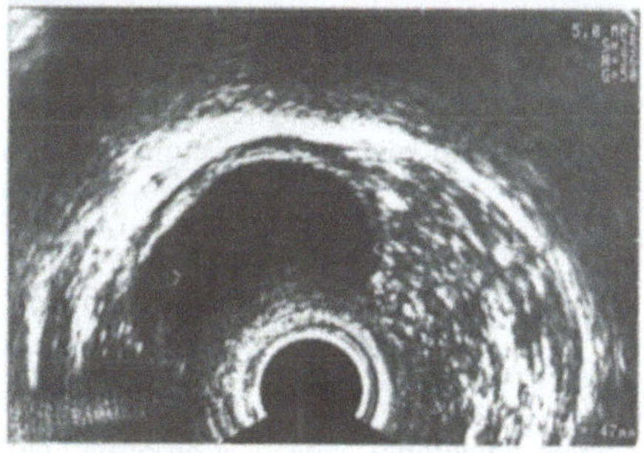

8

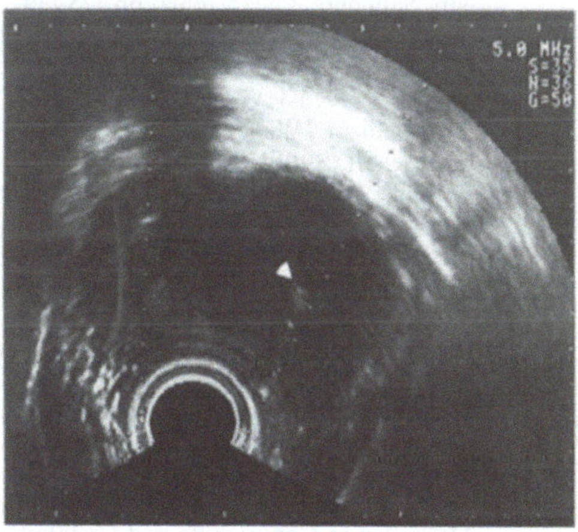


9

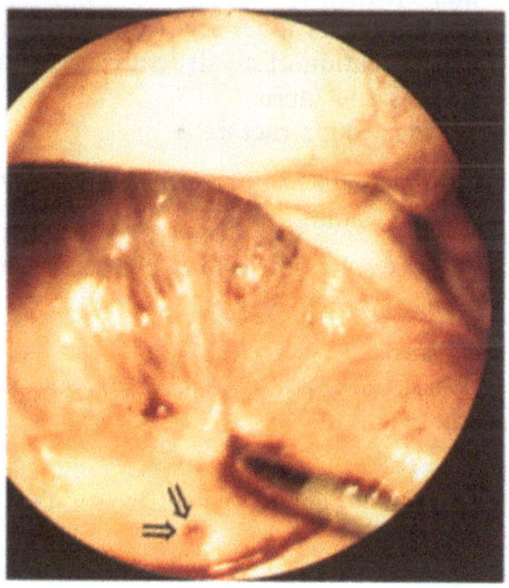

10

**Fig. 5.** Polycystic tumor at the end of vagina after hysterectomy, cyclical bleeding

**Fig. 6.** Hyperdense structures behind uterus towards Douglas' cul-de-sac

**Fig. 7.** Ovarial endometriosis in a 17-year-old girl, before operation; confirmed by histology

**Fig. 8.** Echo-free, solitary cyst

**Fig. 9.** Echo-free, solitary cyst with puncture needle

**Fig. 10.** Laparoscopic picture after puncture; location of puncture marked by *arrows*

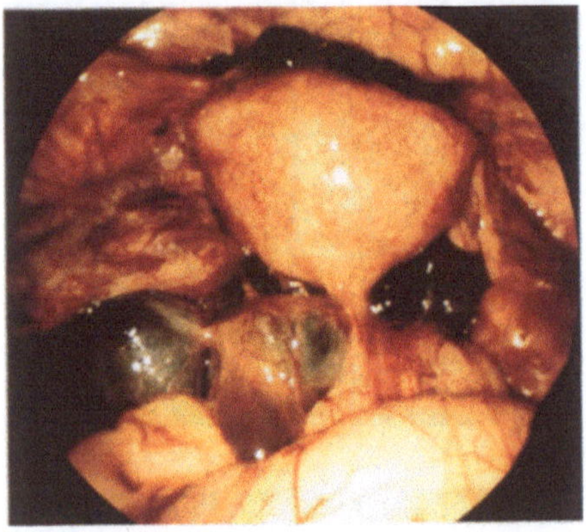

**Fig. 11.** Ultrasonically diagnosed endometriosis (Fig. 5) and laparoscopic picture

**Table 2.** Ultrasound program "Endometriosis"

| |
|---|
| 1. Preoperative screening |
| – exact topographical description of findings and their measurements |
| – description of inner segmental structures |
| – exact location of all structures of interest |
| – ovaries and uterus |
| – bladder wall structure |
| 2. Comparison of ultrasound with laparoscopic findings |
| 3. Four-weekly ultrasound control during drug therapy |
| 4. Ultrasonically guided vaginal puncture (selected cases, cytological and histological evaluation) |

checked by laparoscopy (Fig. 11). Short-term ultrasound controls evaluate the effect of drug therapy.

In our multiphase therapy of endometriosis vaginal sonography has proved to be valuable in the evaluation of the effect of drug therapy as well as in the preoperative staging.

To establish our findings, we correlate the ultrasound results with laparoscopy. In selected cases we use ultrasonically guided puncture by vaginal probe (Table 2).

## References


1. Clifford SL, Lyons EA, Lindsay DJ, Ballard G (1988) Normal anatomy of the female pelvis. In: Callen PW (ed) Ultrasonography in obstetrics and gynaecology, 2nd edn. Saunders, Philadelphia, pp 375–392
2. Degenhard F (1988) Atlas der vaginalen Ultraschalldiagnostik. In: Schneider J, Weitzel H (eds) Gynäkologie und Geburtsmedizin. Wissenschaftliche Verlagsgesellschaft, Stuttgart, S 52–63
3. Grischke EM, Dietz HP, Jeanty P, Schmidt W (1986) Eine neue Untersuchungsmethode: Perineal Scan in der Geburtshilfe und Gynäkologie. Ultraschall Med 4:154–161
4. Hansmann M, Hackeloer B-J, Staudach A (1985) Ultraschalldiagnostik in Geburtshilfe und Gynäkologie. Lehrbuch und Atlas. Springer, Berlin Heidelberg New York Tokyo, S 367–376
5. Popp LW (1986) Endosonographische Methoden in Geburtshilfe und Gynäkologie. In: Popp LW (ed) Gynäkologische Endosonographie. Ingo Klemke Verlag, Quickborn, S 35–50

# Laparoscopic Laser Surgery for Endometriosis – Eight Years' Experience

C. Sutton [1]

Laparoscopic laser surgery was first performed in the United Kingdom at St. Luke's Hospital in Guildford in October 1982. The initial idea was inspired by observations on cervical healing following colposcopic treatment of cervical intraepithelial neoplasia (CIN) with the carbon dioxide laser. Since most of the cell debris was removed as smoke in the laser plume, healing occurred with no contracture or anatomical distortion, and 6 weeks later healing was usually so perfect that even with the magnification provided by the colposcope it was difficult to see the site of the original laser surgery. This was in marked contradistinction to the scarring and fibrosis caused by electrodiathermy and cold-knife cone biopsy. For these reasons it was felt that the laser would be the ideal surgical modality to use endoscopically, especially when fertility was a prime consideration.

The carbon dioxide laser provides a high degree of precision when vaporizing tissue, and there is very little thermal damage beyond the impact crater. Once the necessary skills have been acquired diseased tissue can be removed accurately to within 100 μm of a surface, thus making it possible to remove endometriosis from the surface of the ureter, bladder, bowel and major vessels without causing damage to the underlying structures [7]. This makes it ideal for the cytoreduction of endometriosis and for the division of adhesions close to the bowel or the fallopian tubes [6].

We have treated 260 patients with pelvic pain thought to be due to endometriosis who have been followed up between 1 and 7 years; 167 (72%) were better, 58 were not helped, 24 were pain free initially but relapsed after 9–18 months and 11 were lost to follow-up. Among 85 patients with endometriosis and infertility the overall pregnancy rate was 60%, and of those with no other subfertility factors 75% became pregnant, most of them within 8 months of the laser laparoscopy. Interestingly, the success rate was virtually the same regardless of the severity of the disease on the American Fertility Society scale. Laser laparoscopy appears to be an effective treatment for endometriosis, given that the average pregnancy rate following danazol therapy is only 37%, with a monthly fecundity rate never higher than 0.068 [3], which is not a great deal more than would be anticipated with expectant management alone [2]. The advantage of laser laparoscopy lies in the ability to treat the disease at the same time as the diagnosis is made and to allow the patients to attempt conception without delay.

[1] St. Luke's Hospital, Dept. of Gynecology, 8 Waterden Road, Guildford, Surrey, United Kingdom.

Lasers in Gynecology
Edited by G. Bastert and D. Wallwiener

In patients with severe dysmenorrhoea the uterosacral ligaments can be vaporised at the point where they insert into the back of the cervix. This results in destruction of the Lee-Frankenhäuser plexus and deprives the uterus of much of its sensory nerve supply. Of 100 consecutive patients with congestive dysmenorrhoea associated with endometriosis, 85 reported considerable pain relief, although the first postoperative period was often worse because of the oedema around the nerve bundles as the laser crater healed. The results with primary dysmenorrhoea were less impressive, with only 73% of patients gaining relief [4].

Adhesiolysis can be carefully and precisely performed with the carbon dioxide laser, but we advocate the use of a backstop if lysing bowel or peritubal adhesions to prevent damage to distal structures once the adhesions have parted. In a small series of 25 patients who had peritubal adhesions that were distorting their tubal anatomy but whose tubes were patent on hysterosalpingography and normal salpingoscopy, 20 (80%) became pregnant, usually during the subsequent three cycles. Another group benefitting from laser adhesiolysis were those women who had a thick band of adhesions, usually tethering the bowel, as a result of previous abdominal surgery. The division of peritubal or periovarian adhesions did not appear to alter pain unless they were associated with endometriosis [5].

Advanced laparoscopic procedures such as the removal of intraovarian endometriomas and the conservative treatment of unruptured ectopic pregnancy are hampered by the absorption of carbon dioxide laser energy by blood, thus rendering it ineffective. In these situations we have found flexible fibre lasers to be much more satisfactory and have recently started using the potassium titanyl phosphate (KTP/532) laser. This is a frequency-doubled YAG laser producing emerald-green light, which works well in the presence of blood [1].

We have performed laser laparoscopic surgery on over 850 patients, and apart from minor problems such as surgical emphysema and wound haematomas we have not had any injuries due to the intraperitoneal use of laser energy. It is a highly cost-effective form of surgery, apart from the initial capital outlay for the laser itself, and allows many operations that are traditionally treated by major surgery to be performed on a day case or overnight stay basis with rapid recovery and early return to work or domestic activity.

## References


1. Daniell JF (1989) Fibreoptic laser laparoscopy. In: Sutton CJG (ed) Bailliere's clinical obstetrics and gynaecology, vol 3. pp 545–562
2. Olive DL, Haney AF (1986) Endometriosis-associated infertility: a critical review of therapeutic approaches. Obstet Gynecol Surv 41:538–555
3. Seibel MM, Berber MJ, Weinstein FG, Taylor ML (1982) The effectiveness of danazol on subsequent fertility in minimal endometriosis. Fertil Steril 38:534–537
4. Sutton CJG (1989) Laser uterine nerve ablation. In: Donnez J (ed) Laser operative laparoscopy and hysteroscopy. Nauwelaerts, Leuven, pp 43–52
5. Sutton CJG (1989) Carbon dioxide laser laparoscopy in the treatment of endometriosis. In: Sutton CJG (ed) Bailliere's clinical obstetrics and gynaecology, vol 3. pp 499–523
6. Sutton CJG, Hill D (1990) Laser laparoscopy in the treatment of endometriosis. A 5-year study. Br J Obstet Gynaecol 97:181–185
7. Wilson EA (1988) Surgical therapy for endometriosis. Clin Obstet Gynecol 31:857–865

# Ovarian Cysts: Cystectomy, Fenestration, Punction (PCOD)

D. WALLWIENER [1], J. GAUWERKY, and G. BASTERT

Besides their application in endoscopic tubo-ovarian reconstructive surgery, laser-assisted techniques can be successfully employed for ovarian surgery as well. Three main interventions may be considered: (a) cystectomy, (b) fenestration, and (c) puncture for polycystic ovarian disease (PCOD).

## Cystectomy

Laparoscopic cystectomy is the procedure of choice for cysts and endometriomas (Fig. 1). Use of the HF-electrode to split the ovarian capsule seems the simplest and most efficient procedure, since vascularization is minimal in this area. The carbon dioxide laser and the Nd:YAG contact laser may also be employed. After aspiration of the fluid it is essential to remove the mucosal lining of the cyst in order to prevent recurrence. This may be performed by sharp or blunt ablation, by laser vaporization, or by stripping the lining of the cyst from the remaining ovary using two instruments to provide traction and countertraction [9].

In our collective of patients the laser has not proved to be the optimal preparation instrument, since exact recognition of tissue layers was deficient. Sharp dissection of the cyst lining from the ovarian capsule through all layers of tissue is best performed with finest laparoscopic scissors.

## Fenestration

Cystectomy of larger non-neoplastic ovarian cysts or very small ones (Fig. 2), especially when the ovaries are embedded in adhesions, is technically difficult, often impossible. In these cases a fenestration can be performed. It seems the optimal method of managing follicular, luteal, serous, germinal inclusion, and endometriotic cysts. In our department the carbon dioxide and the Nd:YAG contact laser have proved their usefulness for this procedure. The laparoscopic fenestration of ovarian cysts must also be combined with ablation or removal of the cyst lining to prevent recurrence (Table 1); this is in accordance with the results of others [9]. Cysts treated by this method heal well and rarely recur. The potential risk of spreading cancer, inherent to the procedure, may be reduced to a minimum by proper selection of patients and careful performance of the surgical intervention.

[1] Universitäts-Frauenklinik, Abt. für Geburtshilfe und Gynäkologie, Voßstr. 9, W-6900 Heidelberg, FRG.

Lasers in Gynecology
Edited by G. Bastert and D. Wallwiener

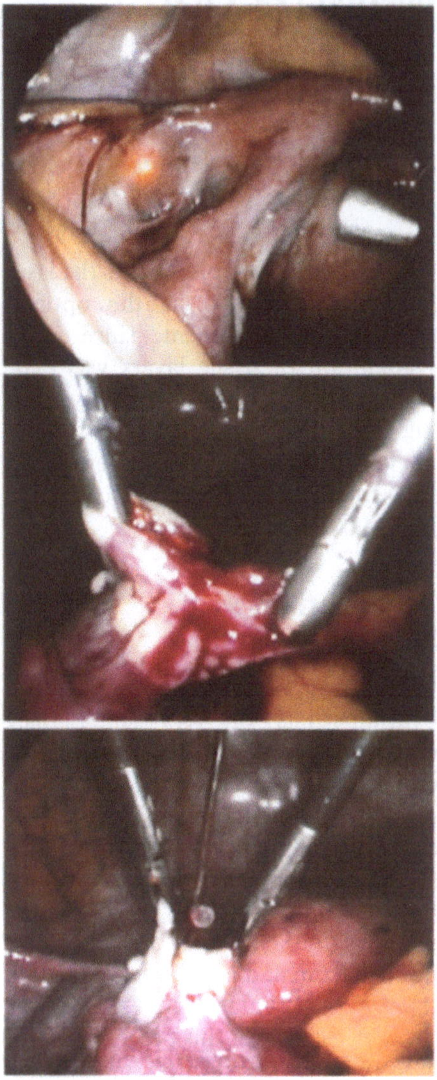

Fig. 1a–c. Ovarian cystectomy: **a** Splitting of the ovarian capsule; **b** Stripping of the lining of the cyst; **c** Reformation by application of tissue adhesive

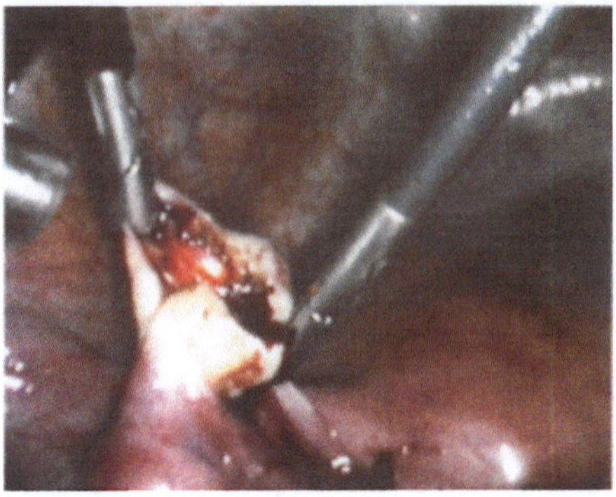

**Fig. 2.** Fenestration of a small ovarian cyst and vaporization of the mucosal lining

**Table 1.** Ovarian surgery: fenestration ($n$ = 15); coagulation of the cyst lining using the $CO_2$ laser versus no removal (modified after Hasson [9])

| | No. of patients | Recurrence |
|---|---|---|
| Fenestration with removal or ablation of cyst wall | 15 | 3/15 |
| Fenestration without removal or ablation of cyst wall | 3 | 2/3 |

# Polycystic Ovarian Disease (PCOD)

Polycystic ovary disease was first described in 1935 by Stein and Leventhal [14] and is characterized mainly by amenorrhea secondary to anovulation and enlarged ovarian mass. Higher levels of luteinizing hormone and testosterone, and a higher free androgen index, as well as associated hirsutism, obesity, and oligomenorrhea have been observed [6]. It is also a cause of subfertility. Its etiology is unclear and its pathophysiology uncertain.

To date various treatment modalities have been used but it is still debatable which is the treatment of choice. The usual approach is the administration of clomiphene citrate or human menopausal gonadotropin (hMG) hMG/human chorionic gonadotropin (hCG); only if hormone therapy fails is a surgical approach indicated. Ovarian surgery is aimed at inducing ovulation by removing intraovarian steroidal and nonsteroidal substances that inhibit ovulation [1].

Traditional conservative open ovarian wedge resection, as first performed by Stein and Leventhal [14], is seldom now. During this procedure a pie-shaped wedge is removed from the ovarian capsule to the ovarian hilium, and the ovary is reconstructed, but the effect is only temporary and the risk of pelvic adhesion formation after this procedure per laparotomy is high [2, 15, 16]. Yet relatively high fertility rates of up to 77% are reported in the literature [10].

Laparoscopic ovarian resection has largely replaced wedge resection. It may be performed by electrocautery or with the laser. Resection by means of multiple biopsy is also an alternative. Common to all these procedures is that the duration of the beneficial response is related to the size of the resected or destroyed ovarian cortex. Postoperative adhesions are reduced.

Application of the laser is a rather new procedure and only a few studies are available as yet. Daniell [4] used the carbon dioxide and later the KTP and argon laser for laparoscopic vaporization and draining of all visible subcapsular follicles on both ovaries of patients with PCOD, and he drilled randomly placed craters into the ovarian stroma. Depending on the size of the polycystic ovary, he placed 25–40 vaporization sites. In his collective, 56% of patients conceived within 6 months of laparoscopy. Keckstein et al. [11] presented a technique of carbon dioxide laser vaporization to evacuate the atretic follicular cysts and destroy androgen-producing tissue, followed by complete vaporization of the interior wall of the opened cyst. In their collective of 19 patients (16 infertility patients) they achieved an overall conception rate of 44%. They also treated 11 patients with the Nd:YAG laser in a noncontact technique, three of whom became pregnant. Yet they observed readhesions at a second-look laparoscopy. However, these can be avoided [5] by careful rinsing of the vaporization areas to remove residual carbonized debris. Other authors [17] describe performing laparoscopic therapy of PCOD with the Nd:YAG laser with a contact-type probe (wedge resection) in six infertility patients, three of whom conceived subsequently. A similar procedure was employed by Kojima et al. [12], who treated 12 women with Nd:YAG laser contact irradiation using a cutting-type probe, performing bilateral ovarian wedge resection. Seven of these patients achieved pregnancy, and another three experienced the return of a normal menstrual cycle.

At our department the technique of multiple puncture of the polycystic ovary with the contact Nd:YAG laser is preferred over electrocautery or the carbon dioxide laser. Multiple and deep punctures may be performed with this technique, with little damage to adjacent tissue and almost no bleeding (Fig. 3).

Our own studies, as well as a literature survey [3, 4, 7–9], show completely heterogeneous results concerning postoperative ovulation and duration of the beneficial response to ovarian surgery. It should therefore be applied only in patients who do not respond to hormone therapy.

We may conclude that laser-assisted preparation techniques (carbon dioxide and the Nd:YAG contact laser) are well suited for fenestration of ovarian cysts or endometriomas, an ablation of

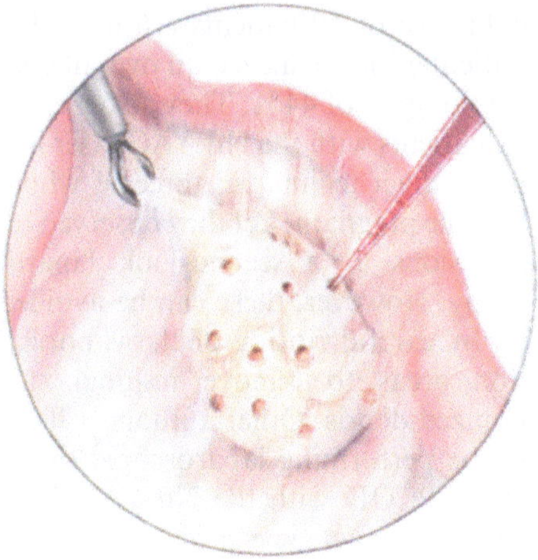

a

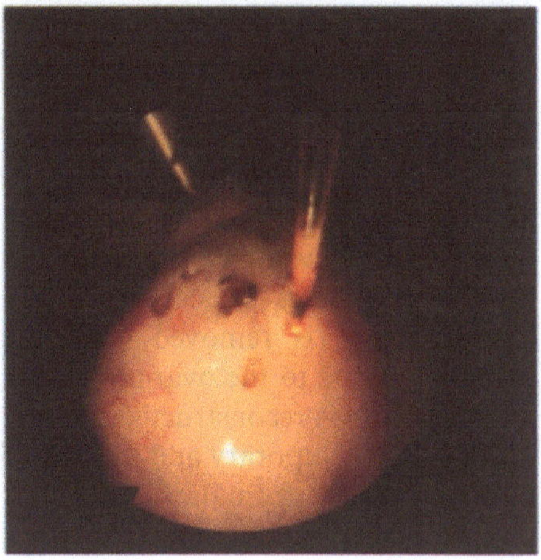

b

**Fig. 3.** **a** Laser puncture of the polycystic ovary; **b** puncture with the Nd:YAG laser "bare fiber"

the cyst lining being mandatory to prevent recurrence.

Regarding puncture of polycystic ovaries, the laser is an elegant technique and seems quite effective. Nevertheless, a longer follow-up is needed to investigate the interval to recurrence as compared with that following classical wedge resection. In our experience, however, the laser presents no advantages over sharp preparation or HF electrosurgery in ovarian cyst extirpation (cystectomy).

An ablative therapy in the sense of an oophorectomy or partial ovarectomy can by no means be performed with the laser. In these cases a loop ligature procedure as described by Semm and Mettler [13] is preferable.

# References


1. Aakvaag A, Gjonnaess H (1985) Hormonal response to electrocautery of the ovary in patients with polycystic ovarian disease. Br J Obstet Gynaecol 92:1258–1264
2. Buttram VC, Vaquero C (1975) Post-ovarian wedge resection adhesive disease. Fertil Steril 26:874–876
3. Campo S, Garcea N, Caruso A et al. (1983) Effect of celioscopic-ovarian resection in patients with polycystic ovaries. Gynecol Obstet Invest 15:213–222
4. Daniell JF, Miller W (1989) Polycystic ovaries treated by laparoscopic laser vaporization. Fertil Steril 51:232–236
5. Donnez J, Fernandez C, Willems T, Casanas-Roux F (1987) Experimental ovarian $CO_2$ laser surgery. In: Donnez J (ed) $CO_2$ laser in intraepithelial neoplasia and in infertility. Nauwelaerts Printing, Leuven
6. Eden JA, Place J, Carter GD, Jones J, Alaghaband-Zadeh J, Pawson ME (1989) The diagnosis of polycystic ovaries in subfertile women. Br J Obstet Gynaecol 96: 809–815
7. Gjonnaess H (1984) Polycystic ovarian syndrome treated by ovarian electrocautery through the laparoscope. Fertil Steril 41:20–25
8. Greenblatt E, Casper RF (1987) Endocrine changes after laparoscopic ovari-

an cautery in polycystic ovarian syndrome. Am J Obstet Gynecol 156: 279–285

9. Hasson HM (1989) Ovarian surgery. In: Sanfilippo JS, Levine RL (eds) Operative gynecologic endoscopy. Springer, Berlin Heidelberg New York, pp 87–106
10. Hjortrup A, Kehlet H, Lockwood K et al. (1983) Long-term clinical effects of ovarian wedge resection in polycystic ovarian syndrome. Acta Obstet Gynecol Scand 62:55–57
11. Keckstein J, Börchers K, Wolf A, Rossmanith W, Steiner R (1989) Is laparoscopic laser surgery an alternative treatment for polycystic ovarian disease? In: Donnez J (ed) Laser operative laparoscopy and hysteroscopy. Nauwelaerts Printing, Leuven, pp 95–107
12. Kojima E, Yanagibori A, Otaka K, Hirakawa S (1989) Ovarian wedge resection with contact Nd:YAG laser irradiation used laparoscopically. J Reprod Med 34:444–446
13. Semm K, Mettler L (1980) Technical progress in pelvic surgery via operative laparoscopy. Am J Obstet Gynecol 138: 121–127
14. Stein IF, Leventhal ML (1935) Amenorrhea associated with bilateral polycystic ovaries. Am J Obstet Gynecol 29:181–191
15. Toaff R, Toaff ME, Peyser MR (1976) Infertility following wedge resection of the ovaries. Am J Obstet Gynecol 124:92–96
16. Weinstein D, Polishuk WZ (1975) The role of wedge resection of the ovary as a cause for mechanical sterility. Surg Gynecol Obstet 141:417–418
17. Yanagibori A, Hojima E, Ohtaka K, Morita M, Hirakawa S (1989) Nd:YAG laser therapy for infertility with a contact-type probe. J Reprod Med 34:456–60

# Advantages of Laser Application in Gynecological Endoscopy

R. P. LUEKEN [1] and A. GALLINAT

There are many advantages to invasive endoscopic treatment compared with conventional surgery. In treating diseases, retaining organs, there is minimal trauma for the patient. Hospitalization of patients is dramatically reduced, the average hospital stay being less than 24 h. The spectrum of indications is continually broadening. Some examples are cholecystectomy, hysterectomy, or laparoscopically guided hysterectomy, as well the preparation of retroperitoneal lymph nodes. These procedures are established methods and proof of what DeCherney foresaw in 1985: that laparoscopic surgery would replace reconstructive pelvic surgery [9].

Several types of lasers are in use for gynecological application: the $CO_2$, the Nd:YAG, the argon, and the KTP laser. They differ in their wavelengths between 532 and 10600 nm. The vaporization effect of the $CO_2$ laser makes it highly suitable for cutting, while the predominating effect of the Nd:YAG laser is coagulation. This special coagulation effect of the YAG laser with a high penetration depth makes it highly suitable for special hysteroscopic indications. Lasers are instruments with a highly predictable tissue effect, i.e. a precisely reproducible tissue destruction. The tissue effect is achieved by coagulation, vaporization or by circumstrict tissue excision. In surgical laparoscopy the $CO_2$ laser predominates, while for hysteroscopic purposes the Nd:YAG laser is most effective. The following examples will show the advantages of laser treatment over conventional treatment using scissors, unipolar current, or endothermic energy. The main questions are: Is endoscopic laser treatment required, effective, and economic, and is the spectrum of indications widened by this new technique?

## Endometriosis

At an high incidence rate, endometriosis is a cause of sterility and dysmenorrhoea. The apparent increase in the number of cases of pelvic endometriosis must be explained by the increased acceptance of laparoscopy as well by increased experience on the part of surgeons [34, 40]. Keye gives an overview of several laser types used to treat endometriosis in cases of sterility [29]. Lomano [31] reported on his experience using the Nd:YAG laser, and Daniell [7] wrote about the use of the KTP 532 laser in this kind of disease. In general, the instrument of choice for the vaporization or excision of endo-

[1] Zentrum für Gynäkologische Endoskopie, Altonaer Str. 59–61, W-2000 Hamburg 36, FRG.

Lasers in Gynecology
Edited by G. Bastert and D. Wallwiener

metriosis as well as the treatment of endometriomata of the ovaries is the $CO_2$ laser.

Implants of endometriosis in our department are treated with a $CO_2$ laser with an output of about 10 W and a focus of about 1 mm in diameter, using an exposure time of 0.1 s. The smoke produced is simultaneously evacuated via a microfilter using the integrated smoke cleaning system of the laparoscopic insufflation machine. The filtration volume of 4 l/min enables a quick, precise vaporization and visibility is not disturbed. Prior to $CO_2$ laser vaporization, the depth of the pouch of Douglas often is filled with Ringer's lactate solution, which works automatically as a backstop. Simultaneous spillage is essential to remove carbonized tissue. In 1986 Olive showed that the pregnancy rate after conservative treatment in endometriosis in 1623 patients was 50% [37]. The pregnancy rate drops continuously with higher levels of endometriosis from 62% to 40%. In this study the effect of a preoperative and postoperative treatment with danazol was not considered [49].

The evaluation of pregnancy rates after treatment with the $CO_2$ laser is rendered more difficult because different authors use different schemata for the classification of endometriosis: the classical one: mild, medium, or severe, or, especially in the past few years, the AFS classification. While Kelly and Roberts [27] reported the lowest pregnancy rate in a stage-2 AFS at 43%, Nezhat and Feste achieved rates of up to 75% [17, 36]. There are other reasons for sterility in addition to endometriosis: the pregnancy rate in 425 women with an isolated endometriosis factor was about 70%; in 324 women with several factors which could cause sterility it was 36%. In the same way, the pregnancy rate after laser treatment decreases with the increase in length of time following treatment for sterility [33]. In a retrospective study Chong compared the pregnancy rates following medical treatment, $CO_2$ laser treatment, and the combination of both. Using danazol the pregnancy rate is 48.9%, using the $CO_2$ laser 44.6%; it is 51.4% after a combination of laser and danazol treatment [3]. In contrast, Donnez reported a pregnancy rate of 61% in cases of light endometriosis and one of 40% in cases of severe endometriosis.

As the diagnosis of small implants is rendered more difficult by preoperative medical treatment which causes shrinking of such implants, a preoperative hormonal treatment should be recommended only for reduction of larger ovarian endometriomata [16]. The results of treatment of pelvic endometriosis with $CO_2$ laser via laparoscopy are better than those using laparotomy or medical treatment [38]. The use of the $CO_2$ laser to vaporize endometriotic implants is safe and effective, and it enlarges the spectrum of indications; in particular, critically localized implants can now be easily treated. The results are superior to those of long-term medical treatment with severe side effects or to those with laparotomy.

## PCOD Syndrome

Fifty-five years ago Stein and Leventhal [45] reported about the syndrome of polycystic ovarian disease (PCOD). Over the years, bilateral wedge resection of the ovaries was the therapy of choice. In nearly 100% of cases adhesions were found following this kind of surgery [2, 43, 47]. In 1984 Gjonnaess first reported on the laparoscopic treatment of PCOD [18]. While Greenblatt [21] used a monopolar current in 1987 and Sumioki [46] described mechanical biopsies in 1988, in the past several years laparoscopic laser treatment has increased [5, 22, 24, 26]. The conception rate following $CO_2$ laser vaporization rose to 44% [26]. Three of 11 patients became pregnant after treatment with the Nd:YAG laser in noncontact technique. We treated eight patients using the Nd:YAG laser contact technique, vaporizing the ovaries at the interior and posterior walls at about 12 points. Within the first 6 months following treatment, four of these patients conceived. The therapeutic value of laparoscopic treatment of PCOD at the moment seems to be only temporary, however.

## Tubal Sterility

In cases of tubal sterility microsurgical techniques are in competition with endoscopic ones. In 1983 Kelly and Roberts first reported their experience using the $CO_2$ laser in gynecological microsurgery [28]. The pregnancy rate for patients with thin-walled hydrosalpinx with mucosal folds is 26%; when a salpingoneostomy is performed using the $CO_2$ laser the results are not superior [13]. For patients with tubo-ovarian or tuboperitoneal adhesions, microsurgery is being replaced by endoscopic techniques with the same excellent result of a more than 50% pregnancy rate. The enlargement of the laparoscope, as well as the precise $CO_2$ laser beam, makes possible exact preparation even directly at the bowel and the ureters, while the neighboring tissue is not affected. The fallopian tubes are handled with atraumatic forceps to keep the tissue under tension. The operating field is rinsed; bleeding is recognized easily under water and can be treated immediately. Coagulation is performed with a defocused laser beam or using a bipolar microforceps. A special indication for the use of the $CO_2$ laser is ampullary stenosis. Using the palpator or a hook instrument, this stenosis can be easily vaporized and therefore automatically dilated. Donnez reported a pregnancy rate of 61% in women with fimbrioplasty [11, 12].

## Tubal Pregnancy

The conservative treatment of tubal pregnancies is varied. In 1980, Bruhat first reported about laparoscopic treatment [1]. After application of a vasoconstrictive agent in the mesosalpinx he prefers the mechanical use of needles and scissors, and performs a linear salpingotomy using monopolar current for cutting [42]. For hematosalpinx the salpingotomy covers an area of 10–20 mm [9, 34, 40]. After the tubal lumen is opened, trophoplastic tissue is generally oozes out by itself and can be removed with an atraumatic forceps or using an endoirrigator. Histological examination is essential, as are postoperative controls of beta-HCG. The salpingotomy is left open; we do not suture it. Having achieved total hemostasis, a Robinson drainage is not necessary. In cases of persistent or rising beta-HCG titers, a second laparoscopy or systematic treatment with methrotrexate is required. At the moment, laparoscopic treatment of ectopic pregnancies is the treatment of choice. Lübke, in Berlin, has shown very impressively how the ratio of laparotomy to laparoscopy for ectopic pregnancies has changed in the past 9 years [32]. While in 1985 he treated in 93% of ectopic pregnancies by laparotomy, in 1989 a laparotomy was required for only 3%. Donnez carried out a follow-up study after conservative treatment of unruptured ectopic ampullary tubal pregnancy via laparotomy or laparoscopy and reported an intrauterine pregnancy rate of 52% and a recurrent ectopic pregnancy rate of 12% when laparotomy was performed; following laparoscopic treatment these rates were 63% and 10% respectively [14]. Other authors who have reported on linear salpingotomy using the $CO_2$ laser are DeCherney, Johns, and Daniell [4, 10, 25]. The application of other laser types is reported by DeCherney [10] (e.g., argon) and Daniell [6] (e.g., KTP). In conclusion, it must be said that most authors prefer laser treatment for this kind of indication and not mechanical instruments such as scissors or the monopolar needle. We began using the Nd:YAG laser for longitudinal salpingotomy, but nowadays we prefer, like Donnez, the $CO_2$ laser for this kind of surgery.

The advantage of treatment can be proved only if there is only one fallopian tube left. In such cases Pouly reports a postoperative pregnancy rate of 46% following laparoscopic treatment [39]. As H. Reich said [41], the future is now; ectopic pregnancies are no longer (except in very special cases) treated by laparotomy.

## Laser Hysteroscopy

The main indications for major intrauterine surgery are

1. Endometrial ablation (laser menolysis)
2. Septum vaporization
3. Myoma coagulation/vaporization

Another indication is intrauterine synechiae, but as this is a very rare finding, it is not listed above. Since 1986 we have been using the Nd:YAG laser for intrauterine surgery, and since then traditional hysteroscopic instruments like flexible biopsy forceps, scissors, or hooks have been used less and less. As we are fully equipped for YAG laser treatment, we do not use the resectoscope, which is very common in other hysteroscopic surgical centers [8, 23, 35, 44, 48]. In some indications – for example, depending on the size and particular localization of a myoma – the use of the resectoscope might be easier and quicker. However, the Nd:YAG laser is the universal tool for handling all kinds of intrauterine surgery. There is no need for unipolar current, which was abandoned in the field of endoscopic gynecological surgery years ago.

The first publication of Goldrath about endometrial ablation was in 1981 [22]. The criteria for this so-called laser menolysis are:

1. Exclusion of intrauterine malignancy
2. Completed family planning
3. Prior hormonal treatment to obtain a thinning of the endometrium (danazol or Zoladex)

Indications for endometrial ablation are first of all medical reasons, i.e., severe cardiovascular disease, diabetes mellitus, patients receiving marcumar or heparin therapy, or therapy-resistent bleeding disorders. The procedure may also be requested by patients with bleeding disorders who want to avoid hysterectomy, and especially following diseases after hysterectomy.

There are several techniques using the bare fiber for YAG-laser coagulation:

1. The noncontact technique is always used at the beginning of the operation for treatment around the tubal ostia and the cornual areas.

2. The withdrawal technique is for coagulation of the anterior, the posterior, and the side walls of the uterine cavity. From the highest point near the fundus, directly at the area coagulated in a non contact technique, the laser fiber is placed at the wall and withdrawn towards the lower uterine segment under yag laser light exposition. One coagulation stripe has a diameter of about 1 mm. Therefore for the treatment of large areas one coagulation stripe is placed parallel the previous one til the whole area (anterior-, posterior and side walls) are laser coagulated. The internal os and the first millimeters at the beginning of the cavity are excluded from treatment to avoid a stenosis. After preoperative treatment with danazol or Zoladex, the thickness of the endometrium is about 1.2 mm. The coagulation effect of the YAG laser (5–6 mm deep) reaches beneath the basal membrane; if there is adenomyosis, this area is treated as well. In our first study in 1987 we treated 62 patients using this technique. The results were amenorrhea in 45% and hypomenorrhea in 52%. All these patients had no more complaints. In two of the first patients who did not receive hormonal treatment owing to our lack of experience, the results were not sufficient. One patient underwent a hysterectomy, and the other required a second laser treatment. Goldrath and Donnez reported amenorrhea in 48% [15, 19], while Loffer reported only 20% [30]. This difference, in our opinion, is due to the use of a fluid distention medium. The advantage is that no smoke is generated, but the cooling effect of a fluid reduces the coagulation effect.

While Goldrath considered the result excellent when amenorrhea was achieved and only good if there was hypomenorrhea, at present we think the best result of endometrial ablation should be hypomenorrhea. We now exclude the lower parts of the uterine cavity from treatment, and thus there must still be some bleeding. Only with this kind of treatment can a cervical stenosis be avoided (no hematometra possible) and the intrauterine situation is better controllable. If we summarize the results of all groups (Goldrath, Donnez, Loffer, and us), amenorrhea and hypomenorrhea were achieved in more than 90% of cases. This means that more than 90% of patients with retractable bleeding disorders can avoid a hysterectomy, which was, and in most hospitals still is, the usual treatment.

There was one light postoperative infection, but this was the only complication. As we use only $CO_2$ hysteroscopy for this kind of procedure, there was no sign of gas embolism. All procedures are performed under continuous intraoperative monitoring of the end-expiratory $CO_2$ using capnography (D. Wallwiener, 1990, personal communication). In all patients with extended surgical laser hysteroscopy an increase of the end-expiratory $CO_2$ up to 10 volume % is seen. It is essential to use a special insufflation machine, so that pressure and $CO_2$ gas flow are limited and there is no risk of a $CO_2$ gas embolism. Typical complications of liquid distention media are fluid overload or an electrolyte imbalance.

This new surgical procedure to produce amenorrhea by laser treatment is performed on an outpatient basis. This new gynecological treatment in cases of benign retractable bleeding disorders reduces the number of hysterectomies performed and thus the hospital stay of 1–2 weeks! This is only one of the various indications in which gynecological laser endoscopy reduces the costs of treatment with a very high economic effect; the benefit to the patient is extraordinary. There is no doubt about the efficiency of gynecological laser endoscopy. For the health and the benefit of the patients, we hope that the use of these new methods will continue to spread.

## References


1. Bruhat MA, Manhes H, Mage G, Pouly JL (1980) Treatment of ectopic pregnancy by means of laparoscopy. Fertil Steril 33:411
2. Buttram VJ, Vaquero C (1975) Post-ovarian wedge resection adhesive disease. Fertil Steril 26:874
3. Chong AP, Keene ME, Thornton NL (1990) Comparison of three models of treatment for infertility patients with minimal pelvic endometriosis. Fertil Steril 53:407
4. Daniell JF (1985) Laser laparoscopy. In: Baggish MS (ed) Basic and advanced laser laparoscopy in gynecology. Appleton-Century-Crofts, Norwalk, CN, pp 343–356
5. Daniell J, Miller W (1989) Polycystic ovaries treated by laparoscopic laser vaporization. Fertil Steril 51:232
6. Daniell JF, Meisels S, Miller W, Tosh R (1986) Laparoscopic use of the KTP/532 laser in nonendometriotic pelvic surgery. Colpos Gynecol Laser Surg 2:107
7. Daniell J, Miller W, Tosh R (1986) Initial evaluation of the use of the potassium-titanyl-phosphate (KTP/532) laser in gynecologic laparoscopy. Fertil Steril 46: 373
8. DeCherney AH (1984) Treatment of irregular menstrual bleeding by hysteroscopic resection of submucous myomas and polyps. In: Siegler AM, Lindemann HJ (eds) Hysteroscopy: principles and practice. JB Lippincott, Philadelphia
9. DeCherney AH (1985) The leader of the band is tired. Fertil Steril 44:299
10. DeCherney AH, Diamond MP (1987) Laparoscopic salpingotomy for ectopic pregnancy. Obstet Gynecol 70:948
11. Donnez J (1987) $CO_2$ laser laparoscopy in infertile women with endometriosis and women with adnexal adhesions. Fertil Steril 48:390
12. Donnez J, Casanas-Roux F (1986) Prognostic factors of fimbrial microsurgery. Fertil Steril 446: 200
13. Donnez J, Nisolle M (1989) $CO_2$ laser laparoscopic surgery. Clin Obstet Gynecol 3:525
14. Donnez J, Nisolle M (1989) Laparoscopic treatment of ampullary tubal pregnancy. J Gynecol Surg 5:157
15. Donnez J, Nisolle M (1989) Laser hysteroscopy in uterine bleeding: endometrial ablation and polypectomy. In: Donnez J (ed) Laser operative laparoscopy and hysteroscopy. p 282
16. Donnez J, Nisolle M, Wayembergh M, Clerkx F, Casanas-Roux F (1989) $CO_2$ laser laparoscopy in peritoneal endometriosis and in ovarian endometrial cyst. In: Donnez J (ed) Laser operative laparoscopy and hysteroscopy. p 58–73
17. Feste J (1985) Laser laparoscopy: a new modality. J Reprod Med 30:413
18. Gjonnaess H (1984) Polycystic ovarian syndrome treated by ovarian electrocautery through the laparoscope. Fertil Steril 41:20
19. Goldrath MH (1985) Hysteroscopic laser surgery. In: Baggish MS (ed) Basic and advanced laser surgery in gynecology. Appleton-Century-Crofts, Norwalk, CN, p 357

20. Goldrath MH, Fuller T, Segal S (1981) Laser photovaporization of endometrium for the treatment of menorrhagia. Am J Obstet Gynecol 140:14
21. Greenblatt E, Casper RF (1987) Endocrine changes after laparoscopic ovarian cautery in polycystic ovarian syndrome. Am J Obstet Gynecol 156:279
22. Grochmal S (1988) Contact Nd:YAG laser superior to $CO_2$ for treatment of ovarian disease. Laser Practise Rep 3:1S-2S
23. Hallez JP, Netter A, Cartier R (1987) Methodical intrauterine resection. Am J Obstet Gynecol 156:1080
24. Huber J, Hosman J, Spona J (1989) Endoskopisch vorgenommene Laserinzision des polyzystischen Ovars. Geburtshilfe Frauenheilkd 49:37
25. Johns DA, Hardie RP (1986) Management of unruptured ectopic pregnancy with laparoscopic carbon dioxide laser. Fertil Steril 46:703
26. Keckstein J, Boerchers K, Wolf A, Rossmanith W, Steiner R (1989) Is laparoscopic laser surgery an alternative treatment for polycystic ovarian disease? In: Donnez J (ed) Laser operative laparoscopy and hysteroscopy. pp 95–107
27. Kelly R, Roberts D (1983) $CO_2$ laser laparoscopy: a potential alternative to danazol in the treatment of stage-I and -II endometriosis. J Reprod Med 28:638
28. Kelly RW, Roberts DK (1983) Experience with the $CO_2$ laser in gynecologic microsurgery. Am J Obstet Gynecol 146:585
29. Keye WR (1986) The present and future application of lasers to the treatment of endometriosis and infertility. Int J Fertil 31:160
30. Loffer FD (1988) Laser ablation of the endometrium. Obstet Gynecol Clin North Am 15:77
31. Lomano JM (1987) Nd:YAG laser ablation of early pelvic endometriosis: a report of 61 cases. Lasers Surg Med 7:56
32. Lübke F (1990) Die laparoskopische Therapie der Extrauteringravidität, Ergebnisse von 1981–1989. Presented at the meeting of the German Society of Gynecology, Berlin, Jan. 1990
33. Martin DC (1986) $CO_2$ laser laparoscopy for endometriosis associated with infertility. J Reprod Med 31:1089
34. Martin DC, Hubert GD, Vander Zwaag R, El-Zeky F (1989) Laparoscopic appearances of peritoneal endometriosis. Fertil Steril 51:63
35. Neuwirth RS (1983) Hysteroscopic management of symptomatic submucous fibroids. Am J Obstet Gynecol 62:509
36. Nezhat C, Crowgey S, Carrison C (1986) Surgical treatment of endometriosis via laser laparoscopy. Fertil Steril 45:778
37. Olive D (1986) Endometriosis. In: De Cherney AH (ed) Reproductive failure. Churchill-Livingstone, New York, p 183
38. Olive DL, Martin DC (1987) Treatment of endometriosis-associated infertility with $CO_2$ laser laparoscopy: the use of one- and two-parameter exponential models. Fertil Steril 48:18
39. Pouly JL, Manhes H, Mage G, Canis M, Bruhat MA (1986) Conservative laparoscopic treatment of 321 ectopic pregnancies. Fertil Steril 46:1093
40. Redwine DB (1987) The distribution of endometriosis in the pelvis by age-groups and fertility. Fertil Steril 47:173
41. Reich H (1989) Ectopic pregnancy. In: Sanfilippo JS, Levine RL (eds) Operative gynecologic endoscopy. Springer, Berlin Heidelberg New York Tokyo, p 83
42. Reich H, Johns DA, Decaprio J (1988) Laparoscopic treatment of 109 consecutive tubal pregnancies. J Reprod Med 33:885
43. Reiter RC, Buttram VJ (1985) Prevention of postoperative adhesions. In: Surgical treatment of the infertile female. Williams and Wilkins, Baltimore, pp 67–88
44. Rock JA, Murphy AA, Cooper WH (1987) Resectoscopic technique for the lysis of a class-V complete uterine septum. Fertil Steril 48:495
45. Stein IF, Leventhal ML (1935) Amenorrhea associated with bilateral polycystic ovaries. Am J Obstet Gynecol 29:181–191
46. Sumioki H (1988) The effect of laparoscopic multiple-punch resection of the ovary on hypothalamo-pituitary axis in polycystic ovary syndrome. Fertil Steril 50:567
47. Toaff R, Toaff ME, Peyser MR (1976) Infertility following wedge resection of the ovaries. Am J Obstet Gynecol 124:92
48. Wamsteker K (1983) Hysteroscopic surgery. In: van der Pas H, van Haerendael B, van Lith DAF, Keith L (eds) Hysteroscopy. MTP Press, Boston, p 165
49. Wheeler J, Malinak L (1981) Postoperative danazol therapy in infertility patients with severe endometriosis. Fertil Steril 36:460

# Antimicrobial Prophylaxis of Perioperative Gynecological Infections

G. MARKLEIN[1]

## Introduction

Antimicrobial agents are indispensable drugs for the causal therapy of clinical manifestations of infectious diseases, but a large proportion of the total amount of antibiotics consumed in the world is used for prophylaxis. At the present time, surgical indications account for the largest amount of antimicrobials being used prophylactically. Antimicrobials are considered prophylactic when they are used in clean or potentially contaminated operations with the aim of preventing infections in the postoperative period. In general, perioperative antimicrobial prophylaxis is a suitable measure for preventing local infections at the operative site, but it does not appreciably influence the development of other postoperative infections such as pneumonia or catheter-associated infections. The development of postoperative infections depends rather on technical, microbial, or patient factors (Table 1). An increasing risk of infection in gynecological surgery is given in cases where several risk factors come together [8]. Postoperative infections such as wound infection, intraabdominal abscess formation, or peritonitis inevitably result in increased morbidity, prolonged hospital stay, and considerably impaired cost-benefit ratios, especially in elderly patients and in patients with preexisting diseases or with reduced antimicrobial defense mechanisms.

In the majority of cases of postoperative abdominal infections, the etiologically responsible pathogens are represented by parts of the microflora resident in the gastrointestinal or genitourinary tract [2, 5]. While the normal human kidney, ureter, urinary bladder, and uterus do not possess an indigenous bacterial flora, the vagina, by

**Table 1.** Common risk factors which may contribute to the development of postoperative infections

1. Technical
   - Hygiene standards
   - Duration of the operation
   - Skill of the operation team
2. Microbial
   - Nature and quantity of potential pathogens
   - Virulence factors (e.g., toxin production)
3. Patient
   - Underlying diseases (e.g., diabetes mellitus, malignant tumor, immunosuppression)
   - Period of life (higher risk in the elderly)
   - Impaired immune defense mechanisms
     - Polymorphonuclear granulocyte / macrophage function
     - Humoral immune function (complement, immunoglobulins)
     - Cellular immune function (T-lymphocytes)

[1] Institut für Medizinische Mikrobiologie und Immunologie der Universität Bonn, Sigmund-Freud-Str. 25, W-5300 Bonn, FRG.

Lasers in Gynecology
Edited by G. Bastert and D. Wallwiener

**Table 2.** Aerobic and anaerobic bacteria isolated from the vagina of healthy women in childbearing age

| | Aerobes | Anaerobes |
|---|---|---|
| Gram-positive rods | *Corynebacterium*<br>*Gardnerella*<br>*Lactobacillus*<br>*Listeria* | *Actinomyces*<br>*Bifidobacterium*<br>*Clostridium*<br>*Eubacterium*<br>*Lactobacillus*<br>*Propionibacterium* |
| Gram-positive cocci | *Enterococcus*<br>*Staphylococcus*<br>*Streptococcus* | *Peptococcus*<br>*Peptostreptococcus* |
| Gram-negative rods | *Escherichia coli* and other *Enterobacteriaceae*<br>*Acinetobacter* | *Bacteroides*<br>*Fusobacterium*<br>*Porphyromonas*<br>*Prevotella* |
| Gram-negative cocci | | *Acidaminococcus*<br>*Veillonella* |

**Table 3.** Bacterial spectrum of nonspecific aerobic/anaerobic polymicrobial infections of the female genital tract (intraabdominal specimens of 49 patients, Bonn 1990)

| Aerobes | No. of strains | Anaerobes | No. of strains |
|---|---|---|---|
| *Escherichia coli* | 14 | *Bacteroides fragilis* | 12 |
| *Proteus* species | 6 | *B. thetaiotaomicron* | 10 |
| *Citrobacter freundii* | 1 | *B. vulgatus* | 5 |
| *Pseudomonas aeruginosa* | 2 | *B. distasonis* | 5 |
| | | *B. uniformis* | 2 |
| *Enterococcus faecalis* | 13 | *B. ureolyticus* | 1 |
| β-haemolytic streptococci | 5 | *Bacteroides* species | 7 |
| *Staphylococcus aureus* | 2 | *Prevotella oralis/bivia* | 9 |
| | | *Porphyromonas gingivalis* | 1 |
| | | *P. asaccharolytica* | 9 |
| | | *Fusobacterium* species | 4 |
| | | *Veillonella* species | 1 |
| | | *Acidaminococcus fermentans* | 1 |
| | | *Propionibacterium* species | 4 |
| | | *Peptococcus* species | 7 |
| | | *Peptostreptococcus* species | 3 |
| Total no. of isolates | 43 | | 81 |

contrast, contains an abundant microflora which is composed of a great variety of aerobes and anaerobes (Table 2). The acidity of this area, caused largely by the metabolic activities of the indigenous anaerobes, seems to be responsible for the extensive stability of the vaginal flora. The pathogens of postoperative gynecological infections may be found in some of these bacteria, often in the polymicrobial-synergistic association of aerobes and anaerobes. The impressive preponderance of gram-negative anaerobes, particularly from the *Bacteroides* group of bacteria, is shown in the microbiological findings of 49 women with unspecific gynecological infections presented in Table 3.

The fact that gynecological operation fields cannot always be adequately decontaminated before surgery does not in general present a serious problem since, as a function of normal human antimicrobial defense mechanisms, slight contamination does not necessarily lead to infectious processes. In certain conditions, however, which are shown in Table 1, it is tempting to assist the defense mechanisms by quantitatively reducing the contaminating bacteria in clean or contaminated operations [3, 9]. Antimicrobial agents which are suitable in preventing postoperative infections in gynecological surgery should meet the following requirements:

1. Bactericidal mechanism of activity with low minimal inhibitory concentrations in vitro and in vivo against the spectrum of possible pathogens
2. High antimicrobial tissue concentrations at the operative site
3. Low binding rate to serum proteins
4. Appropriate biological half-life according to the duration of the operation
5. Metabolic stability
6. Lack of interference with drugs used to induce or to maintain anesthesia
7. Few or no side effects
8. Inexpensive

Antibiotics which satisfy these conditions best belong to the groups of the:

1. Penicillins
2. Semisynthetic penicillins plus betalactamase inhibitors
3. Cephalosporins
4. Nitroimidazoles

## Guidelines for Perioperative Antimicrobial Prophylaxis in Gynecology

While procedures such as colon resection or vaginal hysterectomy carry a relatively high risk of postoperative infections [1, 4, 8], the rate of infection following endoscopic surgery seems to be far lower, obviously as a consequence of the fact that microbial contamination of the incision site in endoscopic procedures is a rarer event than in laparotomy. However, no report has yet been published on a prospective, randomized clinical trial which might suggest the benefit of perioperative antibiotic prophylaxis in endoscopic gynecological surgery.

*The Suspected Pathogens Should Be Defined and the Antibiotics Should Be Active Against Them.* The etiologically responsible agents of postoperative gynecological infections comprise, in principle, all the aerobes and anaerobes listed in Table 2. The compounds used should have sufficient efficacy in vitro and in vivo against both bacterial groups. No single antimicrobial agent available today fulfills these criteria, we have to accept an activity gap in the antimicrobial spectrum of a single agent, or we have to use a combination of antibiotics with complementary activities.

If one accepts an incomplete activity spectrum of a single substance, penicillins or cephalosporins (second generation) are the drugs of choice. The activity spectrum of these cephalosporins against clinically important bacteria is shown in Table 4. The most important disadvantage of these antimicrobials is their wide instability against the betalactamases of clinically

**Table 4.** Medically important bacteria which are predominantly susceptible to second generation cephalosporins

| Gram-positive bacteria | Gram-negative bacteria |
|---|---|
| *Staphylococcus aureus*[a] | *Escherichia coli* |
| *S. epidermidis*[a] | *Klebsiella pneumoniae* |
| *Streptococcus pyogenes* | *Proteus mirabilis* |
| *S. agalactiae* (Group B) | *Neisseria* species[a] |
| *S. pneumoniae* | *Haemophilus influenzae*[a] |
| *Peptococcus* species | |
| *Peptostreptococcus* species | |
| *Clostridium* species | |

[a] Including penicillinase-producing strains

important *Bacteroides* species (especially *B. fragilis* and *B. thetaiotaomicron*). On the other hand, nitroimidazole derivatives in many ways seem to come very close to an ideal antimicrobial agent against anaerobes. For example, they possess a very high level of efficacy against almost all important anaerobes, including those strains which produce penicillinase and cephalosporinase, without clinical development of resistance. For these reasons, the combination of metronidazole with a penicillin or cephalosporin is favored by many authors [4].

In fertility surgery of women who have experienced one event of pelvic inflammatory disease, the possibility of infection due to *Chlamydia trachomatis* has to be taken into consideration. Logically, therefore, in such patients the drug of choice should be doxycycline or erythromycin or one of the recently developed macrolide compounds (e.g., roxithromycin, clarithromycin, or azithromycin). The duration of such prophylaxis has not been systematically evaluated. From the practical point of view, clinics that use postoperative steroids to prevent adhesion should probably extend their antibiotic prophylaxis accordingly.

*The Antibiotic(s) Should Be Present in Effective Tissue Levels at the Time of Incision.* Topical administration of antimicrobial agents seems simple to perform, inexpensive, and free from some of the side effects that can follow systemic administration. Following longer complex procedures with intraabdominal bacterial contamination, antimicrobials applied locally will have minimal effects on the development of postoperative infections. Superficial administration of antibiotics does not lead to sufficient tissue concentrations to kill potential pathogens. Starting antimicrobial prophylaxis in the postoperative phase has also proved to be ineffective. Experimental and clinical studies revealed that the best results were achieved when the perioperative prophylaxis was begun before the contamination occurred [4, 5, 8].

*The Drug Should Be Short-term and Primarily Intraoperative.* The rationale for using prophylactic antibiotics is protection during the operative and the immediate postoperative phase. Extended antibiotic use merely increases the risk of superinfection due to resistant pathogens, but does not reduce the subsequent incidence of septic complications. A regimen appropriate for gynecological surgery is a second generation cephalosporin with sufficient activity against anaerobes of the *Bacteroides fragilis* group such as cefoxitin [1, 4] or without this special activity such as cefazoline, cefuroxime, or cefotiam in combination with metronidazole starting intravenously about 30–60 min before operation [4]. With regard to the combination of cefotiam with metronidazole, the following economical aspects might be of interest.

The bottle containing 2.0 g cefotiam as dry substance is large enough to hold the content of the bottle of 0.5 g metronidazole in aqueous solution. Chemically and physically, this mixture shows compatibility and stability for at least 4 h [7].

*Potent Antibiotics Used for Resistant Pathogens Should not Be Used in Perioperative Prophylaxis.* Widespread use of such agents promotes the development or selection of resistant strains within the hospital microflora and thereby renders them ineffective. Drugs with a narrow spectrum should be preferred, in order to reserve the more potent agents for life-threatening infections that may develop. Thus, aminoglycosides, third generation cephalosporins, and carbapenems are interdicted.

In conclusion, the combination of a second generation cephalosporin with metronidazole, if administered preoperatively, is an optimal means of preventing perioperative gynecological infections. Furthermore, however, to avoid postoperative infections in gynecological as in other types of endoscopic surgery, stringent indications must be regarded [6].

## References

1. Berkeley AS, Freedman KS, Ledger WJ, Orr JW, Benigno BB et al (1988) Comparison of cefotetan and cefoxitin prophylaxis for abdominal and vaginal hysterectomy. Am J Obstet Gynecol 158:706–709
2. Finegold SM, George WL (1989) Anaerobic infections in humans. Academic Press, New York
3. Fischbach F, Voss A, Loos W, Thurmayr R, Graeff H (1988) Perioperative Antibiotika-Prophylaxe bei abdominaler Hysterektomie. Geburtshilfe Frauenheilk 48: 889–892
4. Gjønnaess H (1990) Antimicrobial prophylaxis in gynaecological and obstetric surgery. Scand J Infect Dis [Suppl.] 70:52–67
5. Mandell GL, Douglas RG, Bennett JE (1990) Principles and practice of infectious diseases, 3rd edn Churchill Livingstone, Edinburgh
6. Manegold BC, Götz F, Schreiber HW, Effenberger T (1991) Laparoskopische Appendektomie und Cholezystektomie. Dtsch Ärztebl 88:998–1004
7. Messerschmidt W (1986) Kompatibilität von Cefotiam mit Metronidazol. Krankenhauspharmazie 7:263–265
8. Shapiro M, Munoz A, Tager IB, Schoenbaum SC, Polk BF (1982) Risk factors for infection at the operative site after abdominal or vaginal hysterectomy. N Engl J Med 307:1661–1666
9. Vasanthakumar V, Bhan GL, Perera BS, Taft P (1990) A study to assess the efficacy of chemoprophylaxis in the prevention of endoscopy-related bacteraemia in patients aged 60 and over. Q J Med (N.S.) 75:647–653

# Pelviscopic Reversal of Sterilization

J. GAUWERKY [1], D. WALLWIENER, R. KLOSE, and G. BASTERT

## Introduction

Laparotomy has been replaced by endoscopic procedures for several indications in classical abdominal surgery. It is widely accepted that pelviscopic surgery plays a key role in the therapy of ectopic pregnancy [1, 8, 11], of endometriosis [9], of salpingo-ovariolysis [2, 6], and of salpingostomy [5, 9]. However, the use of endoscopic surgery for tubal reconstruction seems to be limited, especially in cases when microsutures are required for tubal anastomosis.

The basic steps of tubal anastomosis by microsurgery rely on muscular sutures, using 8-0 to 10-0 Vicryl, or equivalent materials. Endoscopic sutures with these fine materials seem to be impossible. Therefore, I have tried to find alternative techniques for tubal anastomosis, applicable endoscopically. This paper deals with basic experiments referring to this topic and with the first applications in human subjects.

[1] Universitäts-Frauenklinik, Abt. für Geburtshilfe und Gynäkologie, Voßstr. 9, W-6900 Heidelberg, FRG.

## Animal Experiments

### Study I: Tubal Anastomosis with Fibrin Glue Versus Microsutures

Forty-one New Zealand White female rabbits were used to study the efficiency of tubal anastomosis by fibrin glue in three groups, compared with conventional anastomosis by microsutures.

Group I: isthmic anastomosis without resection of tubal segments, 20 animals
Group II: ampullary anastomosis without resection of tubal segments, 11 animals
Group III: isthmic anastomosis with resection of 1.5-cm tubal segments, 10 animals

The experimental model has been published earlier [4].

### Results

The results of the fertility studies are summarized in Fig. 1. For all types of tubal anastomoses the glueing technique was as effective as the microsurgical technique. However, the morphological investigations revealed problems with the exact approximation of tissue planes in ampullary anastomoses using fibrin glue.

Lasers in Gynecology
Edited by G. Bastert and D. Wallwiener

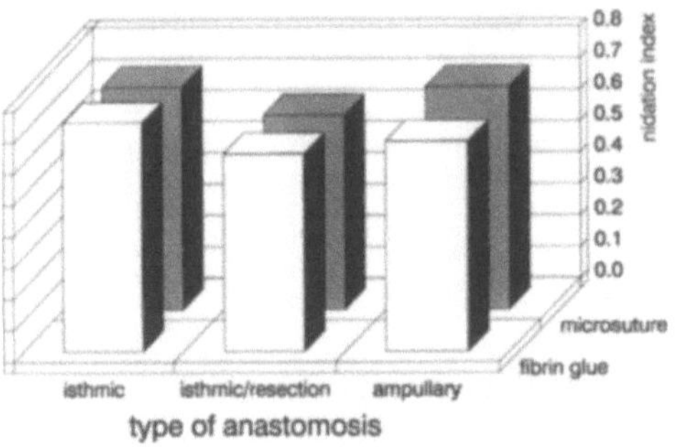


**Fig. 1.** Fertility after tubal anastomoses with fibrin glue and conventional microsutures in rabbits. There is no statistical difference between these two techniques

**Fig. 2. a** Pelvi-trainer for simulation of pelviscopic surgery with rat. **b** The abdomen is opened for endoscopic procedures

## Study II: Experimental Tubal Anastomosis

Twenty female Wistar rats weighing 215 g were used. Anesthesia was performed with pentobarbitone. The abdomen was opened through a midline incision and the uterus exposed. The following steps were carried out endoscopically in a teaching device (Fig. 2) [10]. The central uterine segment was infiltrated with 0.5–1.0 ml of ornipressin (POR-8, 2.5 I.U., 1:10 dilution) to reduce bleeding. Uterine dissection was performed with scissors, and a polyethylene splint was introduced through a separate distal incision of the uterus. A 6-0 seromuscular suture

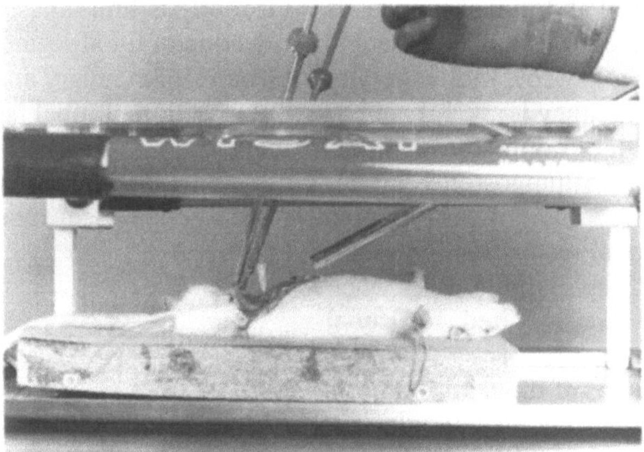

a

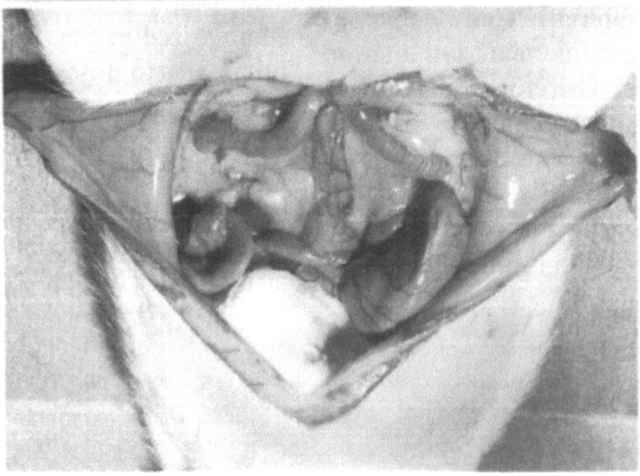

b

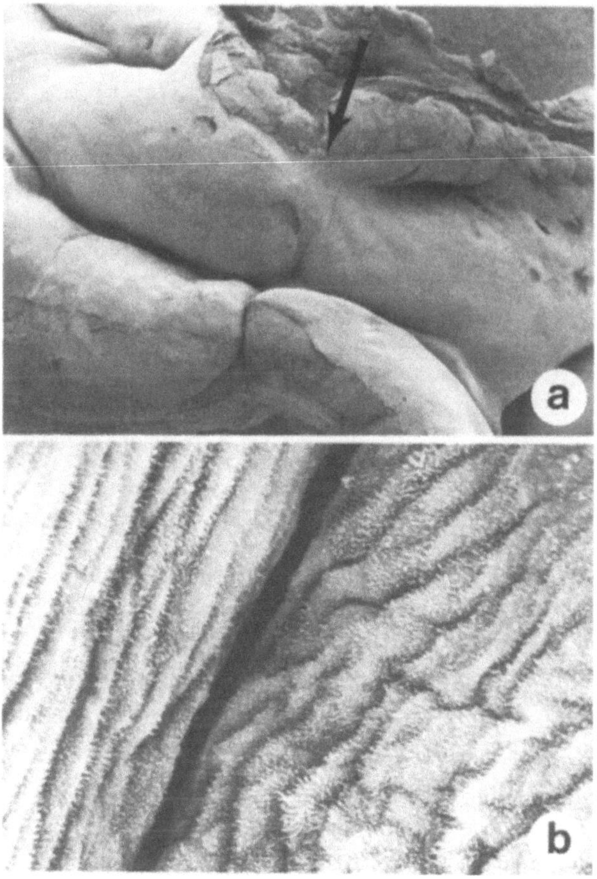


**Fig. 3. a** Scanning electron microscopy of an endoscopic anastomosis of the rat uterus. SEM, × 20 **b** High-power micrograph of **a**. SEM, × 2400

(Prolene) was placed at the antimesenteric border and left untied. Subsequently, the uterine segments were brought together on the polyethylene splint with the aid of atraumatic forceps. The fibrin glue (Tissucol) was placed over the incision site and the edges of the wound were kept tight for about 2 min. After the fibrin became firm the suture at the antimesenteric border was tied and the splint removed. The laparotomy wound was closed in two layers, using 4-0 Vicryl in the peritoneal-fascial layer and 4-0 Mopylene for the skin. Four weeks later the uterus and the adnexa were fixed by aortic perfusion and removed for light microscopy and electron-microscopic examination, following the procedure by Forssmann [3] and Meller [7].

## Results

The scanning electron microscopy of the anastomoses revealed in all cases good healing with very smooth edges. No stenoses or intrauterine adhesions were observed, and the surfaces were

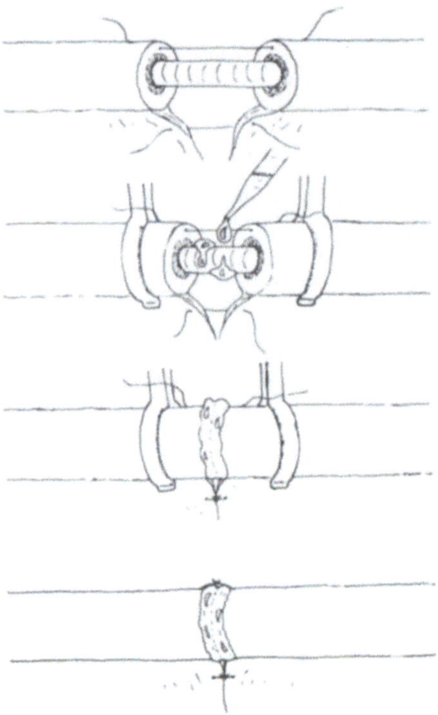

4

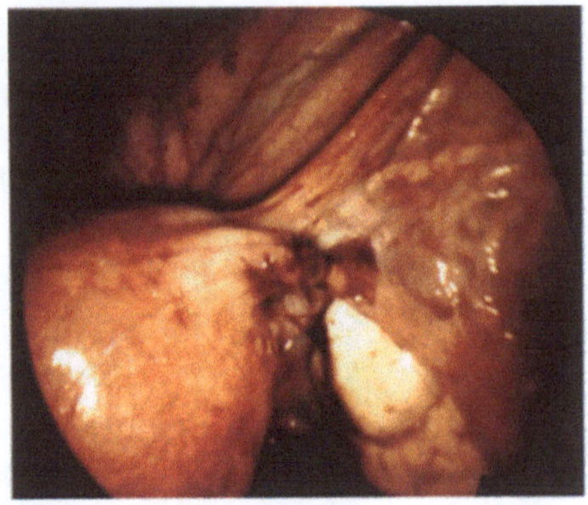

5

**Fig. 4.** Technique of endoscopic tubal anastomosis

**Fig. 5.** Endoscopic tubal anastomosis before application of fibrin glue. In this case two antimesenteric seromuscular sutures were placed

found to be intact (Fig. 3). In particular, the cells within the anastomoses displayed a dense layer of microvilli (Fig. 3b). The analysis of anastomoses by transmission-electron microscopy also revealed intact cellular structures with continuous tissue formation.

In two cases a tapering of the anastomosed uterine wall was noticed. Fistulas, diverticula, or dehiscence were not detected.

## Clinical Study

### Surgical Technique

The surgical technique is demonstrated in Fig. 4. After preparation of tubal stumps and ascending and descending hydropertubation, tubal stumps were approximated with a 6-0 PDS suture of the mesosalpinx. A polyethylene splint was inserted hysteroscopically using a catheter for peridural anesthesia, 0.8 mm in diameter. A 6-0 PDS seromuscular suture was placed at the antimesenteric border. Tubal stumps were approximated over the inlaying splint (Fig. 5) and fibrin glue was applied. After the fibrin clot became firm, the antimesenteric suture was tied.

### Patient Selection

Ten sterilized patients who desired a laparoscopic approach only were included in the ongoing study. Laparoscopic reanastomosis was performed only in cases where an isthmic-isthmic anastomosis seemed to be possible. The endoscopic procedure was performed only on one side. The contralateral tube remained unoperated.

### Results

The preliminary results are summarized in Table 1. Four intrauterine

**Table 1.** Preliminary results of endoscopic tubal anastomoses to reverse sterilization in human beings

| Patient | Age | Patency | Pregnancy |
|---|---|---|---|
| M.L. | 37 | + | abortion, LPD |
| M.G. | 25 | + | term pregnancy |
| M.L. | 38 | + | abortion, LPD |
| H.M. | 39 | + | no, LPD |
| A.D. | 35 | sec. microsurgery | |
| I.E. | 41 | ? | non, LPD |
| U.B. | 30 | + | no |
| C.B. | 36 | ? | no |
| M.Z. | 33 | ? | no |
| H.K. | 34 | ? | no |
| F.A. | 37 | ? | no |

LPD, luteal phase deficiency

pregnancies occurred in the three patients operated first. The second patient had a term pregnancy. In two other cases tubal patency and perfect healing were documented by laparoscopy and hydropertubation or hysterosalpingography. In one case the decision was taken to do a laparotomy because of problems with sufficient approximation of the stumps to create an anastomosis without tension.

## Discussion

In the animal experiments I was able to demonstrate that isthmic tubal anastomoses with fibrin glue are as effective as anastomoses with microsurgical techniques. Also, the morphological studies of endoscopic anastomoses of the rat uterine horn revealed good healing and intact tissues in all cases. Therefore, I decided to go into a clinical trial of endoscopic tubal anastomoses in human subjects. The first results show that this new technique may be an alternative to classical microsurgical reanastomosis in certain cases. This is also the first report of a term pregnancy after endoscopic tubal anastomosis. I believe that endoscopic reversal of sterilization will gain a definite place in reconstructive tubal surgery, but more data are necessary to evaluate the significance of this new method.

## References


1. Bruhat MA, Manhes H, Mage G, Pouly JL (1980) Treatment of ectopic pregnancy by means of laparoscopy. Fertil Steril 33:411–414
2. Bruhat MA, Mage A, Manhes H et al. (1983) Laparoscopy procedures to promote fertility ovariolysis and salpingolysis: results of 93 selected cases. Acta Eur Fertil 14:113–115
3. Forssmann WG (1981) General methods in transmission electron microscopy of the nervous system. In: Heym C, Forssmann WG (eds) Techniques in neuroanatomical research. Springer, Berlin Heidelberg New York, pp 21–39
4. Gauwerky JFH, Kubli F, Forssmann WG (1988) Morphology and fertility after reanastomosis of the rabbit fallopian tube with fibrin glue. Human Reprod 3:327–330
5. Gomel V (1977) Salpingostomy by laparoscopy. J Reprod Med 18:265–268
6. Gomel V (1983) Salpingo-ovariolysis by laparoscopy in infertility. Fertil Steril 40:607–611
7. Meller K (1981) General methods in scanning electron microscopy. In: Heym C, Forssmann WG (eds) Techniques in neuroanatomical research. Springer, Berlin Heidelberg New York, pp 55–66
8. Pouly JL, Manhes H, Mage G, Canis M, Bruhat MA (1986) Conservative laparoscopic treatment of 321 ectopic pregnancies. Fertil Steril 46:1093–1097
9. Semm K (1977) Pelviskopische Chirurgie in der Gynäkologie. Geburtshilfe Frauenheilkd 37:909–920
10. Semm K (1986) Pelvi-Trainer, ein Übungsgerät für die operative Pelviskopie zum Erlernen endoskopischer Ligatur und Nahttechniken. Geburtshilfe Frauenheilkd 46:60–62
11. Semm K, Mettler L (1980) Technical progress in pelvic surgery via operative laparoscopy. Am J Obstet Gynecol 138:121–127

# Suture-free Tissue Sealing: Fibrin Glue

D. WALLWIENER[1], D. POLLMANN, J. GAUWERKY, W. STOLZ, S. RIMBACH, and G. BASTERT

The introduction of surgical laparoscopic techniques into gynecology has made it possible to carry out the majority of reconstructive interventions within the tubo-ovarian functional unit and to remove adhesions or endometric tissue in the vicinity of the internal genitals without having to perform an abdominal section. In the process, the surgical techniques have been optimized by the utilization of lasers, in particular the $CO_2$ laser, as well as by the continued use of the Nd:YAG contact laser.

A further step towards perfecting surgical laparoscopic techniques has been taken with the use of fibrin glue, which is employed for the following purposes:

1. For the sealing of the edges and surfaces of wounds
2. To attain a local hemostatic effect
3. As an adhesion prophylaxis by covering peritoneal damage

The basic advantages of fibrin glue are its exceptional tissue tolerance, its absorbability, its high elasticity, and its adhesive capacity in a wet environment.

## The Technique of Fibrin Sealing

The technique of fibrin sealing is based on the physiologic mechanism of the last phase of blood coagulation. The fibrin glue which we used – Dissucol Duo S – is a two-component glue which is basically made up of highly concentrated human fibrinogen and a highly concentrated thrombin solution. The solidification begins within seconds after application, and fixation is largely completed within a few minutes.

During punctiform application and application on small surfaces, for example, after end-to-end tubal anastomosis, both components are applied with a syringe. The adhesive components can, furthermore, be applied using multichanneled catheters. These catheters are inserted through a 3-mm trocar via a second or third auxiliary puncture near the pubic hair. When larger wound areas are involved the components are sprayed onto the damaged area with the help of $CO_2$ gas. During laparoscopic application using an additional pumping system one must be especially careful to ensure

[1] Universitäts-Frauenklinik, Abt. für Geburtshilfe und Gynäkologie, Voßstr. 9, W-6900 Heidelberg, FRG.

Lasers in Gynecology
Edited by G. Bastert and D. Wallwiener

that an intra-abdominal pressure of 12 mm Hg is not exceeded and that $CO_2$ gas is never applied in a hollow organ, as this could lead to dangerous complications.

The catheter tip should be 2–3 cm from the wound surface. A few seconds after application one can see the fibrin glue turn white and observe its solidification. Too rapid absorption of the fibrin glue is prevented by the addition of an antifibrinolytic agent such as aprotinin.

## Clinical Utilization

There are two basic fields of application for fibrin glue within the framework of surgical laparoscopy:

- For the adhesion of tissue
- For hemostasis or sealing

The following set of indications results:

*1. Tissue closure*
- End-to-end anastomosis after laparoscopic refertilization
- Fimbrial eversion during salpingostomy
- Salpingotomy due to tubal pregnancies
- The shaping of ovaries after extirpation of cysts

*2. Sealing and hemostasis*
- Covering of peritoneal damage
  a) after adhesiolysis
  b) after laser treatment of endometriosis
- Bleeding peritoneal injuries
- Artificial perforation of the uterus

## Ad 1: Tissue Closure

The time interval until absorption of the fibrin glue and completion of the healing procedure is largely dependent upon the thickness of the applied glue; hence, it ought to be applied thinly when wound edges are to be united.

### Laser Laparoscopic Salpingostomy with Fimbrial Eversion

Salpingostomy of a sactosalpinx can be performed with a laparoscopically applied $CO_2$ laser beam. Under chromopertubation pressure the visible or assumed cicatricial lines in the vicinity of the navel of the old scar are opened bloodlessly using the laser; negligible tissue damage is caused. The small fimbrial "lobes" cut in this manner can be everted by application of fibrin glue, thus avoiding a time-consuming and complex eversion suture. Single drops of fibrin glue are applied to the rear of the reconstructed fimbrial "lobes" which are apposed in an everted position until sealed with atraumatic fixation forceps (Fig. 1). If the tubal wall is

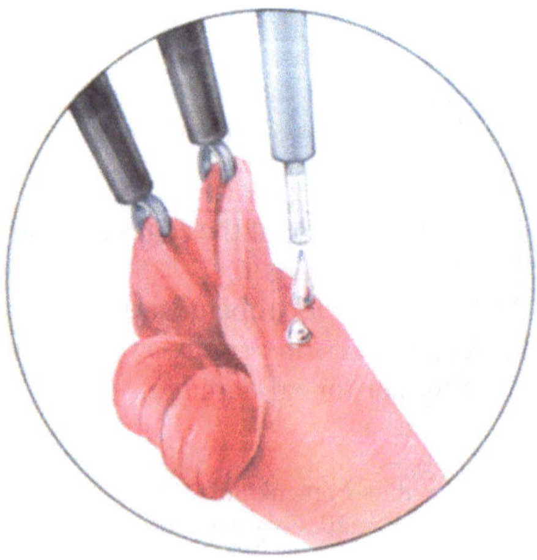

**Fig. 1.** Laser laparoscopic salpingostomy with eversion of the fimbriae

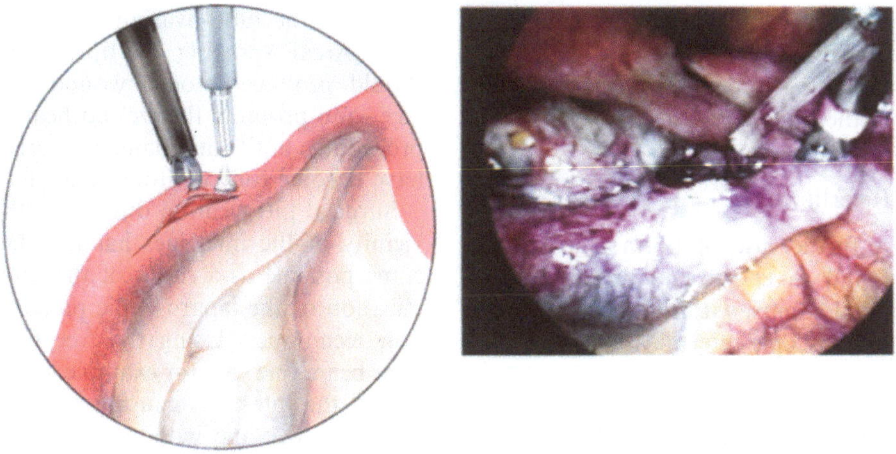

a b

**Fig. 2a, b.** Fibrin sealing of a linear salpingotomy in an isthmo-ampullary tubal pregnancy

thin enough laser eversion in the sense of "flowering" is the therapy of choice.

## Linear Salpingotomy in Isthmo-ampullary Tubal Pregnancy

The use of fibrin glue for the adaptation of wound edges in order to close the lumen is an extremely quick and easy manipulation; it enables one, on the one hand, to avoid having to carry out the complex endoscopic suture needed to close a linear salpingotomy after surgical treatment of a tubal pregnancy and, on the other hand, to prevent having to take the risk of fistulization of widely dehiscent salpingotomy wound edges (Fig. 2).

## Extirpation of Ovarian Cysts and Ovary Shaping

A technically easy operation, but an important intervention as far as an expense-versus-benefit analysis is concerned, is the shaping of the ovary subsequent to the extirpation of ovarian cysts (Fig. 3). In a large number of patients ovarian cysts can be removed and the organ preserved without an abdominal section being necessary.

The ovarian capsule is initially incised prior to the partially sharp and partially blunt extirpation of the cyst together with the cystic wall. After hemostasis the resultant damage is ad-

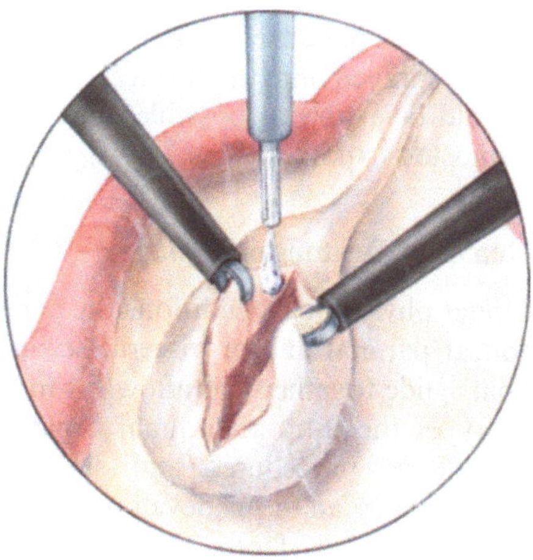

**Fig. 3.** Shaping an ovary using fibrin glue after organ-preserving extirpation of an ovarian cyst

hered with the necessary amount of fibrin glue, either in the sense of a reshaping of the ovary or as a closure of the injury.

### Refertilization

A further use of fibrin glue is to be found in end-to-end anastomosis of tubal stumps within the framework of laparoscopic refertilization [3, 7].

## Ad 2: Hemostasis and/or Sealing

Fibrin glue can furthermore be used for the sealing of surfaces for hemostatic purposes or for the repair of a damaged area.

### Hemostasis

Bleeding surfaces are rinsed with a warm saline solution using a water purifier. The hemostatic effect is promoted by positive intra-abdominal pressure generated by an insufflator. Electrocoagulation or laser coagulation can follow if a temporary hemostasis is not attained by the rinsing and the positive pressure. Thereafter, fibrin glue is sprayed onto the surfaces, which should be as dry as possible, for final hemostasis and sealing.

### Repair of Peritoneal Damage

Under physiologic conditions the peritoneal parietal and/or visceral serous coats glide together and allow for mobility of the intra-abdominal organs. Pathologic procedures and surgical trauma can lead to the development of deperitonized surfaces. These can adhere to one another, and thus cause pain or lead to adhesions which restrict or obstruct the ovum pick-up mechanism and the tubo-ovarian functional unit. Areal spraying of fibrin glue should produce smooth wound surfaces and promote the wound healing procedure until the damaged areas have become fully peritonized again.

Surgical intervention directly in the vicinity of the intestines has also become possible as a result of the optimization of the laparoscopic preparation techniques. Using the free $CO_2$ laser beam one can, for example, operate on the intestines causing minimal tissue damage and without an intestinal lesion resulting. Inevitably, larger injuries develop within the region of the visceral peritoneum.

It is generally not necessary to repair damage within the tubo-ovarian functional unit which has resulted from fine surgical incisions, as in ovariolysis of adhesions caused by endometriosis. However, should more extensive vaporization of the endometriosis result in greater peritoneal damage, sealing of the injury ought to be taken into consideration (Fig. 4).

A further example of the use of fibrin glue is the sealing of artificial perfora-

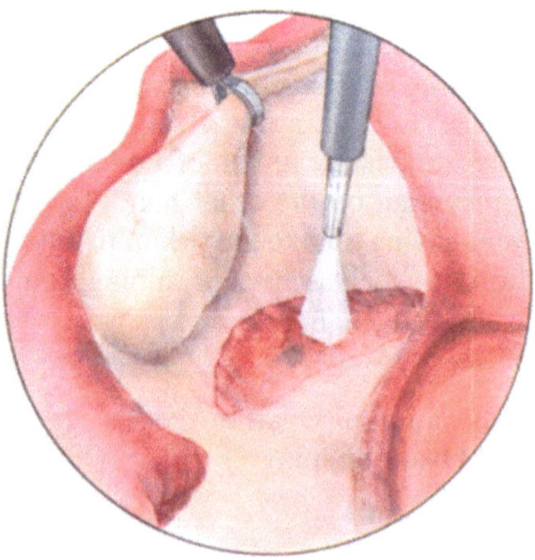

**Fig. 4.** The sealing of peritoneal damage

tions of the uterus. Patients with a perforation of the uterus are in danger, on the one hand, of an ascending infection as a result of germs in the lower genital canal and, on the other hand, of primary or secondary hemorrhaging from the damaged area. A perforation of the uterus can also be sealed using fibrin glue, thus avoiding a laparoscopic suture. After a local bloodless area has been established fibrin glue is applied. A second layer of fibrin glue is then applied on top of the first to seal the area. The uterus is held in this position for 3–5 min to allow the fibrin glue to consolidate and to ensure that the upper layer, applied to seal the damaged area, is appropriately stable.

## Results and Discussion

Table 1 summarizes the application of fibrin glue carried out by our team using various laparoscopic surgical techniques (Table 1). Subject to the initial condition (e.g., of the tubes after a postinfectious alteration), particularly adhesion induction or multiple scar formation within the vicinity of the distal tubes were excluded subsequent to fibrin sealing. In those cases in which the sealing of a peritoneal injury and a second-look laparoscopy had been performed, almost no postoperative adhesions were observed.

**Table 1.** Laparoscopic preparation techniques and fibrin glue ($n = 70$)

| Indication | No. of cases |
|---|---|
| Peritoneal damage | 15 |
| Serous damage (intestines) | 5 |
| Ovary shaping after extirpation of cysts | 7 |
| Linear salpingotomy | 7 |
| Eversion of fimbriae | 10 |
| Refertilization | 23 |
| Perforation of the uterus | 3 |

The postoperative courses of the three cases of sealing of uterus perforations using fibrin glue carried out thus far were completely free of complications.

In 11 extirpations of ovarian cysts with subsequent shaping of the remaining ovary using fibrin glue, no pathologic changes in the newly shaped ovary could be laparoscopically and/or ultrasonographically detected over a longer observation period.

## Summarizing Evaluation

In summary, we may conclude that the optimization of the laparoscopic techniques through the use of fibrin glue, especially within the framework of laser-assisted preparation, has led to a reduction in the duration of an endoscopic intervention.

Moreover, fibrin sealing leads to a physiologic tissue union or sealing of damaged tissue, because human physiologic fibrin is used and no tissue destructive procedures such as traumatic sutures or tissue-damaging traumatization (for example, electrocoagulation or tissue-toxic destruction due to synthetic glue) can exert a negative influence on the healing procedure.

The extremely laborious endoscopic suturing, which many centers cannot perform due to a lack of necessary instruments, is thus often superfluous.

When fibrin glue is used as a method of sealing peritoneal damage, an adhesion prophylactic effect is hoped for in addition to the attained tissue union; this has been experimentally demonstrated by both Waag and Koltai. The relatively high price of fibrin glue, which exceeds the cost of endosuture

**Table 2.** Advantages and disadvantages of fibrin glue

| Advantages | Disadvantages |
|---|---|
| Reduction in operating time | More expensive than suture material |
| Physiologic tissue union and sealing of damage | Necessity of sealing peritoneal damage unclarified |
| Atraumatic operating technique | To date, too little experience and insufficient follow-up data to enable definitive evaluation |
| Easy handling | |
| Adhesion prophylaxis | |

material does, however, make a careful cost-benefit analysis essential in each individual case. The extent to which the reduced operating time compensates for this disadvantage remains to be seen.

Fibrin glue as a means of sealing peritoneal damage is still under discussion, in contrast to the various indications within the range of reconstructive tubal surgery. Convincing evidence of the necessity of sealing damage, particularly after laser intervention, is still lacking. The advantages and disadvantages of fibrin glue are summarized in Table 2.

Due to the crystallization of a range of indications and the resultant advantages, the utilization of fibrin glue does appear to be very promising for surgical laparoscopy. Detailed comparative studies with a suitably long follow-up period must follow, however, to enable a long-term clinical evaluation.

## References


1. Baumann R, Volk M, Taubert H-D, Rücker KJ (1987) Refertilisierung beim Menschen unter zusätzlicher Anwendung von Fibrinkleber. In: Kubli F, Schmidt W, Gauwerky J (eds) Fibrinklebung in der Frauenheilkunde und Geburtshilfe. Springer, Berlin Heidelberg New York London Paris Tokyo, S 73–80
2. de Virgilio C, Dubrow T, Sheppard BB, MacDonald WD, Nelson RJ, Lesavoy MA, Robertson JM (1990) Fibrin glue inhibits intra-abdominal adhesion formation. Arch Surg 125:1378–1382
3. Gauwerky J, Forssmann WG, Kubli F (1987) Fibrinklebung versus Nahtanastomose: experimentelle Untersuchungen zur mikrochirurgischen Refertilisierung. In: Kubli F, Schmidt W, Gauwerky J (eds) Fibrinklebung in der Frauenheilkunde und Geburtshilfe. Springer, Berlin Heidelberg New York London Paris Tokyo, S 61–68
4. Koltai JL, Gerhard A (1990) Intraperitoneal application of fibrinogen gluing in the rat for adhesions prophylaxis. Pediatr Surg 25:71
5. Larsson B, Fianu S, Jonasson A, Rodriquez-Martinez H, Hedström CG, Thorgirsson T (1987) The use of Tisseel (Tissucol) – a two-component fibrin sealant – in operations for fertility as a sealant and for prevention of adhesions: an experimental study and a preliminary clinical evaluation. In: Schlag G, Redl H (eds) Gynaecology and obstetrics – Urology. Springer, Berlin Heidelberg New York London Paris Tokyo, pp 90–94
6. Riss P, Spernol R, Beck A, Schindler K (1986) The use of fibrin glue in experimental tubal surgery. In: Schlag G, Redl H (eds) Gynaecology and Obstetrics – Urology. Springer, Berlin Heidelberg New York London Paris Tokyo, pp 84–89
7. Scheidel PH, Wallwiener D, Wiedemann RA, Hepp H (1982) Experimental anastomosis of the rabbit fallopian tube using fibrin glue. Fertil Steril 38:471–474
8. Waag K-L, Litzinger B (1988) Umkehrung des Fibrinkleberprinzips zur Adhäsionsprophylaxe. Eine experimentelle Studie. Monatsschr Kinderheilkd 136 (8):581
9. Wallwiener D (1982) Experimentelle Anastomose-Technik mit Fibrinkleber an der Kaninchentube. Inaugural dissertation, University of the Saarland

# Hysteroscopy

# The Increasing Role and Future Development of Laser Hysteroscopy

H.-J. LINDEMANN [1]

Hysteroscopy is the name given to the technique of examining the uterine cavity and the cervical canal by means of a special endoscope, known as a hysteroscope. This consists of a straight tube which incorporates a series of optical lenses. The tube varies in diameter between 3 and 7 mm, depending on the nature of the examination to be conducted. It also incorporates a built-in lighting system, which serves to illuminate the interior of the uterine cavity. Using this instrument, or one of the flexible or semiflexible endoscopes which transfer pictures by fibers, the examining physician is able to observe normal conditions or pathological changes inside the uterus.

Apart from attempts at interstitial sterilization of the uterine tubes, hysteroscopy was formerly used exclusively for diagnostic purposes. But today this method is also used in certain forms of therapy. With improved technical possibilities, i.e., better optical lenses, instruments, and light sources, and more knowledge in general based on experience gained in this field, therapeutic procedures have been developed.

In many cases hysteroscopic surgery obviates the need to make incisions in the abdomen; formerly this was the only possibility. A uterotomy is no longer necessary, for example, to remove a septum, an operation traditionally performed with the Strassmann or Tompkins or other similar methods. A pregnancy that follows a successful uteroplasty has to be terminated by cesarean section. This is not necessary after the hysteroscopic removal of a uterine septum.

The hysteroscopic technique can also be used to remove synechiae, submucous myomas, and polyps. To achieve a bloodless operating field administration of POR 8 (Sandoz), a derivative of the pituitary hormone vasopressin, is very helpful. A dosage of 2.5 IU diluted in 50 ml saline is injected in four equal portions into the myometrium. One can practice conventional surgery with a planing knife, scissors, and other sharp instruments or high-frequency current and thermocoagulation instruments, or use one of the latest techniques such as a neodymium:YAG laser.

Hysteroscopy for intrauterine operations should always be the preferred method for gynecologists who have been trained in the latest surgical techniques. The complication rate is lower, and the length of hospitalization is shorter; the patient can be discharged the same day – as an in- and outpatient

[1] Michaeliskrankenhaus, Oderfelderstr. 6, W-2000 Hamburg 13, FRG.

Lasers in Gynecology
Edited by G. Bastert and D. Wallwiener

– or after an overnight stay. There is no doubt that this technique is much less traumatic.

A comparison between patients who underwent a classical operation by laparotomy and those treated by a hysteroscopic procedure is not yet possible, as the number of hysteroscopic cases is still too small and not all the results have been published.

The number of patients who have received laser treatment in hysteroscopy to date is very small. However, the experience gained with this latest technique show the following results, and possibly some advantages.

The bioeffect of a laser light is the optical resonance between the laser beam and the molecules of a tissue. A molecular reaction takes place if photoenergy is absorbed. The wavelength of the incident photon determines the capacity of absorption. The molecule can react in two ways: (a) with fluorescence – this would be the emission of another photon, or (b) with a rise of temperature, which is due to the increase of movements of the molecules.

For hysteroscopic laser surgery we need the effect of coagulation and vaporization of tissue. Depending on the amount of energy absorbed, the heat released by the absorption of a particular laser light will cause: coagulation of the proteins, if the temperature is about 60 °C; thermal necrosis, if the temperature is between 80° and 100 °C; or loss of tissue in a kind of vaporization, if the temperature is above 100 °C.

Laser-caused vaporization produces a large amount of smoke and the difficulty is to deal with it in the small uterine cavity. In order to minimize the disturbing smoke effect, the uterus is distended not with carbon dioxide but with fluids which absorb the smoke, and an Nd:YAG laser is used as it is not in optical resonance with $H_2O$. The $CO_2$ laser is not suitable, for its optical resonance with $H_2O$ prohibits the use of fluid distention media. However, the state of the art is a new technical construction by A. Gallinat and R. Lueken (personal communication), manufactured by the Wolf company. They use carbon dioxide as a dry distention medium for the uterine cavity with its known advantage of allowing a good view, and the $CO_2$ gas pressure blows the smoke out. It should be mentioned that lasers other than the Nd:YAG are also theoretically suitable, such as argon and dye lasers. Their use has not yet been extensively evaluated, however.

The therapeutic indications for the use of an Nd:YAG laser are: removal of myoma, synechia, or polyps, resistant bleeding of the endometrium, hyperplasia endometrii, and transection of a septum – either with coagulation or vaporization. A future diagnostical possibility is the induction of fluorescence in pathological tissues, as in patients with adenomatous hyperplasia or adenocarcinoma.

The advantages and disadvantages of this technique are a matter of controversy, but there is no question about its effectiveness. At the last Congress of the European Society of Hysteroscopy in Taormina, Italy, in June 1990, Neuwirth presented some cases of submucous myoma and other pathological conditions treated by high-frequency current and the laser technique. Postoperative histological examinations of the extirpated uteri revealed no scars or remarkable changes in comparison with results when other techniques were used. Similar results were reported by Tennessee at the Ibero-American

Gynecological Congress in May, 1989, in Rio de Janeiro. Tozzini from Argentina and others criticized the long duration of the procedure, and even they emphasized that the use of laser is not without risk. Perhaps the sapphire technique by Morawski of Homburg (personal communication) will enable us to proceed safer and faster.

## References


1. Neuwirth RS (1976) Excision of submucous fibroids with hysteroscopic control. Am J Obstet Gynecol 126:95
2. Strassmann EO (1966) Fertility and unification of double uterus. Fertil Steril 17:165
3. Tompkins P (1962) Comments on the bicornuate uterus and twinning. Surg Clin North Am 42:1049

# The Role of Operative Hysteroscopy in Fertility Surgery

D. Wallwiener [1], S. Rimbach, D. Pollmann, J. Gauwerky, G. Bastert, and H. Wedler *

Due to optimized endoscopic visualization and improved preparation techniques, hysteroscopy has developed into an effective procedure with increasing applicability in gynecologic fertility surgery for treatment of intrauterine pathology. Though the principle of hysteroscopy has been known for the past 100 years, the importance of the transcervical surgical approach for diagnosis and therapy of diverse uterine and tubal sterility factors has only recently been appreciated. This became possible as a result of improvements in uterine distention methods, development of a reliable high-intensity light source, and introduction of new preparation techniques.

## Hysteroscopes and Ancillary Instruments

Representative for the newly developed generation of diagnostic and operative hysteroscopes, the hysteroscopic equipment (Olympus) used at our department is described below. Figure 1 a shows an instrument generation (Olympus) consisting of diagnostic hysteroscopes with 4- and 5-mm shafts, and of therapeutic hysteroscopes with 7- and 8-mm shaft diameters.

### Diagnostic Hysteroscopes

The 4-mm hysteroscope consists of a 4-mm sheath, and a 3-mm, 30° fore-oblique telescope. This instrument is particularly well suited for application in young women or in patients with a narrow cervical canal, where dilatation is not indicated. The 5-mm hysteroscope consists of a 5-mm sheath and a 4-mm-wide angle telescope, with a 30° angled direction of view. It is the standard rigid hysteroscope for diagnostic procedures. Advantageous is the compatibility of the telescope, which may also be employed in the therapeutic hysteroscopes. Telescope and sheath are easily and safely connected by an automatic snap-in lock. When the instrument is assembled, the 30° telescope is in an upward viewing position and the stopcock can be operated from above (Fig. 1 b).

### Therapeutic Hysteroscopes with Continuous Flow System

Hysteroscopic systems with 7- or 8-mm shaft diameters are available for opera-

* Olympus (Winter und IBE), Hamburg.

[1] Universitäts-Frauenklinik, Abt. für Geburtshilfe und Gynäkologie, Voßstr. 9, W-6900 Heidelberg, FRG.

Lasers in Gynecology
Edited by G. Bastert and D. Wallwiener

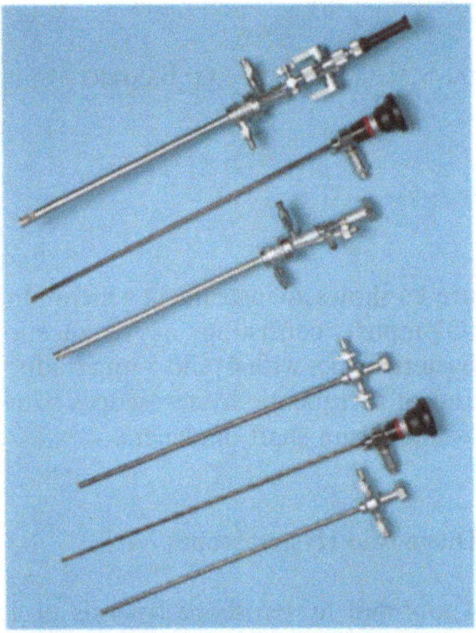
a

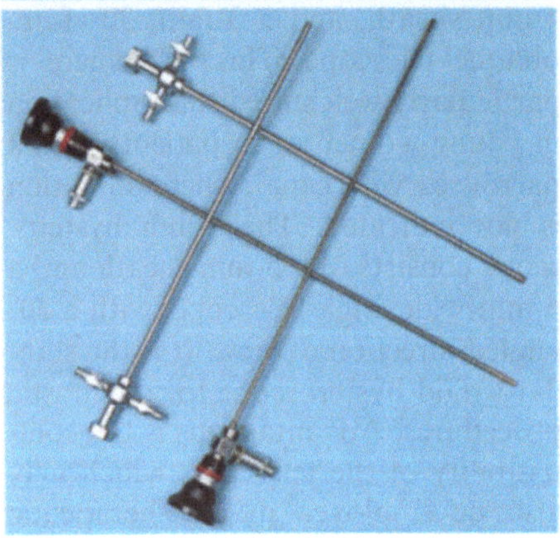
b

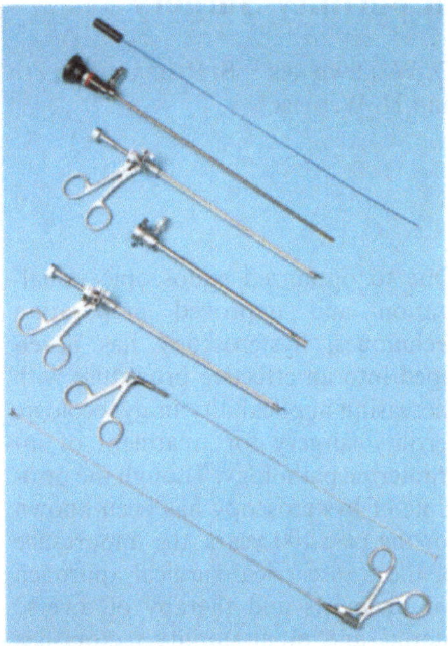
2

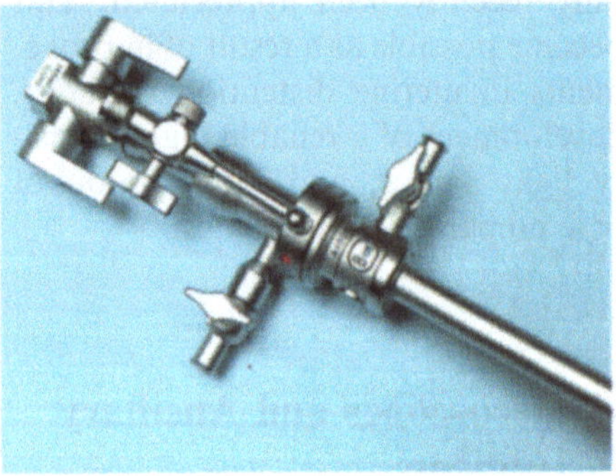
3

**Fig. 1. a** Diagnostic and therapeutic hysteroscopes (Olympus); **b** Diagnostic hysteroscopes (Olympus)

**Fig. 2.** Ancillary equipment for operative hysteroscopy

**Fig. 3.** Hysteroscopic working channel with in- and outflow (valve fluid or $CO_2$ gas), ensuring continuous flow or irrigation

tive procedures. Ancillary instruments such as flexible and semiflexible biopsy and grasping forceps, scissors, electrodes, puncture syringes, catheters, and probes may be used. For both systems additional instruments for laser application over 400- and 600-µm light guides have been developed (Fig. 2). The hysteroscopes shown in Fig. 2 are equipped with a continuous-flow rinsing system. The inflow irrigant or gas is guided directly along the telescope through a "telescope irrigation channel". The irrigant is thus directed to the

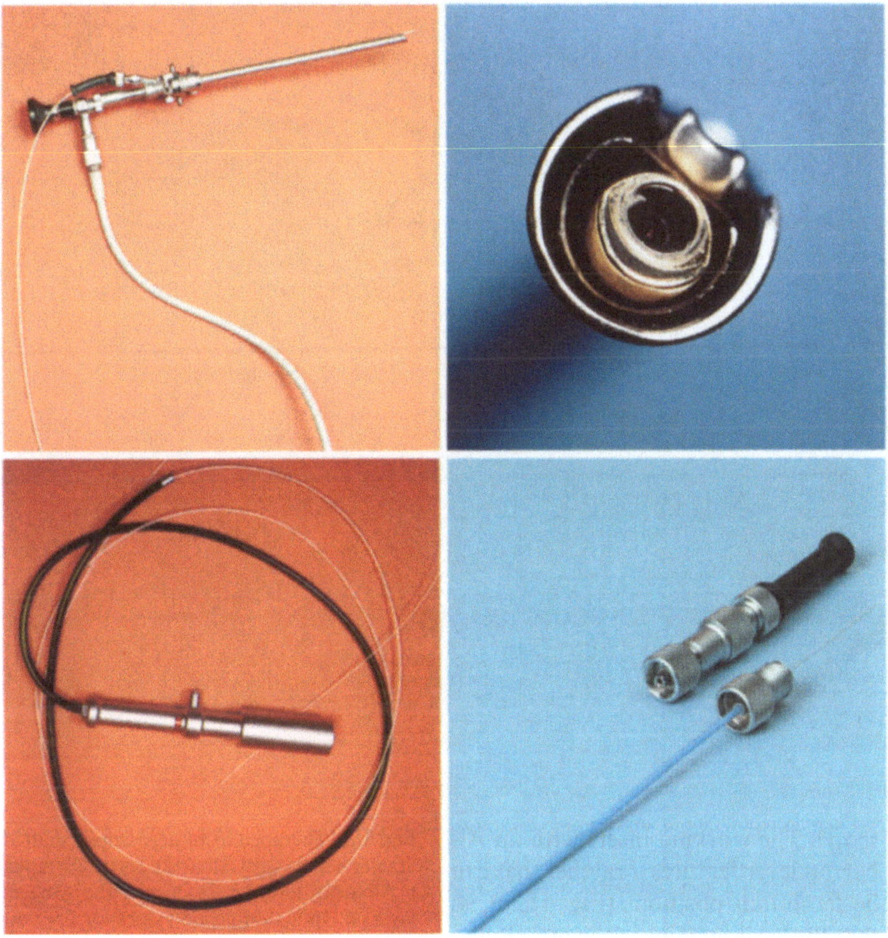

**Fig. 4. a** Operative hysteroscope (shaft diameter 7 mm) with ramp and flexible fiber inserted for Nd:YAG laser application (Olympus); **b** Ramp of the 7-mm therapeutic hysteroscope; **c** Nd:YAG "bare fiber" (quartz glass fiber); **d** Fixation and protection sheath for laser fiber

front lens. Continuous rinsing is a prerequisite for optimal vision. Outflow of liquid or gas is achieved via the sheath (Fig. 3).

*The 7-mm therapeutic hysteroscope* consists of a 7-mm sheath, a working insert with ramp, and a 30° wide-angle telescope 4 mm in diameter. It will accept instruments of 5-French diameter. Semiflexible biopsy and grasping forceps as well as electrodes are available for surgery. It is the ideal "laser hysteroscope," since the flexible fiber with its sheath can be readily passed through (Fig. 4a–d).

*The 8-mm therapeutic hysteroscope* consists of an 8-mm shaft, a working insert with an Albarrán lever, and a 4-mm telescope with either a 30° wide-angle or a 70° angle lens (Fig. 5a). It will accept instruments of 7-French perimeter. The same instruments may be applied as with the 7-mm hystero-

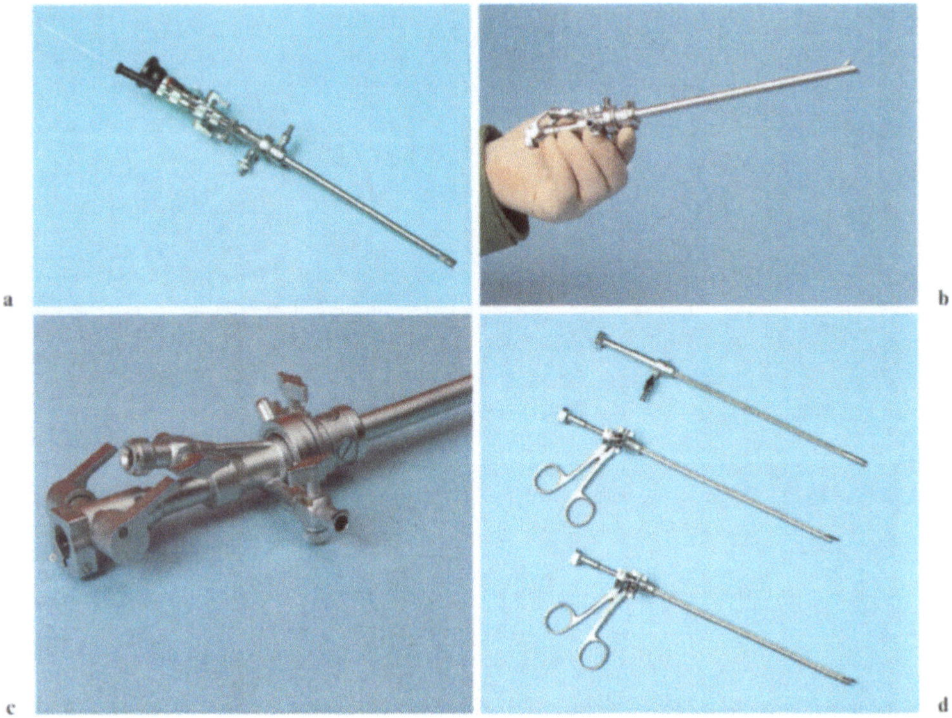

Fig. 5. **a** Therapeutic hysteroscope (shaft diameter 8 mm) with Albarrán lever (Olympus); **b** Albarrán lever; **c** Ratchet of the Albarrán lever; **d** Optical forceps/scissors for operative hysteroscope (8 mm)

scope. The working insert with an Albarrán lever features a ratchet that can be set in four positions (Fig. 5b, c), so that the lever remains in the preselected position. Operating lever and Albarrán lever are positioned at the same angle. The position of the Albarrán lever can be checked at a glance. Instead of the working insert with the Albarrán, optical biopsy forceps and optical scissors can be employed with the 30° telescope (Fig. 5d).

An optical obturator is available for insertion of the instrument. All these optical instruments feature a channel for continuous rinsing of the lens. The 8-mm sheath also allows combination with a working insert with ramp. It this combination, it will accept semiflexible instruments with a perimeter of 7 French.

For both surgical hysteroscopes ancillary equipment for *laser application* via light guides is available: At the instrument inlet both working inserts (with Albarrán lever and with ramp) have the standard screw thread for laser application. This thread connects to light guides with protective tubing or to the new type of accessories used with the bare fiber technique (Fig. 4). A fiber guide which fills up the working channel of either 5 French or 7 French is available for the 7-mm and the 8-mm hysteroscope. This additional covering of the bare fibers of up to 600 μm ensures that the light guide is centered and safely guided to the distal end of the sheath. Thus, the optical surface of the light guide does not come into direct contact with the metal of the endoscope. It also prevents the fiber from getting caught in the working channel. If required, the instrument can be fitted with a fiber fixture that prevents leakage of the distention medium and longitudinal dislocation of the fiber (Fig. 4d).

*The Importance of Video-Endoscopy.* Videoscopy, i.e. the possibility to perform a procedure under indirect vision over a video "chain" (CCD camera adapted to the hysteroscope, image transmission to a video monitor), has revolutionized hysteroscopy, as indeed it has revolutionized the entire field of diagnostic and operative endoscopy in medicine. It presents advantages not only from an ergonomic point of view, but also for documentation and training (Fig. 6).

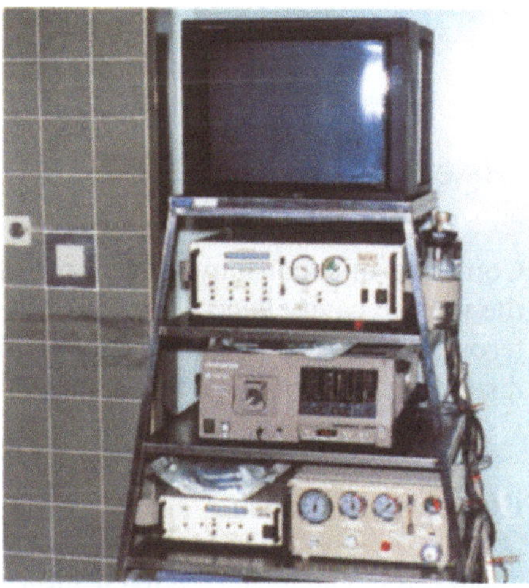

**Fig. 6.** Endoscopic unit (laparoscopic and hysteroscopic insufflators with smoke-suction device, light source, control panel for endoscopic video camera, video recorder)

## New Hysteroscopic Preparation Techniques

The development of new surgical techniques such as laser preparation or HF electroresection via a resectoscope has optimized operative hysteroscopy.

### Hysteroscopic Laser Preparation

For details see chapter on "Lasers in gynecology. An overview", this volume.

#### Operative Hysteroscopy with the Nd:YAG Laser Contact Technique

Due to its wavelength-specific tissue effects the Nd:YAG laser (Fig. 7) has proved best suited for hysteroscopic surgery. By preparation using thin quartz fibers precise fine cutting movements and concomitant hemostasis are possible. Coagulation can be achieved in both contact and noncontact techniques, even if fluid is used as a distention medium. The Nd:YAG laser light is applied over bare fibers via flexible or rigid hysteroscopes.

#### Surgical Procedure

The laser fiber is introduced into the uterine cavity via the working channel. Surgery is carried out with a videoscope or with a lens attached to the telescope, filtering the Nd:YAG beams that may cause damage to the eye.

The fiber must be advanced into the cavity until its tip can be seen hysteroscopically. Thus, damage to or destruction of the optic by direct laser light or heating up is avoided. However, the fiber must not be inserted too deep, since stability is necessary for exact

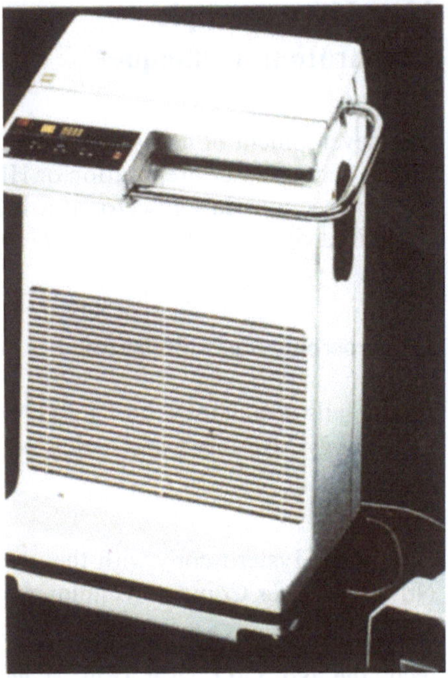

**Fig. 7.** Nd:YAG laser

preparation, and unintended lesions must be prevented.

For resection of a septum under laparoscopic control, the Nd:YAG laser is applied via bare fibers in the contact technique. The laser is set at 40–60 W/s in case of fluid distention and at maximally 20 W/s when $CO_2$ is used.

### Hysteroscopic High-frequency Electroresection

High-frequency electroendosurgery with the resectoscope is also very effective for intrauterine tissue ablation. Continuous-flow resectoscopes with shafts of 7, 8, and 9 mm in diameter are available. Small resectoscopes are well suited for resection of synechiae or adhesions. If the outer shaft diameter is larger than 8 mm, the high-frequency electrodes are of a size that without a considerable increase in operation time endometrial ablation or resection of larger myomas may be performed. Resectoscopes feature a system of an inner sheath and an outer sheath for continuous flow, an HF-electrode, a working element, an obturator, and a telescope. For gynecologic application the urologic resectoscope was somewhat modified, having a round shaft profile, uniform from the distal to the proximal end, so that optimal sealing of the cervical canal is achieved. It is equipped with an outer sheath for continuous irrigation, on which holes are evenly distributed to prevent suction of the posterior uterine wall to the instrument. Accidental pressure increase is thus ruled out.

The advantages of high-frequency endosurgery via resectoscope, namely the availability of material for histologic examination and the good cutting ability are, however, associated with an increased risk of lesion of blood vessels (Fig. 8a–c).

## Distention

Complete and constant distention of the uterine cavity is necessary for hysteroscopic procedures. Gas and fluid distention media are currently in use.

### $CO_2$ Gas Insufflation

Technique: The hysteroscope is inserted and $CO_2$ gas is insufflated for cavum distention at a maximal flow rate of 100 ml $CO_2$/min to prevent $CO_2$

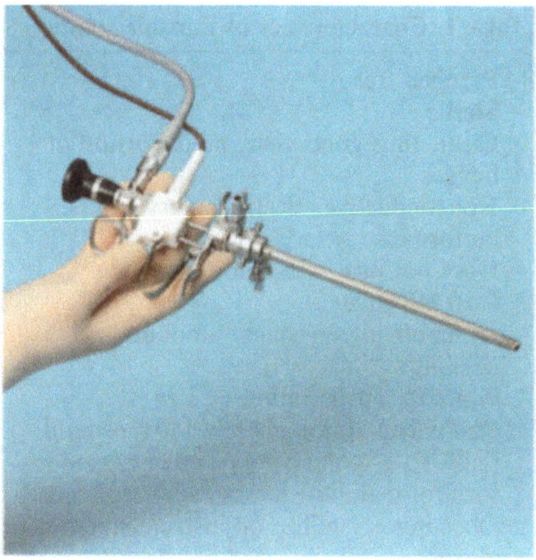
a

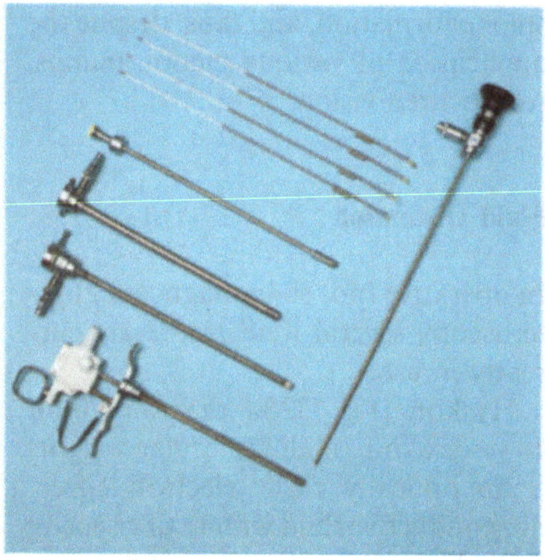
b

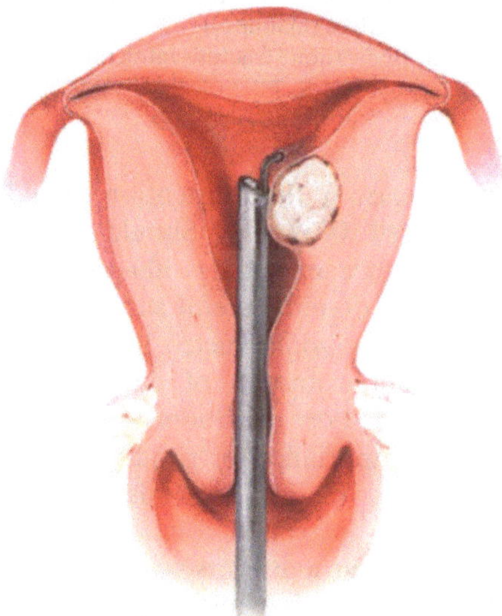
c

**Fig. 8. a**, **b** Hysteroscopic electroresectoscope; **c** Submucous myoma: resection with the HF resectoscope

embolism. Rubin [36] was the first to use $CO_2$ as a distention medium. Visualization is excellent, and videoscopic techniques can be applied. However, the surgeon must be aware of the potential risks of the procedure, e.g., gas intravasation, possibly with secondary hypercarbia, acidosis, or cardiac arrhythmia, if security precautions are not observed. Though in Germany, Lindemann [29] and Gallinat [20], have reported no complications in large series of operative hysteroscopies performed with $CO_2$ insufflation, a final evaluation is not yet possible, since major, even lethal, complications have been reported by some American hysteroscopists [2, 16], who therefore advocate the sole use of fluid distention.

In case of laser preparation, better vision is achieved with fluid distention. $CO_2$ distention leads to increased smoke formation, and thus, despite the application of various suction pumps, to obscured vision.

### Fluid Distention

In operative (not so in diagnostic) hysteroscopy several fluid media are currently in use:

Hyskon [17] (32% dextran 70 in 10% dextrose, high molecular weight) is an optically clear, electrolyte-free, nonconductive fluid with high viscosity and a high refractive index. It does not mix with blood and thus is particularly suited to laparoscopic-hysteroscopic combination procedures. Its high viscosity is a disadvantage. Complications reported in the literature range from anaphylactic reactions to dextran (1% occurrence) [5], dyspnea, and adult respiratory distress syndrome to coagulopathy [19, 25]. Hyskon, however, is not licensed in Germany.

Low-viscosity fluids (5% or 10% dextrose in water, NaCl, dextran 4% or 6%) are also well suited. Vision is clear at the beginning but gradually diminishes, since these media do mix with blood. This is, however, only valid for fluid distention as performed to date, no continuous rinsing system being available. With the newly developed continuous flow system, however, debris is constantly removed from the uterine cavity.

NaCl solutions cannot be employed for electrosurgical interventions.

Fluid distention presents risks as well. An exact monitoring of in- and outflow, electrolyte determination, and careful anesthesiologic surveillance are mandatory. Irrespective of the fluid used, exact pressure control and an exact fluid and electrolyte balance are of utmost importance. Intrauterine pressure in case of fluid distention should not exceed 100 mm Hg. Should a negative balance exceeding 300–500 mm Hg occur despite careful surveillance, this must be counteracted by prophylactic administration of diuretics. A continuous-flow irrigation system is of utmost importance for good vision by removing blood and debris (S. de Block, personal communication, 1990). At our unit Purisole (Fresenius, Bad Homburg) is employed, consisting of 0.54% mannitol and 2.7% sorbitol (Table 1).

**Table 1.** Characteristics of Purisole solution

- Pyrogen free
- Sterile
- Clear, thus good view, no distortion of images
- Nonconductive: electrocautery can be performed
- Does not mix with blood
- Cost-effective
- Available in convenient amounts and packing
- Glucose concentration of 3.24%, osmolarity approximately 180 mosmol/l (half-iso-osmolar); iso-osmolar sorbitol-mannitol solutions are not suited
- Does not crystallize on instruments or gloves
- Early elimination by metabolization (sorbitol) combined with an osmotic diuretic effect (mannitol)
- Contraindication: fructose intolerance

## Indications

The more hysteroscopies are performed as routine diagnostic procedures, the more often intrauterine factors for the etiopathogenesis of infertility are detected, authors reporting between 11% and 60% [8]. The most frequent are:

1. Uterine malformations, mainly müllerian fusion defects
2. Submucous intrauterine myoma
3. Intrauterine synechiae

## Malformations of the Uterus

### Etiology

Müllerian fusion defects, in 80% of cases a septate uterus [18], can be associated with normal fertility, infertility, or habitual abortion. These lesions may range from a minimal deformity of the fundus uteri to complete agenesis of the müllerian ducts [39]. The classification of Strassmann may serve as basis for the identification of the various forms of congenital müllerian anomalies [37] (Table 2).

**Table 2.** Müllerian fusion defects (classification by Strassmann [37])

*I. Symmetric double formations*

A. Externally divided uterus: two separate uterine bodies
   1. Arcuate uterus
   2. Uterus bicornis unicollis
   3. Uterus bicornis bicollis
   4. Uterus didelphys (2 corpora, 2 cervices)
B. Externally unified uterus but two cavities
   1. Uterus septus (reaching the internal os)
   2. Uterus subseptus (not reaching the internal os)

*II. Asymmetric double formations*

### Importance

The causal relation of a septate uterus with infertility has not been accounted for in the past. However, DeCherney [10] found in his study collectives an occurrence rate of 5% for müllerian fusion defects, and Dunselman [37] stated that "the malformation can no longer be regarded as a rarity" (Table 3).

**Table 3.** Müllerian fusion defects: incidence

| | | |
|---|---|---|
| 1924 | 0.014% | Smith[a] |
| 1962 | 0.31% | Zabriskie[a] |
| 1984 | 5% | DeCherney [10] |

"The malformation can no longer be regarded as a rarity" (Dunselman[a])

[a] Cited in [37]

### Diagnosis

A filling defect in the hysterosalpingogram carried out for all patients within the routine infertility workup must be further visualized by diagnostic hysteroscopy and often combined laparoscopy. Most commonly found are a bicornuate uterus, a submucous fibroid, or intrauterine synechiae. A rudimentary uterine horn, another form of müllerian anomaly, can be detected laparoscopically and its connection to the main cavum concomitantly examined hysteroscopically.

### Conventional Surgical Techniques

In the past, correction of intracavitary pathology was accomplished by large transabdominal procedures, namely uterotomies or metroplasty [37] (Table 4). These invasive procedures are associated with the risk of intraperitoneal adhesion or scar formation, possibly resulting in tubal occlusion, as well as with the risk of infection and bleeding. Furthermore, a cesarean section is indicated after metroplasty. Pregnancy rates of 75% after metroplasty compare favorably with the spontaneous pregnancy rate (nor surgical correction of the anomaly) of 4%.

**Table 4.** Müllerian fusion defects: techniques of metroplasty

| Technique | Year |
|---|---|
| *Resection through the cervix* | |
| Ruge and Schroeder[a] | 1884 |
| *Resection through the fundus* | |
| Transverse incision (Strassmann[a]) | 1952 |
| Vertical incision (Jones [26]) | 1957 |
| Modified incision (Bret-Palmer[a]) | 1968 |

[a] Cited in Strassmann [37]

For many decades, transcervical resection of septa was attempted with fine scissors, or later with electrodes. In a retrospective study Fayez [19] proved hysteroscopic metroplasty to be superior to the transdabdominal approach based on cost and morbidity considerations, but also on reproductive outcome. He reported on pregnancy rates of 84% (87% deliveries, no ectopic pregnancy, 13% abortions) after correction with hysteroscopic rigid scissors, and on 71% pregnancies (70% deliveries, 10% ectopic pregnancies, 20% abortions) after a Tompkins procedure. Many authors report on hysteroscopic septum resection [1, 10, 11, 28, 33, 40].

### Optimized Hysteroscopic Surgical Procedures

1. Nd:YAG contact technique (Fig. 9a–e) (See chapter "Laser in Gynecology: An Overview".)
2. HF electroendosurgery with the resectoscope

### Postoperative Treatment

Various methods are utilized to prevent adhesion formation. Most frequently, estrogens are administered postoperatively to regenerate the endometrium, or an intrauterine device (e.g., Nova-T, Schering, Hamburg) is inserted for a certain time. However, no statistically relevant results are yet available.

### Success Rate

Some authors [11] have reported on good results after electroresection with the resectoscope (term pregnancy rate of 80%). For Wamsteker [42], the resectoscope and the laser are alternative devices, and it is mostly a matter of personal preference. In a collective of 34 patients we attained a pregnancy rate of 75% (2 years' follow-up) after Nd:YAG laser resection of intrauterine septa (Table 5). The precondition for a surgical approach was associated fertility disorders (at least one abortion). No complications occurred. All patients could be delivered spontaneously, with the exception of one with a duplex uterus. Donnez [14] had a 43% pregnancy rate 1 year after Nd:YAG laser therapy.

## Intrauterine Synechiae

### Etiology

Post-traumatic intrauterine adhesions, a partial or complete Asherman syndrome, can mostly be traced back to

**Table 5.** Müllerian fusion defects: operative hysteroscopy for septate uterus in 34 cases using the Nd:YAG laser contact technique

| | |
|---|---|
| Preconditions for surgery: | |
| Infertility | 1. habitual abortions<br>2. no endocrinologic dysfunction |
| Complications: | none |
| Pregnancy rate: | 75% (follow-up 2 years) |
| Mode of delivery: | spontaneous delivery<br>1 cesarean section (Uterus duplex) |

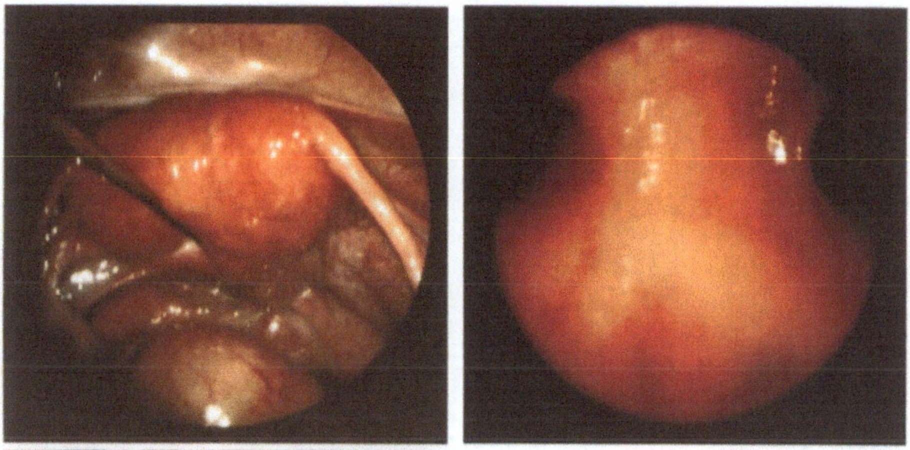

a b

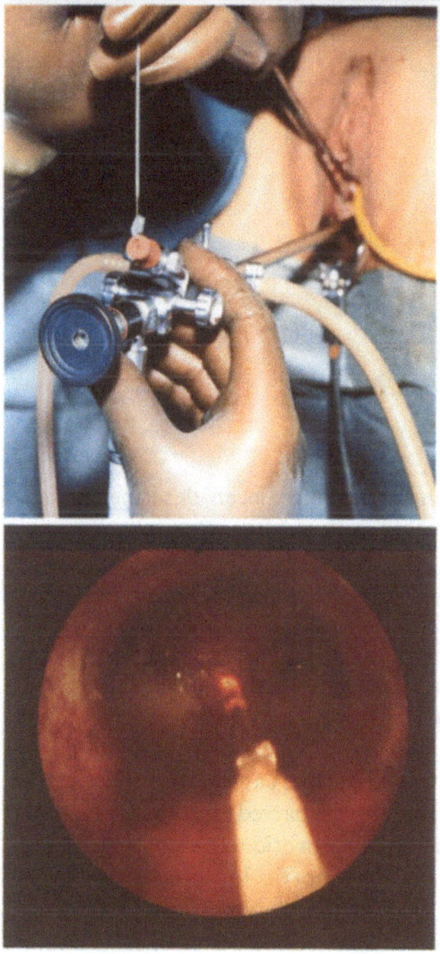

c e

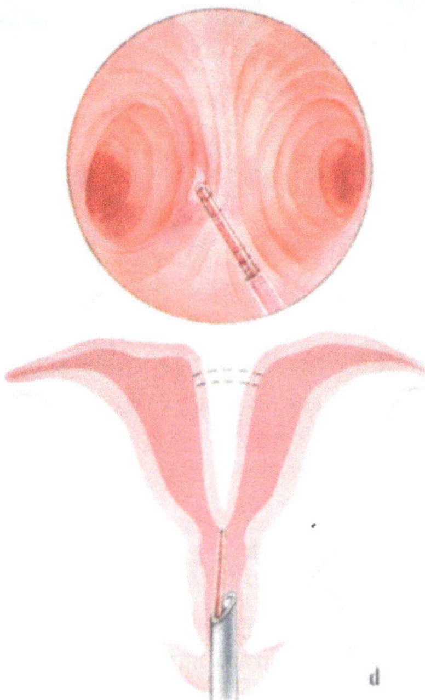

d

**Fig. 9. a** Septate uterus with median raphe (laparoscopic aspect); **b** Septum uteri (hysteroscopic aspect); **c** Laser hysteroscope with "bare fiber" in situ (Wolf, Knittlingen); **d** Laser hysteroscopic excision of a uterine septum; **e** Excision of a small septum with the laser "bare fiber," $CO_2$ distention and hysteroscopic smoke evacuation

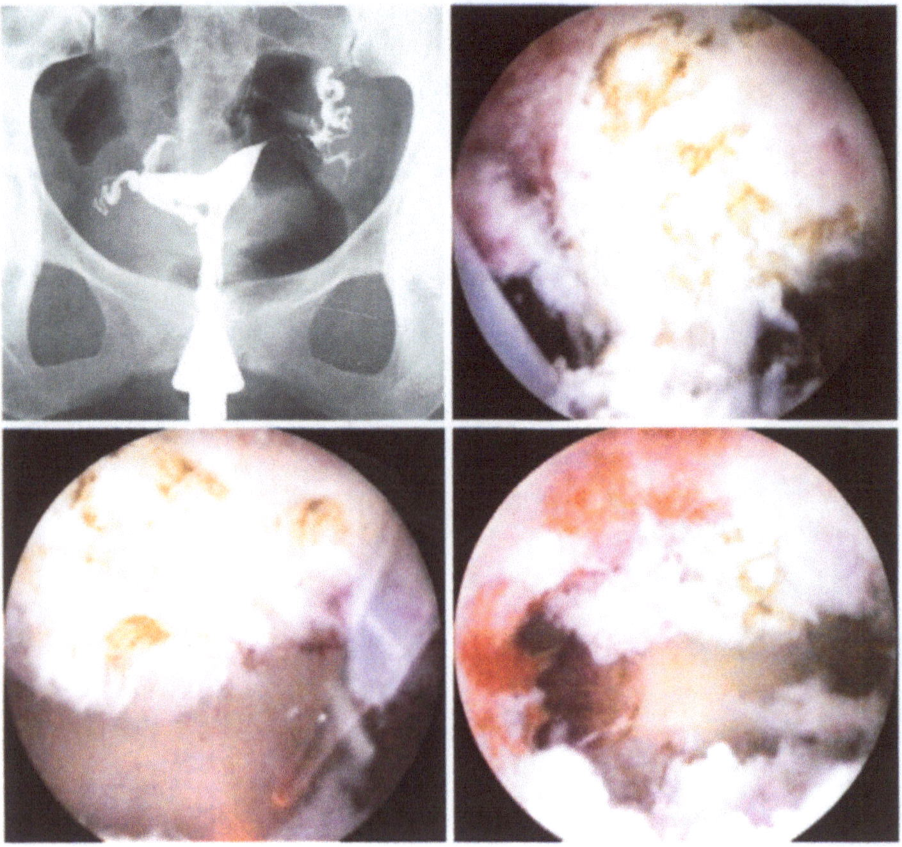

a b c d

**Fig. 10 a–d.** Intracavitary synechiae: **a** hysterosalpingogram; **b–d** synechiolysis

intracavitary manipulations on the endometrium after a pregnancy: postpartum or postabortion curettage may result in scar formation or induction of adhesions, due to the destruction of the basal layers of the endometrium.

## Importance

The spectrum of symptoms for intrauterine adhesions covers hypo- to amenorrhea, infertility, and recurrent abortion. In cases of habitual pregnancy losses, implantation or placentation anomalies have been described. Associated pain is often very severe.

## Diagnostics

For diagnosis of intrauterine adhesions a hysterosalpingogram (HSG, Fig. 10) and hysteroscopy are performed. The HSG shows a slender, fragmentated, or twisted uterine cavity; through the hysteroscope, scarred endometrial alterations or fibrotic tissue may be identified. Diagnostic hysteroscopy is thus essential to establish the exact nature and extent of the adhesions.

By their surface appearance, adhesions may be classified into endometrial, myofibrous, and connective tissue

[21]. Extensive adhesions within the uterine cavity may lead to severe intrauterine pathology, a compartmentation of the cavum, or a partial or complete Asherman syndrome.

### Therapy

Endometrial adhesions can be broken up by slight pressure with the shaft of the hysteroscope. Myofibrous and connective tissue adhesions may, in rare cases, be divided by blunt dissection, but usually sharp preparation by means of electrosurgery or laser application is necessary. For intracavitary synechiolysis laser application seems superior to all other techniques (see septum resection, above).

The more extensive the adhesions, the more difficult the procedure and the higher the risk of a *via falsa*. A sufficient distention of the uterine cavity facilitates the surgical intervention. Simultaneous laparoscopic control is mandatory in these cases. If the laparoscopic light source is shut off, the laser pilot beam can often be seen diaphonoscopically through the uterine wall. In cases of compartmentation or partial Asherman syndrome an intraoperative hysteroscopic hysterosalpingogram can be useful to establish exactly the intracavitary preparation area/surface and reduce the risk of perforation (K. Wamsteker, personal communication, 1990).

## Postoperative Treatment (See Septum)

### Success Rates

Pregnancy rates of up to 75% have been achieved in patient collectives with no other causes of infertility [34]. Donnez [12] reported a pregnancy rate of 50% after laser therapy, and normalized bleedings in 75%. In our collective of infertility patients the pregnancy rate after laser adhesiolysis was only 20%, mainly due to the fact that the patients had severe adhesions, often necessitating a combined hysteroscopic-laparoscopic approach.

**Table 6.** Classification of submucous myomas

1. Small (diameter < 1–2 cm)
2. Medium-sized (diameter < 2–3 cm)
3. Large (diameter > 3–4 cm)
   a) sessile, extending intramurally
   b) pedunculated

## Submucous Myomas

### Etiology and Importance

Though many theories exist, much is still unknown regarding the etiology of myomas [7, 35]. They arise from a single neoplastic cell derived from the smooth-muscle elements of arterioles, as well as directly from the myometrium [24]. Their growth is influenced by a variety of factors, above all by increased estrogen.

Among the various forms of uterine myomas, the submucosal ones with intraluminal extension may cause uterine bleeding, or infertility, often due to pathologic implantation of the embryo. They may also lead to premature pain, or severe dystocia. Our experience has shown that diagnosis and management of submucous fibroids should differentiate between myomas of various sizes and localization (Table 6).

## Diagnosis

An enlarged uterus and/or a filling defect in the uterine area detected on the hysterosalpingogram may be evidence of a submucous myoma. Diagnostic hysteroscopy is essential in these cases for further clarification of the diagnosis. The tumor protruding deeply into the cavum uteri, often covered by a typically colored flat endometrium, may be directly visualized by hysteroscopy. A differentiation from endometrial polyps is also possible. The tumor can be exactly localized and the size determined. Intraluminal extension cannot be estimated, however.

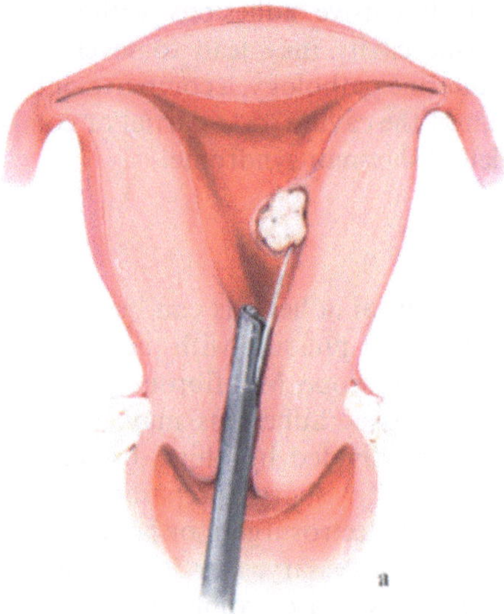

a

## Therapy

Pedunculated and sessile submucous myomas can be managed hysteroscopically. Hysteroscopic myomectomy can be performed both with the HF electrode and with the laser. Very small submucous myomas can be coagulated, but the absence of tissue sampled for histology and the relatively high risk of recurrence due to residual myomatous cells in the base are arguments against such a procedure. Therefore, laser contact enucleation of the myomatous tissue for smaller fibroids and laser morcellation for larger ones seems to be the therapy of choice.

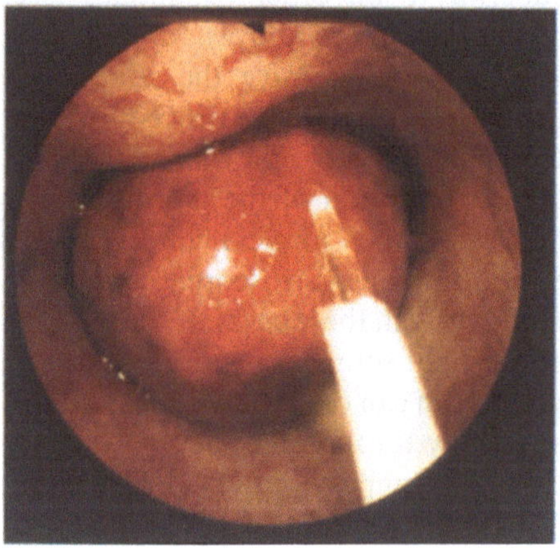

b

**Fig. 11. a** Resection of a myoma with the Nd:YAG contact laser via "bare fiber"; **b** Submucous, pedunculated myoma: attempt at resecting the peduncule with the Nd:YAG contact laser via "bare fiber" (diameter of the fibroid too large for a laser procedure)

For larger fibroids, especially sessile ones, this procedure is time consuming, however, and often impossible due to impaired handling of the laser fiber in the hysteroscopic shaft. For sessile fibroids a combined laser-HF therapy, consisting in removal of the bulk with the resectoscope and vaporization of the base with the laser, seems the treatment of choice. Wamsteker (personal communication, 1990) advocates that for larger sessile submucous fibroids, HF electroendosurgery with the resectoscope is the most effective procedure, and tissue for histologic evaluation is produced. In cases of pedunculated submucous myomas, a stalk that is easily accessible may be cut with the laser hysteroscopically; otherwise, laparoscopy has to be performed (Fig. 11a, b).

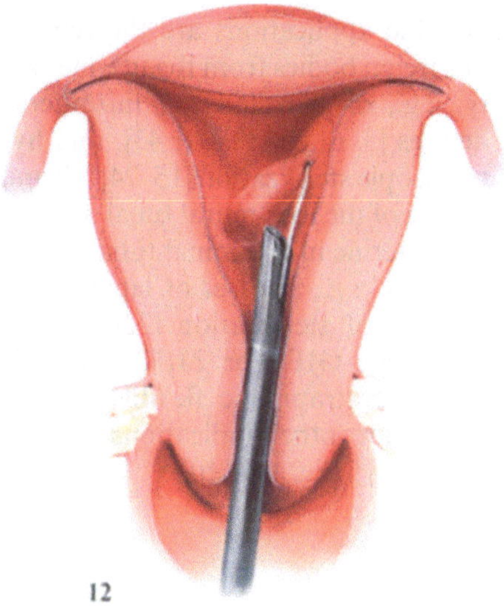

12

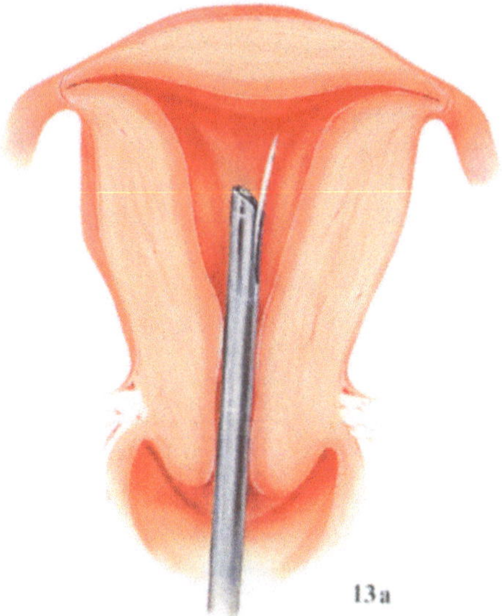

13a

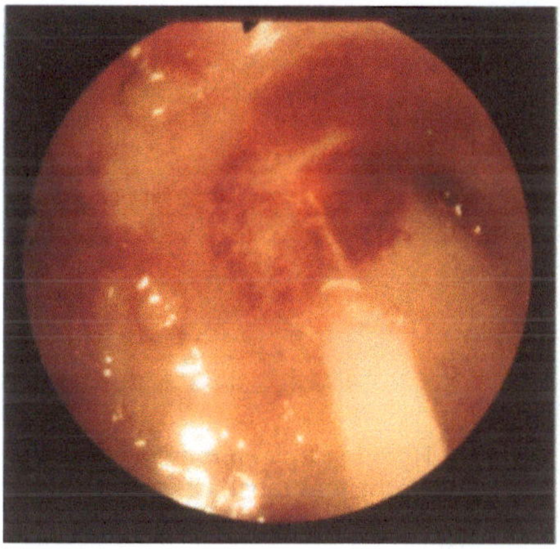

b

**Fig. 12.** Removal of a polyp by means of the "bare fiber" Nd:YAG contact technique

**Fig. 13. a, b.** Endometrial ablation with the Nd:YAG contact laser

## Success Rates

Donnez and co-workers [14] successfully (control hysteroscopy or hysterosalpingography 6 weeks after surgery) performed Nd:YAG laser myomectomy in 60 patients (fibroid, mainly in the uterine cavity) after preoperative GnRH therapy (reduction in size approximately 30%). For larger myomas they advocate a two-step therapy: 8 weeks GnRH treatment, partial laser myomectomy, 8 weeks hormone therapy, and subsequent second-look hysteroscopy. In any case, a preoperative biopsy is essential before laser procedures.

In our collectives smaller myomas were successfully removed by means of the Nd:YAG contact bare-fiber technique in almost all cases. The same management proved unsatisfactory for larger lesions, however (no pregnancy). Hysteroelectroresection seems more ef-

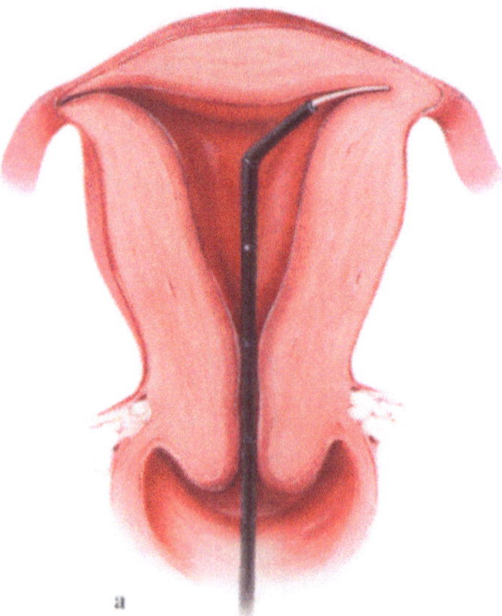

a

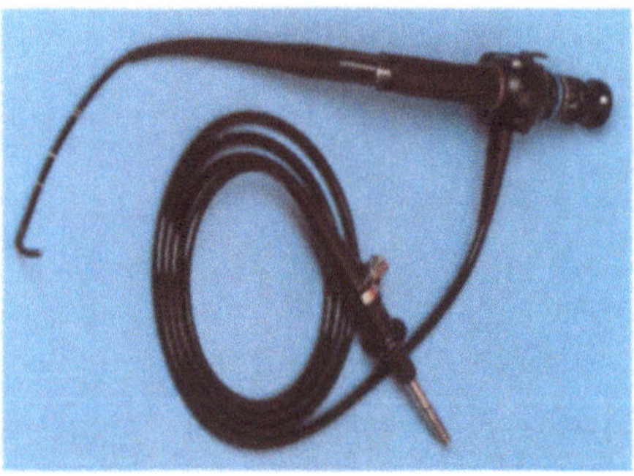

b

**Fig. 14. a** Hysteroscopic splinting/catheterization of the proximal tubal aspect via flexible hysteroscope (Olympus); **b** Flexible hysteroscope (Olympus)

ficient. For final evaluation, larger series of patients and a longer follow-up are necessary.

### Other Indications

Hysteroscopy is an effective approach to the management of polyps (Fig. 12) or to endometrial coagulation or ablation (Fig. 13) [4, 9, 30, 31, 32]. Other possible applications are the removal of foreign bodies from the uterine cavity, e.g., of so-called lost intrauterine devices [3, 6, 23, 38, 41, 43], and hysteroscopic sterilization [15, 24]. Examination of the intramural tubal sections, transuterine intraluminal tubal manipulations, e.g., in cases of laparoscopic reversal of sterilization [22], or hysteroscopic GIFT or ZIFT (especially with the atraumatic flexible hysteroscope) are other indications for hysteroscopy (Fig. 14a, b).

## Conclusions

Within the framework of the newly developed concept of "minimal invasive surgery" less invasive hysteroscopic procedures have replaced laparotomy or uterotomy. A further advantage is that spontaneous vaginal delivery is possible in most cases after a hysteroscopic procedure.

Optimal technical equipment, the continuous awareness of the risks inherent to hysteroscopic procedures, disregard of which may even be lethal, and intensive endoscopic training of the surgeon are mandatory.

## References


1. Assaf A, Serour G, Elkady A, Agizy el H (1990) Endoscopic management of intrauterine septum. Int J Gynecol Obstet 32:43–51
2. Baggish MS, Daniell JF (1989) Catastrophic injury secondary to the use of coaxial gas-cooled fibers and artificial sapphire tips for intrauterine surgery: a report of five cases. Lasers Med Surg 9: 581–584

3. Bartfai G, Barad DH, Kaali SG, Feinman M (1988) Video hysteroscopic removal of "lost" intrauterine devices during the first trimester of pregnancy. J Reprod Med Obstet Gynecol 33:877–878
4. Bent AE, Ostergard DR (1990) Endometrial ablation with the neodymium:YAG laser. Obstet Gynecol 75:923–925
5. Borten M, Seibert CP, Taymor ML (1983) Recurrent anaphylactic reaction to intraperitoneal dextran 75 used for prevention of postsurgical adhesions. Obstet Gynecol 61:755
6. Burmucic R, Mayer HO, Kometter R (1987) Diagnosis and treatment of lost intrauterine devices. Wien Med Wochenschr 137:104–107
7. Buttram VC, Reiter RC (1981) Uterine leiomyomata: etiology, symptomatology, and management. Fertil Steril 36:433–445
8. Cornier E (1982) La fibroscopie en gynecologie: la fibro-hysteroscopie et al fibrotuboscopie. Nouv Presse Med 11:2841–2843
9. Daniell J, Tosh R, Meisels S (1986) Photodynamic ablation of the endometrium with the Nd:YAG laser hysteroscopically as a treatment for menorrhagia. Colpo Gynecol Laser Surg 2:43–46
10. DeCherney AH (1984) Hysteroscopic management of müllerian fusion defects. In: Siegler AM, Lindemann HJ (eds) Hysteroscopy. Principles and practice. Lippincott, Philadelphia, p 204
11. DeCherney AH, Russell JB, Graebe RA, Polan ML (1986) Resectoscopic management of müllerian fusion defects. Fertil Steril 45:726
12. Donnez J (1989) Instrumentation. In: Donnez J (ed) Laser operative laparoscopy and hysteroscopy. Nauwelaerts, Leuven, pp 207–221
13. Donnez J, Nisolle M (1989) Operative laser hysteroscopy in mullerian fusion defects and uterine adhesions. In: Donnez J (ed) Laser operative laparoscopy and hysteroscopy. Nauwelaerts, Leuven, pp 249–261
14. Donnez J, Gillerot S, Nisolle M, Bourgonjon D, Clerckx F (1989) Laser hysteroscopy in submucous fibroids. In: Donnez J (ed) Laser operative laparoscopy and hysteroscopy. Nauwelaerts, Leuven, pp 263–276
15. Donnez J, Malvaux V, Nisolle M, Casanas F (1989) Hysteroscopic sterilization. In: Donnez J (ed) Laser operative laparoscopy and hysteroscopy. Nauwelaerts, Leuven, pp 289–298
16. ECRI (1989) Fatal gas embolism associated with intrauterine laser surgery. Health Devices 18:325–326
17. Edström K, Fernström I (1970) The diagnostic possibilities of a modified hysteroscopic technique. Acta Obstet Gynecol Scand 49:307
18. Fayez J (1986) Comparison between abdominal and hysteroscopic metroplasty. Obstet Gynecol 68:399
19. Fayez JA (1989) Diagnostic and operative hysteroscopy. In: Sanfilippo JS, Levine RL (eds) Operative gynecologic endoscopy. Springer, Berlin Heidelberg New York Tokyo, pp 251–262
20. Gallinat A (1984) Carbon dioxide hysteroscopy: principles and physiology. In: Siegler AM, Lindemann JH (eds) Hysteroscopy. Principles and practice. Lippincott, Philadelphia, pp 45–47
21. Hamou J (1981) Microhystéroscopie: Mises à Jour Collège Francais. Vigot Press, Paris
22. Hamou J, Gasparri F, Scarselli GF et al. (1984) Hysteroscopic reversible tubal sterilization. Acta Eur Fertil 15:123
23. Hepp H (1977) Zum Problem des "verlorenen" Intrauterinpessars. Geburtshilfe Frauenheilkd 37:653–659
24. Hosseinian AH, Morales WA (1983) Clinical application of hysteroscopic sterilization using uterotubal junction blocking devices. In: Zatuchni GI, Shelton JD, Goldsmith A et al. (eds) Female transcervical sterilization. Harper & Row, Philadelphia, p 234
25. Jedeikin R, Olsfanger D, Kessler I (1990) Disseminated intravascular coagulopathy and adult respiratory distress syndrome: life-threatening complications of hysteroscopy. Am J Obstet Gynecol 162:44–45
26. Jones HW (1981) Reproductive impairment and the malformed uterus. Fertil Steril 36:137
27. Levine RL (1989) Myomectomy. In: Sanfilippo JS, Levine RL (eds) Operative gynecologic endoscopy. Springer, Berlin Heidelberg New York Tokyo, pp 133–139
28. Lindemann H (1985) Hysteroscopy for the transcervical resection of the septum uteri. XI World Congress of Obstetrics and Gynecology. Berlin, October, Abstract 08, 28, 04

29. Lindemann HJ, Mohr J, Gallinat A et al. (1976) Der Einfluß von $CO_2$-Gas während der Hysteroskopie. Geburtshilfe Frauenheilkd 36:153
30. Loffer FD (1987) Hysteroscopic endometrial ablation with the Nd:YAG laser using a nontouch technique. Obstet Gynecol 69:679–82
31. Lomano JM, Feste JR, Loffer FD, Goldrath MH (1986) Ablation of endometrium with neodymium-YAG laser: a multicenter study. Colpo Gynecol Laser Surg 2:4
32. Magos AL, Baumann R, Turnbull AC (1989) Transcervical resection of endometrium in women with menorrhagia. Br Med J 298:1209–1212
33. March CM, Israel R (1987) Hysteroscopic management of recurrent abortion caused by septate uterus. Am J Obstet Gynecol 834:42
34. March CM, Israel R, March AD (1981) Hysteroscopic management of intrauterine adhesions. Fertil Steril 36:455
35. Robbins SL, Cotran RS (1984) Leiomyoma. The pathogenic basis of disease, 3rd edn. Saunders, Philadelphia, p 1136
36. Rubin IC (1925) Uterine endoscopy, endometroscopy with the aid of uterine insufflation. Am J Obstet Gynecol 10:313
37. Strassmann EO (1966) Fertility and unification of double uterus. Fertil Steril 17:165–176
38. Tadesse E, Wamsteker K (1985) Evaluation of 24 patients with IUD-related problems: hysteroscopic findings. Eur J Obstet Gynecol Reprod Biol 19:37–41
39. Tompkins P (1962) Comments on the bicornuate uterus and twinning. Surg Clin North Am 42:1049
40. Valle R, Sciarra J (1986) Hysteroscopic treatment of the septate uterus. Obstet Gynecol 67:253
41. Wagner H, Schweppe KW, Beller FK (1983) Fragmentation von Intrauterinpessaren als Komplikation bei der Extraktion. Geburtshilfe Frauenheilkd 42:123–126
42. Wamsteker D, de Blok S (1990) HF-Electrosurgery vs. laser in hysteroscopy. Paper, International Workshop on Operative Endoscopy and Laser, June 22–23, Heidelberg
43. Zakin D, Stern WZ, Rosenblatt R (1981) Complete and partial uterine perforation and embedding following insertion of intrauterine devices. II. Diagnostic methods, prevention, and management. Obstet Gynecol Surv 36:401–417

# The Application of the Nd:YAG Laser in $CO_2$ Hysteroscopy

R. P. LUEKEN [1], A. GALLINAT, and C. P. MÖLLER

## Introduction

For more than 50 years, $CO_2$ has been a proven medium for distention of the uterine cavity in gynecological endoscopy. While for laparoscopy only gaseous media such as air, $CO_2$, or $N_2O$ are used, there are also liquid distention media for hysteroscopy. The intrauterine pressure and flow conditions were first investigated by Gauss in 1928 [8], but it was Lindemann [11] who, following the initiative of Menken [17], developed a gaseous distention method for inspection of the uterine cavity in the late 1960s. This method has become increasingly important during the past few years.

Hysteroscopy in particular is indicated in the diagnosis of intrauterine bleeding, or in the thorough research of reasons for intrauterine sterility and infertility. The combination of this proven, and in the meantime worldwide accepted, surgical method of hysteroscopy with the newly introduced laser technique in gynecological endoscopy provides a number of new approaches to therapy which we will describe in this report.

[1] Zentrum für Gynäkologische Endoskopie, Altonaer Str. 59–61, W-2000 Hamburg 36, FRG.

## Methods and Results

Hysteroscopy as a routine examination in menometrorrhagia and sterility has been used at our center for gynecological endoscopy since 1984. The increase in outpatient surgery including hysteroscopy is shown in Fig. 1. We prefer $CO_2$ hysteroscopy because it provides a fast overview and $CO_2$ is a physiological gas. Furthermore, there are no allergic reactions and no intra-abdominal spillage. For the insufflation a special apparatus like the Metromat must be used, which generates a limited maximum flow of 70 ml/min at a limited pressure of 150 mm Hg [14]. With the Metromat, there are no side effects, no danger of gas embolism, no change

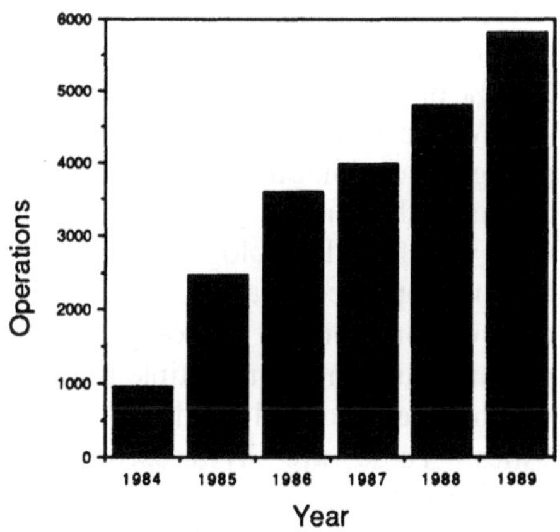


**Fig. 1.** Outpatient operations performed at the center for gynecological endoscopy

Lasers in Gynecology
Edited by G. Bastert and D. Wallwiener

**Table 1.** Endoscopic findings of bleeding disorders in 1519 analysed cases

| Disorder | *n* |
|---|---|
| Polypoid changes | ? |
| Polyps | 201 |
| Myomata | 129 |
| Carcinoma | 62 |
| Synechiae | 17 |
| | 409 (26.9%) |

**Table 2.** Intrauterine pathologies in 137 of 937 patients examined hysteroscopically for sterility

| Finding | *n* |
|---|---|
| Cervical stenosis | 42 |
| Isolated/multiple polyps | 39 |
| Myomata | 36 |
| Septa | 11 |
| Synechiae | 9 |
| | 137 (14.6%) |

in blood gas analysis, and there are no constrictions on the duration of surgery. With limitations there is no danger for the patient; this has been proven by clinical research [12, 13]. Liquid distention media are somewhat more complicated to handle, and with a patent uterotubal system the liquid flows via the tubes into the abdominal cavity. If large quantities of liquid are used fluid overload may result and disturbances of the electrolyte balance may follow, the most serious consequence of which is pulmonary edema. The indications for $CO_2$ hysteroscopy are menometrorrhagia, sterility/infertility, and problems with intrauterine contraceptive devices.

We found intrauterine changes which gave evidence of menometrorrhagia in 26.9% of 1519 cases analyzed (Table 1). We have left out the polypous changes shown in other statistics [5, 20, 21], as all the various typical cyclic patterns were found histologically without hysteroscopic correlation. The only ones of considerable pathological importance are single or mutiple polyps, which have been evaluated in our statistics. These are hysteroscopically clearly different from myomas with respect to shape, color, and vascularization.

The D + C is still the most common type of surgery performed in operative gynecology. Should hemorrhaging persist after a D + C with inconspicuous histological findings, an endoscopic clarification is required in our opinion. Very often, large intrauterine myoma are found that are accessible only by endoscopic therapy.

A complete sterility check includes hysteroscopy. We examined 937 patients hysteroscopically for sterility and found intrauterine pathologies in 137 (14.6%; Table 2). We also found patients with isolated or multiple polyps almost equal in number to those with myomas, followed by those with intrauterine septa. These conditions are responsible for habitual abortions, but also for non-implantation. The uterus septus, not the uterus bicornis, is the reason for infertility or sterility. As neither by hysteroscopy nor by hysterography can a uterus septus be differentiated from a uterus bicornis, laparoscopy can be performed; a non-invasive method is sonography, by which the various double malformations can be clarified [6]. The septate uterus is accessible to hysteroscopic operative therapy, whether conservative with the scissors or with the Nd:YAG laser [3, 19, 22].

The therapy-resistant, hormone-refractory, intractable menometrorrhagia of preclimacteric women is surely a difficult problem for the gynecologist.

Laser menolysis in such cases can reduce or eliminate menstrual bleeding while retaining the organ and avoiding major surgery. The D + C produces only short-term or even no success, and no pathological changes are seen histologically. Very often the hysterectomy is the last resort and entails a certain amount of danger for the patient. The same is true for patients with coagulation disturbances, or for women who carry a high operative risk for hysterectomy such as diabetics, those susceptible to thrombosis, and risk groups 3 and 4 of the ASA classification. For these women, therapeutically induced amenorrhea is advisable via laser menolysis or endometrium ablation, as described in detail below.

Conservative hysteroscopic operations with a scissors and rigid angular instruments often present difficulties; the base of an intrauterine myoma is very often found in the wall and cannot be excised because of the danger of hemorrhage, and perforations are overlooked very easily. The aim is to produce a smooth surface of the uterine cavity.

For intrauterine surgery two main techniques are currently accepted [2, 10, 18, 23], one using the resectoscope and one using laser. The indications for hysteroscopy using the Nd:YAG laser are (a) endometrial ablation (menolysis), (b) vaporization of septa, and (c) vaporization of myoma.

The physical features of the Nd:YAG laser enable the precise destruction of specific tissue. The collimated light is not absorbed by liquid. The penetration depth is 6 mm. The uterus has a wall thickness of 1–2 cm and is hollow organ, which makes laser therapy undangerous. As the laser light is conveyed via a flexible quartz fiber, the Nd:YAG laser is regarded as ideal for gynecological endoscopy. If a gaseous distention medium is used, laser energy of 40–60 W is normally sufficient. If liquid distention media are used the laser should generate a capacity of up to 100 W. For 4 years we have been using an Nd:YAG laser, type Medilas 40–60N produced by the MBB Company, which is extremely compact and is mobile in the operating room. The apparatus is extremely easy to handle as disposable probes are now available. When an intrauterine myoma is found which is accessible to laser therapy, the probe is led into the working channel of the scope after calibration of the laser probe. The pilot laser is switched on and then the myoma is vaporized step by step with 40 W of energy. For vaporization we use the touch as well as the nontouch technique. We work with a coated Teflon quartz-glass fiber 0.6 mm in diameter (outer diameter 1.4 mm). If the probe tip comes into contact with the mucous membrane debris or with burnt particles it may become blunt and the concentration of laser light is no longer sufficient. The fibers can be shortened or resharpened, or shortened only at the head. The disadvantage of $CO_2$ hysteroscopy with simultaneous YAG laser application is the generation of smoke by vaporization of a large volume of tissue. In fact, the smoke is made turbulent and condensed by the permanent presence of $CO_2$ gaseous fluid; however, if the smoke remains some time in the fundus the operation will be prolonged and an operating time of more than 1 h for large myomas is necessary. Sometimes several probes are needed. When liquid distention methods are used no smoke is generated, but a known disadvantage is fluid overload. It should be mentioned that fluid overload does not

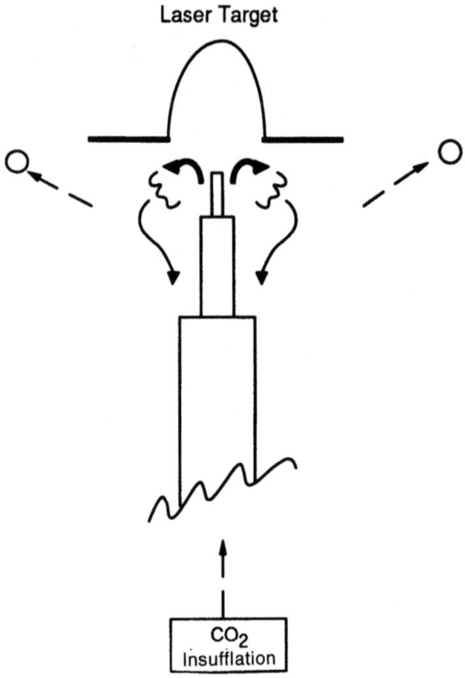


**Fig. 2.** $CO_2$ insufflator with smoke evacuation for hysteroscopic surgery using Nd:YAG laser

occur when both tubes have been closed by sterilization or other changes. Together with the Richard Wolf Co. we have developed a $CO_2$ insufflator which at the same time filters the smoke out of the distending gas by means of a separate circulation system [7]. With guided insufflation of maximally 70 ml/min the smoke is simultaneously filtered out at a circulation capacity of 1 l gas per min (Fig. 2). With this newly developed apparatus the operating time can be reduced. Another advantage is that the tip is cooled and the smoke is pressed aside on the spot as the gas is insufflated via the laser probe, and then filtered via a separate cleansing channel.

The therapy-resistant periclimacteric hemorrhage or the permanent menometrorrhagia of a patient with a coagulopathy and the need to avoid the operative risk for women who need a hysterectomy have produced a completely new indication – the so-called photovaporization of the endometrium. This was reported first in 1981 [9]. In the meantime, this method has been published in the United States by Lomano [16] and Loffer [15], in France by Cornier [1], and in Switzerland by Dequesne [4]. We ourselves have experience with laser menolysis. However, for ablation of the endometrium other procedures have also been developed, in particular the resection of the endometrium by the electric sling. As monopolar electricity is simultaneously used to stop hemorrhage and to improve cutting efficiency, here again only a liquid distention medium can be employed. Another method is thermocautery by means of a roller ball which is led over the endometrium. With all three methods the Asherman-Fritsch-syndrome is obtained, but due to the side effect of the necessary liquid distention medium (fluid overload) using the resectoscope as well as the roller balls, these two methods are not performed by us. However, substantial costs to the public health services could be saved by avoiding hysterectomy. All hysteroscopic laser operations described are outpatient treatments. This means a shorter convalescence, reduced trauma, and fewer costs.

Prior to laser menolysis, a malignancy must be definitely excluded by hysteroscopy and a D + C. Family planning should be completed, as sterility will inevitably follow from the subsequent amenorrhea. As the endometrium thickness, depending on the cycle, is between 1 and 4 mm and the penetration depth of the Nd:YAG laser is 4–6 mm, it is essential to carry out a pretreatment to thin the endometrium,

**Table 3.** Duration of laser surgery

| Scope length (cm) | Range (min) | Mean (min) |
|---|---|---|
| < 8 | 15–30 | 21 |
| 8–12 | 30–60 | 45 |

such as administration of danazol or a GnRH analogue Zoladex. After the patient has taken 600 mg danazol per day for 6 weeks an average thickness of the endometrium of 1.2 mm is obtained. On inspection of the cavity the endometrium then presents itself as consistently white and bleeds only very little if touched. After induction of the laser probe beginning at the fundus, the endometrium is vaporized/coagulated with an output of 40 W. Any endometrial islands (adenomyosis) located under the basic membrane are coagulated at the same time. The menolysis is carried out using three different techniques: the touch technique for the ostia, air-brushing at the fundus, and "dragging" up to the ostium internum of the cervical canal, which should not be treated in order to avoid cervical cicatrization. The length of the operation depends on the size of the uterine cavity. With a probe length of less than 8 cm the average time of operation is 21 min; with a probe longer than 8 cm it can be up to 60 min (Table 3). These average times are for operations carried out without the smoke filter pump.

In 53% of patients thus treated complete amenorrhea followed; in 41.5% a clear hypomenorrhea has been reported, together with an absolute decline of menstrual complaints. There were two failures: In one case a hysterectomy has since been performed; the patient rejected a second application of the laser. The other patient underwent laser hysteroscopy of the endometrium a second time. As the operation is extremely painful to the patient under local anesthesia due to the heating of the myometrium and the peritoneum above, we prefer to use general anesthesia for this type of surgery. Only one intrauterine septum has been lasered with the patient under local anesthesia.

We have treated 16 large intrauterine myomas with laser therapy alone. While those of 3–4 cm could be vaporized in one session, the size and poor view of a myoma that completely fills the cavity does not always allow for treatment in one session. For one patient vaporization of a myoma of 8 cm filling the total cavity required three sessions. Two patients experienced relatively heavy bleeding during vaporization; even simultaneous coagulation of the vessels by laser could not prevent it. The bleeding was stopped conservatively by intrauterine tamponade and the patients were discharged 12 h later. In all these cases a planned hysterectomy was avoided. In one case, two fibroids 2 cm in diameter were vaporized. At a second look 3 months later, one of these had grown again to double its size.

With advances in neurosurgery and in hysteroscopic laser therapy, new instruments have been developed, especially for the treatment of intramural myomas, where the laser probe is led directly into the center of the myoma via a metal head and the myoma is melted from the inside.

## Conclusion

Contrary to the situation in the United States, further development of intrauterine laser therapy is not favored in the Federal Republic of Germany because (a) hysteroscopy is not generally accepted and taught at medical schools, and (b) the Nd:YAG laser is relatively expensive to purchase and to maintain. When a laser plant is planned, a decision has to be taken by the hospitals and outpatient clinics as to which laser for which purpose should be purchased. We are convinced that the Nd:YAG laser is applicable for intrauterine therapy.

## References


1. Cornier E (1986) Traitement hystérofibroscopique ambulatoire des métrorragies rebelles par laser Nd:YAG. J Gynecol Obstet Biol Reprod 15:661
2. Daniell JF (1989) Hysteroscopic laser surgery. In: Sanfilippo JS, Levine RL (eds) Operative gynecologic endoscopy. Springer, Berlin Heidelberg New York, pp 263–271
3. Daniell JF, Osher S, Miller W (1987) Hysteroscopic resection of uterine septa with visible light laser energy. Colpo Gynecol Laser Surg 3:217–220
4. Dequesne J (1987) Traitement hystéroscopique des hemorragies utérines au Nd:YAG laser. Rev Med Suisse Romande 107:405
5. Deutschmann C, Lueken RP, Lindemann HJ (1983) Hysteroscopic findings in postmenopausal bleeding. In: Siegler HM, Lindemann HJ (eds) Hysteroscopy: principles and practice. Lippincott, Philadelphia
6. Fedele L, Ferrazzi E, Dorta M, Vercellini P, Candiani GB (1988) Ultrasonography in the differential diagnosis of "double" uteri. Fertil Steril 50:361
7. Gallinat A, Lueken RP, Möller CP (1989) A new smoke cleaner in $CO_2$ hysteroscopy for Nd:YAG laser application. Proc. 18th annual meeting American Association of Gynecologic Laparoscopists, Washington DC, Sept. 15–19
8. Gauss CD (1928) Hysteroscopie. Arch Gynakol 133:18
9. Goldrath M, Fuller T, Segal S (1981) Photovaporization of endometrium for the treatment of menorrhagia. Ann J Obstet Gynecol 140:14
10. Hallez JP (1986) Function endoscopique des fibromyomes uterin. Une nouvelle technique. In: Netter A, Gorins P (eds) Actualités gynecologiques. Masson, Paris
11. Lindemann HJ (1977) The use of $CO_2$ in the uterine cavity for hysteroscopy. Int J Fertil 17:221
12. Lindemann HJ, Gallinat A (1976) Physikalische und physiologische Grundlagen der $CO_2$-Hysteroskopie. Geburtshilfe Frauenheilkd 36:729
13. Lindemann HJ, Mohr J, Gallinat A et al. (1976) Der Einfluß von $CO_2$-Gas während der Hysteroskopie. Geburtshilfe Frauenheilkd 36:153
14. Lindemann HJ, Gallinat A, Lueken RP (1979) Metromat. A new instrument for producing pneumometra. J Reprod Med 23:73
15. Loffer FD (1987) Hysteroscopic endometrial ablation with the Nd:YAG laser using a non-touch technique. Obstet Gynecol 69:679
16. Lomano JM, Feste JR, Loffer FD, Goldrath MH (1986) Ablation of endometrium with neodymium: YAG laser: a multicenter study. Colpo Gynecol Surg 2:4
17. Menken FC (1967) L'endocerviscopie. Bull Soc Sci Med Grand Duche Luxemb 104:97
18. Neuwirth RS, Amin HK (1976) Excision of submucous fibroids with hysteroscopic control. Ann J Obstet Gynecol 126:95
19. Perino A, Mencaglia T, Hamou I, Cittadini E (1987) Hysteroscopy for metroplasty of uterine septa: report of 24 cases. Fertil Steril 48:321
20. Peterson WF, Novak ER (1956) Endometrial polyp. Obstet Gynecol 8:40–49
21. Taylor PJ, Gomel V (1986) Endoscopy in patient with abnormal vaginal bleeding. In: Gomel V, Taylor PJ, Yuzpe AA, Rioux JE (eds) Laparoscopy and hysteroscopy in gynecologic practice. Year Book Medical, Chicago
22. Valle RF, Sciarra JJ (1986) Hysteroscopic treatment of the septate uterus. Obstet Gynecol 67:253
23. Wamsteker K (1983) Hysteroscopic surgery. In: van der Pas H, Hernendael BV, von Lith J, Keith L (eds) Hysteroscopy. MTP Press, Boston, p 165

# HF Electrosurgery Versus Laser in Hysteroscopy

K. WAMSTEKER[1] and S. DE BLOK

## Introduction

Intrauterine hysteroscopic diagnosis has opened up completely new avenues for endosurgical treatment of intrauterine disorders. Newly developed instruments and special adaptions of endosurgical instruments from other disciplines provide for a still rapidly expanding range of surgical possibilities in the uterine cavity.

The most important methods for intrauterine surgery are electroendosurgery with the resectoscope and Nd:YAG laser endosurgery; some of the techniques have already been used for a long time in urology and gastroenterology. The resectoscope is equipped with a sliding loop for high-frequency (HF) electrosurgery with an HF generator. Nd:YAG laser endosurgery can be performed with flexible or rigid hysteroscopes; the laser light is transmitted via quartz fibers.

## Instruments

The resectoscopes are equipped with a 0°, 12°, or 30° foreoblique telescope, a working element with the resection loops, and an outer sheath. The outer sheath can be single (24 F) for a single-flow system or double (27 F) for a continuous flow system. Different electrodes can be used: a loop, a knife, or a roller-ball electrode. The HF electrosurgery generator must provide both cutting and coagulating current.

For Nd:YAG laser surgery a 4.9 mm flexible fiberoptic hysteroscope with tip-bending capabilities and a combined 6 F working/irrigation channel can be used. With a rigid endoscope an operating hysteroscope has to be used. This is equipped with a working bridge with or without a deflection device and a 7 F working channel for the laser fiber. The working bridge is covered by an outer sheath of 7–8 mm.

## Distension Media

For HF electrosurgery the uterine cavity must be distended by a nonconducting fluid. This can be a low-viscosity fluid (LVF) like sorbitol 4% or glycine 1.5% or a high-viscosity fluid (HVF) like 32% dextran 70. Special pressure systems have to be used, especially with LVF. Dextran cannot be used with coagulation because it becomes caramelized, resulting in insulation of the electrode.

Hysteroscopic laser endosurgery can be performed with $CO_2$ gas or LVF dis-

[1] Hysteroscopy Training Center, Spaarne Hospital Haarlem, PO Box 1644, 2003 BR Haarlem, The Netherlands.

Lasers in Gynecology
Edited by G. Bastert and D. Wallwiener

tension of the uterine cavity. With $CO_2$ gas a special smoke evacuation system has to be used. In both techniques special emphasis must be put on safety precautions, because of the intravasation of the distention medium. Fluid overload with serious circulatory, pulmonary, and electrolyte disturbances has occurred, with fatal accidents. There have also been some deaths because of $CO_2$ embolus formation, due to improper equipment or systems for insufflation. The procedures must be limited in time and in the amount of distention medium used.

## Advantages and Disadvantages of HF Electroendosurgery and Laser Endosurgery

Both types of endosurgery have advantages and disadvantages.

### Advantages

The equipment needed for HF electroendosurgery is generally available in the operating room and the procedure can be used for rather extensive intrauterine pathologies. Another advantage of HF surgery is that the resected tissue can always be removed from the uterine cavity for histology (except with endometrial coagulation).

Laser endosurgery produces almost no bleeding and is a rather clean procedure. Cutting and coagulation can be performed very precisely.

### Disadvantages

With HF electroendosurgery large underlying vessels may be transected during tissue resection, which causes bleeding and necessitates increased intrauterine pressure. This again leads to increased intravasation of the distention medium. The tissue "chips" produced by resection have to be removed, which can be cumbersome.

The equipment for laser endosurgery, on the other hand, is expensive; the effective depth of the laser is uncertain, and it produces tissue carbonization. The procedure can be rather time consuming, depending on the technique used. With endometrial ablation the tissue remains in the uterine cavity, and this can produce prolonged discharge. Laser endosurgery appears not to be suitable for extensive intrauterine pathologies.

## Procedures

The most important procedures in which HF electroendosurgery and laser surgery are competitive are myomectomy of submucous fibroids, transection of uterine septa, and endometrial ablation.

The techniques for laser "myomectomy" are still being developed. Small fibroids are usually vaporized easily, but this does not produce tissue for histology. The possibilities of laser treatment of large submucous fibroids are still uncertain.

Almost all submucous fibroids can be resected with the resectoscope and electrosurgery; with large fibroids, however, two sessions are sometimes necessary. The tissue removed can be examined histologically.

Uterine septa can be transected excellently with either the electrosurgical knife or laser surgery. The two techniques are almost equal in both performance and results.

For endometrial ablation different techniques are applied.

HF electrosurgery techniques include complete or partial resection, complete or focal roller-ball coagulation, and a combination of resection and coagulation. The resection should not be too deep in the myometrium to prevent bleeding of large vessels. With resection the endometrium "chips" have to be removed from the uterine cavity and can be examined histologically. With roller-ball coagulation the tissue remains in the uterus or is removed with a curette.

Laser surgery techniques include the "dragging" touch technique, the "pointwise" or "blanching" non-touch technique for complete or focal ablation. The Nd:YAG laser penetrates the endometrium and exerts its maximum coagulating effect in the superficial layer of the myometrium with coagulation of the large radial arteries (front scatter). Backscatter produces its effect in the more superficial layers (popcorn effect). In laser ablation the coagulated endometrial tissue remains in the uterine cavity or is removed with a curette.

Although the results of the majority of the different techniques employed appear to be good, a long-term follow-up covering more than 5 years is not yet available, and comparative studies need to be performed in order that their results, advantages, and disadvantages can be evaluated. Several points remain to be settled:

- Necessity for and type of preoperative medication
- Postoperative management
- Incidence and extent of postoperative intrauterine adhesions
- Long-term results

## Conclusion

Both HF electroendosurgery and laser surgery seem to be very effective for the transection of uterine septa and endometrial ablation. The application of one or the other technique today appears to be more a matter of personal preference.

For the resection of larger submucous fibroids especially and of large endometrial polyps electrosurgery appears to be more useful than laser surgery. Small endometrial polyps and uncomplicated intrauterine adhesions should preferably be treated with conventional or laser endosurgery and not primarily with HF electrosurgery.

# Pilot Study on Transcervical Resection of the Endometrium

R. BAUMANN [1], A. L. MAGOS, and SIR A. C. TURNBULL

Unacceptably heavy or prolonged menstrual blood loss is a common complaint of women in their reproductive years and an important cause of iron deficiency anaemia in up to 25% of women [20]. In most cases the increase in menstrual blood loss cannot be explained; neither local abnormalities such as endometrial polyps and fibroids, nor systemic disorders such as hypothyroidism and blood dyscrasias are found [4]. Traditionally, patients are treated with drugs for symptomatic relief after a normal curettage. If medical treatment fails, they are referred for hysterectomy. In fact, poorly controlled menorrhagia has become the primary indication for hysterectomy, which is the most commonly performed major operation for women during their reproductive years [10].

A simpler operation to treat patients with abnormal menstruation was proposed by DeCherney and Polan in 1983, using a cystoscope-resectoscope to destroy the endometrium by surgical diathermy. Their patients were unfit to undergo hysterectomy because of anaesthetic risk factors or blood dyscrasias. After electrosurgery, most of them either became amenorrhoeic or had slight spotting for up to 5 years of follow-up [7].

Encouraged by these early reports, we decided to undertake a prospective pilot study of transcervical hysteroscopic resection of the endometrium as an alternative to hysterectomy in fit women with menorrhagia.

## Patients and Methods

We recruited 16 women from the gynaecological outpatient department at the John Radcliffe Hospital in Oxford. All patients complained of unacceptably heavy and prolonged periods (Table 1). Fourteen of them had tried unsuccessfully to control their blood loss with various drugs, including progestogens, danazol or mefenamic acid. The uterus was in no case larger than in a pregnancy of 8 weeks. All but four women had undergone diagnostic curettage or endometrial aspiration biopsy within the previous 12 months, showing no pathology. The women

**Table 1.** Characteristics of 16 parous women treated with transcervical resection of the endometrium

| Characteristic | Mean | Range |
|---|---|---|
| Age (years) | 42.7 | 28–54 |
| Length of periods (days) | 11.2 | 5–30 |
| Length of cycle (days) | 26.8 | 14–30 |
| Duration of menstrual complaints (months) | 33.9 | 4–72 |

[1] Abt. für Gynäkologie, Clemenshospital, Düesbergweg, W-4400 Münster, FRG.

Lasers in Gynecology
Edited by G. Bastert and D. Wallwiener

were eager to avoid hysterectomy but appreciated that the proposed treatment might fail. They were advised that the operation might not guarantee sterility.

At the time of operation eight women were taking drugs to reduce their menstrual blood loss; these were stopped postoperatively. We did not prescribe hormones preoperatively to ensure a thin endometrium at the time of surgery and did not confine the operation to a specific phase of the patient's cycle. Twelve of our patients opted for total endometrial resection aiming at amenorrhoea; four women preferred partial resection to preserve light periods. In this pilot series, general anaesthesia was used in all patients. The procedures were carried out in the lithotomy position with an unmodified urological resectoscope fitted with a cutting loop, two sheaths for continuous irrigation and suction (outer diameter 26 F) and a 4-mm forward-oblique telescope (Karl Storz, Tuttlingen, FRG). We used blended diathermy current of 100 W for cutting and 50 W for coagulation (Valleylab force II, United Kingdom). The uterine cavity was distended with 1.5% glycine solution at a pressure of 100 mm Hg in the glycine reservoir and a pressure of 80 mm Hg in the suction bottle. Glycine in- and outflow was carefully monitored throughout the procedure.

A video camera was attached to the eyepiece of the resectoscope to allow operating from a monitor. We first inspected the uterine cavity for focal lesions or any suspicious areas. We then started the resection in the cornual region, followed by the fundus, the posterior wall, the lateral walls of the uterus, and finally the anterior wall. This approach facilitated systematic resection of the endometrium. In order to remove deep glandular crypts, we aimed at resecting 1 or 2 mm of myometrium as well. We continued the operation up to the upper part of the endocervical canal for total resection. In case of partial resection, a 0.5-cm cuff of endometrium was left above the internal cervical os. Resected debris was removed from the cavity with polyp forceps or conventional curettes, and all material was sent for histological examination. After resection, the cavity was carefully checked for haemostasis and residual endometrium was removed. A single intravenous dose of cefuroxime was given as antibiotic prophylaxis. In two women, resection was followed by laparoscopy for sterilization. We discharged our patients as soon as they felt well enough and reviewed them in the clinic after 6 weeks and 3 and 6 months. We assessed menstrual symptoms before and after resection both retrospectively and prospectively. We asked the patients to fill in menstrual calendar forms rating their vaginal bleeding, vaginal discharge and lower abdominal pain daily on a scale from 0 to 3 (0 = none, 1 = mild, 2 = moderate, 3 = severe). Data comprised at least one or two menstrual cycles before surgery and follow-up for a maximum of 6 months in this pilot series. Outpatient hysteroscopy and endouterine biopsy were carried out 3 months after resection in 13 of the patients.

## Results

We completed 15 of the 16 operations successfully, including one case in which a 3-cm submucous fibroid was resected along with the endometrium.

The failed procedure was in a woman with an acutely retroverted and retroflexed uterus who sustained a small perforation in the anterior wall requiring laparotomy. No damage to bowel or other pelvic organs was found and the resection was abandoned. In the 15 successful cases histological examination of the resected tissue confirmed normal endometrium with underlying myometrium; evidence of adenomyosis was found in three specimens. The mean duration of the operation was 41 min (range 20–100 min); it was reduced to 28 min (range 20–35 min) for the last five resections. The mean volume of glycine required for distension and irrigation was 7.1 l (range 2.3–20.9 l). The mean difference between glycine in- and outflow at the end of surgery was 523 ml (range 200–1000 ml). There was a tendency to lower readings, meaning less glycine absorption in patients with previous sterilization. Clinical estimates of blood loss during the operation ranged from 100 to 300 ml. The only postoperative complication was low-grade fever in one case, which resolved spontaneously within 24 h. Thirteen patients were discharged from hospital the day after the operation, the remaining three with successful procedures staying two days. The one woman who required laparotomy left hospital after 6 days. Immediately after surgery, uterine bleeding was light to moderate and it persisted longer than 10 days in only two cases. Postoperative bleeding was followed by inoffensive vaginal discharge for a mean of 17.5 days (range 5–34 days). The women were able to return to domestic activities after a mean of 1.9 weeks (range 0.5–4 weeks); 10 of the 11 who were employed resumed their work after a mean of 2.2 weeks (range 1–3 weeks).

Six of the 11 patients who underwent total endometrial resection became amenorrhoeic during the 6-month follow-up period. Of the remaining five, one had only a single episode of bleeding and four reported regular, but shorter and lighter periods. Partial resection also led to lighter menstruation of shorter duration. Interestingly, menstrual pain improved with endometrial resection, although none of our patients had complained of severe dysmenorrhoea before treatment. Four of the eight women still menstruating after resection reported completely painless bleeding. Analysis of the daily ratings from the menstrual calendars confirmed the reports, the duration of bleeding, the amount of blood loss, and menstrual pain all being improved throughout follow-up after total and partial resection (Figs. 1 and 2). Women who continued to menstruate after the operation experienced a shortening of the episodes of monthly bleeding as well as a reduction in the mean daily loss.

Outpatient hysteroscopy was performed in 13 cases 3 months after surgery, showing that the uterine cavity had not been completely obliterated as in Asherman's syndrome. We saw pale and translucent tissue in the upper part of the uterine cavity obscuring one or both tubal ostia in nine cases. In the four women with partial resection, a narrow ring of normal-looking endometrium was present above the internal os. Endometrium from endouterine specimens in these patients was reported normal on histological examination. There was no clear relation between histological findings and postoperative menstrual history. Microscopic deposits of endometrium were confirmed in three of the five women who had become amenorrhoeic after surgery

1
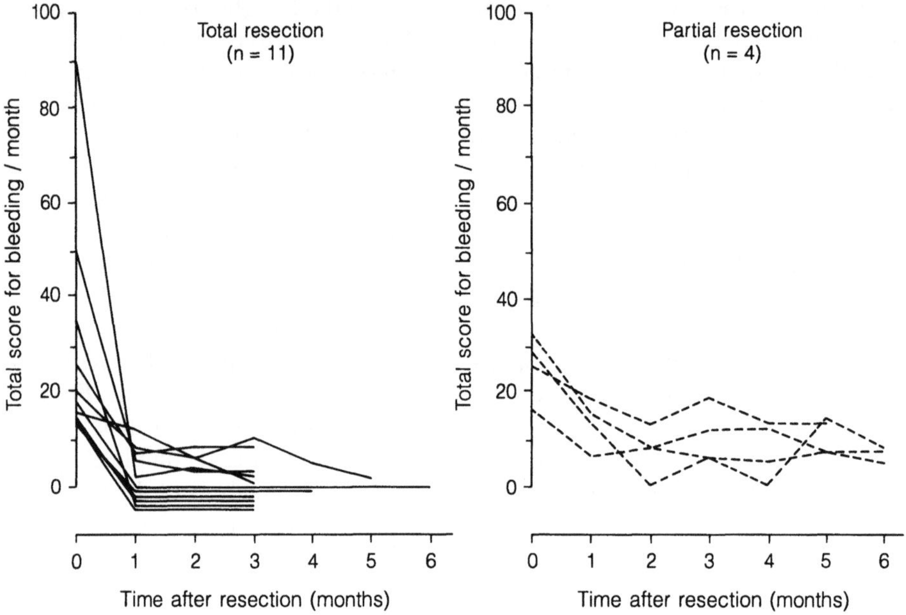


2
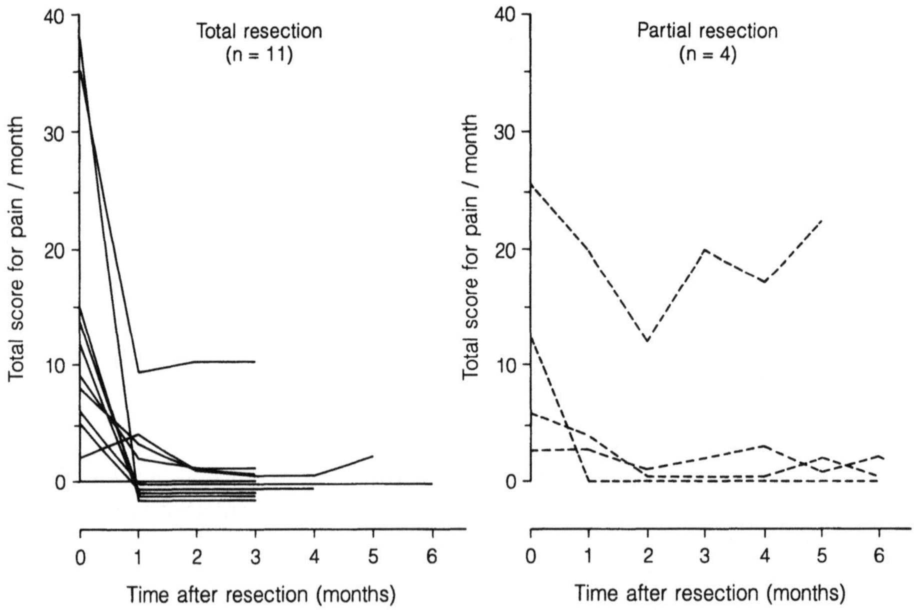


**Fig. 1.** Effect of total and partial endometrial resection on total amount of menstrual bleeding each month. Blood loss was scored daily on a scale of 0–3 (none to heavy)

**Fig. 2.** Effect of total and partial endometrial resection on total amount of menstrual pain each month. Pain was scored daily on a scale of 0–3 (none to severe) and in seven of the eight who continued to menstruate. Multinucleate giant cells of a foreign-body type and histiocytes were described in three specimens and myometrium in one. There was no evidence of squamous metaplasia or atypia in any of the endouterine biopsies.

Twelve of the 15 patients with technically successful procedures were pleased with the results of the operation. The remaining three, two of whom were among the first patients treated, were disappointed in spite of improvements in their periods. Two of these three patients requested hysterectomy; one of them presented for vaginal hysterectomy 4 months after her transcervical resection. Macroscopic examination of the operative specimen showed a uterine cavity lined by greyish-pink, slightly translucent tissue of 1 mm average thickness, with obstruction of the cavity in several places. Histological assessment confirmed the presence of thin endometrium throughout, with occasional points of granulation tissue.

## Discussion

The idea of destroying the endometrium to cure menorrhagia is more than 20 years old. Cryocautery and cytotoxic chemicals were used inside the uterine cavity but did not lead to satisfactory long-term results [11, 24]. Recently, hysteroscopic laser ablation of the endometrium has been described. Since Goldrath et al. published their report in 1981 [12], several studies have confirmed the efficacy of this treatment option. Laser ablation of the endometrium was described as an attractive alternative for outpatient management of menstrual problems [5]. Its main disadvantages seem to be the expensive equipment required, the absence of a histological specimen and longer operative times compared with electrosurgery [1]. As opposed to laser units, the resectoscope is relatively cheap and widely available in hospitals with urological departments. Besides endometrial resection [17], removal of submucous fibroids and uterine septa has been successfully carried out using electrosurgical instruments [6, 18]. Our pilot study showed that otherwise healthy women with prolonged, heavy, or frequent periods may be candidates for transcervical resection of the endometrium as an alternative to hysterectomy. Partial or complete excision of the endometrium with a diathermy loop led to a considerable and consistent reduction in the duration of menstrual bleeding, the amount of blood flow per cycle, and menstrual pain in all women and to amenorrhoea in half the patients after total resection. Our results seem to be comparable to those obtained in studies on laser ablation [16].

As menstrual benefits of endometrial resection persisted for the entire follow-up, they cannot be explained by a simple curettage effect. Typically, menstrual blood loss is reduced only in the first period after dilatation and curettage, but increases again in the following months [14]. This is due to the great regenerative powers of the endometrium. Complete excision with destruction of the basal layer, however, is able to prevent future growth [21]. In a series of 100 patients followed up for 12 months after total or partial endometrial resection, 50% of all those undergoing total transcervical resection became and remained amenorrhoeic (A. L. Magos, R. Baumann, A. C.

Turnbull, unpublished data). In a subgroup of 27 patients, menstrual blood loss was measured before and up to 12 months after surgery. The mean menstrual loss of women who were still menstruating postoperatively was reduced by 75% compared with preoperative measurements (A. L. Magos, R. Baumann, M. P. Rees, A. C. Turnbull, unpublished data). As to why partial endometrial resection reduces menstrual blood loss, we assume that overall reduction in endometrial surface area may be the responsible factor. Although objective assessment of menstrual blood loss in untreated women with a complaint of heavy periods does not correlate with endometrial surface area [4], the correlation may well exist in cases of extensive endometrial ablation.

A further argument in favour of new approaches to treating menorrhagia may be derived from a critical look at hysterectomy and its current role in the management of heavy periods. In 1985, 18600 hysterectomies were carried out in England for menstrual disorders [8]; no histological abnormality is found in over half of such operative specimens, according to a recent audit [13]. Hysterectomy for benign conditions carries a mortality risk of 6.0 per 10000 [23]. A large American study reported a short-term morbidity rate of 24.5 per 100 for vaginal, and 42.8 per 100 for abdominal hysterectomy [9]. Worrying potential long-term effects associated with hysterectomy include an increased risk of cardiovascular disease [3], premature ovarian failure [22], and psychosexual dysfunction [19]. In the light of these problems associated with hysterectomy, as well as financial and social implications, it seems more than attractive to develop and evaluate a minor operative procedure such as endometrial ablation as an alternative to hysterectomy for menorrhagia.

Our results of transcervical resection of the endometrium are preliminary. Long-term follow-up and further studies with objective measurements of menstrual blood loss will be necessary. The problem of potential malignant change within the uterine cavity after resection must be addressed. More immediately, the operative risks associated with endometrial resection should be studied in detail, with special attention given to the risks of haemorrhage, infection, uterine perforation and fluid overload by excessive absorption of solutions for distention and irrigation. A recent study of glycine absorption during transcervical resection confirmed the real risk of rapid and severe hyponatraemia (120 mmol/l) if the glycine difference between in- and outflow exceeds 1000 ml [2]. The risk of uterine perforation may be reducible by using flexible hysteroscopes and a laser unit rather than a rigid resectoscope with diathermy in case of an acutely retroverted uterus. With regard to intrauterine electrosurgery, its similarity to transurethral resection of the prostate may stimulate closer cooperation between gynaecologists and urologists, who have been in the forefront of endoscopic surgery for over 40 years.

As a final remark, may we point to the fact that transcervical resection of the endometrium has already proved to be a beneficial and successful alternative in high-risk patients when hysterectomy was undesirable, dangerous or impossible [15]. Provided further studies confirm the promising preliminary results, minimally invasive endoscopic surgery for endometrial ablation could well become a routine option in the concept of treating menorrhagia.

## References


1. Baggish MS, Baltoyannis P (1988) New techniques for laser ablation in high-risk patients. Am J Obstet Gynecol 159:287–292
2. Baumann R, Magos AL, Kay JDS, Turnbull AC (1990) Absorption of glycine irrigating solution during transcervical resection of endometrium. Br Med J 300:304–305
3. Centerwall BS (1981) Premenopausal hysterectomy and cardiovascular disease. Am J Obstet Gynecol 139:58–61
4. Chimbira TH, Anderson ABM, Turnbull AC (1980) Relation between measured menstrual blood loss and patient's subjective assessment of loss, duration of bleeding, number of sanitary towels used, uterine weight and endometrial surface area. Br J Obstet Gynaecol 87:603–609
5. Cornier E (1986) Traitement hysterofibroscopique ambulatoire des metrorragies rebelles par laser Nd:YAG. J Gynecol Obstet Biol Reprod (Paris) 15:661–664
6. DeCherney AN, Russell JB, Graebe RA, Polan ML (1986) Resectoscopic management of mullerian fusion defects. Fertil Steril 45:726–728
7. DeCherney AH, Diamond MP, Lavy G, Polan ML (1987) Endometrial ablation for intractable uterine bleeding: hysteroscopic resection. Obstet Gynecol 70:668–670
8. Department of Health and Social Security and Office of Population Censuses and Surveys (1985) Hospital in-patient enquiry, 1985. HMSO, London, table P1(MB no.4)
9. Dicker RC, Greenspan JR, Strauss LT et al. (1982) Complications of abdominal and vaginal hysterectomy among women of reproductive age in the United States. Am J Obstet Gynecol 144:841–848
10. Dicker RC, Scally MJ, Greenspan JR et al. (1982) Hysterectomy among women of reproductive age. Trends in the United States, 1970–1978. JAMA 248:323–327
11. Droegemueller W, Makowski E, Macsalka R (1971) Destruction of the endometrium by cryosurgery. Am J Obstet Gynecol 110:467–469
12. Goldrath MH, Fuller TA, Segal S (1981) Laser photovaporization of endometrium for the treatment of menorrhagia. Am J Obstet Gynecol 140:14–19
13. Grant JM, Hussein IY (1984) An audit of abdominal hysterectomy over a decade in a district hospital. Br J Obstet Gynaecol 91:73–77
14. Haynes PJ, Hodgson H, Anderson ABM, Turnbull AC (1977) Measurement of menstrual blood loss in patients complaining of menorrhagia. Br J Obstet Gynaecol 84:763–768
15. Lockwood GM, Magos AL, Baumann R, Turnbull AC (1990) Endometrial resection when hysterectomy is undesirable, dangerous or impossible. Br J Obstet Gynaecol 97:656–658
16. Loffer FD (1988) Laser ablation of the endometrium. Obstet Gynecol Clin North Am 15:77–89
17. Magos AL, Baumann R, Turnbull AC (1989) Transcervical resection of endometrium in women with menorrhagia. Br Med J 298:1209–1212
18. Neuwirth RS (1978) A new technique for an additional experience with hysteroscopic resection of submucous fibroids. Am J Obstet Gynecol 131:91–94
19. Richards DH (1974) A post-hysterectomy syndrome. Lancet 2:983–985
20. Rybo G (1966) Clinical and experimental studies on menstrual blood loss. Acta Obstet Gynecol Scand 45 [Suppl]:1–23
21. Schenker JG, Polishuk WZ (1973) Regeneration of rabbit endometrium following intrauterine instillation of chemical agents. Gynecol Invest 4:1–13
22. Siddle N, Sarrel P, Whitehead MI (1987) The effect of hysterectomy on the age of ovarian failure: identification of a subgroup of women with premature loss of ovarian function and literature review. Fertil Steril 47:94–100
23. Wingo PA, Huezo CM, Rubin GL, Ory HW, Peterson HB (1985) The mortality risk associated with hysterectomy. Am J Obstet Gynecol 152:803–808
24. Zipper J, Medel M, Pastene L, Rivera M (1969) Intrauterine instillation of chemical cytotoxic agents for tubal sterilisation and treatment of functional metrorrhagias. Int J Fertil 14:280–288

# Hysteroscopic Surgery in the Outpatient Clinic

H. van der Pas[1] and T. Vancaillie

## Introduction

The term "outpatient" basically emphasizes the fact that the patient can rise and leave immediately after the intervention; still it should be given a much broader meaning. There is no need at all for doing a hysteroscopic intervention exclusively in a hospital; it can just as well be done in the gynecologist's office. There are several reasons why hysteroscopy is still not a generally accepted office procedure. These reasons lie with the patient as well as with the gynecologist and with the technical equipment at our disposal. Outpatient hysteroscopy is a highly specific gynecological examination that, besides thorough training, also requires highly adapted equipment.

Since all of my hysteroscopic outpatient interventions were done by means of $CO_2$ distention, all of them were conducted with the patient under local anesthesia, for I am convinced of the greater safety of this method as distinct from intervention under total anesthesia. It appears that in all cases where serious complications have occurred, sometimes even resulting in death, the patients were under total anesthesia. A patient under local anesthesia breathes spontaneously; in case of an overdose of $CO_2$ gas in the bloodstream, she will automatically, by means of hyperventilation, try to eliminate this inconvenience. This autocorrection precludes the possibility of serious complications occurring during ambulant hysteroscopy.

[1] Dept. of Obstetrics and Gynecology, St. Elisabeth Hospital, 2300 Turnhout, Belgium.

## Preparation of the Patient

When the operation is carried out as an ambulant and/or outpatient procedure, it is preferable that the patient not have an empty stomach, to prevent vagal reactions. Fearful patients can take an anxiolyticum the night before the operation, or possibly in the morning. In order to avoid spasms during the operation, a prostaglandin synthetase inhibitor is administered 0.5–1 h before the intervention. The patient is asked to empty her bladder before the hysteroscopic procedure is begun. The success of the intervention is directly proportional to the preceding conversation. It is absolutely necessary that the patient know what precisely is going to happen, how much time the operation will take, more or less, and what exactly will be observed.

It is obvious that during the whole intervention the patient has to be comfortable, as far as this is possible in gynecological position. It is of great im-

Lasers in Gynecology
Edited by G. Bastert and D. Wallwiener

portance that during the intervention the gynecologist explain to the patient step by step what is going on precisely, so that the patient need not be frightened when one or another maneuver is carried out.

My personal experience has taught me that the presence of the husband or of another close relative of the patient during the intervention usually implies more advantages than disadvantages. Most patients do not need any escort to go home, where they can resume their normal activities right away. Sometimes slight spasms occur, comparable to menstruation pain.

## Method

As soon as the patient is in gynecological position, and after vaginal examination to determine the position of the uterus, a speculum is inserted. In order to determine whether the cervical canal has to be dilated or not, a semi-rigid plastic urethral catheter with a diameter of 4 mm is introduced into the ostium internum uteri to the point where the resistance of the ostium internum can be felt. Advancing it a little bit more will show whether the catheter can also penetrate this part without any difficulty. If this is the case, then dilatation is completely unnecessary. Next, the vagina will be thoroughly disinfected with Isobetadine, which is not only a bactericide but also antimycotic. It is important to ask the patient if she is allergic to iodine, for if she is, she will also be allergic to Isobetadine. If necessary, paracervical block is administered. A mixture of 7.5 ml Xylocaine 1% with adrenaline 1/100000 with 22.5 ml scandocaine 1% is used as the analgesic. In order to get maximum efficiency it is important to inject very superficially, this means 0.3 cm beneath the vagina wall, at an angle of 5 and 7 o'clock.

As soon as the hysteroscope has entered the uterus the $CO_2$ flow should be reduced to a maximum of 40 ml/min or until the minimum pressure accomplishes a good distension of the uterus wall.

Passage of the hysteroscope through the ostium internum can cause slight pain for the patient, possibly followed by a vagal reaction. The patient should be told that this might happen and warned right before the penetration begins. Further penetration into the uterine cavity must be carried out very slowly and very carefully, for touching the endometrium can lead to bleeding. As soon as possible the left and right ostium tubae are located to provide a clear orientation within the uterine cavity.

## Material

Intrauterine laser surgery, just as the more conventional surgery, requires certain circumstances in order to be fully practicable. It is for instance important that:

1. The uterine walls be distended as much as possible
2. The $CO_2$ blown into the uterus is kept from escaping via the cervical canal
3. Any point in the uterine cavity can be located precisely
4. The surgical instruments can be directed to this point several times in succession
5. The hysteroscope can be moved forward millimeter by millimeter

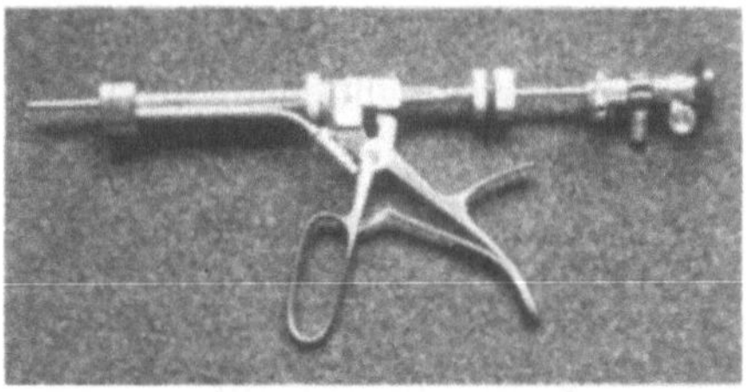

**Fig. 1.** The self-attaching hysteroscope for use with one hand, developed for outpatient hysteroscopy

We use a self-attaching single-hand hysteroscope (Fig. 19) which is extremely suited for outpatient hysteroscopy. The portio-adaptor adheres firmly to the cervix on one side by suction; on the other it is joined by a bayonet catch to a handle through which the hysteroscopic shaft is advanced. Consequently, once firmly attached, the hysteroscope forms a unity with the uterine body.

The portio-adaptor, sucked on by means of vacuum pressure at the portio, prevents even the smallest $CO_2$ leak via the cervical canal. Thus, the uterine walls remain maximally distended. The forward movement or, in other words, the progression in the sagittal plane, can be fixed on any given point by means of a rack system. The circular movement, or the turning movement in the transverse plane, can equally be pinpointed by means of a screw system.

When the two planes transect in a way that a fixed point from the one coincides with a fixed point from the other, then the intersection of both planes is a set point in the uterine cavity to which any conventional instrument as well as any laser beam can be directed.

The complete maneuver can be executed with one hand, the other being free to handle the instruments for surgery advanced through the workshaft. The scope is moved into the ostium externum, guided by its own light intensity. The portio-adaptor attaches automatically. The speculum is carefully removed. Step by step, the hysteroscope is pushed through. Each time the grip is completely closed, the scope advances by 0.5 cm. In reaction to the supply of $CO_2$ under pressure, the cervical canal opens.

## Use of Laser Energy

To assess the usefulness of laser energy in hysteroscopy we need to ask the following question: "What are the deficiences of hysteroscopy, as it is actually applied for diagnosis and treatment, without laser?"

### Diagnosis

Hysteroscopy in essence allows the evaluation of shape, color and consistency of a specific lesion. This type of information is usually sufficient to differentiate between polyps, fibromyomas, congenital anomalies, inflammatory lesions, and exophytic adenocarcinoma with good accuracy. Where hysteroscopy fails quite often is in the differentiation of glandular hyperplasia, adenocarcinoma in situ, and endophytic adenocarcinoma. This is the field where research should be conducted to increase the diagnostic value of hysteroscopy with laser energy.

## Therapy

Gynecologists are so used to performing a hysterectomy for any plausible reason, that very few are aware of the potential of hysteroscopy for the treatment of various conditions, many of them on an outpatient basis: extraction of foreign bodies, synechiolysis, polypectomy, submucous myomectomy, septum resection, sterilization, and ablation of the endometrium. The current trend in hysteroscopic surgery is to investigate whether these techniques can be improved by using laser energy. Whether laser energy indeed improves the technique or the prognosis is debatable, and some hard data are required.

However, the array of interventions can be expanded: treatment of intramural myomas, adenomatous hyperplasia, adenocarcinoma of the endometrium (in particular cases), intramural polyps, and intramural occlusion of the fallopian tube. The introduction of laser in hysteroscopy should be an incentive to expand the current possibilities of hysteroscopic surgery.

Laser represents a major victory in medicine and mainly in endoscopy; we have finally learned to control wavelengths and the intensity of light energy. This is the real revolution in medicine: the mastery of light energy, but we should not regard laser as a replacement for electrosurgery!

The photobiologic properties of human tissue in general and of genital tissues in particular are not yet fully understood. These properties can be divided into three categories:

1. Luminescence (i.e., fluorescence)
2. Photochemical reactivity (i.e., induction of specific nonenzymatic reactions)
3. Thermodynamic effects (i.e., transformation of light energy into thermal energy)

Currently, the bioeffect of lasers which has boosted its widespread use is the thermodynamic effect. Properties such as precision and minimal scarring have indeed proven beneficial in the treatment of cervical dysplasia, eye surgery, and microsurgery. These properties, from a theoretical point of view, will have less impact on intrauterine surgery and none on hysteroscopic diagnosis.

The use of laser energy for diagnosis of conditions such as adenomatous hyperplasia or endophytic adenocarcinoma is in the early phase of research. A team of gynecologists and physicists at the University of San Antonio, Texas, has demonstrated that these conditions can be differentiated by laser-induced fluorescence, at least in a laboratory setup (T. Vancaillie, personal communication). It will take years of research, however, to bring this technology to the office of the private practitioner.

Laser energy in therapeutic hysteroscopy is theoretically applicable for the following surgical procedures: myomectomy, polypectomy, endometrial ablation, and septum resection. All these interventions are performed to date using monopolar current or simply by mechanical means (scissors, forceps), but they can be accomplished either by direct application of the laser beam or indirectly, by heating a sapphire with the laser beam.

There are no studies that compare prospectively the use of laser with the established electrosurgical techniques; most authors merely state that these interventions can be done. It is our experience that the laser does not bring any advantage in treatment of uterine septa or (strictly) submucous myomas.

Concerning endometrial ablation, there is a theoretical advantage of the laser over electrosurgery: safety. Indeed, the destruction caused by the laser is predictable, whereas due to the unpredictable pattern of the electrical current in the monopolar mode, unwanted destruction of extrauterine structures can occur. However, the use of laser for endometrial ablation also presents some potential hazards (e.g., fluid overload). We are currently performing a prospective study comparing endometrial ablation with Nd:YAG laser versus the resectoscope. So far, there is no indication that there will be any difference, but the follow-up period is not long enough.

An application of laser energy still to be investigated is the destruction of intramural or partly intramural fibromyomas. The current technique (when hysterectomy is not desired) is to "shave" or "peel" off the portion of the myoma that is protruding into the cavity, leaving the rest in situ. It is evident that such a technique is suboptimal and carries a high recurrence rate. Laser energy theoretically offers the possibility of in-depth destruction of such a myoma. We have done this with small myomas ($<1$ cm) with apparently good success, i.e., disappearance of the symptoms.

Our efforts should be aimed at opening new perspectives in hysteroscopic surgery. Some of the possible applications are: photocoagulation of localized malignancy in young or medically disabled persons and treatment of intramural disease, such as intramural tubal polyps and intramural occlusion of the fallopian tube.

# Cervical Dilatation by $PGE_1$ Analogue Application Prior to Hysteroscopy

S. RIMBACH [1], D. WALLWIENER, T. RABE, D. POLLMANN, and G. BASTERT

## Introduction

Since the beginning of the century, Hegar rods have been used for mechanical cervix dilatation prior to transcervical interventions. The new techniques of diagnostic and operative hysteroscopy both require cervix dilatation. Long experience with Hegar rods for other procedures has shown that forced mechanical dilatation with forces up to 60 N [1, 19, 24] carries a high risk of traumatization in about 4%–5% of patients [6, 17], followed by isthmocervical insufficiency in further pregnancies [8, 9, 14, 21], with a special risk for nulliparae and multiparac with scarred cervices [8]. Mechanical stretching of the cervical tissue leads to measurable liberation of endogenous prostaglandins, as shown by in vitro studies [10, 13]. This fact generated the idea of using exogenous prostaglandins for softening and dilatation of the cervix.

Today in OB/GYN, prostaglandins are used with success in the pregnant uterus to induce labor, as well as for induction of abortion in the first trimester. Compared with mechanical dilatation, prostaglandin pharmacological cervix dilatation in abortion patients proved to have fewer complications [14] and caused less discomfort in the postoperative recovery phase, which suggests the assumption of less operative trauma [17].

In order to determine the applicability of prostaglandins, in our study using Gemeprost,* a $PGE_1$ analogue, for cervical dilatation prior to hysteroscopy, we investigated (a) how frequently mechanical dilatation for hysteroscopy was necessary at all, and (b) whether sufficient dilatation might be achieved by Gemeprost application in the nonpregnant uterus as well, thus avoiding the risk of possible mechanical traumatization.

## Material and Methods

In 35 infertility patients we applied a new prostaglandin $E_1$ analogue, Gemeprost, for cervical dilatation prior to hysteroscopy. The 1-mg vaginal suppository was administered 1–3 h prior to the procedure. The suppository was positioned in the fornix posterior of the vagina. Stored at $-10\,^{\circ}C$, it must be opened directly before application and administered at room temperature. The insertion of the suppository is simple and was easily done by the nursing staff.

[1] Universitäts-Frauenklinik, Abt. für Geburtshilfe und Gynäkologie, Voßstr. 9, W-6900 Heidelberg, FRG.

* Cergem, Nourypharma [11].

Lasers in Gynecology
Edited by G. Bastert and D. Wallwiener

**Table 1.** Retrospective analysis of 100 sterility patients: necessity of cervix dilatation for diagnostic and operative hysteroscopy in relation to the hysteroscope shaft diameter

| | *n* | Hysteroscope shaft diameter (mm) median (range) | Mechanical dilatation | |
|---|---|---|---|---|
| | | | Necessary | Not necessary |
| | | *n* (%) | *n* (%) | *n* (%) |
| Diagnostic hysteroscopy | 70 | 5 (all) | 5 (7) | 65 (93) |
| Operative hysteroscopy | 30 | 10 (8–12) | 30 (100) | - |
| Total | 100 | 5 (5–12) | 35 (35) | 65 (65) |

Patients with allergic asthma, glaucoma patients, patients with an infection of the vagina or cervix, and women with circulation disorders and prostaglandin allergy were excluded from the study.

## Results

A retrospective analysis of 100 infertility patients was made. The necessity of cervical dilatation differed significantly in the two collectives of diagnostic and operative hysteroscopy (Table 1). In 70 cases diagnostic hysteroscopy had been performed. Shaft diameters of 5 or 7 mm had been used. In 7% of the cases mechanical dilatation had been necessary. In 30 cases operative hysteroscopy had been performed. Here, shaft diameters had been 8, 10, or 12 mm. Dilatation had been necessary in all cases.

In our pilot study collective of 20 diagnostic hysteroscopies no further mechanical dilatation was necessary after Gemeprost application (Table 2). Neither was additional mechanical dilatation needed before operative hysteroscopy ($n = 15$).

Though the effect of Cergem administration lasted 12 h, and optimal cervical softening and dilatation occurred between 3 and 6 h after administration, we obtained the best results after 1 h. In diagnostic hysteroscopy the generally used 5- or 7-mm shafts did not close the cervical canal sufficiently after dilatation. Gas was lost between the dilated cervix and the shaft. In operative hysteroscopy, however, the larger shafts employed sufficiently closed the cervical canal, so that pressure was high enough.

Side effects of Cergem mentioned in the user's manual are vaginal bleeding, uterine pain, and gastrointestinal disorders such as nausea, vomiting, and diarrhea, which are rarely severe; possible are headache, sedation, muscle fatigue, disorder of the central nervous temperature regulation, tachycardia, decrease in blood pressure, bronchioconstriction, thoracic pain, anaphylaxis. In our study uterine pain was recorded in some cases, but no analgetics were required.

## Discussion

Since the end of the 1960s, prostaglandins have been used clinically in OB/GYN. Introduced for induction of labor or abortion, they have now been investigated for cervical dilatation as well [12]. The use of prostaglandins for cervical softening and dilatation could

**Table 2.** Prospective analysis of 35 sterility patients: effect of Gemeprost administration prior to diagnostic or operative hysteroscopy on cervical canal diameter, the necessity of additional mechanical dilatation, and technical feasibility, related to time of application

| | Administration of Gemeprost 2 h preoperatively | | | | |
|---|---|---|---|---|---|
| | *n* | Diameter of the cervical canal (intraoperative) median (range) | Mechanical dilatation necessary | Hysteroscopy could be performed *n* (%) | Hysteroscopy could not be performed (pressure loss) *n* (%) |
| Diagnostic hysteroscopy | 20 | 10 (7–12) mm | - | 4 (20) | 16 (80) |
| Operative hysteroscopy | 15 | 10 (7–12) mm | - | 15 (100) | - |
| | Second intervention: Gemeprost application 1 h preoperatively | | | | |
| Diagnostic hysteroscopy | 16 | 5 (3[a]–8) mm | - | 16 (100) [b]$p < 0.01$ | - |
| Operative hysteroscopy | - | - | - | - | - |

[a] Small shaft diameter, but good softening
[b] Significant in the chi-square test versus values after 3 h

prevent traumatization by mechanical cervix dilatation in up to 5% [6, 17, 19].

First used for labor induction, prostaglandins were thought to imitate natural cervical dilatation during childbirth. Therefore, natural $PGE_2$ and $PGF_2$ were tested for preoperative softening of the cervix, but frequent gastrointestinal side effects, such as diarrhea, vomiting, fever, and labor-like pain occurred. A further disadvantage was the delayed onset of their effect [4, 17].

To avoid these typical side effects, prostaglandin analogues were developed and different modes of administration were investigated. Prostaglandin analogues are more selective than natural prostaglandins and more efficient. There are substantial differences between $PGE_2$ and $PGE_1$ derivatives.

The first prostaglandin analogue investigated in order to achieve cervical dilatation was Sulproston (a $PGE_2$ derivative). It had to be administered locally and systemically. By injection of 25 µg intramurally, or 750 µg i.m., cervical dilatation of up to more than Hegar 8 with remarkable cervical softening was achieved. Thus, hysteroscopy could be performed without anesthesia [15, 22]. Minimal bleeding occurred in some of the cases, but no severe side effects were noticed. Systemically injected, more side effects such as bleeding, uterine contractions, nausea, and vomiting were registered, sometimes requiring therapy. Four hours after the administration of Sulproston no effect was observed. Dilatation was achieved as late as 12 h after injection.

It was concluded that Sulproston might allow hysteroscopy without anesthesia, and that local administration was better than systemic because of fewer side effects and equivalent efficiency [15], but the first generation of prostaglandins of the $PGF_2$ and $PGE_2$ type had many disadvantages. They had to be injected, the time span between injection and dilatating effect was long, the dilatation effect was brief, and side effects were still frequent, even increasing with time after the administration [3]. Consequently, they were rarely employed [5].

With prostaglandins of the third generation, such as Gemeprost, local administration in the form of suppositories became possible [19]. Animal experiments on monkeys showed good tolerance by skin and mucosa, whereas typical systemic prostaglandin side effects occurred only at 40 times the human dosage [5]. Clinically, Gemeprost was primarily used to induce abortion in first-trimester pregnancy patients. In a controlled study, cervical dilatation of 10 mm (median) was achieved, compared with 7 mm in the placebo group [17]. Gemeprost was found to be more efficient as regards cervical softening and opening of the cervical canal than Sulproston, and fewer side effects were noticed [3, 7]. The $PGE_1$ derivative can thus be classified as more selective, producing fewer systemic side effects such as pain or gastrointestinal disorder. Also, the administration as a suppository is much easier than injection, as in the case of Sulproston [17]. The exact mechanism of action of these prostaglandin analogues is still not known, but probably a change of the fibroblast activity plays the central role, which leads to changes in proteocollagen fibers, reticulum fibers, mucopolysaccharide envelopes and geomatrix at the cervix [17]. Thus, the stretch module raises about 50% [20]. The fact that a cervical dilatation of 7 mm was reached in a control group led, however, to the hypothesis that the mechanism might possibly be a combination of the PG-analogue effect and the effect of manual stretching during vaginal examination to administer any suppository, which might lead to liberation of endogenous prostaglandins as well [18]. Compared with Laminaria, Gemeprost proved to be more efficient [3].

The good experiences with Gemeprost administration in first-trimester abortion patients led to the study of efficiency and tolerability in nonpregnant women as well [18]. In a group of premenopausal and postmenopausal women, 3 h after administration of the Gemeprost suppository a significantly better dilatation of the cervical canal was achieved than in the control placebo group. In 32% of the patients some contraction in the form of pain in the uterus area was reported, but this did not require any analgetics. No severe side effects were observed. The authors concluded that Gemeprost in general is efficient, well tolerated, and suited for preoperative cervical softening in pre- and postmenopausal women. Despite atrophy and a changed hormone situation, postmenopausal women reacted similarly to premenopausal women. Other authors [2, 16, 23] confirmed these observations and did not notice any severe gastrointestinal or other side effects either.

Based on these data showing the efficiency and tolerability of Gemeprost when used locally as a suppository, we investigated the usefulness of Gemeprost prior to hysteroscopy, diagnostic and operative. In our study good cervical dilatation was achieved, as already described in the literature. But despite the advantage of easy handling, Gemeprost did not prove to be applicable for diagnostic hysteroscopy, because after softening and dilatation by Gemeprost the shafts did not close the cervical canal sufficiently for the hysteroscopic procedure. For operative hysteroscopy, however, where larger shaft diameters of 8, 10, and sometimes up to 12 mm were used, the problem of gas loss did not occur and occlusion of the cervical canal was sufficient. Best results were obtained 1 h after administration of the suppository.

To conclude, based on the results of our study, we recommend the use of a prostaglandin $E_1$ analogue such as Gemeprost for cervical dilatation before operative hysteroscopy where it has proven to be efficient and safe, reducing the danger of traumatization of the cervical canal.

## References


1. Antony GS, Fisher J, Cuuts JRT, Calder AA (1982) Forces required for surgical dilatation of the pregnant and nonpregnant human cervix. Br J Obstet Gynaecol 89:913–916
2. Chen JK, Elder MG (1983) Preoperative cervical dilatation by vaginal pessaries containing $PGE_1$ analogue. Obstet Gynecol 62:339
3. Christensen NJ, Bygdeman M, Géen K (1983) Comparison of different prostaglandin analogues and laminaria for preoperative dilatation of the cervix in late first-trimester abortion. Contraception 27:51
4. Craft I (1973) The use of prostaglandin pessaries prior to vaginal termination. Prostaglandins 3:377
5. Crowshaw K (1983) Comparison of the pharmacological and biochemical properties of ONO 802 (Cervagem) and the naturally occurring prostaglandins. In: Karim SMM (ed) Cervagem. A new prostaglandin in obstetrics and gynaecology. MTP Press, Lancaster, pp 1–14
6. Edström KGM (1975) Early complications and late sequelae of induced abortion. Review of the literature. Bull WHO 52:123
7. Fylling PR, Jerve F (1983) Experience with prostaglandins for therapeutic abortion in Norway. Acta Obstet Gynecol Scand [Suppl] 113
8. Goeschen K (1989) Behandlung mit Prostaglandinen in Geburtshilfe und Gynäkologie. Enke, Stuttgart
9. Grünberger W, Riss P. (1979) Isthmocervikale Insuffizienz nach vorangegangener Cervixdilatation und Cürettage. Wiener Med Wochenschr 14:390
10. Hillier K, Coad N (1982) Synthesis of prostaglandins by the human cervix in vitro during passive mechanical stretch. J Pharm Pharmacol 34:262–263
11. Informationsschrift (1989): Cergem Firmenbroschüre, Nourypharma GmbH, Norderstedt
12. Karim SMM (1983) Clinical applications of prostaglandins in OBGYN. In: Karim SMM (ed) Cervagem. A new prostaglandin in obstetrics and gynaecology. MTP Press, Lancaster, pp 15–34
13. Kurzrok R, Lieb CL (1930) Biochemical studies of human semen. Proc Soc Exp Biol Med 28:268
14. Lauersen NH, Graves ZR (1984) Menstrual induction and cervical priming prior to evacuation. In: Toppozada M, Bygdeman M, Hafez ESE (eds) Prostaglandins and fertility regulation. MTP Press, Lancaster, pp 91–106
15. Lueken RP, Lindemann HJ (1978) Lokale and systemische Sulproston Anwendung zur Cervixdilatation bei nicht Schwangeren. International Sulproston Symposium, Vienna, November 1978. Schering AG, Berlin, p 157
16. Prasad RNV, Lim C, Wong YC et al. (1978) Vaginal administration of 16, 16-dimethyl-trans-delta2-PG $E_1$ methyl-ester (ONO 802) for preoperative cervical dilatation in first-trimester nulliparous pregnancy. Singapore J Obstet Gynecol 9:69–71
17. Rabe T, Helk J, Kiesel L, Zwick J, Runnebaum B (1985) Anwendung eines neuen Scheidenzäpfchens: Prostaglandin $E_1$ Analog Gemeprost zur Cervixreifung vor Schwangerschaftsabbrüchen im ersten Trimester. Geburtshilfe Frauenheilkd 6: 393–401
18. Rabe T, Willinger G, Kiesel L, Runnebaum B, Kubli F, Mösche R, Nobakht N (1985) Study of 16, 16-dimethyl-trans-delta2-prostaglandin $E_1$-methylester vaginal suppository for cervical dilata-

tion in premenopausal and postmenopausal women. Adv Contraception 1:91–102
19. Rath W, Kuhn W (1987) Die Anwendung von Prostaglandinen zur Schwangerschaftsbeendigung im ersten Trimenon. Gynakol Praxis 11:113–126
20. Rath W, Adelmann-Grill BC, Schauer A, Kuhn W (1988) Physikalische, morphologische und biochemische Untersuchungen zur prostaglandininduzierten Cervixreifung im ersten Trimenon. In: Haller U, Kubli F, Husslein P (eds) Prostaglandine in Geburtshilfe und Gynäkologie. Springer, Berlin Heidelberg New York Tokyo, pp 58–70
21. Richardson JA, Dixon G (1976) Effects of legal termination on subsequent pregnancy. Br Med J 1:1303
22. Ulbrich J, Bartels H (1978) Klinische Ergebnisse mit Sulproston. International Sulproston Symposium, Vienna, November. Schering AG, Berlin, p 63
23. Welch C, Elder MG (1982) Cervical dilatation with 16, 16-dimethyl-trans-delta2-prostaglandin $E_1$-methylester vaginal pessaries before surgical termination of first trimester pregnancies. Br J Obstet Gynaecol 89:849–852
24. Zahradnik HP, Beyor J, Schillfart D et al. (1979) Are Hegar's rods omissible? Geburtshilfe Frauenheilkd 39:43–45

# Distention Media

A. Gallinat [1]

Hysteroscopic examination of the uterine cavity and the cervical canal requires distention of these spaces. Various media can be used to distend the uterine cavity: one can choose between liquid media and a gaseous one, $CO_2$. While for diagnostic hysteroscopy all media can be used, for surgical hysteroscopy there are special indications for and limitations to the various distention media.

## Carbon Dioxide

For many reasons we perform exclusively $CO_2$ hysteroscopy. The advantages of this dry medium are:

1. It provides the best view.
2. Hysteroscopy is performed clean and very fast.
3. $CO_2$ is a physiological gas, and there are no allergic reactions.

The visibility during $CO_2$ hysteroscopy is not disturbed by particles, blood or mucous. These are pressed against the uterine walls because of the intrauterine gas pressure and cannot swim through the distending medium, as they can when a liquid medium is used.

When $CO_2$ is the distending medium it is essential to use a special insufflation apparatus like the Metromat (Richard Wolf Company, Knittlingen). This machine works with a fixed insufflation pressure and a limited $CO_2$ insufflation flow. This guarantees a safe hysteroscopic procedure [3]. Using such an insufflation machine the maximum insufflation pressure is approximately 150 mm Hg and the maximum flow rate is limited to below 100 ml/min. Within this fixed range $CO_2$ gas insufflation is safe to use. Approximately 10–20 s after $CO_2$ insufflation a good view is obtained, while the pneumometra builds up very quickly. $CO_2$ gas with a refraction index of 1 gives the best-quality view of all the media [2].

For routine $CO_2$ hysteroscopy taking about 2 min, about 100 ml $CO_2$ gas is used. For intrauterine surgery, which can last 1 h or longer, there is no electrolyte imbalance, no fluid overload, and no intra-abdominal spillage, all of which can occur when liquid media are used. The complication rate using liquids for this kind of surgery is up to 10%, the worst complication being pulmonary edema.

$CO_2$ is a physiological gas to which there are no allergic reactions, in comparison, for example, to dextran.

Using $CO_2$ as the insufflation medium intravasation does occur, as with a liquid medium. However, due to its

[1] Zentrum für gynäkologische Endoskopie, Altonaer Str. 59–61, W-2000 Hamburg 36, FRG.

Lasers in Gynecology
Edited by G. Bastert and D. Wallwiener

buffer capacity, the blood can absorb large quantities of $CO_2$. Furthermore, $CO_2$ is the gas with the highest diffusion rate. The transport of even large quantities of $CO_2$ succeeds because the intravenous $CO_2$ is completely eliminated by the lungs in the first round of circulation.

$CO_2$ for uterotubal persufflation was introduced by Rubin in 1925 [6]. In 1947 he reported the results of 80000 carbon dioxide tubal persufflations by 380 different authors with no untoward sequelae [7]. Experiments have shown that complications occurring at higher flow rates range from cardiac arrhythmias and tachypnea, caused by more or less severe acidosis, up to cardiac arrest [3].

Intrauterine surgery was performed at the beginning of the 1970s by mechanical means, i.e., with scissors and biopsy forceps, but new techniques have been developed in the past few years. The main indications for major intrauterine surgery are:

1. Endometrial ablation
2. Septum resection
3. Myoma coagulation/resection

All of these procedures are performed today using the YAG laser or a modified resectoscope. For surgery with the resectoscope, liquid distention media are required, while with the YAG laser both $CO_2$ and a liquid can be used. The only disadvantage of $CO_2$ for distention is the smoke production during surgery.

## Liquid Distention Media

For liquid distention both high- and low-viscosity fluids are in use.

### Hyskon

High-viscosity fluids have the advantage of not mixing with blood [5]. No general spillage is therefore required. Most popular for fluid hysteroscopy, especially when a surgical hysteroscopy is performed, is 30% dextran [1]. Dextran has become the most used distention medium in the USA. Due to its high viscosity the intravasation rate is small. With hyskon and other liquids there is spillage through the open fallopian tubes into the peritoneal cavity. However, the passage of cells through the fallopian tubes appears to be very limited. The greatest disadvantage of hyskon is the danger of anaphylactic shock caused by an allergic reaction. This reaction occurs in 1:10000 cases [4].

When hyskon is used for distention, the instruments must be cleaned in hot water directly after hysteroscopy to avoid a caramelization, which would destroy the endoscopes.

Hyskon is an electrolyte-free fluid which is not conductive. Therefore, it can be used for diagnostic purposes as well as for electroresections. Its reflection index is 1.4.

### Low-viscosity Fluids

Low-viscosity fluids are saline solution, dextrose (5%–10%), dextran solution (5%–10%), and glycine 1.5%. As these fluids mix with blood and mucous, higher amounts of fluids are required. Special electric hydraulic pumps or blood-pressure cuffs are used in order to maintain a constant distention. For surgical procedures using electrotechniques only the nonconductive fluids can be used. With low-viscosity fluids there is a relatively high amount of fluid intravasation. For these kinds of intrauterine surgery up to 10–15 l of fluid are reported. Therefore, input/output measurement is essential. If the absorption rate becomes too high the surgical intervention must be stopped.

New distention media have been developed in the past few years, and for surgical hysteroscopy glycine is now being used more and more. Glycine was used earlier in urology for cystoscopy. It was noticed that in some cases after intravasation of relatively small amounts of glycine an irreversible encephalitis, so-called glycine encephalitis, was seen. Therefore, this substance was abandoned by urologists years ago!

In any case, when liquid distention media are used, electrolytes must be checked during hysteroscopic procedures and afterwards, and blood pressure and diuresis must be controlled.

In the past 15 years hysteroscopy has become increasingly important. Today it is an established method for diagnostic as well as for therapeutic purposes. At several centers, special surgical hysteroscopic techniques are in use. As we use the YAG laser, we employ $CO_2$ exclusively to establish pneumometra, because it is the only method without any complications. Other experienced centers (Wamsteker, de Blok) use $CO_2$ for diagnostic hysteroscopy too. As they use the resectoscope for surgical procedures, however, they prefer hyskon for distention in these cases.

## References

1. Edstroem K, Fernstroem I (1979) The diagnostic possibilities of a modified hysteroscopic technique. Acta Obstet Gynecol Scand 49:327
2. Lindemann HJ, Gallinat A (1976) Physikalische und physiologische Grundlagen der $CO_2$-Hysteroskopie. Geburtsh Frauenheilkd 36:729
3. Lindemann HJ, Mohr J, Gallinat A, Buros M (1976) Der Einfluß von $CO_2$-Gas während der Hysteroskopie. Geburtsh Frauenheilkd 36:153
4. Maddi V, Wyso EM, Zinner EN (1969) Dextran anaphylaxis. Angiology 20:243
5. Menken FC (1971) Ein neues Verfahren mit Vorrichtung zur Hysteroskopie. Endoskopie 3:200
6. Rubin IC (1925) Uterine endoscopy, endometroscopy with the aid of uterine insufflation. Am J Obstet Gynecol 10:313
7. Rubin IC (1947) Utero tubal insufflation. C.V. Mosby, St. Louis

# Problems in Operative Gynecological Endoscopy – The Anesthesiological Point of View

G. HUNDT [1], R. CONRADI, and H. BÖHRER

In general, laser laparoscopy and laser hysteroscopy can be performed with the patient under local, regional, or general anesthesia. Since pathophysiological changes caused by lithotomy and Trendelenburg's position as well as by pneumoperitoneum and pneumometra significantly alter respiratory function, general anesthesia with muscle relaxation and controlled mechanical ventilation represents the appropriate anesthetic technique for these major endoscopical procedures.

## Pathophysiological Changes During Operative Gynecological Endoscopy

### Effects of Positioning

Laser laparoscopy requires a 15°–20° Trendelenburg position, often in combination with a lithotomy position. During Trendelenburg's position the abdominal viscera move cephalad, compromising diaphragmatic excursions and leading to a decrease in functional residual capacity (FRC) and total vital capacity (TVC). Tidal volume is reduced by up to 20% under general anesthesia and spontaneous breathing. In addition, an increase in pulmonary blood flow and in pulmonary compliance further impair respiratory function.

Young, healthy women undergoing minor endoscopical operations may maintain respiratory homeostasis when under spontaneous or assisted ventilation. However, in the case of a major procedure, which is common in laser surgery, only controlled mechanical ventilation via a cuffed endotracheal tube can counteract the effects of positioning [8].

### Pathophysiological Changes Following Pneumoperitoneum

Air, nitrous oxide ($N_2O$), and carbon dioxide ($CO_2$) are used as gases to establish pneumoperitoneum. Recently, $CO_2$ has become most widely used because of its reduced risk of gas embolism when compared with $N_2O$ and air [6]. The following considerations on the effects of pneumoperitoneum refer to $CO_2$ as the insufflation medium.

*Resorption of $CO_2$.* An increase in arterial $CO_2$ tension ($PaCO_2$) of about 10 mm Hg is generally described in the current literature when laparoscopy with $CO_2$ pneumoperitoneum is performed and the patient is under general

[1] Klinik für Anaesthesiologie, Universität Heidelberg, Im Neuenheimer Feld 110, W-6900 Heidelberg, FRG.

Lasers in Gynecology
Edited by G. Bastert and D. Wallwiener

anesthesia with a constant minute volume during controlled ventilation. With patients breathing spontaneously, increases in $PaCO_2$ up to 75 mm Hg are seen. The highest $PaCO_2$ values should be expected following the release of the pneumoperitoneum, because there is an increase in $CO_2$ transport after reperfusion of the smaller vessels that are closed during the phase of pneumoperitoneum [14].

*Mechanical Alterations of the Respiratory System.* There is a decrease in FRC and TVC of about 20%, which is similar to the changes following the lithotomy and the Trendelenburg position. The compliance of the chest is reduced by 30%, due primarily to a decrease in the compliance of the chest wall [14].

*Hemodynamic Changes.* The hemodynamic alterations are minor until the intraperitoneal pressure reaches 30 mm Hg. Above this pressure, which is achieved only very rarely, there might be cardiovascular incidents following pneumoperitoneum. These may also be caused by vasovagal reflexes or gas embolism [14].

*Gas Embolism.* $CO_2$ embolism during operative laparoscopy with a low intraperitoneal pressure of 10–20 mm Hg is a rare event. In 1975, Phillips et al. reported on 15 possible events of $CO_2$ embolism during 113253 laparoscopies [11]. There are also some case reports of cardiovascular incidents where gas embolism could not be excluded [14]. Since the $CO_2$ laser and the Nd:YAG laser require no cooling in laser laparoscopy, the risk of $CO_2$ embolism is the same as in conventional operative laparoscopy.

## Pathophysiological Changes During Pneumometra

The pathophysiological effects on the respiratory and cardiovascular system following pneumometra are less marked than those of pneumoperitoneum. Since laser hysteroscopy is commonly carried out in combination with laparoscopy, the anesthesiological point of view is similar.

The risk of gas embolism is higher during laser hysteroscopy, which is performed under a high intrauterine pressure. The discussion about the risk of $CO_2$ embolism is as old as $CO_2$ hysteroscopy itself. The risk seems to be low when flow- and pressure-limited insufflator systems are used. In 1979, Lindemann described 5220 $CO_2$ hysteroscopies without any case of $CO_2$ embolism [7]. In the anesthesiological literature there are only two case reports of such an event [4, 10].

The problem is completely different in laser hysteroscopy with the Nd:YAG laser, as high amounts of gas are used for cooling the sapphire tip of the laser [2, 3]. Several cases of gas embolism were reported in the United States in the past when gas was used for cooling or distention in laser hysteroscopy. Thus, laser hysteroscopy in the USA is allowed only with fluid distention of the uterus and without gas cooling [1–3, 5].

In our hospital, we use $CO_2$ distention without cooling of the laser tip. Instead, the tip of the laser probe is cut off each time. Even under these conditions, we were able to detect two cases of $CO_2$ embolism with the precordial Doppler device during ten laser hysteroscopies. Both patients presented with unspecific cardiovascular changes that made $CO_2$ embolism likely. One patient showed a tachycardia and neg-

**Table 1.** Detection of $CO_2$ embolism during laser hysteroscopy using precordial Doppler device and capnography

| | |
|---|---|
| Laser hysteroscopy | $n = 10$ |
| Age in years | $33.5 \pm 8.5$ |
| Weight in kg | $63.8 \pm 20.8$ |
| Tachycardia and ECG changes | 1 |
| $F_{ET}CO_2$ increase > 0.5 vol% | 1 |
| Gas bubbles detected by precordial Doppler | 2 |
| Oxygen saturation | $98 \pm 2$ |

**Table 2.** Total intravenous anesthesia (TIVA)

| | | |
|---|---|---|
| Premedication: | midazolam | 7.5 mg per os |
| Induction: | droperidol | 2.5–5 mg |
| | alfentanil | 15 µg/kg |
| | vecuronium | 0.1 mg/kg |
| | propofol | 2–2.5 mg/kg |
| Maintenance: | alfentanil | as required |
| | vecuronium | as required |
| | propofol | 0.1–0.2 mg/kg/min |
| Ventilation: | oxygen in air | |

ative ST segments, the other had a sudden increase in end-tidal $CO_2$ concentrations ($F_{ET}CO_2$) of greater than 0.5 vol% (Table 1). These changes are not serious on their own, but they demonstrate that there is a risk of gas embolism during laser hysteroscopy that makes extended monitoring mandatory.

# Monitoring

Measurement of $F_{ET}CO_2$ and application of a precordial Doppler device should be added to routine monitoring such as ECG, blood pressure measurement, and pulse oximetry.

### Capnometry

Measuring $F_{ET}CO_2$ provides us with two types of information. On the one hand, it is a means to assess the adequacy of minute ventilation, which is important as higher amounts of $CO_2$ may have to be exhaled. On the other hand, a sudden decrease in $F_{ET}CO_2$ is a very specific sign of gas embolism in general, whereas in the case of $CO_2$ embolism there may be an initial sudden increase in $F_{ET}CO_2$ greater than 0.5 vol% [12].

### Precordial Doppler

In addition to capnometry, the precordial Doppler is considered to be one of the most specific methods for detecting gas embolism [9]. Thus, gas bubbles as small as 0.1 ml can be detected over the right heart. Since gas embolism may lead to serious complications, including cardiovascular reactions requiring resuscitation, the precordial Doppler is essential for an early diagnosis of gas embolism.

# Choice of Anesthesia

The recommended anesthetic technique is general anesthesia with muscle relaxation and controlled mechanical ventilation, since impairment of pulmonary function during laser laparoscopy and laser hysteroscopy under spontaneous breathing may lead to hypoxia and hypercapnia, followed by respiratory acidosis, cardiac arrhythmias and hyper- or hypotension, up to cardiovascular collapse [8]. Mild hyperventilation to an $F_{ET}CO_2$ of 3.8–4.2 vol% is necessary to promote exhalation of the $CO_2$ resorbed from the pneumoperitoneum.

Supplementation of $N_2O$ should be avoided, because $N_2O$ intensifies the pulmonary hypertension caused by $CO_2$ embolism [13]. We recommend total intravenous anesthesia (TIVA) with propofol, fentanyl, or alfentanil, and relaxation with vecuronium or atracurium (Table 2). A nasogastric tube should be inserted to empty the stomach and avoid regurgitation, pulmonary aspiration, or gastric perforation.

## References


1. Alexander GP (1990) Gas/air embolism associated with intrauterine laser surgery. FDA Safety Alert
2. Baggish MS, Daniell JF (1989) Death caused by air embolism associated with Nd:YAG-laser surgery and artificial sapphire tips. Am J Obstet Gynecol 16:877–878
3. Baggish MS, Daniell JF (1989) Catastrophic injury secondary to the use of coaxial gas-cooled fibers and artificial sapphire tips for intrauterine surgery. Lasers Surg Med 9:581–584
4. Gomar C, Fernandez C, Villalonga A, Nalda MA (1984) Carbon dioxide embolism during laparoscopy and hysteroscopy. Ann Fr Anesth Reanim 4:380–382
5. Kelly J, Coleman S (1989) Fatal gas embolism associated with intrauterine laser surgery. ECRI Problem Reporting System 9:325–326
6. Kunkler A, King H (1949) Comparison of air, oxygen and carbon dioxide embolization. Ann Surg 149:95–99
7. Lindemann H-J (1979) $CO_2$ hysteroscopy today. Endoscopy 2:94–100
8. Lucas LF, Rigor BM (1989) Anesthesia. In: Sanfilippo JS, Levine RL (eds) Operative gynecologic endoscopy, 1st edn. Springer, Berlin Heidelberg New York
9. Michenfelder JD, Miller RH, Gronert GA (1972) Evaluation of an ultrasonic device (Doppler) for the diagnosis of venous air embolism. Anesthesiology 36:164–167
10. Obenhaus T, Maurer W (1990) $CO_2$-Embolie bei Hysteroskopie. Anaesthesist 39: 243–246
11. Philips JM, Keith D, Hulka B, Keith L (1976) Gynecologic laparoscopy in 1975. J Reprod Med 16:105–117
12. Shulman B, Aronson MB (1984) Capnography in the early diagnosis of carbon dioxide embolism during laparoscopy. Can Anaesth Soc J 31:455–459
13. Steffey EP, Johnson BH, Eger EI (1980) Nitrous oxide intensifies the pulmonary arterial pressure response to venous injection of carbon dioxide in the dog. Anesthesiology 52:52–55
14. Wurst H, Finsterer U (1990) Pathophysiologie und klinische Aspekte der Laparoskopie. Anaesth Intensivmed 31: 187–197

# Anesthesiological Considerations Concerning the Choice of Fluid for Irrigation During Hysteroscopy with the Nd:YAG Laser

G. HUNDT[1], H. BÖHRER, and D. WALLWIENER

Because the use of $CO_2$ for distention during hysteroscopy with the Nd:YAG laser includes the risk of $CO_2$ embolism (1–3), fluid irrigation represents an alternative method for distention. In urology, different fluids for irrigation such as Ringer's solution, NaCl 0.9%, or a solution of mannitol 0.54% and sorbitol 2.7% are in common use. Similar to the conditions during transurethral resection of the prostate (TURP), the uterus consists of highly vasculated tissue. Concluding from TURP experiences, we want to give a short summary of the possible complications that are imminent when fluid distention is used for hysteroscopy. Further research is needed to investigate this problem.

## Pathophysiological Changes Due to the Influx of Irrigation Fluid into the Circulation

### Saline Solutions

*NaCl 0.9%.* The solution of NaCl 0.9% is isotonic to plasma and therefore may be used for every endoscopical procedure, except when electrocautery or electroresection is needed. Because ot its isotonicity, normal saline solution will not cause hyponatremia following resorption into the circulation [5]. The influx of greater amounts of NaCl 0.9% may cause hypokalemia, which leads to a higher risk of cardiac arrhyrhmias.

*Ringer's Solution.* As Ringer's solution contains $K^+$ 4.0 mEq/ml and $Ca^{++}$ 4.5 mEq/l it should be preferred to NaCl 0.9% in order to avoid hypokalemia following rapid absorption.

### Nonelectrolyte Solutions

Since the beginning of endoscopy many different nonelectrolyte solutions such as aq dest, glucose 5%, glycine 1.5%, and various concentrations and combinations of sorbitol and mannitol have been used. Nonelectrolyte solutions are mandatory when electrocautery and electroresection are needed. For technical and hemostaseological reasons, the combination of sorbitol 2.7% and mannitol 0.54% seems to be most useful [5].

During hysteroscopical laser surgery the combination of sorbitol 2.7% and mannitol 0.54% is preferred because it assures the best optical conditions. Every nonelectrolyte irrigation solution may cause hyponatremia, hemodilution, and water intoxication following

[1] Klinik für Anaesthesiologie, Universität Heidelberg, Im Neuenheimer Feld 110, W-6900 Heidelberg, FRG.

Lasers in Gynecology
Edited by G. Bastert and D. Wallwiener

influx into the circulation. During anesthesia for urological endoscopy this phenomenon has been known as the TUR syndrome [4].

## TUR Syndrome

The TUR syndrome is caused by rapid absorption of hypotonic fluid via the opened prostatic veins. It results in hyponatremia (serum sodium less than 120 mEq/l) and hypervolemia, with subsequent cardiovascular alterations. Confusion, restlessness, or muscle rigidity may occur as early signs which are caused by cerebral edema and disturbed blood chemistry. Acute left- or right-sided cardiac failure, myocardial infarction, and pulmonary edema may further complicate the situation [4].

## Anesthesia for Laser Hysteroscopy with Fluid Distention

As confusion and restlessness are the first signs of the influx of nonelectrolyte solution, regional anesthesia should be preferred if laser hysteroscopy is carried out without previous laparoscopy. This allows early recognition of the symptoms. A central venous catheter should be inserted, and the finding of an increased central venous pressure will reveal massive fluid absorption in an early phase. Frequent blood analyses should be done to detect hyponatremia. Accurate monitoring of the fluid balance, including irrigative and intravenous fluids, is essential.

## References


1. Alexander GP (1990) Gas/air embolism associated with intrauterine laser surgery. FDA Safety Alert
2. Baggish MS, Daniell JF (1989) Death caused by air embolism associated with Nd:YAG laser surgery and artificial sapphire tips. Am J Obstet Gynecol 161:877–878
3. Baggish MS, Daniell JF (1989) Catastrophic injury secondary to the use of coaxial gas-cooled fibers and artificial sapphire tips for intrauterine surgery. Lasers Surg Med 9:581–584
4. Browne B (1989) Anaesthesia of urological procedures: current practice. Curr Opin Anaesth 2:735–740
5. Juricic D, Al-Naieb Z, Hoffmann CF, Engelmann U (1989) Spüllösungen in der Endoskopie. Uroscop 1:3–17

# Laser and Endoscopical Technology News

# Nd:YAG Laser System for Contact Surgery with Tissue-Effect Control

G. HAUPTMANN [1] and F. FRANK

The different interaction mechanisms of Nd:YAG laser irradiation with biological tissue in noncontact application have been adequately described. Extinction of laser irradiation in biological tissue depends on the laser parameters (power density, irradiation time, and wavelength) as well as on the specific tissue properties and occurs within the range of some centimeters [1]. If a beam of 10 W is transmitted through an optical fiber 365 μm in diameter, with a divergence of 19° and held at a distance of 30 mm from the tissue, a power density of approximately 10 $W/cm^2$ is achieved, which is sufficient to coagulate the tissue. Placing the fiber in direct contact with the tissue increases the power density 800 times. Such a high power density heats the tissue so strongly that within a short time carbonization, and finally vaporization occur. This effect is reinforced by the fact that a much smaller tissue volume is affected by laser light in contact application than in noncontact application. Due to these processes, cutting with the Nd:YAG laser in contact with the tissue becomes possible.

Problems arise when tissue is prepared with a bare optical fiber using the conventional laser systems with constant power settings, as the distal fiber tip is contaminated with tissue residues. On the one hand, this improves the cutting efficiency because of increased absorption; on the other hand, it can damage the fiber when not in contact with the tissue. The latter causes a problem particularly in endoscopic procedures where the spatial orientation is difficult and the user is often not able to control the exact distance of the bare fiber from the target tissue during the whole cutting process.

Several attempts have been made to eliminate this problem by developing techniques that use sapphire probes. However, these methods are technically expensive and require detailed instruction of the user. In addition, the necessary cooling of the sapphire probes carries the risk of gas embolism for the patient.

Our investigations indicated that a constant temperature at the fiber-tissue interface guarantees a uniform cutting effect and makes the thermal stress which can damage the fiber tip controllable. We therefore developed a tissue-effect controlled Nd:YAG laser system. During the process of cutting biological tissue using the Nd:YAG laser, visible light is produced as a result of burning and carbonization processes at the fiber-tissue interface. There is a correlation between light intensity and the temperature at the flat end of the fiber. This correlation per-

[1] MBB-Medizintechnik, Applikationsforschung, Postfach 801168, W-8000 München 80, FRG.

Lasers in Gynecology
Edited by G. Bastert and D. Wallwiener

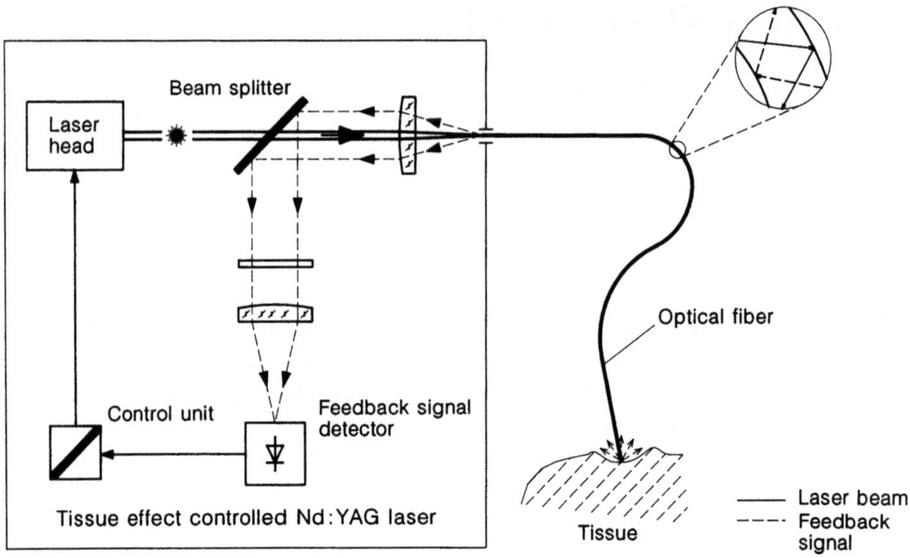


**Fig. 1.** Nd:YAG laser system with an optical fiber for contact surgery with tissue-effect control. The feedback signal is transmitted simultaneously to the Nd:YAG laser through the optical fiber

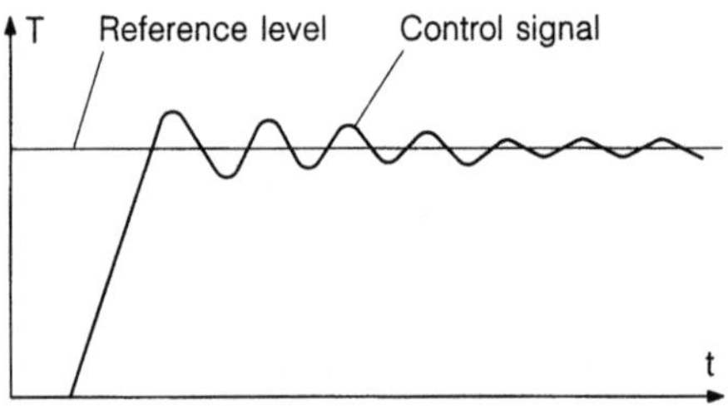


**Fig. 2.** Procedure for controlling the light guide tip temperature

mits light intensity to be used as a control parameter for the temperature at the fiber-tissue interface. In this way we have a surgical laser system based on an Nd:YAG laser which can be used in contact application with optimal cutting efficiency.

The light produced during the burning process is transmitted back through the therapy fiber to the laser system and is separated by means of a beam splitter from the beam path of the Nd:YAG laser. This feedback-signal is focussed by means of a lens to a detector and converted into an equivalent voltage. A control unit with an integrated microprocessor uses this signal as a control parameter for the distal laser output (Fig. 1). Therefore, the visible light produced at the end of the light guide during the cutting process is kept at a constant intensity. Control is effected by comparing the reflected beam with a reference value which can be adjusted by the user according to the operating needs (Fig. 2). By varying the light intensity and the maximum obtainable laser power, different cutting parameters can be achieved.

Such a tissue-effect controlled laser system offers the following advantages:

- The optical fiber is protected against thermal damage by limiting the temperature at the distal fiber end.
- The laser energy required for cutting is optimized with regard to cutting

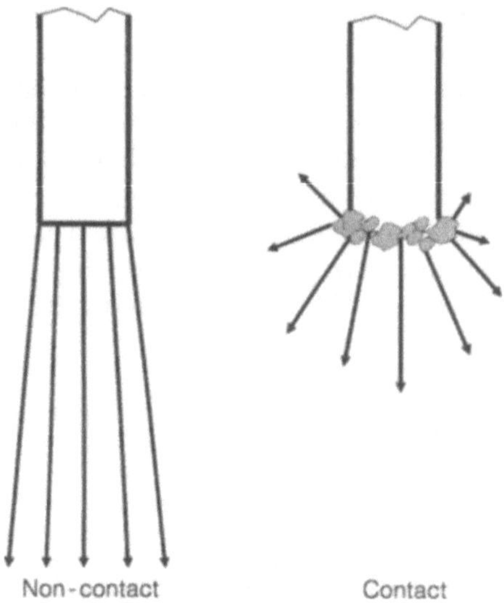


**Fig. 3.** Beam characteristics of a standard bare fiber for noncontact surgery and a contaminated fiber for contact surgery

speed, so that a maximum of energy is available without the optical fiber being damaged.
- When the fiber is not in contact with thc tissue, laser power is automatically reduced within fractions of a second, which prevents the fiber from being damaged.
- Cooling the fiber tip is not necessary.
- Only a small amount of smoke develops as a result of the defined, narrow incision.
- A uniform coagulation necrosis develops at the border of the vaporization zone which is due mainly to the homogeneous distribution of Nd:YAG laser radiation in tissue [2], but also to the heat conduction process in connection with the rapid carbonization.

During the cutting process, at certain points on the fiber tip surface, temperatures develop for a very short time that reduce the viscosity of quartz glass, to a degree that carbon and tissue particles burn into the fiber material. Due to these melting processes the fiber is contaminated, such that the carbon particles sticking to the fiber tip absorb part of the laser energy and convert it into heat. The resulting higher temperature of the burnt-in fiber as compared with the noncontaminated fiber causes immediate car-

**Fig. 4a, b.** Vaporization and coagulation zone after tissue-effect controlled incision: **a** schematic; **b** histological findings in pig uterus

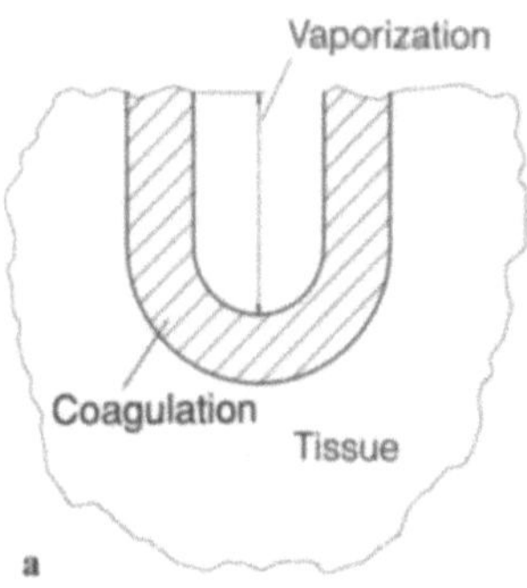


a

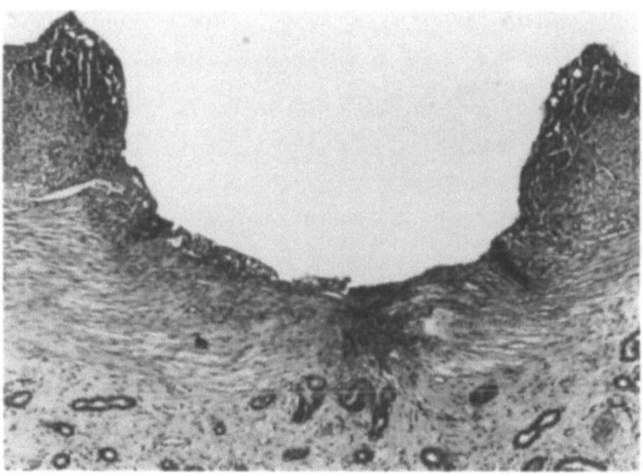

b

bonization and vaporization of the tissue, as is required for a high cutting efficiency.

Contamination also changes the beam characteristics of the fiber and reduces the power density in an axial direction. The result is a homogeneous coagulation necrosis along the border of the incision (Fig. 3).

With the laser system mediLas 4060 fibertome (manufacturer: MBB-Medizintechnik GmbH) such a tissue-effect controlled system has been realized, offering the possibility of setting the temperature at the distal fiber tip at three levels (approximately 700 °C, 800 °C, 900 °C). This range of variation is sufficient to achieve an optimal cutting effect according to the different kinds of tissue and to produce a lesion with a vaporization depth and coagulation zone according to the therapeutic requirements (Fig. 4).

In in vitro experiments on specimens of pig uteri using a constant cutting speed of 2.5 mm/s, vaporization depths between 0.4 mm and 0.9 mm, and coagulation zones between 0.2 mm and 0.6 mm were achieved, depending on the setting of the fiber tip temperature.

## References

1. Boulnois JL (1986) Photophysical process in recent medical laser developments: a review. Lasers Med Sci 1:47
2. Keiditsch E, Hofstetter A, Zimmermann I, Stern J, Frank F, Babaryka I (1985) Histological investigations to substantiate the therapy of bladder tumors with the neodymium-YAG-laser. Laser Med Chir 1:19

# An Intelligent Surgical Laser – Science Fiction or a Pathbreaker for Future Technical Developments?

D. WALLWIENER[1], D. POLLMANN, S. RIMBACH, and G. BASTERT

A basic disadvantage of all laser preparation techniques, one which is common to all endoscopic preparation techniques used in surgical intervention, cannot be denied: the subjective control of the energy output by the constant energy emission from the preparation instrument. During contact preparation the energy output is preset at a constant level and must be preselected by the surgeon, as is the case with all electrosurgical and laser surgical preparation techniques. The output cannot be influenced by tissue-specific properties or by the extremely irregular surface configuration of the tissue often encountered in microsurgery. Contact laser preparation has further shown that application of laser without direct tissue contact can result in an overheating of the distal quartz fibers until they smoulder, even up to a point where the blazing hot, distal quartz glass tip falls off.

Thus the question arises whether "nonsubjective" control of the thermal tissue effect attained with the preparation techniques is not possible by means of electronic feedback and consecutive automated control of the energy level. This is applicable not only to endoscopic tissue preparation but also to contact laser preparation of external tissue.

An "intelligent" laser system can be distinguished by the fact that is capable of directly measuring the tissue effect or, if this is not possible, of measuring a closely related parameter with ensuing calculation of the tissue result using a preincorporated function.

"Coagulation" and "resection" are not measurable quantities; hence, only derived quantities can be taken into consideration to control the course of treatment. These can be roughly divided into optic and acoustic techniques.

The interaction of the laser with biological tissue changes the visually perceivable properties of the tissue. The light emitted from the laser used for treatment purposes or from a separate diagnosis laser is differently scattered; its polarization properties are changed and the tissue becomes fluorescent. Theoretically, all these parameters can be reported back to the laser apparatus by the "treating" fiber and used for regulation purposes.

The development of entirely new objective, automated, electronically controlled and preprogrammable output ratios, enabling the control of the energy dose during tissue preparation by means of feedback on the basis of the thermal tissue effects and other parameters within split seconds, has made it possible to produce a new generation of tissue preparation systems with the

[1] Universitäts-Frauenklinik, Abt. für Geburtshilfe und Gynäkologie, Voßstr. 9, W-6900 Heidelberg, FRG.

Lasers in Gynecology
Edited by G. Bastert and D. Wallwiener

help of the laser technique. This leads to the possibility of developing a new generation of preparation systems with unforseeable potentials, not only for the Nd:YAG laser but also for other surgical laser systems or even other tissue preparation techniques.

This enables tissue preparation with "intelligent" calculation of the required energy level; the preparation possibilities are then independent of the tissue-specific properties and the microsurface configuration of the different types of tissue, primarily the irregular tissue surface usually encountered during microsurgical preparation, as well as being independent of the water content of the tissue. Moreover, they are also independent of the surgeon's training, and thus enable a completely safe preparation technique which is free of potential damage.

This is of exceptional importance, particularly for endoscopic preparation: the two-dimensional view and visual appraisal of the tissue to be prepared using the currently available preparation systems makes a power output adapted to the respective tissue and operative technical situation impossible.

## Initial Experimental Surgical and Clinical Experience with the Nd:YAG Laser Contact Technique with "Effect Control"

The automated output control during Nd:YAG laser contact preparation by means of distal light measurements via the quartz glass fiber was reliably tested in animal experiments. Thereafter, the mediLas-fibertome Nd:YAG laser, with a temperature-controlled cutting system, was experimentally tested on beagles and introduced into a pilot study within the framework of surgical laparoscopy for distal tubal pathology.

The laser system utilized is an optoelectronic system enabling temperature-controlled cutting of tissue with naked fiber tips. A special sensor within the machine measures the temperature at the fiber tip during the cutting process. The laser output is automatically regulated in the preselectable temperature ranges so that predetermined fiber tip temperature remains constant or is not exceeded.

During the cutting procedure the optoelectronic regulating system can be switched on, in addition to the conventional continuous-wave Nd:YAG laser. The laser output can be set between 10 and 40 W in three ranges by the measurement of the cutting temperature at the fiber tip.

## Method

In vivo material was gathered within the framework of laser laparoscopic linear uterotomies on beagles to determine the incision depth and width morphometrically as well as the extent of the zone which had been thermally damaged by the laparoscopic cutting when using the Nd:YAG laser with tissue-effect control.

The basic setting chosen for the laser was 20 W using the continuous-wave system; all three settings were tested. The quartz glass fiber utilized had a diameter of 360 µm. A laparoscopic cutting velocity of approximately 1 cm/s was selected.

The tissue was fixed in a 5% formalin solution and then embedded in paraffin, once the animals used for the experiment had been killed. Morphometry was carried out using the morphometric method described in the chapter "Tissue Effects by Different Laser Techniques".

A total of 100 incisions were evaluated per power setting. The average cutting width was found to be 420 μm and the average cutting depth 700 μm. An average thermal destruction zone of 200 μm was determined.

## Clinical Utilization

The fibertome was utilized for laser laparoscopic revisions of distal tubal pathology within the framework of a clinical pilot study.

## Results

A satisfactory laparoscopic techniques was attained when using the temperature-regulated Nd:YAG laser with special bare fibers for the contact preparation technique during laparoscopic intervention. At all preselected output levels, melting of the fiber tip was avoided by means of the optoelectronic measuring system, despite occasional utilization of the laser without tissue contact. The laser output was automatically reduced to a noncritical level within fractions of a second shortly before the fiber tip would have begun to melt. During precise preparation, adhesion of the fiber tip to the tissue was avoided due to the regulated cutting temperature.

Morphometrical evaluation of the standardized laser tissue sections showed that, in addition to the described advantages of the system, the small thermally damaged zone is not larger during automatic laser control with an optimized cutting effect than during preparation with a nonautomatically controlled laser output.

The development of this Nd:YAG laser with automatic tissue-effect control appears to have been a successful step in the development of a new laser generation.

## References

See chapter Hauptmann and Frank (pp 247–250 in this volume).

# SharpLase Sculpted Fibers for High-precision Contact Surgery

M. SLATKINE [1] and Z. ROZENBERG

## Introduction

A highly flexible surgical tool that performs comparable to a precise monopolar electrosurgical unit without its possible risks has long been the desire of general and other surgeons. The recent flourishing of laparoscopic minimal invasive procedures [3] has made the development of such a tool almost mandatory due to possible risks of fatal bowel perforation with electrocautery. SharpLase conical fibers provide an excellent cutting tool for minimal invasive procedures without the above-mentioned possible risks of electrocautery (see Fig. 1).

Incisions are achieved with good hemostasis. The thermally affected zone is as small as 0.5 mm. Moreover, the fibers act only upon actual contact, thereby eliminating the possibility of damage to adjacent or forward tissue. Ease of use is comparable to or better than that of monopolars. Simultaneous cutting and irrigation are possible. Here we briefly present the operating principles of the conical SharpLase fibers and elucidate the origin of their inherent safety. We also describe a hemispherical contact fiber for surface ablation (such as endometrial ablation).

## Operating Principles of Contact Conical Nd:YAG SharpLase Fibers

The operating principles of conical SharpLase fibers are presented in

[1] Sharplan Laser Industries, PO Box 13135, 49100 Tel Aviv, Israel.

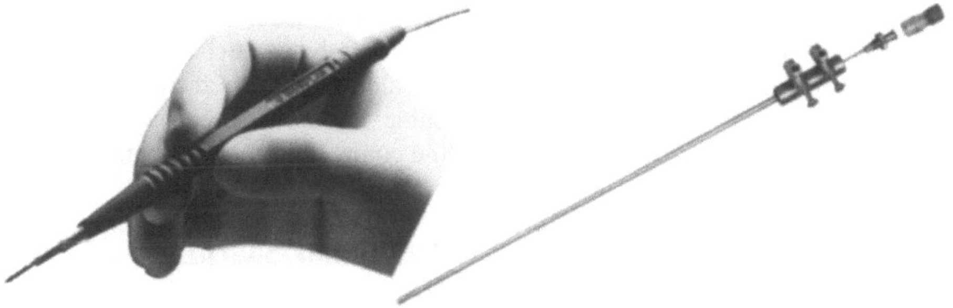

a b

**Fig. 1. a** Conical SharpLase fiber with a handle for freehand contact surgery. **b** Irrigation/aspiration probe designed to accept SharpLase fibers for laparoscopic surgical procedures

Lasers in Gynecology
Edited by G. Bastert and D. Wallwiener

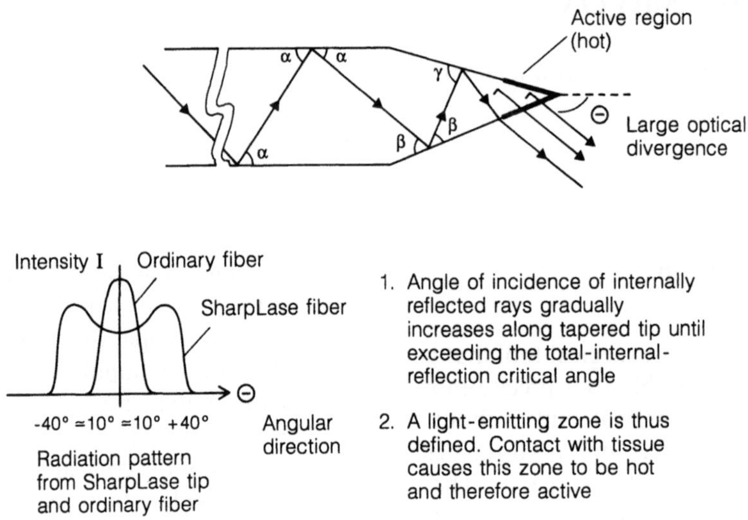


**Fig. 2.** Optics of a conical SharpLase fiber

Figs. 2 and 3. Nd:YAG laser energy is focused on a 3-m long silica optical fiber of numerical aperature 0.4. The angle of incidence of internally reflected rays gradually increases along the 10° tapered end of the fiber until it exceeds the total internal reflection critical angle. An optical leakage emitting zone is thus created. Contact of the leaking zone with tissue causes this zone to become hot and therefore active. At 15–40 W Nd:YAG laser power levels, the sharp conical leaking zone reaches tissue ablation temperature, thus enabling dissection with a tissue-affected zone of less than 0.5 mm. Cutting speed can be increased by increasing the laser power level. Fast incisions in a liquid environment can be achieved at power levels up to 60 W. For lysis of adhesions in minimal invasive laparoscopic surgical procedures, power levels as low as 15 W may be used.

A remarkable feature of conical SharpLase fibers is the very large angular divergence of the optical radiation emitted from the distal sharp end (80°) of the fibers. This provides clinical safety, since subjacent tissue will essentially not be affected immediately after lysis of adhesions although laser emission may be continued for a short time. Figure 4 elucidates the origin of this fiber safety. A simple calculation [2] shows that the temperature of subjacent tissue located at a distance greater than 3 mm from the fiber will not reach more than 30 °C rise at 40 W of power and 1 s irradiation time on a single tissue location. Practical incision conditions are far below these, thus insuring no thermally irreversible denaturation of tissue. As depicted in Fig. 5, the conical SharpLase fiber is suitable for coagulation by contact with bleeding tissue along the tapered end of the fiber. Table 1 presents the specifications of the conical SharpLase fibers.

## Additional SharpLase Fibers

Two additional contact fibers are frequently used in surgical procedures:

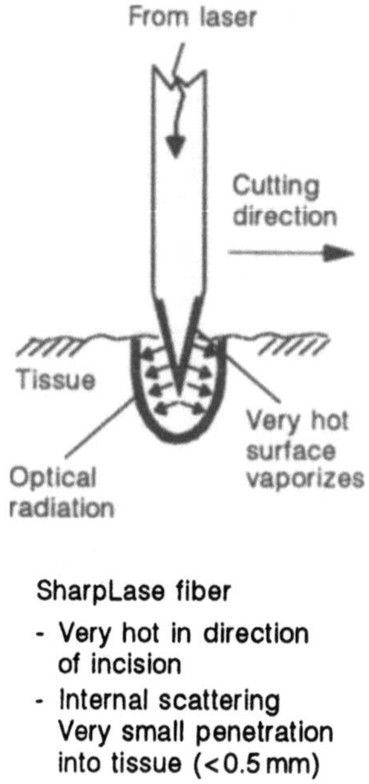


a

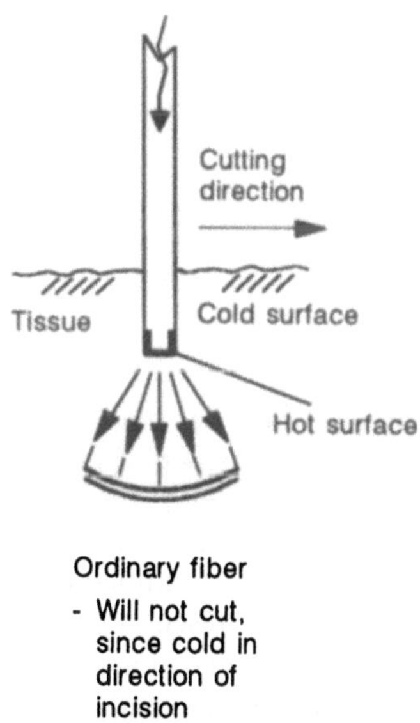


b

**Fig. 3a, b.** High incision efficiency of conical SharpLase fibers. **a** Conical SharpLase fiber; **b** ordinary flat fiber for comparison

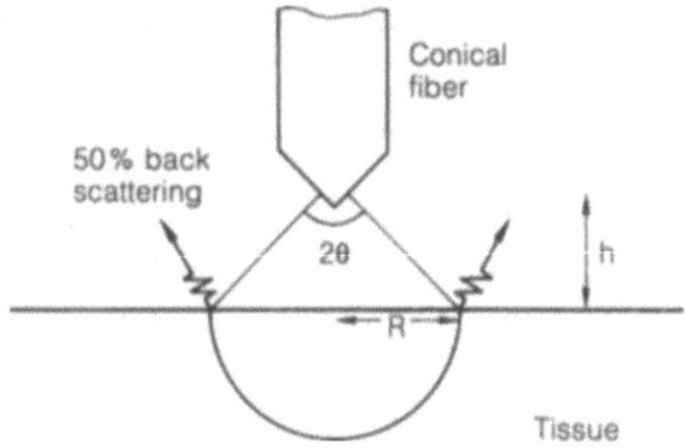


**Fig. 4.** Temperature rise of tissue for noncontact irradiation is a source of safety. Assume an irreversible thermal effect for $\Delta T > 65\,°C - 37\,°C$, extinction coefficient $\sim 0.6\,cm^{-1}$ [1], power input of 40 W (two times higher than in normal procedure), and 1 s irradiation (longer than normal). Calculations show lack of thermal effect for $2\theta \geqq 50°$ at $h \geqq 4$ mm. Since SharpLase fibers are produced with $2\theta \geqq 80°$, thermal effect on tissue is achieved at a distance of less than $\sim 3$ mm from tissue

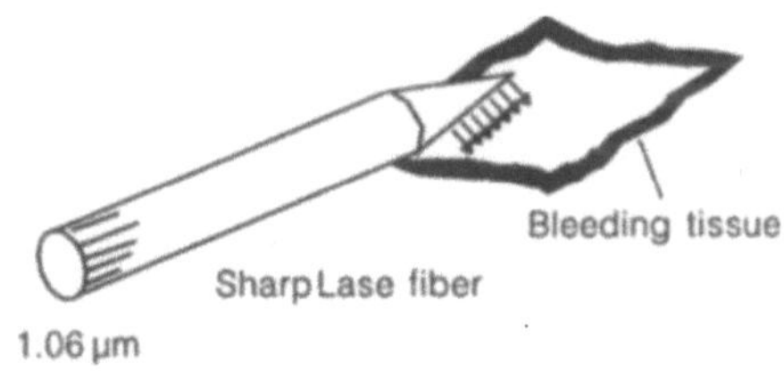


**Fig. 5.** Coagulation with a conical SharpLase fiber

The first is a hemispherical fiber (800 μm and 1000 μm), for coagulation and ablation (Fig. 6). Hemispherical fibers are currently used mainly in arthroscopy (menisectomy), urology (bladder neck contracture), and hysteroscopy (endometrial ablation) [4].

The second is a "dual-effect" flat fiber which cuts and coagulates. This fiber (400 μm, 600 μm) is suitable for incisions of thin adhesions. Its forward radiation pattern enables coagulation of excessive bleeding in case of emergency. The fiber may be used for laparoscopic cholecystectomy. For other

**Table 1.** Specifications of SharpLase conical fibers

| | Dimensions | | |
|---|---|---|---|
| Core diameter | 1000 μm | 800 μm | 600 μm |
| Tapering angle | 10° | 10° | 10° |
| Overall diameter | 1.2 mm | 1 mm | 0.92 mm |
| Total length | 3 m | 3 m | 3 m |
| | Power range | | |
| Cutting mode<br>(fluid environment) | 10–40 W<br>30–70 W | 10–30 W<br>30–60 W | 10–30 W<br>30–60 W |
| Coagulation mode | Up to 40 W | Up to 40 W | Up to 40 W |
| Beam divergence | 80° | 80° | 80° |
| Tensile torque threshold | 600 g × cm | 500 g × cm | 250 g × cm |
| "Burn in" time<br>(tissue dependent) | Very fast –<br>< 0.2 s | Very fast –<br>< 0.2 s | Very fast –<br>< 0.2 s |
| Cooling | Not required | Not required | Not required |

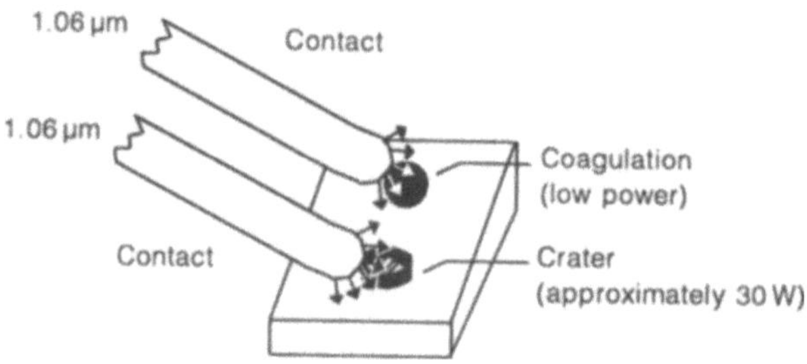


**Fig. 6.** Vaporization/coagulation with a hemispherical SharpLase fiber

laparoscopic procedures where subjacent tissues are critical, the conical fiber is recommended.

## Instrumentation and Applications

The contact SharpLase fibers may be inserted in a variety of surgical probes, as depicted in Fig. 7. The most frequently used probe is the irrigator/aspirator for all laparoscopic procedures. Other probes are employed for free-hand general surgery, neurosurgery, arthroscopy and otorhinolaryngology.

In Table 2 we present a comparison between conical SharpLase fibers and electrosurgical units used in laparoscopic minimal invasive surgical procedures. The comparison clearly indicates the advantage of Nd:YAG conical fibers over electrocautery. Conical fibers will eventually play a key role in the success of the minimal invasive surgery revolution.

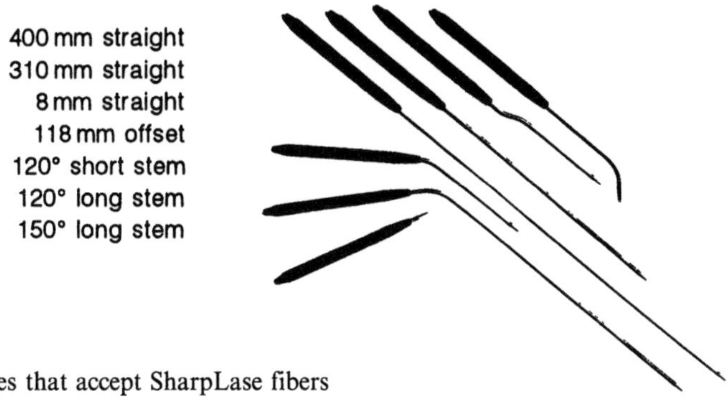


**Fig. 7.** Probes that accept SharpLase fibers

**Table 2.** SharpLase contact fibers versus electrocautery

| Nd:YAG SharpLase fibers | Electrocautery |
|---|---|
| High precision and high-speed cutting capability; excellent hemostasis; 0,5 mm thermally affected zone | Bipolar – imprecise, slow cutting<br>Monopolar – seldom used in abdomen |
| Can be easily adjusted to the plane of dissection, facilitating for example removal of gall bladder from liver bed | Difficult to adjust to plane of dissection, particularly with the bipolar probe |
| Irrigation during lasing provides a clear field of view throughout the procedure | Irrigation is virtually impossible |
| | Possible risks: |
| – Conical SharpLase Fibers act only upon actual contact with tissue, eliminating adverse effects on adjacent or forward tissue<br>– Fully predictable tissue effect<br>– No electrical currents | Monopolar – risk of uncontrolled arcing and possible current flow through internal organs (fatal in bowels) to grounding electrode |
| Ease of use: | |
| The SharpLase fiber is easier to use than the bipolar electrocauter unit. Required training is minimal and does not cover any new conceptual techniquel. Fast cutting | More difficult to obtain reasonable cutting quality, and is slower |

# References

1. Carruth JAS, McKenzie AL (1986) Medical lasers. Adam Hilger, Bristol, pp 66–72
2. Rozenberg Z, Slatkine M (1992) Laser contact fibers as an alternative to electrocautery in microsurgery. IEEE Transactions on Biomedical Engineering (in press)
3. Rubio PA, Rowe G, Feste JR (1989) Endoscopic laser cholecystectomy. Houston Med 5:124
4. Silber N, Servadio C (1992) Neodymium: YAG laser treatment of bladder neck contracture following prostatectomy. Lasers in Surgery and Medicine (in press)

# Endoscopic Surgery with the Argon Laser in Gynecology

M. KORELL[1]

Endoscopic surgery has found wide applications in gynecology and is now emplied in many fields of medicine. The advantages of such surgery are the avoidance of large scars, the rapid recovery time, and less pain. The addition of laser ($CO_2$, argon, and neodym YAG laser) to the operative instrumentarium created great expectations. It has been suggested that the laser, by means of its unique properties, may improve the success rate of gynecologic infertility surgery.

Based on our clinical experience and the data from the literature, we evaluate the use of Argon laser in endoscopic surgery.

## Use of Argon Laser in Endoscopy

### Adhesiolysis

Adhesions among genital organs and, for instance, the bowel are common finding in gynecologic endoscopy. With the exception of very extensive adhesions, which require laparotomy, both a fiber and a sapphire probe can be used for adhesiolysis. During cutting it is important that the tissue is taut, especially in the case of thick adhesions. Otherwise the laser instrument is less "sharp" and the coagulation effect predominates.

### Endometriosis

One of the major advantages of the argon laser in infertility surgery is that the absorption of the argon laser beam coincides with the pigmented endometriotic implants. This specific absorption makes the argon laser most suitable for the therapy of endometriosis.

### Ectopic Pregnancy

Endoscopic surgery became the standard therapy of tubal therapy to avoid laparotomy and preserve tubal function. The required longitudinal salpingotomy to remove the gestation can easily be done with the argon laser using the contact technique. The use of the bare fiber is recommended because the necrotic zone is smaller than with the sapphire probe. This meets the requirements of reconstructive surgery with the aim of minimal trauma.

[1] Klinikum Großhadern, Abt. für Geburtshilfe und Gynäkologie, Marchioninistr. 15, W-8000 München 70, FRG.

Lasers in Gynecology
Edited by G. Bastert and D. Wallwiener

### Salpingostomy

The reconstructive surgery of hydrosalpinx can be performed with the monopolar microneedle or a laser. It is a matter of debate whether the results of endoscopic salpingostomy equal those of microsurgery. In severe cases of tubal damage, there is no doubt that the pregnancy rate is not determinated by the surgical method. In the case of mild tubal pathology, i.e., intact mucosa and thin tubal wall, one can achieve pregnancies in over 50% of the cases by using microsurgical methods. There has been no definite proof that endoscopy can offer the same results. In our opinion salpingostomy should be reserved for centers with great experience.

### Myomectomy

Because of its ideal combination of cutting and coagulation capacity, the argon laser is most suitable for myomectomy. The risk of intensive bleeding during this operation is thus reduced. The additional local injection of vasopressin may be helpful. Laparotomy is nevertheless often still necessary to remove large intramural myoma.

## Results of Laser Compared to Conventional Surgery

The use of lasers in infertility surgery was connected with hopes of less tissue trauma, more rapid healing, and better results. Only a few studies have compared the outcomes of laser and conventional surgery [1–4], and they have not detected any significant difference in the results.

The different physical qualities of a laser beam and of electrocautery cause equal effects on biological tissue. The energy is transformed into heat, and the result depends on energy density. Thus, one must not expect different clinical results from lasers and electrocautery. This has to be taken in account when evaluating the role of laser technology.

Studies comparing laser endoscopy with conventional laparotomy should be suspected. It is undisputed that the advantages for the patients are mainly from the endoscopic techniques and not lasers. It is thus no surprise that no study has proven any significant differences regarding cutting with laser beam or conventional methods.

Nevertheless, the unique properties of laser technology are beneficial for infertility surgery. Table 1 summarizes the different abilities of laser surgery compared with alternative methods and illuminates its indispensable value. No other technology offers such different possibilities – cutting, coagulation, and vaporization. One can expect problems only in the case of cutting thick tissue that is not taut; then there is a larger necrotic zone than with electrocautery and, especially, scissors.

## Advantages of the Argon Laser

The high cost of laser equipment normally requires the choice of one system. Thus, it is important to choose the one with greatest variety of possibilities. Based on different absorption characteristics, $CO_2$, argon, and neodym YAG lasers have different tissue effects. While the neodym YAG laser offers deep coagulation, the $CO_2$ laser

**Table 1.** Laser uses compared to alternative methods

| Method | Dissection | Vaporization | Coagulation | Endoscopy |
|---|---|---|---|---|
| Conventional surgery | + | – | – | (+/–) |
| Electrocautery | + | – | + | + |
| Infrared coagulation | – | – | + | – |
| Laser surgery | + | + | + | + |

**Table 2.** Effects of different laser systems

| Laser | Vaporization | Dissection | Coagulation | Specific absorption | Flexible fibre | Touch technique |
|---|---|---|---|---|---|---|
| $CO_2$ | ++ | + | +/– | – | – | – |
| Argon | + | + | + | + | + | + |
| NeodymYAG | – | + | ++ | – | + | + |

can vaporize tissue. The argon laser combines both effects. Argon and neodym YAG lasers can be used in flexible fiber systems and allow cutting in contact technique. The advantages of both over the $CO_2$ laser include the reduction of smoke and better hemostasis. Furthermore, the argon laser provides a specific absorption by endometrial implants. Table 2 displays the different qualities of $CO_2$, argon and neodym YAG lasers. Because of its universal applicability, the argon laser seems to be the best choice.

## Conclusions

The unique properties of the laser offer great possibilities and create high expectations as to its use in infertility surgery. In the future the role of lasers can be increased, for example by the use of photosensitizers. The argon laser is ideally suited for endoscopic surgery of ectopic pregnancy, adhesiolysis, and many other applications because of its ability to cut and coagulate simultaneously. The laser is not, however, a cure-all. The development of endoscopic surgery signifies great progress for patients, and in reasonable hands the laser can be used to great advantage in gynecologic surgery.

## References

1. Daniell JF, Diamond MP, McLaughlin DS, Martin DC, Feste J, Surrey MW, Friedman S, Vaughn WK (1986) Clinical results of terminal salpingostomy with the use of the $CO_2$ laser: report of the Intraabdominal Laser Study Group. Fertil Steril 45(2):175
2. Keckstein G, Wolf A, Wittek S (1989) Tubenzustand nach pelviskopisch operierter (konventionell versus Laser) Tubargravidität. Arch Gynecol Obstet 245(1–4): 416
3. Tulandi T, Farag R, McInnes RA, Gelfand MM, Wright CV, Vilos GA (1984) Reconstructive surgery of hydrosalpinx with and without the carbon dioxide laser. Fertil Steril 42(6):839
4. Tulandi T (1986) Salpingo-ovariolysis: a comparison between laser surgery and electrosurgery. Fertil Steril (1986) 45(4):489

# Use of Holmium Laser in Gynecological Endoscopic Surgery

J. DEQUESNE [1] and D. SUMNER

## Introduction

Over the past decade, the development of new endoscopic techniques has revolutionized the treatment of intrauterine and intra-abdominal gynecological pathology. Bruhat et al. [1] introduced the $CO_2$ laser in laparoscopy, and in 1980, Goldrath [7] was the first to use the Nd:YAG laser in hysteroscopic surgery. Since 1985, the Clinic Cecil Laser Center has evaluated each new laser development medical applications in the field of gynecology [4]. The aim of this ongoing research is to assay each laser in the hope of finding a polyvalent, multitask system.

However, there are still limitations to the use of lasers in gynecological surgery. The lasers used in most clinical trials required cumbersome delivery systems and failed to precisely ablate tissues. For example. the carbon dioxide laser (wavelength 10.6 µm) is capable of precise ablation because the wavelength is strongly absorbed by water, and the resulting thermal damage covers less than 100 µm, but the major limitation preventing clinical use is the lack of an optimal fiber delivery system. Furthermore, carbon dioxide laser energy cannot be transmitted through a liquid environment.

Similarly, the noncontact Nd:YAG laser has been controversal because ablation depth cannot be controlled precisely and heat produced by the continuous-wave irradiation produces a large zone of tissue damage. The visible wavelength lasers such as argon and KTP require pigmented tissue for absorption, and the area of tissue damage is also much larger than that with carbon dioxide lasers.

Recently, a pulsed near infrared laser based on the rare earth element holmium has been developed that operates at a wavelength of 2.1 µm. The beam is transmitted through a quartz fiber and works equally in a gas or a liquid environment. In addition, the laser radiation is absorbed by most biological tissues, which allows for precise tissue removal.

Up to now, only animal experimentation with holmium laser, has been reported [8], and the tissue damage seems to be very similar to that caused by $CO_2$ laser (Table 1). Ninety-two endoscopic procedures (hysteroscopy and laparoscopy) have allowed us to make a preliminary evaluation of the in vivo possibilities of the Ho-YAG lasers compared with the conventional laser used in gynecological endoscopic surgery.

[1] FMH Gynécologie et obstétrique, Chemin des Croix-Rouges 16, 1007 Lausanne, Switzerland.

Lasers in Gynecology
Edited by G. Bastert and D. Wallwiener

**Table 1.** Characteristics of the holmium and $CO_2$ lasers compared

| Holmium | $CO_2$ |
|---|---|
| "Solid" medium | "Gas" medium |
| Flexible fiber delivery | Articulated arm delivery |
| Depth of thermal damage = 0.4–0.6 mm | Depth of thermal damage = 0.2 mm |
| "Wet" environment | "Dry" environment |
| Ablating and coagulating | Vaporizing and coagulating |

# Material and Method

### Laser

A flashlamp-pumped, pulsed holmium-yttrium-aluminum-garnet laser (Coherent Medical, Palo Alto, CA) was used in our surgery. The laser produces 250-μs pulses at the wavelength of 2.1 μm. The pulse repetition rate can be varied from 5 to 20 Hz, and the pulse energy can be varied from 0.5 to 2 J. Pulse-to-pulse energy variation is less than 5%. A red helium laser (5 mW) produces the aiming beam; transmission is done by a 400-μm quartz fiber.

### Hysteroscopy

In order to treat benign endometrial pathology, the technique already described with Nd-YAG laser [2, 3] was used, using the noncontact technique for coagulation and the contact technique for cutting (Fig. 1). However, the pulsed mode (cavitation effect) was used for better fragmentation of hard lesions such as myomas and synechiae, and the shallow penetration depth was favorable for precise surgery close to the tubal ostium.

The distension medium used was glycine 1.5%. The power selected for fibroids was 1.5 J per pulse and the pulse rate 10 pulse/s, following trials with other parameters.

### Laparoscopy

Conventional laparoscopic surgical techniques were used. The Ho-YAG laser was introduced by second puncture with a Nezat probe, allowing suction-aspiration of gas and liquid. The specific effect of the holmium laser allows us to enlarge the conventional indications for laparoscopic surgery (see Table 3). The surgery was performed in both $CO_2$ and Ringer's lactate medium. The power and the laser pulse rate were adapted to these different indications. All equipment for conventional laparoscopic and hysteroscopic surgery was present (including Nd:YAG/KPT and $CO_2$ laser) in case of failure during surgery with the Ho/YAG laser, but it was never needed.

# Results

The short follow-up time has not allowed us to demonstrate a better efficiency of treatment with regard to bleeding, sterility, and pain. However, control hysteroscopy has been used for anatomical evaluation in the early follow-up (Table 2), except in cases of subtotal ablation. These endoscopic evaluations seem to indicate that anatomical recovery is faster for fibroid surgery; healing is better in the cornual areas as with cervical stenoses and lateral synechiae. Furthermore, the coagulation effect was evaluated as being as efficient as that of the Nd:YAG laser.

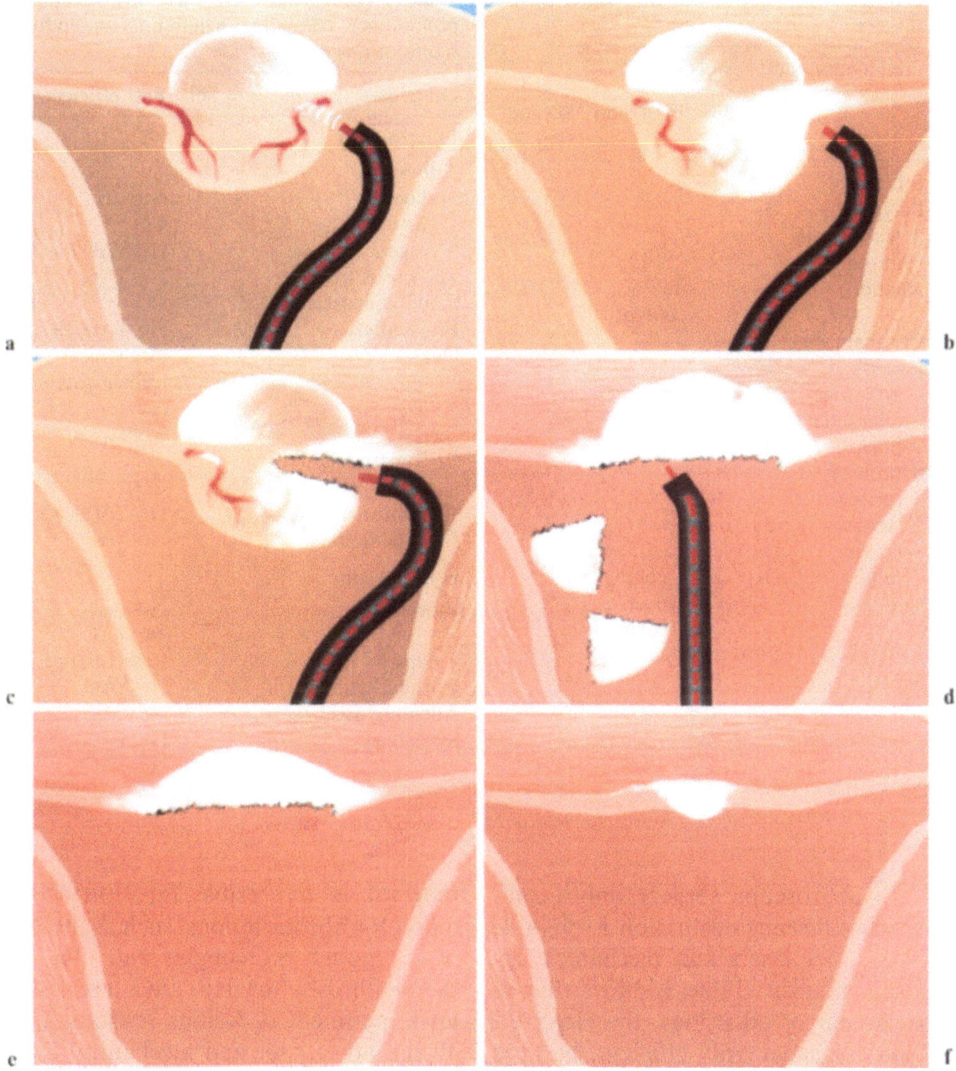

**Fig. 1a–f. Myolysis in hysteroscopic Nd:YAG laser surgery. a** Banding of the flexible hysteroscope with fiber inserted; coagulation of vessels (20 W, noncontact) to avoid fluid intravasation and bleeding. **b** Deep coagulation (60–70 W) in non contact technique. **c** Fibroid dissection (60–70 W) in contact technique. **d** Deep coagulation (70–90 W) in noncontact to induce myolysis. **e,f** Myolysis occurs between 6 and 8 weeks after laser treatment.
Second-look hysteroscopy is then needed for evaluation and possibly second-step surgery

**Table 2.** Ho-YAG laser in hysteroscopic surgery

| Preliminary good results (HSC) | $n = 36$ | First step | Second step |
|---|---|---|---|
| Fibroids | 9 | 7 | 2 |
| Polyps | 3 | 3 | |
| Adenomyosis (subtotal ablation) | 4 | no control | |
| Cornual lesions: | | | |
| – synechiae | 3 | 3 | |
| – interstitial polyps | 4 | 4 | |
| Synechiae: | | | |
| – cervical stenoses | 5 | 5 | |
| – lateral | 3 | 1 | 2 |
| – fundic | 3 | 1 good<br>1 over-correction | 1 |
| Septa | 2 | 1 | 1 |

**Table 3.** Ho-YAG laser/other instruments: preliminary compared evaluation in endoscopic surgery

| Diagnosis | $n = 56$ | Best efficiency |
|---|---|---|
| Laparoscopy | | |
| Adhesions | | |
| stage 1 | | scissors |
| stage 2 | 15 | $CO_2$ |
| stage 3 | | Ho-YAG |
| Subserous fibroids | 7 | Ho-YAG |
| Salpingoneostomy | 5 | $CO_2$ |
| Fimbriolysis | 7 | Ho-YAG/$CO_2$ |
| Ovarian cysts | 4 | mechanical instruments (scissors + bipolar) |
| Endometriosis: | | |
| – peritoneal | 6 | $CO_2$ |
| – endometrioma | 5 | Ho-YAG/$CO_2$ |
| Fibrotic lesions | 7 | Ho-YAG |
| Hysteroscopy | | |
| Fibroids | | Ho-YAG/ Nd:YAG |
| Polyps | | Resectoscope |
| Cornual lesions | | Ho-YAG |
| Adenomyosis | | Nd:YAG |
| Synechiae | | Ho-YAG/ Nd:YAG |

In laparoscopic surgery, only a compared efficiency evaluation is possible with other lasers and mechanical instrumentation. Table 3 shows that in treatment of adhesions, the Ho-YAG laser is particularily efficient, because of its coagulating and cutting effect. Furthermore, tubo-ovarian dissection in cases of sterility is possible with very little tissue damage immersion technic. With subserous fibroids, the Ho-YAG laser has been more efficient than the other lasers, including Nd:YAG and KTP lasers, for both coagulating and cutting purposes. The most impressive indication, however, appears to be endometriosis, as we now have the possibility of treating encapsulated endometriomas when conventional capsula removal is impossible [5]. Furthermore, for fibrotic lesions, such as those of the pouch of Douglas and uterosacral fibrosis, the Ho-YAG laser allows treatment of lesions resistant to all the other laser and mechanical instruments (cavitation effect).

## Discussion

Since 1985, we have been establishing the Cecil Laser Center in Lausanne, Switzerland, focusing on new laser developments, principally in the field of gynecological endoscopic surgery.

Since May 1985, we have developed hysteroscopic laser surgery principally with the Nd:YAG laser [2,3]. The

KTP laser was used for evaluation, but no advantage over the Nd:YAG laser was found, excepting the possibility of vaporization of the tissues and easier handling of disposable quartz laser fiber.

The argon laser has been abandoned in the meantime, as it is underpowered and there are visual difficulties during surgery caused by the laser shutter.

Preliminary evaluation on uterine tissue in vitro was done by Duffy [6]. Our preliminary evaluation of the Ho-YAG laser shows a better sculpturing effect with less tissue damage than that caused by the Nd:YAG, argon, and KTP lasers. This allows for precise surgery around the tubal ostia and during hysteroscopic sterility procedures.

The pulsed mode is principally advantageous with fibroids and sclerotic lesions such as synechiae, as it allows for precise and quick fragmentation of the lesion and permits immediate removal of the interstitial part of the fibroids, which might otherwise have to be treated by a secondary myolysis using Nd:YAG (Fig. 1a–f).

In laparoscopic surgery we have generally used the Bruhat-Manhes mechanical instruments and, since 1985, the $CO_2$ laser. However, the Nd:YAG/KTP laser has been used in contact techniques (including that with the sapphire tip). In our experience, the KTP laser should be used only for removal of very thick vascularized adhesions; the $CO_2$ laser is suited for all other indications.

The new holmium laser on the market can be regarded as a $CO_2$ laser with the advantage of the quartz laser fiber delivery system, working equally well in liquid and gas media and with better, more efficient, coagulation during major destruction (subserous myomectomy, fibrotic endometriosis).

The disadvantage of the holmium laser, however, is the slightly larger area of tissue damage than that produced by the $CO_2$ laser worse cutting effect also there is liquid and blood projection due to the pulsed mode, and the materials are expensive.

This inconvenience is compensated for by the possibility of operative usage in a larger range of different medical specialities, including orthopedics [9].

## Conclusion

Preliminary evaluation of the Ho-YAG laser in gynecological endoscopy shows it to be of interest in hysteroscopy, principally when benign endometrial pathology is found (fibroids, synechiae) and for the treatment of cornual pathology. With regard to endometrial ablation, however, a longer follow-up is needed for evaluation.

In laparoscopy, the Ho-YAG laser allows for enlargement of the indications in endoscopy (fibrotic endometriosis, fibromyomas) technical improvements must be done before it may be regarded as a $CO_2$ laser improved by the fiber delivery system and the transmission through fluids.

## References

1. Bruhat M, Mage G, Manhes M (1979) Use of the $CO_2$ laser via laparoscopy. In: Kaplan I (ed) Laser surgery, III. Proc. Third International Society for Laser Surgery. International Society for Laser Surgery, Tel Aviv, p 275
2. Dequesne J (1987) Hysteroscopic treatment of uterine bleeding with the Nd:YAG laser. Lasers Med Sci 2:73

3. Dequesne J (1990) Traitement hystéroscopique des métrorragies et de l'infertilité au laser Nd:YAG (5 ans d'expérience). Arch Gynecol Obstet 247:52–59
4. Dequesne J, De Grandi P (1990) Utilisation laparoscopique du laser $CO_2$ dans le traitement de l'endométriose. Med Hygiène 48: 1354–1360
5. Donnez J, Nisolle M, Karaman Y, Wayembergh M, Clerckx F, Casanas-Roux F (1990) $CO_2$ laser laparoscopy in peritoneal endometriosis and in ovarian cyst. J Gynecol Surg
6. Duffy S, Sharp F, Ginsberg R (in press) Holmium:YAG laser: preliminary observations of laser-tissue interaction using human uterus. Northern General Hospital, Sheffield
7. Goldrath MH, Fuller TA, Segal S (1981) Laser photovaporization of endometrium for treatment of menorrhagia. Am J Obstet Gynecol 104:14
8. Rosenberg C, Tadir Y, Braslavsky D, Fisch B, Karni Z, Ovadia J (in press) Endometrial laser ablation in rabbits: a comparative study of three laser types
9. Trauner K, Nishioka N, Patel D (1990) Pulsed holmium: yttrium-aluminium-garnet (Ho:YAG) laser ablation of fibrocartilage and articular cartilage. Am J Sports Med 18:3

# Flexible Fiber in Gynecology

J. Bouquet de Joliniere [1], J. B. Dubuisson, B. Tessier, and M. Levardon

It has been about 15 years since the first $CO_2$ lasers were used in laparotomy during tubal microsurgery operations. It was the Clermont school of Bruhat [1] which pioneered in this field and demonstrated the usefulness of the laser beam through an operating microscope to obtain a precise cut and microhemostasis.

In 1980, Pouly [2] succeeded in transmitting the beam during a laparotomy through a suprapubic handpiece in a rigid tube. Later, Daniell [3] attempted to repeat this directly through a bayonet laparoscope.

Concomitantly, the advent of the YAG laser with its highly specific wavelength led to the first use of fibers, in particular for operational hysteroscopy, though in a very special configuration: "touch surgery". Indeed, the tissular effects are very different from those obtained with the $CO_2$ laser, very quickly limiting the therapeutic indications for the YAG laser.

In October 1988, our team used the first flexible $CO_2$ laser fiber, on the basis of a portable laser with a power of 40 W achieving very interesting therapeutic results in ten cases of peritoneal endometriosis.

[1] Hospital Beaujon, 100, Boulevard Général Leclerc, 92118 Clichy, France.

## Physical Principles of Lasers

### Physical Process of Gas – Carbon Dioxide ($CO_2$)

The active environment of a $CO_2$ laser generally consists of a mixture of carbon dioxide, nitrogen ($N_2$) and helium (He). Each of these three constituents plays a crucial role. The $CO_2$ laser can thus provide infrared emissions in wavelengths of 9–11 μm. It can also put out power ranging from a few fractions of a watt to several tens of kilowatts. In the medical field we use powers of a fraction of a watt to 100 W. The efficiency of the $CO_2$ lasers is remarkable, reaching up to 30%.

At first sight, open-tube $CO_2$ lasers seem the most simple: a tube, mirrors, a mixture provided by external cylinders, and electrodes generating a discharge. However, these lasers are cumbersome, and this technique has been replaced by sealed lasers.

Sealed lasers apply the same principle as the open-tube $CO_2$ laser, but the mixture is kept inside a sealed tube. New developments with these lasers have enabled the overall dimensions to be reduced to a few tens of centimeters, with excellent efficiency. Today, reverse catalysts regenerate the $CO_2$, and lifetimes can be as long as 10000 h.

If the internal diameter of the $CO_2$ tube is reduced to a few millimeters, a waveguide is obtained by capillary ef-

Lasers in Gynecology
Edited by G. Bastert and D. Wallwiener

fect, generally consisting of alumina or BeO. The $CO_2$-$N_2$-He mixture can be excited by a DC electric discharge or by a strong microwave field. Enormous progress has been achieved in this area, particularly in France, and we now have lasers with powers of 100 W for lengths of 60 cm and with efficiencies of 15%.

### Physical Processes of a Solid-state Laser (Nd:YAG)

The atomic system uses trivalent ions of a rare earth known as neodymium (Nd 3+) present as dopers in an yttrium aluminum garnet (Y Al 0), with the dominant emission at 1.0641 μm. The pumping and supply systems are similar in all respects to those used for ruby lasers.

Recently, the Nd:YAG laser has been given a new shape, that of an elongated bar or "slab." The slabs, the ends of which are cut to be Brewster angle, enable the laser to operate at a very high mean power.

These lasers operate both continuously in the pulsed mode, with pulse durations ranging from about 1 ps to 1 ns.

### Means of Beam Transmission

*Nd: YAG.* The Nd:YAG beam is transmitted via an optical fiber. In general, this is a multimodal single fiber with an index jump comprising a silica core and an external sheath with an index less than that of silica.

The useful diameter is from 200 to 600 μm and the doped silicone optical sheath has a diameter slightly greater, with the plasic sheath increasing the outer diameter to 1 mm. However, the emergent beam at the distal end of the fiber is less coherent than a beam sent via an articulated arm.

*$CO_2$.* Until now, the $CO_2$ beam was transmitted via a reflecting articulated arm. However, recent tests with special infrared fibers have enabled several tens of watts to be sent over more than 1 m. A French patent on these new fibers was granted in April 1989, providing surgeons with a $CO_2$ optical fiber.

## Comparative Tissular Effects

These fibers are of interest primarily in endoscopy, for which the $CO_2$ laser was poorly suited. By exercising great ingenuity, engineers have made it possible for the beam to penetrate; using rigid endoscopes it was very difficult to fire and the ergonomics were limited. The thermal effect of the $CO_2$ laser combined with optical fibers will revolutionize therapeutic endoscopy. Major research is presently being carried out in France on the laser-tissue interaction for the Nd:YAG and $CO_2$ lasers. The $CO_2$ laser would seem to be the ideal instrument for human tissues. The Nd:YAG appears to entail disadvantages owing to its wavelength, since the great proportion of photons not absorbed by the tissues can create undesirable effects at a distance.

Another process of the surface action of the laser is the action of the electric field accompanying the beam. A sinusoidal electromagnetic wave is associated to each photon, consisting of an electric field and a magnetic field. Owing to the spatial coherence of the laser, all the photons are in phase, and at each point the value of the fields is cumulative. It follows that the electromagnetic field linked to the beam can attain considerable values in the range of $10^7$–$10^{12}$ V/m.

The electric fields prevailing in the atoms, binding electrons and nuclei, are of the same order. It is hence understandable that the laser shots can destroy the entire molecular organization, resulting in ionization, breakage of bonds, and the appearance of free radicals.

These effects take on first-order importance when ultrashort (nanosecond) pulse lasers with high peak powers are used. Microplasma is created, and the extension of the plasma will create a very powerful shock wave in the environment, capable of mechanically lacerating the tissues. The effects of the laser on the cellular system cannot be ignored, since these mechanical effects can be highly detrimental in the short term. They can be transmitted by nervous or osteoconduction to sites remote from the point of generation and produce remote lesions, in particular bleeding.

## Choicc of Laser

One must define the therapeutic indications in the light of the tissular effect desired. The choice of the type of laser will govern the results, depending on whether one decides to cut, to evaporate a lesion to destroy it while leaving the neighboring tissues intact, or to use a simple thermal effect as part of a seroserous eversion or peritoneal plasty.

It is fundamental to introduce the concept of power density at tissue level. Its domestication will optimize the safety of the surgical act, enabling the best possible result to be obtained at the lesion treated. Power density depends on:

1. The *power set* at the laser, provided that beam control is correct so as to be sure to have the same power at the tissues. (This is not the case for most lasers available on the market. For some of them, one must allow for 15%–20% loss.)
2. The *distance of the shot*, considering the *focal point*. It is by the combination of focusing-defocusing that the depth of the shot at the tissues and the area of thermal distinction will be varied.
3. The *time* and *mode* of the shot: the power density obviously depends on the pulse frequency for repeated shots. It is in this type of configuration that the super-pulse delivers high peak powers at tissue level for a low set power and a very brief shot time.

Our work also consists in assessing two other fundamental concepts without which the surgeon would not be able to control this power density output: (a) the effective power (availability of beam) at the tissues and (b) the ergonomics of the equipment. Until recently, the $CO_2$ beam could be transferred into abdominal cavity only via stiff tubes about 20 cm long, through a lens with a focal length of 125–300 mm. Accordingly, the surgeon could never be certain that the beam was perfectly focused, since it might reflect several times inside the tube. At the exit, the telltale red helium-neon spot no longer corresponded to the photon beam, creating a risk if there was accidental damage to the nearly tissues.

In addition, these laparoscopic handpieces are mounted on articulated arms, the internal mirrors of which increase the radius. These arms are often heavy or bulky and difficult to handle, with a reduction of the moments of rotation in the three axes. This is hardly compatible with the accuracy needed to evaporate and destroy lesions through the abdominal wall at a total distance of about 30 cm. Lastly, defocusing of the radius is almost impossible to achieve without shifting the suprapubic trocar.

Our revolutionary technique offers maximum ergonomics. We have introduced a flexible $CO_2$ laser fiber 2.4 mm in diameter in the suprapubic position, through an adaptor (the BTL, a special handpiece for laparoscopy), providing (a) direct access to the lesion, (b) a focal point a 2 cm, and (c) a 350-μm spot. It is very easy to defocus, and surgical movements become easy and precise. There is no problem of size, since the 70-cm-long fiber is attached to the laser outlet.

This laser can be carried in a 25-kg case, offering a power of 45 W in a sealed tube set on a console of adjustable height. With it we have performed about 20 peritoneal endometriosis evaporations, ten adhesiolyses and two neosalpingostomies. Used in combination with video optical fibers, the $CO_2$ laser fiber will provide high power for evaporation of intracavitary myomas in a dry atmosphere through an instrument, the BTH a special handpiece for hysteroscopy, introduced via the cervix. The purpose is to avoid complicated and lengthy manipulations of the Nd:YAG and the danger of endoscopic resection by diathermy.

## References


1. Bruhat MA, Mage G, Manhes H, Pouly JL (1979) Use of the $CO_2$ laser via laparoscopy. In: Kaplan I (ed) Laser surgery. Proc. 3rd Intl. Congress on Laser Surgery, Tel Aviv. Jerusalem Academic Press, pp 235–238
2. Bruhat MA, Mage G, Manhes H, Pouly JL (1981) Utilisation du laser $CO_2$ par coelioscopie. Rev Fr Gynecol Obstet 76:397–399
3. Daniell JF, Herbert CM (1984) Laproscopic salpingostomy utilizing the $CO_2$ laser. Fertil Steril 41:558–563

# Possibilities of Combination of Carbon Dioxide and Nd:YAG

K. DINSTL [1]

In general surgery, the $CO_2$ laser and the Neodym:YAG laser have been used so far. The $CO_2$ laser has an excellent cutting property and the Nd:YAG laser a coagulating, vaporizing quality, but neither of them was completely satisfying for single use. For obvious reasons, these two systems have been combined. After unsatisfying experiments with technical problems [1], a combilaser was constructed. The Combilaser from the Lasermatic company is the first laser which combines these two systems.

At our institute we tested the system with the $CO_2$ output up to 27 W (Nd:YAG laser 75 W), which turned out to be too little for general surgery. We are now working with a stronger system which puts out $CO_2$ laser performance up to 50 W and Nd:YAG laser performance up to 80 W. The aiming beam is produced by the helium-neon laser (2 mW). The combi beam can be conducted only through a rigid system, but both kinds of laser can be applied alone. A flexible light guide is available for the Nd:YAG beam. It was expected that the application of the $CO_2$ and the Nd:YAG laser together would combine the effects of both kinds of laser. Together with the well-known advantages of the $CO_2$ laser – excellent cutting ability, little postoperative wound infection, and little postoperative wound pain – the blood clotting was improved in this combination. On the other hand, it was expected that, for instance, deep malignant rectal stenosis might be opened more quickly and better with the addition of the $CO_2$ laser to the Nd:YAG beam.

We performed 228 operations with this combilaser. In 82 cases there was palliative application in anorectal surgery (stenosed rectal cancer or local recurrence after deep anterior resection, piles, rectal or anal polyps, condylomas, and anal carcinomas; Table 1). Moreover we operated on 21 patients with breast cancer, on 14 patients with soft tumors, on 59 with condylomas, verrucae, and pilonidal fistulas, and performed four liver resections and one pancreas resection. The Combilaser was applied in eight cases of Dupuytren's contracture as well as in five cases of gangrene of the leg.

**Table 1.** Combilaser in used anorectal surgery ($n = 82$)

| Indication | No. of patients |
|---|---|
| Anal carcinoma | 11 |
| Rectal carcinoma (palliative) | 30 |
| Rectal polyps | 12 |
| Hemorrhoids | 18 |
| Anal condyloma | 11 |

[1] Krankenhaus Rudolfstiftung, 1. chirurgische Abt., Juchgasse 25, 1030 Wien, Austria.

Lasers in Gynecology
Edited by G. Bastert and D. Wallwiener

**Table 2.** Operations performed with the Combilaser ($n = 228$)

| | |
|---|---|
| Anorectal surgery | 82 |
| Surgery of breast cancer | 21 |
| Liver resection (wedge) | 4 |
| Pancreas resection | 1 |
| Lymphadenectomies | 3 |
| Soft tissue tumors | 14 |
| Condylomas, verrucae, tatoos, pilonidal fistulas | 59 |
| Dupuytren's contracture | 8 |
| Resection of gangrene (diabetes) | 5 |
| Endoscopic application (YAG) | 20 |
| ENT operations | 11 |

With the contact sapphire tip we carried out lymphadenectomies in three cases, and in 20 cases we performed an endoscopic operation with the YAG fiber. Within otorhinolaryngology surgery this laser was used 11 times. Table 2 gives an overview of our applications.

The following indications result for the use of the Combilaser in general surgery:

1. Anal surgery
2. Rectal surgery (palliative)
3. Surgery of soft tissue and superficial tumors
4. Surgery of parenchymatous organs (liver, spleen, kidney, female breast)

## Discussion

Without any doubt, the application of the Combilaser represents progress in field of laser surgery, though problems of wound healing occurred at first which can be exemplified by the surgery of hemorrhoids:

We used the $CO_2$ laser in 205 cases, the Nd:YAG laser in 16 cases, and the Combilaser in 20 cases. It became evident that the application of the Nd:YAG laser alone must be stopped, because of the extreme wound necrosis and the strong wound pains that persisted for up to several weeks and disturbed the patients. The $CO_2$ laser was employed because it caused less postoperative wound pain and proved to be satisfactory. Our impression of the Combilaser was similar to that of the Nd:YAG laser at the beginning because we chose a very strong performance of the Nd:YAG component (20–25 W). We achieved excellent results concerning blood clotting and postoperative pain only when we used the Nd:YAG laser in combination at a maximum power of 5 W. Therefore, we applied only a little additional Nd:YAG power with the combi beam in order not to delay wound healing after the operation of small, superficial tumors as well.

Some progress was observed in liver resection compared with the use of the $CO_2$ or the Nd:YAG laser alone. The application of the combibeam for liver resection requires temporary obstruction of the blood supply to the liver (wedge resections of the liver).

The larger vessels still have to be ligated in the conventional manner. Lymphadenectomy with the contact sapphire tip was quite useful and the result was acceptable.

In summary, the Combilaser has the following advantages compared with the use of either of its components alone:

1. Better cutting than with the Nd:YAG laser alone (especially in palliative surgery)
2. Better coagulation than with the $CO_2$ laser alone
3. Sealing of the lymph vessels with the combination is more likely than with the $CO_2$ beam alone.

The disadvantages are:

1. The $CO_2$ beam and the combi beam are not transferable through flexible fibers alone.
2. Carbonization is stronger than with the $CO_2$ beam alone (causing a possible delay of wound healing)
3. One cannot be sure of the thermal penetration depth with the noncontact method.

In general, however, this instrument represents progress in general surgery using laser.

## References

1. Sultan RA, Fallouh H, Lefebvre-Vilardebo M, Ladouch-Badre A (1986) Separate and combined use of Nd:YAG and carbon dioxide lasers in liver resections: a preliminary report. Laser Med Sci 1:101

# Smoke Evacuation in Endoscopic Laser Surgery

A. GALLINAT [1] and R. P. LUEKEN

In the field of gynecological surgery there are several types of lasers in use. The great advantage is that all these lasers can be used in gynecologic endoscopic surgery, and with these new methods the number of classical operations performed, from laparotomies, over hysterotomies, up to hysterectomies, can be reduced.

The various laser types differ in their wavelengths. This implies a different physical behavior as well as a different biophysical reaction seen as the effect on tissue.

High laser energy projected onto tissue generates more or less profuse amounts of smoke, depending on exposure time and output. In a closed cavity, depending on its volume, vision is obscured with in a short time. To reduce or do away with working gaps smoke evacuation is necessary. For purposes of endoscopic gynecologic surgery the Nd:YAG laser and the $CO_2$ laser are the most interesting and most commonly used.

## Hysteroscopic Surgery

For many reasons the $CO_2$ laser – so familiar in the field of gynecology – cannot be used in hysteroscopic surgery. In contrast to the $CO_2$ laser, the Nd:YAG laser has virtually ideal characteristics for hysteroscopic application:

1. The normal standard hysteroscope (7-mm hysteroscope with working channel) can be used.
2. The thin quartz-glass fiber that conducts the laser beam can be passed down through the normal working channel.
3. Thanks to the physical properties with the predominating coagulating effect, only small amounts of smoke are generated, which obscure vision but little or only temporarily.
4. The depth of penetration of the Nd:YAG laser and this physical property in particular makes it highly suitable for intrauterine application – is approximately 5–6 mm.

Due to the physical characteristics of the Nd:YAG laser a liquid as well as a gaseous distention medium for hysteroscopy can be used. The advantage of a liquid medium is that no smoke is generated. However, because of the lack of an ideal liquid distention medium we use $CO_2$ exclusively for hysteroscopy. Use of a liquid is negated by the large amounts of fluid required. In a high percentage of cases, these large amounts of fluid lead to serious complications, such as severe electrolyte imbalance, fluid overload with circula-

[1] Zentrum für gynäkologische Endoskopie, Altonaer Str. 59–61, W-2000 Hamburg 36, FRG.

Lasers in Gynecology
Edited by G. Bastert and D. Wallwiener

tory disregulations, and pulmonary edema. Dextran, in particular, has produced severe allergic reactions. A liquid distention medium also requires a higher laser output. Only $CO_2$ used with a proven insufflation apparatus has no side effects; higher wattage ensures the generation of steam.

The power output required for a minor intrauterine operation is between 20 and 25 W. In the case of more major interventions such as the treatment of submucosal myomas or septa 30–40 W are needed to achieved vaporization. Up to an output of approximately 30 W only small amounts of smoke are generated that hardly obscure the view. At higher outputs, however, enough smoke is generated to obscure the view until it has been cleared again by continuous $CO_2$ gas insufflation. The carbon dioxide is usually insufflated via the scope and flows from the cervix toward the fundus. If the target area is located within the upper part of the uterine cavity, the smoke collects here before condensing in the regions of the cornua. In many cases vision is improved when $CO_2$ insufflation is effect via the laser light guide. The $CO_2$ issues from the tip of the light guide, despelling the smoke to the side and keeping the view unobstructed. However, at higher outputs a considerable amount of smoke is generated, hindering the view.

To avoid longer breaks in the operating procedure a smoke evacuator for continuous gas circulation was constructed[2]. Essential in developing such a machine is that the $CO_2$ insufflation not be allowed to exceed 100 ml/min. Under normal uterotubal conditions the $CO_2$ flow is about 30–40 ml/min. The smoke generated by hysteroscopic Nd:YAG laser treatment is aspirated. Under pressure- and flow-controlled conditions this new machine clears the smoke by suction via a microfilter and simultaneously replaces the $CO_2$, preventing a collaps of the pneumometra. That means that the gas insufflation from the Metromat[3] takes place within its fixed physiological limits while at the same time the intrauterine gas volume is cleared of the smoke by the apparatus. In this new machine these two completely different systems are separated and – most importantly – there is no interference of the smoke evacuator with the $CO_2$ gas insufflation.

For intrauterine laser surgery we use a coated Teflon probe with an internal diameter of the quartz-glass fiber of 0.6 mm. The $CO_2$ is insufflated via the laser probe. The gas is delivered to the same area where the smoke is produced by the tissue vaporization. The smoke is blown aside; visibility remains clear. While laser surgery continues, the distending $CO_2$ of the pneumometra is cleared by the smoke evacuator. The smoke-containing gas is aspirated via the working channel of the scope and reinsufflated via the laser probe. The problem for hysteroscopic smoke evacuation was the relatively small volume of the uterine cavity of only 20–30 ml. The thick myometrium requires a high intrauterine pressure, which means that the smoke evacuator needs a pressure-controlled gas circulation system.

During hysteroscopic surgery, a collaps of the pneumometra should be avoided. In contrast to procedures in other cavities, a collaps of the pneu-

[2] $CO_2$ Cleaner: Richard Wolf Company, Knittlingen.

[3] Metromat: Richard Wolf, Knittlingen.

mometra during uterine surgery immediately leads to bleeding. Even limited bleeding and possible mucus production render continuous hysteroscopic surgery more difficult. Because of the very small intrauterine volume (see above), the exchange of the laser probe, for example or simply an unintentional movement of the nipples of the endoscope shaft would lead to a collaps of the pneumometra.

Different methods are available for intrauterine surgery, but among them only the Nd:YAG laser is the universal tool, for all types of indications.

## Laparoscopic Surgery

Most experience in gynecologic laparoscopy has been gained with the $CO_2$ laser, which has now been used for many years. The Nd:YAG laser, too, can be used for laparoscopic surgery, and in some indications it has advantages over the $CO_2$ laser; the problem of smoke evacuation in laparoscopic laser surgery is technically much easier with the Nd:YAG laser, as it generates less smoke than the $CO_2$ laser.

There are different techniques in use, the simplest being to introduce via an additional puncture a tube through which the smoke can escape. This tube can be moved by the assistant close to the location of smoke production, and most of the smoke is expelled by the intra-abdominal pressure. With the insufflation machine the same amount of $CO_2$ must be replaced. A better effect is obtained using the endoirrigation unit for smoke aspiration. This unit is required for laparoscopic surgery anyway. The smoke can be aspirated directly at the area of the laser target. This aspiration volume is much greater and the view remains good. The replacement technique requires greater amounts of $CO_2$. The strong effect of anesthesia on the intra-abdominal pressure leads to failure of automatic pressure-driven $CO_2$ delivery machines for this use. Therefore, an assistant should make sure that spillage and $CO_2$ replacement with this method are within the same range.

For frequent laparoscopic surgery a special insufflation machine with a closed system was constructed[4]. This universal apparatus has in the middle the normal $CO_2$ delivery system. The left part of the machine is the endoirrigation unit for spilling and aspiration and on the right side is the continuous $CO_2$ evacuation system with a microfilter. After the normal pneumoperitoneum has been established at the beginning of surgery, the intraabdominal $CO_2$ is aspirated and cleared via the microfilter, and the same amount of $CO_2$ is replaced simultaneously. Normally, the working volume at laparoscopy is 3–5 l. The clearance volume is 4 l/min. In other words, the entire intra-abdominal $CO_2$ content is exchanged every minute. In this closed system exactly the aspirated $CO_2$ amount is replaced. The smoke-containing $CO_2$ is not released in the operating area.

The continuous $CO_2$ exchange makes laparoscopic surgery much easier. Working gaps are nearly eliminated. Therefore, even the use of the $CO_2$ laser in laparoscopic surgery creates no problems of vision, and continuous laser surgery is made possible.

[4] Richard Wolf Company, Knittlingen.

# HF Electrosurgery in Operative Gynecologic Endoscopy

S. DE BLOK [1] and D. J. HEMRIKA

## General Aspects of Electrosurgery

### Introduction

High-frequency alternating current has been used in general surgery for many decades to achieve hemostasis by coagulation or to perform cutting procedures with concomitant sealing of small vessels. In theory, a wide range of frequencies can be used to achieve the same electrosurgical effects, but there are some practical restrictions. Frequencies below 100 kHz cause nerve and muscle stimulation, resulting in the risk of electrocuting the patient. Electromagnetic waves in the megahertz range are difficult to confine to conducting wires because of the phenomena of inductance and capacitance. Due to these limitations, HF generators used in electrosurgery usually operate at frequencies between 200 and 800 kHz.

### Unipolar and Bipolar Equipment

Electrical energy can be applied to human tissues with either unipolar or bipolar instruments. In unipolar electrosurgery a small active electrode delivers the energy at the desired spot, allowing the current to flow through the body to a relatively large neutral ground plate attached to the patient's buttocks or thigh. The electrical current will take the "way of least resistance" to reach the ground plate. The electrical resistance of tissue is determined by its electrolyte content: in abdominal surgery the intestines, the urinary bladder, and the ureters all contain fluid with a high electrolyte content and are therefore excellent electrical conductors, making them prone to unintended electrical injury. Unipolar current can be used for either cutting or coagulation. Unipolar coagulation is most effective in a dry operating field because excess fluid will cause the current to disperse in all directions, thereby decreasing the local effect. With bipolar instruments both the active and the neutral electrode are contained in the same probe, and current will flow only through the tissue that is grasped between the tips of the electrodes (in bipolar electrosurgery it is essential to have a short distance between the electrodes: when both tips touch, the system is short-circuited and no current will flow!). Because current flows between the tips of the electrodes and not between the electrode and the patient, bipolar instruments are poor cutting tools. For the same reason, however, they have excellent coagulat-

[1] O. L. V. Gasthuis, 1e Oosterparkstraat 179, 1091 HA Amsterdam, The Netherland.

Lasers in Gynecology
Edited by G. Bastert and D. Wallwiener

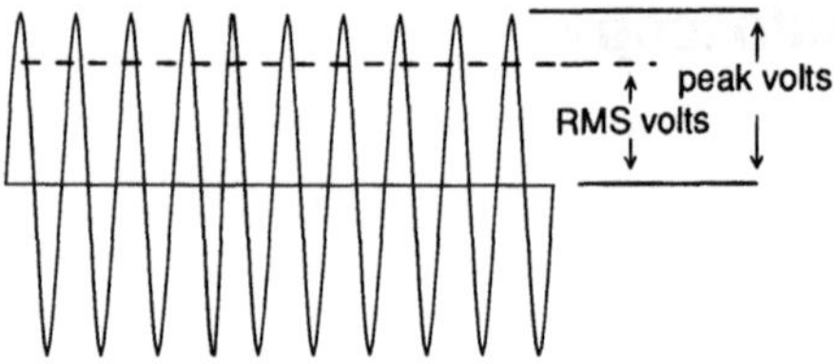


Fig. 1. HF cutting current

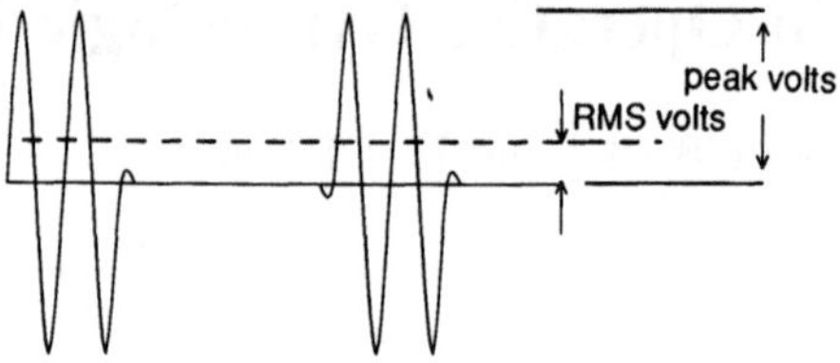


Fig. 2. HF coagulating current

ing properties, especially in a wet operating field (the electrolyte fluid between the tips facilitates the flow of current). Precise coagulation of small bleeding vessels can even be accomplished under constant irrigation, as is practiced in microsurgical procedures.

## Cutting and Coagulating Current

Electrosurgical equipment can generate either cutting or coagulating current. Cutting current is characterized by a continuous sine wave alternating at the frequency of the generator. The amplitude of the sine waves (peak voltage) determines the power output (expressed in Watts) of the current. Because of the on-off nature of alternating current the effective average voltage is lower than the peak voltage and is expressed as the root mean square voltage (RMS; Fig. 1).

Coagulating current consists of short bursts of sine waves with a relatively long pause between bursts. At equal peak voltages a coagulating current has only about 30% of the power of a cutting current, because the sine wave is turned off most of the time (Fig. 2). To put it differently: the RMS voltage of coagulating current is much lower. The ratio between peak voltage and RMS voltage is called the crest factor.

Modern HF generators used in surgery are able to produce pure cutting waveforms (crest factor 1.4) and pure coagulating waveforms (crest factor ± 8), as well as intermediate waveforms with increasing crest factors to produce cutting currents with more hemostatic effects.

## Biological Effects of HF Current

Depending on the type of waveform used and the power applied, three different effects on tissue can be achieved: desiccation, cutting, and fulguration (Fig. 3).

*Desiccation* is achieved by passing current through tissue. Due to the electrical resistance of the tissue, heat is generated, which will slowly drive the water out of the tissue. To achieve this the electrode must be kept in good contact with the tissue and low power must be used (< 30 W); otherwise a fulgurating effect will be achieved.

For desiccation any type of current can be used (cutting or coagulating current) in monopolar mode, but the bipolar forceps is best designed for desiccation. At power settings of 25–50 W the bipolar forceps will seal vessels as large as the ovarian artery through the mechanism of desiccation.

*Cutting* requires a continuous sine wave with adequate peak voltage pow-

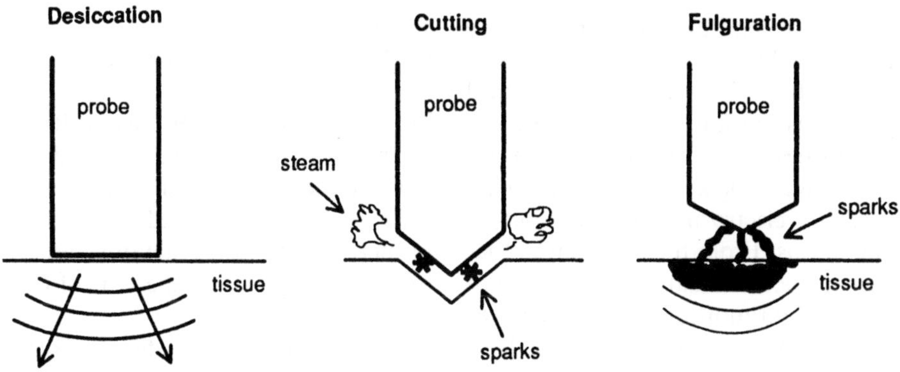


Fig. 3. Various biological effects of HF current

er to create sparking to the tissues that are to be divided. In electrosurgical cutting the electrode should not touch the tissue. Due to the high peak voltage the air gap between the electrode and the tissue becomes ionized and a spark jumps from the electrode to the tissue.

A spark is densely packed with electrical energy, creating intense heat on the spot where it hits the tissue. Intracellular water will boil instantly and will be converted to steam, making the cell explode. Almost all energy will be conducted in the steam, thereby preventing damage to adjacent cells.

Pure cutting current (creast factor 1.4) will produce a nice, clean cut without a thermal effect on the cut edge, resulting in a freely bleeding incision. To achieve better hemostasis while cutting with an electrosurgical "knife" the crest factor can be increased in blended current. The higher the crest factor, the more fulguration and charring of the cut edges will occur.

*Fulguration* of tissue is attained by the short intermittent bursts of sine waves typical for coagulating current. To achieve pure fulguration the electrode should not be in contact with the tissue. Because the sparks produced by this type of waveform are dispersed more widely, and because of the pause between each burst, heating of cellular water is more gradual and will not result in formation of steam that severs tissue as in the cutting mode. Instead, the cells are slowly dehydrated. Due to the high peak voltage, current will continue to flow with increasing desiccation, resulting in the formation of eschar.

Coagulation is a combination of desiccation and fulguration. When the coagulating electrode is brought in contact with tissue desiccation will take place first. As the tissue dries out, its electrical resistance increases until the electrical contact is broken. Now sparks will jump to the nearest moist tissue, as in fulguration.

# Electrosurgery in Operative Laparoscopy

## Introduction

Electrosurgical instruments are indispensable in microsurgical operations performed by laparotomy. As these operations tend to be performed more and more by laparoscopy, it stands to reason that the same electrosurgical principles are applied in laparoscopic techniques. However, some important facts have to be kept in mind to safely perform electrosurgical laparoscopic operations. From the early days of laparoscopic sterilization, when monopolar fulguration of the oviducts was performed, many reports of inadvertent damage to intestines, bladder, and ureters have been published. Surgeons who oppose the use of electrosurgery in operative laparoscopy are eager to recall these unfortunate events.

Some complications of monopolar sterilization can be attributed to poor technique, like a poorly insulated electrode making direct contact with intestines. Other mishaps are caused by a lack of understanding of the laws of physics that apply to electrical current: when the tube is fulgurated or coagulated at its midisthmal portion it may well become electrically isolated from the uterus and the mesosalpinx. When a second burn is applied distal to the first, current will be forced to flow distally to the fimbrial end. When one of the fimbriae touch, for instance, the terminal ileum, thermal injury to the ileum can result. At high power settings, even sparking from the distal fimbrial end can occur, resulting in serious necrosis of intestinal structures.

With the introduction of bipolar desiccation of the tubes with adquately isolated forceps, sterilization procedures became much safer. With these historical facts in mind, it is possible to perform electrosurgical operations by laparoscopy in a safe and sensible way.

## Equipment for Laparoscopic Electrosurgery

To assure maximal safety peak voltages of the current used and power delivered at the tissue should be limited. This can be guaranteed by specific properties of the HF generator, the size of the electrodes, and the type of current applied.

The American Association of Gynecologic Laparoscopists has defined limits for electrosurgical generators to be used in laparoscopic procedures: the maximum peak voltage should not exceed 600 volts and the maximum power must be limited to 100 W (at impedances of 200–500 ohms). This means that not all standard operating room electrosurgical units are suited for laparoscopy.

In our department we use the Valleylab Force 2. This generator allows the operator to calibrate the power output of the monopolar and bipolar outlets on a continuous scale from zero to maximum power. The preset values are clearly depicted by digital displays. This makes the apparatus suitable for general surgery as well as for microsurgical operations.

Yet another way to minimize the power applied is the choice of electrodes. Remember that the smaller the electrode, the less power is needed to achieve a tissue effect.

The electrodes used in laparoscopic surgery should be well insulated to prevent inadvertent damage to intra-abdominal structures like the bowel. Insulation of the outer metal trocar shaft is not necessary, because even when current is conducted to the trocar sleeve the large area of contact with the abdominal wall will result in a very low current density, so thermal injury will not occur.

Keeping the electrode free of eschar is another way of minimizing the power used. When tissue effects are unsatisfactory, the electrode should always be cleaned and checked for erratic insulation before the power is turned up!

As was stated in the previous section on general aspects of electrosurgery, the choice of the electromagnetic waveform has dramatic consequences for the peak voltage and the power delivered at the cellular level. High peak voltages increase the risk of sparking; therefore, it is often wise to concede to blended cutting current for monopolar coagulating jobs instead of using pure coagulating current.

**Specific Electrosurgical Procedures**

Lysis of adhesions by electrodissection is accomplished by using a 1-mm microsurgical electrode. With filmy, loose adhesions a pure cutting current is used at 40–80 W. When adhesions are dense and vascularized a blended current should be used to assure hemostasis. Alternatively, small vessels can be occluded with bipolar coagulation preceding electrosurgical division.

For electrosurgical adhesiolysis using monopolar cutting current it should always be kept in mind that the current will spread beyond the trajectory of the cutting needle. Therefore, a safe margin of $\pm$ 2 mm between the cut edge and the structure to be divided should always be respected. In this respect, the electrosurgical cutting needle is inferior to surgical scissors and even to $CO_2$ laser.

When performing a salpingotomy for conservative laparoscopic treatment of ectopic pregnancy a blended current should always be used (crest factor 2–3) to ensure adequate hemostasis, especially when no vasoconstrictive agent like vasopressin is used. Typical power settings for this type of blended current would be between 10 and 30 W.

Coagulation of bleeding vessels is best accomplished by bipolar desiccation. In microsurgical procedures like lysis of tubo-ovarian adhesions or terminal tubostomy for hydrosalpinx, power settings as low as 15–30 W will suffice. Sealing of bleeding vessels is ideally accomplished under constant irrigation (e.g., by the Aquapurator from Wisap).

When coagulating bigger vessels higher power settings are needed (25–50 W). In these circumstances it is essential to compress the vessel between the tips of the bipolar forceps to assure complete desiccation of the vessel wall. Adherence to this principle is vital, because otherwise a partially occluded artery will not be able to withstand the systolic blood pressure and will start to bleed again during the operative procedure (or, even worse, postoperatively).

*Coagulation* or fulguration of endometriotic implants can be performed with unipolar electrodes. The major drawback of electrosurgery, i.e., poor depth control, is best demonstrated in these cases.

We prefer using a bipolar forceps to coagulate small endometriotic lesions. Larger implants can be treated with monopolar coagulation, but only when localized in anatomic regions where depth control is not a major concern (e.g., not in the vicinity of the ureters or on the sigmoid colon). To minimize the risk of sparking to adjacent organs a blended cutting current at 10–30 W is to be preferred to a pure coagulating current. When coagulating current is used the power should always be kept below 10 W.

For laparoscopic treatment of endometriosis we usually employ endocoagulation (Endocoagulator according to Semm by Wisap) or the $CO_2$ laser.

### Conclusion

Reich [4] states that "electrosurgery is the mainstay of microsurgical laparotomy today," and in our opinion, this holds true for laparoscopic surgery as well. When the fundamentals of electrosurgery are clearly understood by the operator and when preventive principles are strictly adhered to, electrosurgical instrumentation can be an integral part of laparoscopic operative procedures.

## Hysteroscopic HF Electroendosurgery

### Introduction

Diagnostic hysteroscopy has become an indispensable technique in modern gynecology to evaluate intrauterine pathology. Similar to other disciplines in medicine, the endoscopic recognition of pathology with hysteroscopy has initiated the development of specific hysteroscopic endosurgical therapeutic procedures. Rigid as well as flexible endoscopes used in other disciplines have been introduced to gynecology and modified into specific hysteroscopic endosurgical tools. The introduction of the urologic electrosurgical resectoscope, already used for many decades for transureteric prostatectomy, has been the most promising development.

The hysteroscopic resectoscope of today, modified into a rigid hysteroresector, has found application as the major instrument for therapeutic hysteroscopic HF electroendosurgery. The hysteroresector can be equipped with an inward-sliding loop, a roller ball coagulator and an electrosurgical knife. The hysteroscopic endoresectoscope can be used only with an HF generator that can deliver true cutting, true coagulation, and blended current. Each specific therapy requires extensive knowledge of electroendosurgery.

### Resectoscope, Procedures

The hysteroresector that has found the most extended application has a diameter of 9.5 mm (27 French). A resector with a smaller 7 mm (21 French) has been developed, but intrauterine pathologies usually require the 9.5-mm hysteroresector. Resectoscopes for hysteroscopic endosurgery consist of an inner and an outer sheath through which a continuous flow of the distention medium with regulated pressure is possible.

The tip of the outer sheath has holes for outflow, and in the tip of the inner sheath a ceramic part provides cutting of tissue parts with the inward-sliding loop.

Tubing can be connected to the stopcock-closed sheaths so that inflow and outflow can be monitored constantly. Monitoring of distention fluid intravasation in hysteroscopic electroendosurgery has turned out to one of the major safety precautions to be taken.

Only nonconductant electrolyte-free media like sorbitol 4%, glycine 1.5%, or hyskon can be used as distention media.

4
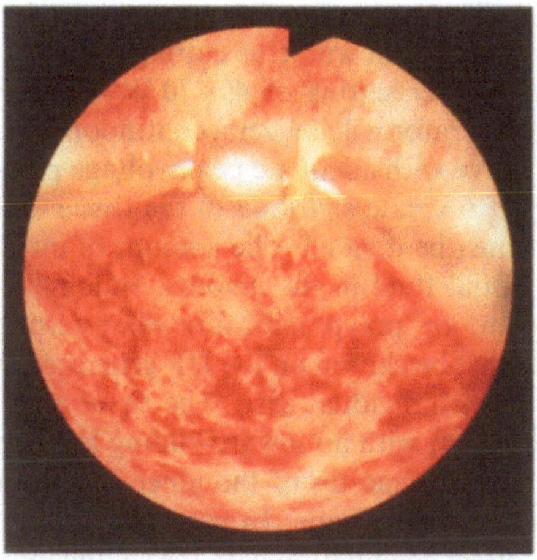

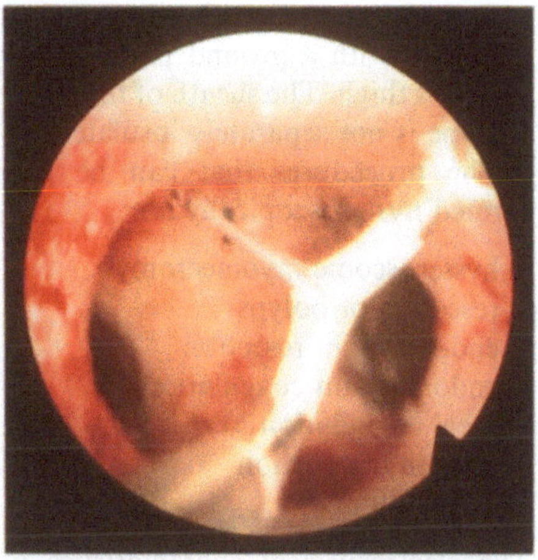
5

6
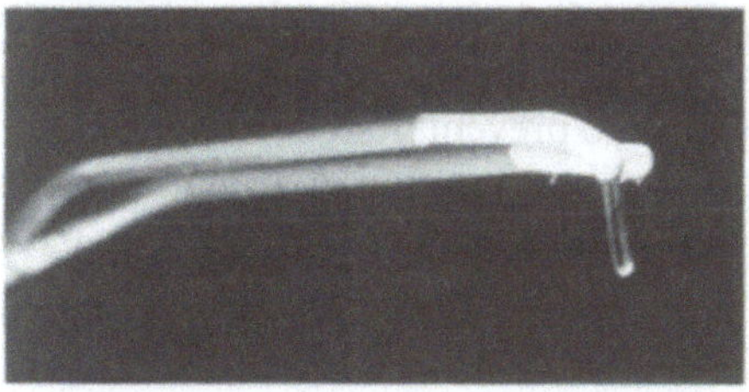

7
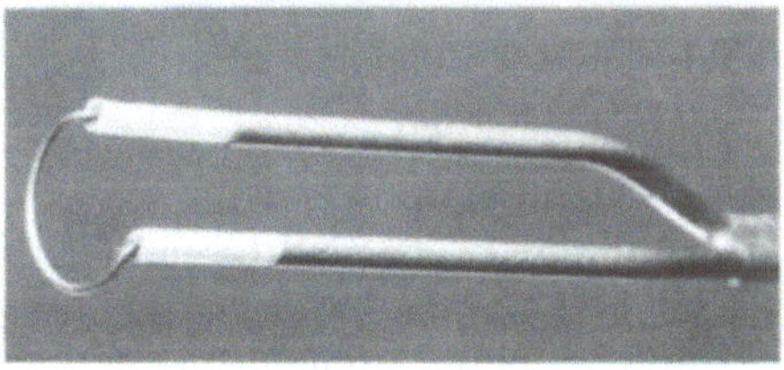

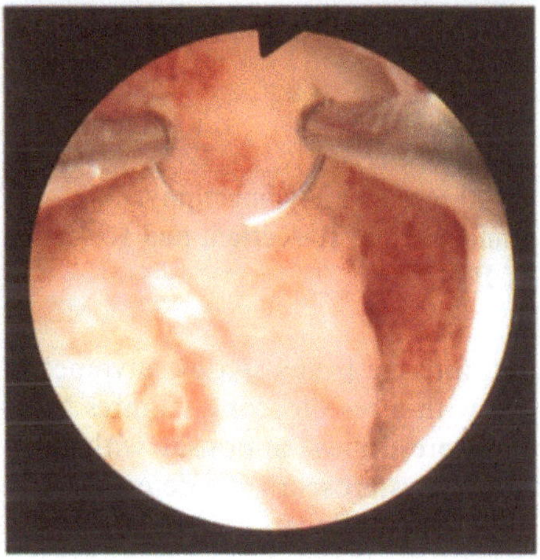
8

**Fig. 4.** Intrauterine coagulation with roller ball coagulation electrode in fundal and right cornual area

**Fig. 5.** Uterine septum preoperatively; 9.5-mm continuous-flow hysteroresector with dissection knife electrode (Olympus)

**Fig. 6.** Dissection knife electrode of the 9.5-mm continuous-flow hysteroresector (Olympus)

**Fig. 7.** Looped-wire electrode for 9.5-mm continuous-flow hysteroresector (Olympus)

**Fig. 8.** Looped-wire electrode being used to cut a submucous myoma with a 9.5-mm continuous-flow hysteroresector (Olympus)

The HF generator is connected to the working element, to which the working electrodes are attached. The telescope is 4 mm, 12° oblique to provide a clear vision of the tubal ostia in the cornual areas.

Several working electrodes can be inserted into the resectoscope, such as a cutting-knife electrode, a roller-ball coagulator electrode, and an inward-sliding loop for cutting and coagulation, depending upon the procedure that must be performed (Figs. 4–8).

Procedures are done in the monopolar mode with a ground plate on the patient's thigh. The sheath of the resectoscope is not separately insulated.

The procedures that can be performed hysteroscopically are:

- hysteroscopic myomectomy
- resection of polyps
- endometrium resection
- endometrium coagulation
- focal coagulation
- septum transection
- synechotomy

The electrosurgical characteristics of these procedures will be discussed separately.

## Hysteroscopic Myomectomy, Endometrium Resection, and Resection of Polyps

Intrauterine hysteroscopic myomectomy with a resectoscope requires a minute presurgical appraisal with diagnostic hysteroscopy to determine the size, number, location, and intramural extension of the myoma.

Because of the rich vascularization and the firm tissue quality of myomas, cutting must be used alternately with coagulation. For this reason, only low-viscosity fluids such as sorbitol 4% or glycine 1.5% can be used for distention. The dextrose particles of hyskon will immediately caramelize on the cutting loop, and hyskon is thus contraindicated as a distention medium. For endometrium resection the patient must be pretreated with danatrol. A thin endometrium will allow cutting through the basal membrane into the superficial myometrium. If the patient is not pretreated, more cuts are needed to reach this layer.

The myoma, endometrium, or polyp must be cut with true cutting current at a power setting of 40–70 W. The HF generator will deliver a continuous sine wave with adequate peak voltage power to create sparking to the tissue that must be divided. The amount of interstitial fluid in the myoma, endometrium, or polyp indicates what setting must be chosen on the HF generator in each individual case. The tissue parts are cut with an inward-sliding loop on the ceramic tip of the inner sheath of the resectoscope. In case of bleeding, direct coagulation with the sliding loop can be done immediately with true coagulation current at proximately 30–50 W to prevent fluid overload through direct intravasation of distention medium. With a blended current tissue parts stick to the loop.

## Endometrium Coagulation, Focal Coagulation

Coagulation to destroy the endometrium can be done through the basal membrane in the myometrium by thermal heating and protein denaturation. This can be done with the roller ball electrode with true coagulation from the HF generator at 30–50 W. The biological effect of fulguration will be obtained.

Roller ball coagulation can be done only with a nonconductant electrolyte free low-viscosity fluid distention medium such as sorbitol 4% or glycine 1.5%. Again because of caramelization, hyskon cannot be used.

At the power settings mentioned, tissue effects up to 4 or 6 mm into the myometrium can be obtained. Usually the procedure is quick, because through the sliding/rolling of the electrode on the pretreated thin endometrium, the vascular architecture underneath the basal membrane is destroyed.

The patient may notice vaginal discharge for 3–4 weeks as a result of deep energy-induced necrosis caused by protein denaturation (protein denaturation occurs at 60 °C). Depth control is not as critical in the uterus as it is in laparoscopy. However, it is advised that electrosurgery should be done under laparoscopic control, since leakage of current to adherent intestines in the thin cornual areas can cause major complications. Focal coagulation of vascular anomalies can be performed in patients with intermenstrual bleeding and can be done even with local anesthesia in an office setting with the small hysteroresectoscope.

## Septum Transection

The hysteroscopic treatment of uterine septa has developed from a cutting loop resection of the septum into a needle-knife electrode transsection. A knife transsection with the resectoscope can be done with a cutting power of approximately 50 W using hyskon as well as low-viscosity fluid distention.

A septum usually contains no vessels and will rarely cause any bleeding during HF electroendosurgery.

With the needle-knife electrode the septum should be transsected under laparoscopic or ultrasonic control to prevent fundal uterine perforation.

## Conclusion

HF electroendosurgery using the resectoscope has become the major method of hysteroscopic therapeutic endosurgery. Since hysteroscopy as a diagnostic tool is now being introduced into many training programs for gynecologists, it is to be expected that the number of hysteroscopic surgical procedures will expand rapidly. Amongst the advantages of electroendosurgery are the avoidance of laparotomy, hysterectomy, or hysterotomy in many different cases. A true knowledge of the HF equipment, the resectoscope and biological effects are the indispensable basis for safe application of these promising, almost established techniques in gynecological endoscopy.

# References

1. De Cherney A et al. (1984) The management of intractable uterine bleeding utilizing the cystoscopic resectoscope. In: Siegler A, Lindemann HJ (eds) Hysteroscopy, principles and practice. Lippencott, Philadelphia, pp 140–142
2. Neuwirth RS (1983) Hysteroscopic management of symptomatic submucous fibroids. Obstet Gynecol 62:509–511
3. Perino A, Cittachini E, Hamou JE et al. (1985) Hysteroscopic management of uterine septa. Acta Eur Fertil 16:331–335
4. Reich H (1990) Treatment of pelvic masses and myomectomy. In: Balmaceda JP, De Cherney A (eds) New developments in endoscopic and laser surgery. Course syllabus, p 126
5. Semm K (1970) Operative manual for endoscopic abdominal surgery. Yearbook Medical, Chicago

# Importance of Video Simulation Training for Operative Endoscopy

A. PIER [1] and F. GÖTZ

Those who have attentively observed the development in abdominal surgery during the past 2 years have seen astonishing progress – not only in Germany. Advances have been made toward a less radical method of operation for the patient; these have culminated in the idea of "maximal noninvasive surgery," i.e., minimal invasive surgery [7]. The starting point of this development is laparoscopic diagnosis and the endoscopic-pelviscopic operating technique in gynecology, developed by Palmer in Paris, Semm in Kiel, and Frangenheim in Constance [3, 4, 6]; this operating technique has been continuously improved through arduous work.

Laparoscopy is a technically perfect procedure for diagnosis and offers the possibility of performing – with the help of special instruments – endoscopic-laparoscopic operations immediately after the diagnosis is made. Thus, more and more surgeons are now realizing the superb possibilities offered by the idea of minimal invasive surgery.

The following procedures can be performed laparoscopically and have already become routine in many gynecological hospitals: removal of ectopic pregnancy, ovarian cystectomy, myomectomy, and oophorectomy.

To date, two of the most frequent procedures in abdominal surgery can be performed endoscopically-laparoscopically: appendectomy and cholecystectomy. These methods of operation have been proved and tested in a number of hospitals in Germany, France, Switzerland, and the United States. It now appears that the field of application can be extended to hernia repair, vagotomy, and bowel surgery.

What qualifications are required of the surgeon? Even for a surgeon who possesses great experience in conventional abdominal surgery and gynecology, qualification involves a thorough training in this new technique of endoscopic surgery. The laparoscopic equipment must be mastered and a basic knowledge of the instruments and apparatus must be acquired.

What possibilities exist today?

1. The surgeon can cooperate with clinics or centers where the endoscopic-laparoscopic techniques of operation are thoroughly demonstrated and where he or she can also be trained under the tutelage of experienced surgeons. (Training courses are planned in our Department for Laparoscopy starting in July 1991).
2. The surgeon can train on a model (the so-called simulation trainer) with corresponding training modules and prepared organs: e.g., liver

[1] Chirurgische Klinik, Abt. für laparoskopische Chirurgie, RWTH Aachen, von Werthstr. 5, 4048 Grevenbroich, FRG.

Lasers in Gynecology
Edited by G. Bastert and D. Wallwiener

with gallbladder and bile ducts; caecum with appendix; gynecological organs.

We introduce here a simulation trainer with which we are in a position to teach the skills needed and to perform – as realistically as possible – the basic steps of laparoscopy. Training with unipolar and bipolar current as well as the use of a laser instrument is ensured. The main feature of the simulation trainer developed and already tested by us is the possibility of creating a pneumoperitoneum.

Thus, an important point of the training program with regard to reality is fulfilled. Further features of the simulation trainer, shown in Fig. 1, are listed in Table 1. The torso form of the simulation trainer provides a direct impression of the patient. The different

**Table 1.** Characteristics of the simulation trainer

1. Torso form
2. The human abdominal skin is simulated with a special tissue.
3. Insertion of the optic trocar in "Z"-Track Entry Method is possible.
4. All safety tests can be performed.
5. "Abdominal wall" bulges when the pneumoperitoneum is created.
6. Creation of an inadvertent preperitoneal emphysema is possible.
7. The light interior ensures an optimal illumination of the situs.
8. Additional lateral opening allows for changing of training modules.
9. A pit for training of skill and for cleaning purposes has been stamped into the bottom plate.
10. A deep pit holds the different training modules.
11. A cholangiography can be performed intraoperatively by means of an additional special module.
12. Irrigation is possible.
13. Use of laser instruments is possible.

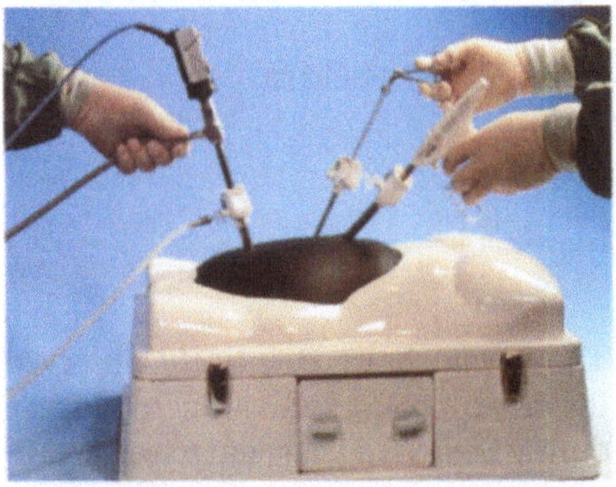

1

2

**Fig. 1.** Endoscopic simulation trainer

**Fig. 2.** Lifting the abdominal wall and inserting the Veress needle

trocars can be inserted – as in reality – into the pelvi-trainer according to the chosen method of operation. A lifting of the abdominal wall for insertion of the Veress needle and the trocars is possible (Fig. 2).

## Training Program

After creation of the pneumoperitoneum and insertion of the trocars, the surgeon has to become used to the new perspective of the operative field; unlike direct eye visualization of the anatomy he must now work from a video screen presenting a two-dimensional

**Table 2.** Summary of the Training program

| | |
|---|---|
| *Phase 1* | |
| 1 | Creation of pneumoperitoneum including safety tests |
| 2 | Insertion of trocar accommodating the scope |
| 3 | Placement of working trocars |
| *Phase 2* | |
| 1 | Diagnostic all-round view on the screen |
| 2 | Grasping and coordination exercises |
| 3 | Training modules 1 and 2 |
| *Phase 3* | |
| 1 | Exercises for appendectomy procedures (module 3) |
| 2 | Cholangiography (module 4) |
| 3 | Intracorporeal suture and knot-tying techniques (module 5) |
| *Phase 4: Exercises with Animal Specimens* | |
| 1 | Adhesiolysis (unipolar/bipolar/scissors technique) |
| 2 | Cholecystectomy (unipolar/bipolar/scissors technique) |
| *Phase 5: Use of Laser Instruments* | |
| 1 | Simple exercises regarding the functioning of laser instruments |
| 2 | Adhesiolysis |
| 3 | Cholecystectomy |

**Table 3.** Safety tests

| | |
|---|---|
| 1. | Injection of NaCl solution |
| 2. | Aspiration |
| 3. | Sip test |
| 4. | Checking the regulator for intraabdominal pressure |

image and he must develop a whole new (three-dimensional) depth perception and eye-hand coordination of the instruments.

We have established a Training Program wherein the different exercises are based on one another, and where the degree of difficulty increases. The training course is divided into five phases. Phase 5 is provided for departments that work with a laser instrument (Table 2). With the help of a video camera, all exercises are transmitted onto a screen.

It is possible for the surgeon to perform several exercises by himself. However, we recommend a training like the operative situation within a team of two to maximally three surgeons.

## Phase 1

### Creation of the Pneumoperitoneum

When the Veress needle is inserted the corresponding safety tests ([1–4], Table 3) can be performed before creating the pneumoperitoneum. If, for example, the insufflation needle has not been placed correctly, the inadvertent creation of a preperitoneal emphysema is possible.

The insufflation of 6 l/min with a selected pressure of 12–14 mm Hg can be controlled optically by observation of the bulging "abdominal wall" (Fig. 3) and manually by percussion; moreover, this can be controlled by observing the regulator for internal abdomi-

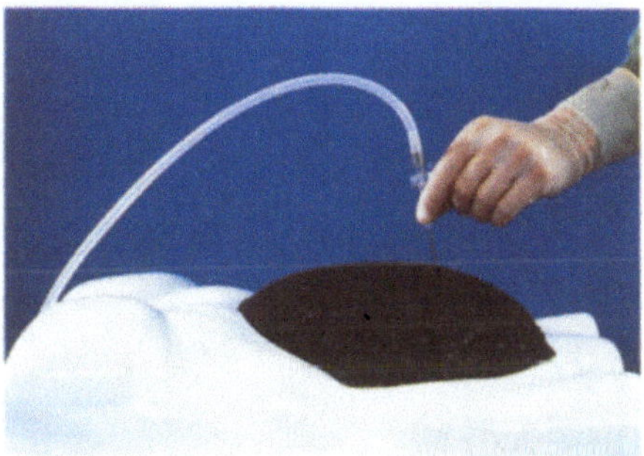

**Fig. 3.** "Abdominal wall," tense after insufflation

nal pressure of the high-flow electronic insufflator.

### Insertion of the Trocar Acommodating the Scope

The insertion of the trocar accommodating the scope can be performed straight (single-layer material) or with the "Z" Track Entry Method according to Semm [6]. This will be possible due to the synthetic material of the abdominal wall which is multilayered and elastic. If necessary, this synthetic material can be changed in a quick and uncomplicated way with the help of a snap closure.

### Trocar Placement

According to the real operative situation, the surgeon can choose the correct position of the working trocars by palpating the abdominal wall under visual control either from the outside or from the intra-abdominal side.

The trocars are inserted after a stab incision has been made or directly under visual control (Fig. 4). Like in the "live situation," the abdominal wall reacts to the force necessary for trocar entry and regains its normal form afterwards. The assistant can fix the trocars and facilitate the insertion of the working instruments for the surgeon or operate the video camera, when required.

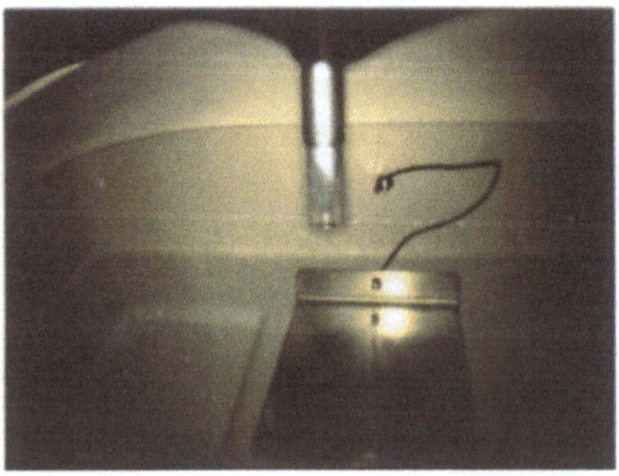

**Fig. 4.** Insertion of trocars under visual control

## Phase 2

### Diagnostic All-round View

Like in a live operation, the surgeon first gets a diagnostic panoramic view with the help of the video camera; he can familiarize himself with the interior of the simulation trainer on a monitor. The optimal illumination depends on the movement of the camera and on the chosen light intensity of the cold light source. Fadings and reflections must be avoided by correct focusing.

### Grasping and Coordination Exercises

For trocar placement, see Fig. 5 (1, 2). Skill is first trained with matches and peas. Six to eight matches are placed within the pit which has been stamped into the bottom plate; the pit has a diameter of approximately 8 cm. With the help of a grasping forceps these matches can be put individually into a matchbox, each with the percussion cap pointing in the same direction. This exercise becomes more difficult when the matchbox is not fixed to the bottom plate; it may slip away if the matches are not placed correctly. We recommended that the surgeon operate the camera himself during this exercise and also during the next one. The assistant should only fix the trocars and help in inserting the instruments.

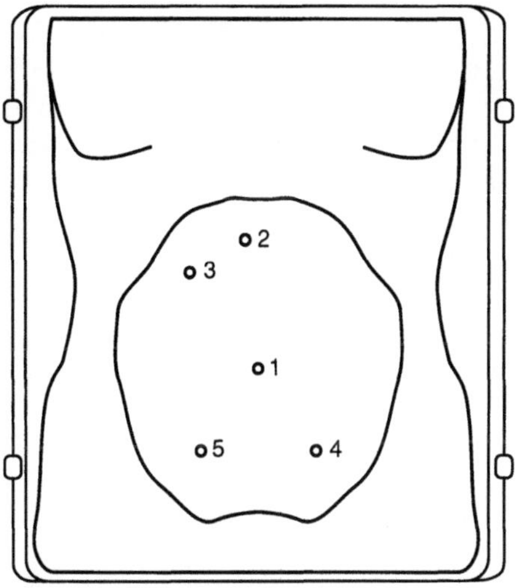


**Fig. 5.** Positioning of trocars 1–5. *1*, Optic trocar 11 mm; *2*, Working trocar 11 mm; *3*, Working trocar 5.5 mm; *4*, Working trocar 5.5 mm; *5*, Working trocar 11 mm

Instead of matches, 8–10 peas are now put into the pit. These can simulate lost gallstones or mini-dissecting pledgets. The trainee must also grasp the peas with the grasping forceps and put them into the corresponding matchbox.

During these two exercises, care must be taken that the camera follows the fixed objects until they are put into the matchbox. Thus, the surgeon gets the feeling for the coordination of camera lead and manipulation of an instrument with the performance of gripping, grasping, and searching movements. Before the surgeon begins with the next training point, he should repeat these skill exercises several times.

## Training Modules 1, 2

For trocar placement, see Fig. 5 (1, 2, 3). The following exercises train, above all, the coordination and the sense of touch with two instruments. For these exercises, five 9 × 15 cm training modules are available. Figure 5 shows how the trocars have to be placed for these exercises.

The first training module is fixed in the prepared space in the bottom – a Plexiglas element containing six fillister-head screws arranged in a circle. The screws are 5 cm long. Additional screw sockets have been screwed on to two screws at head level (Fig. 6).

Three to five elastic bands have been placed in the working pit. Now the surgeon tries – with the help of *one* grasping forceps that has been inserted through point 1 in Fig. 5 to grasp an elastic band and to stretch it over two screw heads facing each other. With this exercise the surgeon gets the feeling for a well measured traction and leverage. In addition, the correct placing of loops is trained.

Next, a *second* grasping forceps is inserted into the simulation trainer through point 2. Now the elastic band should be stretched over the two screw sockets (screwed on to the screw heads) so that it can be placed exactly in the second screw thread.

In this exercise it is best if the assistant operates the camera and at the same time manipulates the trocars. The surgeon should only concentrate on the handling and coordination of the grasping forceps. If need be, a second assistant can fix the trocars.

Module 2 consists of several elastic bands that have been stretched longitudinally (Fig. 7). These bands are led through two eyes and are fixed at each end with a bent hook. These bands represent tissue structures which have to be prepared or to be clipped in order to be dissected afterwards. Here, again, it is best if the surgeon only handles the instruments while the assistant guides the camera.

6 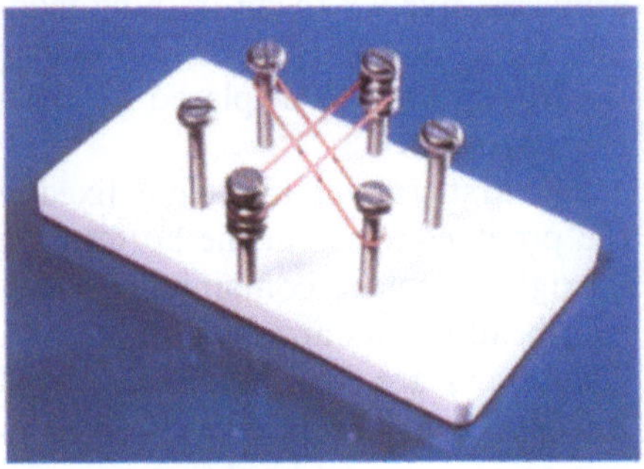

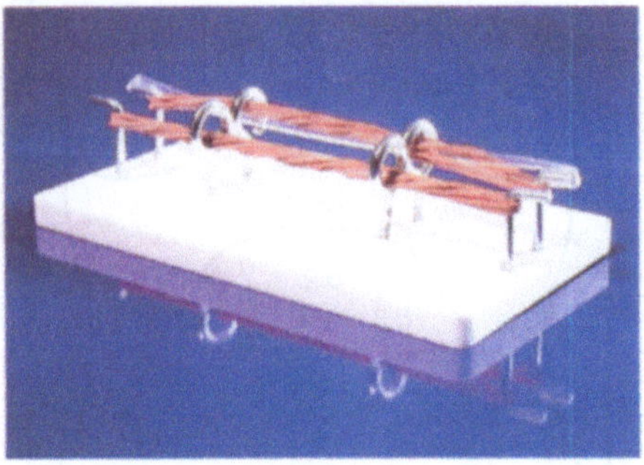 7

**Fig. 6.** Training module 1

**Fig. 7.** Training module 2

With a so-called dissecting forceps the skill in dissecting can be trained for the exposure of tissue. The dissecting forceps is pushed into the stretched elastic bands. The individual loops can be excellently exposed by spreading of the jaws (Fig. 8). Depending on the skill of the individual, it is also possible to grasp an elastic band and to ligate an area of 1–2 cm on both sides with two endoclips; then this will be dissected with scissors and removed through an extractor.

**Fig. 8.** Preparing elastic bands

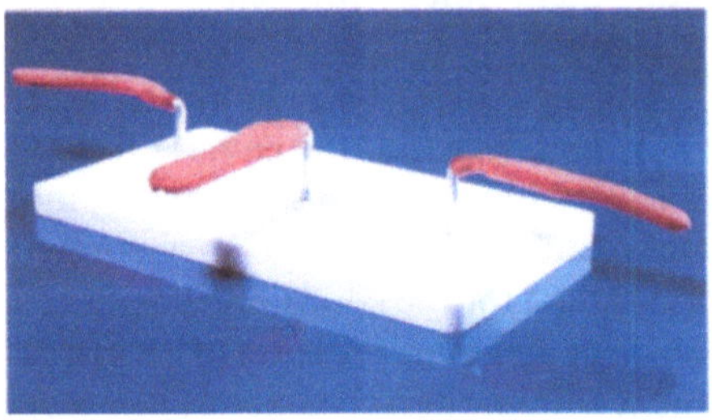

**Fig. 9.** Training module 3 (appendix)

## Phase 3

### Exercises for Appendectomy Procedure – Module 3

In order to simulate the individual steps of an appendectomy procedure, module 3 is placed in the simulation trainer. It consists of a plate with three appendix foam models with a lateral kinking at the base (Fig. 9). The model is fixed with an adhesive closure in the right lower quadrant of the simulation trainer.

*Operating Procedure.* The Roeder loop carrier with the Roeder loop is inserted into the 5.5 mm working trocar that has been placed in the left lower quadrant (point 4 in Fig. 5) of the simulation trainer. As shown in Fig. 5, the third working trocar accommodating the appendix extractor is inserted un-

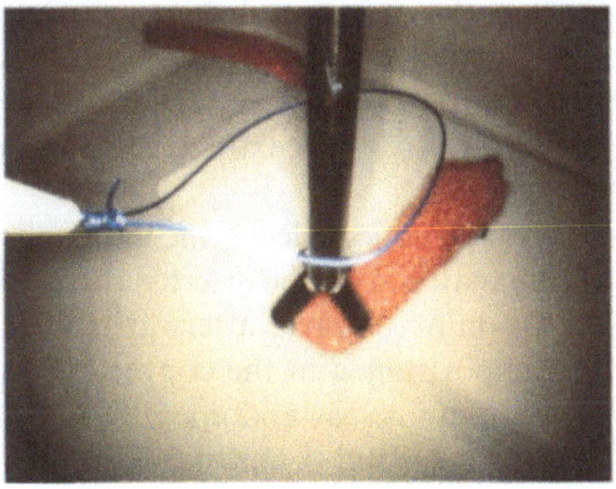

10

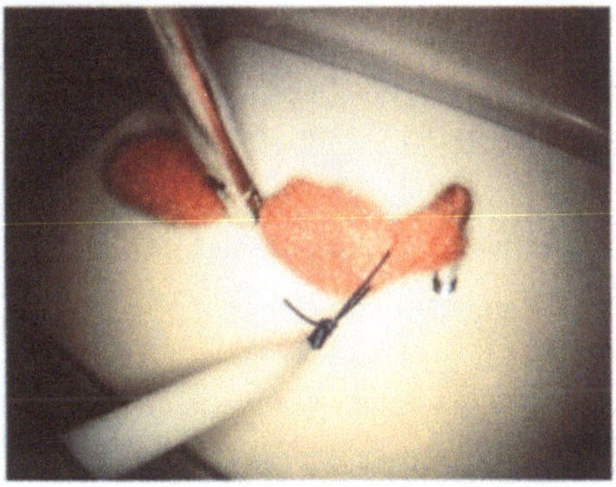

11

Fig. 10. Grasping the top of the appendix

Fig. 11. Placement of the Roeder loop

der visual control into the right lower quadrant (point 5). Now, also under visual control, the Roeder loop is advanced directly above the first appendix model, until the loop can – like a lasso – be pulled over the grasping forceps that has been inserted through the appendix extractor (Fig. 10).

The tip of the appendix is grasped and drawn into the appendix extractor. After reaching the breaking point, the loop carrier can be advanced in order to ligate near to the base of the appendix (Fig. 11).

The extracorporeal suture is cut right behind the breaking point; the carrier device and the delivery system can be removed from the 5.5 mm trocar. With scissors that have been inserted through the same trocar, the suture is cut near to the base at an appropriate distance of 1–1.5 cm from the knot. In this case it is helpful if the assistant tightens the suture slightly.

**Cholangiography – Module 2**

For trocar placement, see Fig. 5 (1, 2, 3). With module 2 an intraoperatively performed cholangiography can be simulated (Fig. 7). The unit to train the dissecting technique with module 2 is used, as described before. Several elastic bands are stretched longitudinally; however, they surround a soft infusion tube which has a diameter of approximately 2 mm.

*Operating Procedure.* According to the trocar placement shown in Fig. 5, first a grasping forceps is inserted through point 2 and scissors through point 3. The elastic bands, which have been stretched longitudinally, are dissected in order to expose the "cystic duct." Now, micro-dissecting scissors are introduced through point 3 and a transverse incision is made on the cystic duct. The cholangiocatheter with a diameter of 1–2 mm is inserted through trocar 2 into the simulation trainer with the help of the Roeder loop carrier. The catheter is fixed with a grasping forceps (point 3), advanced to the incision, and inserted into the duct. In this case it is helpful to grasp the tip of the catheter 1.5–2 cm at the tangential free end. This facilitates the intubation of the infusion tube (Fig. 12).

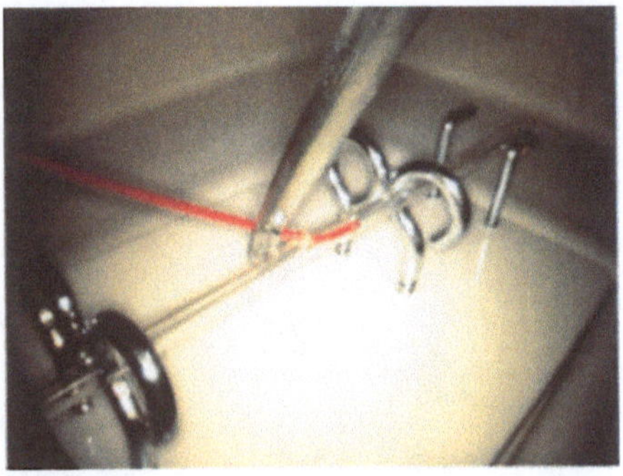

**Fig. 12.** Intubation of a cholangiography catheter

### Intracorporeal Suture and Knot-tying Technique – Module 4

For trocar placement, see Fig. 5 (1, 2, 3). With module 4 the intracorporeal suture placement can be simulated. The module has the same size as the others already described; it consists of two clamping devices holding two foam models approximately 5 × 8 cm. These foam models are fixed such that they join at both sides at a distance of 5 mm and in an angle of approximately 45° (Fig. 13). In order to ensure "realistic" suturing, we recommend the use of a suture with a curved needle (e.g., HS 27). The needle fits exactly through the 10-mm appendix extractor sleeve.

*Operating Procedure.* In order to insert the suture material into the abdominal cavity, the following steps have to be taken.

A 5-mm suture holder is inserted through a 10-mm appendix extractor sleeve using a reducer or a converter. The strand, approximately 10 cm in length, is grasped at the end and pulled out of the extraction sleeve. Then the needle holder is again inserted into the extraction sleeve, and care must be taken that the suture end does not slide into the sleeve. Now the needle is grasped tangentially and completely drawn into the extraction sleeve (Fig. 14).

The extraction sleeve including the suture material can be inserted through an 11-mm trocar; afterwards, the needle positioned in the converter can be inserted into the abdomen. With the help of a second 5-mm needle holder that has been inserted through the 5.5 mm working trocar, the needle is grasped. One foam model is fixed with the second needle holder, and the first suture can be placed on the foam model (Fig. 15). It is possible to secure the suture ends with a knot or with a clip (Fig. 16).

This exercise requires certainly a thorough training, and we therefore recommend that these steps be repeated several times.

## Phase 4 – Exercises with Animal Specimens

### Adhesiolysis (Unipolar/Bipolar/Scissors Technique)

#### Unipolar Dissection

For trocar placement, see Fig. 5 (1, 2, 3). Adhesiolysis or dissection of tissue can be trained on an animal specimen, for example a chicken. We recommend marking an area of approximately 4 × 4 cm for each surgeon in case several are working on the same specimen.

Before the chicken is placed in the trainer, the supplied neutral electrode has to be connected to the plug situated at the front end of the simulation trainer. Afterwards, the trainer is closed in

13 

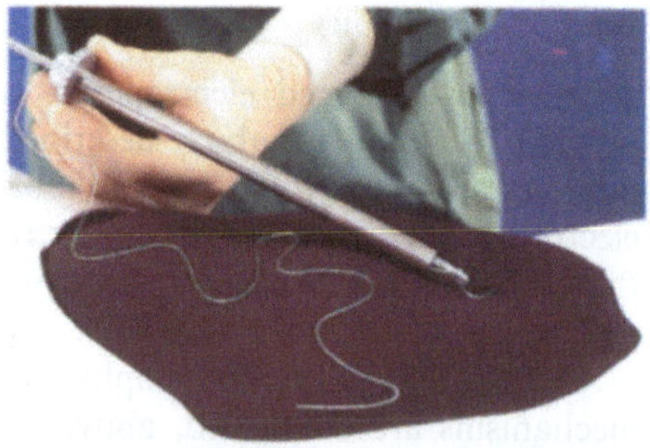 14

15 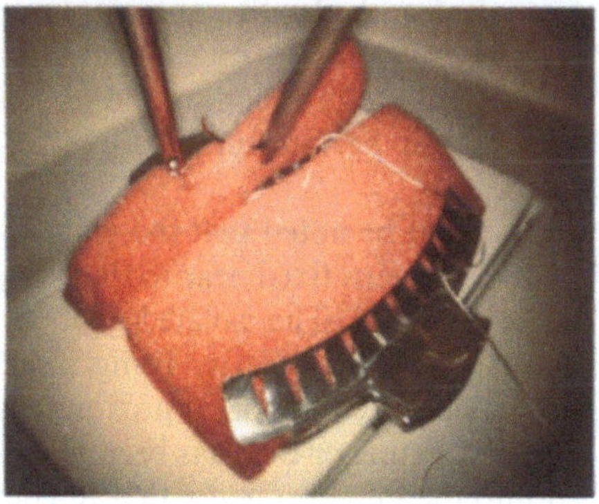

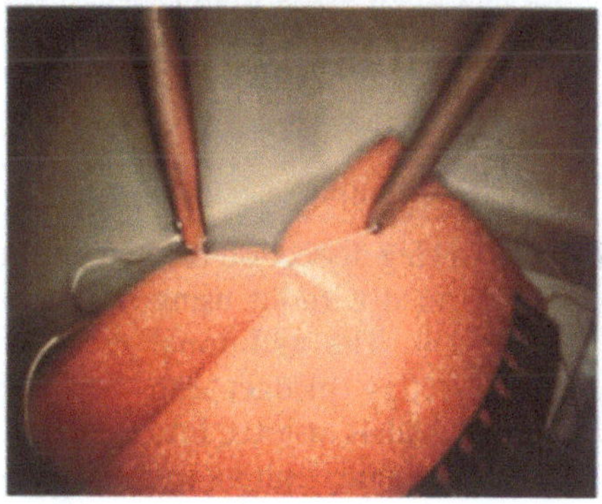 16

**Fig. 13.** Training module 4 (suture technique)

**Fig. 14.** Procedure for insertion of sutures

**Fig. 15.** Demonstration of suture techniques

**Fig. 16.** Intracorporal knotting

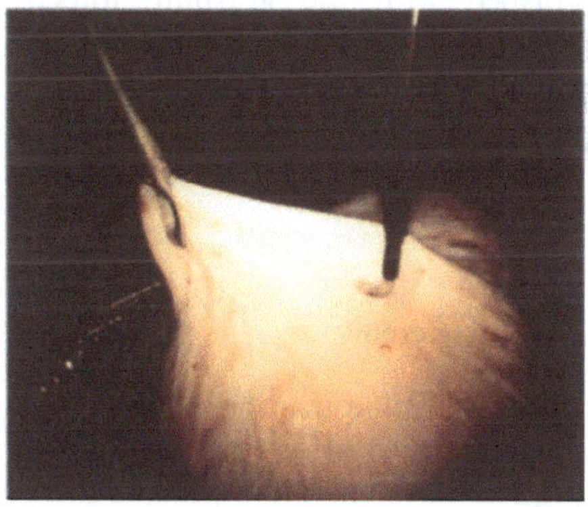

**Fig. 17.** Preparation with the monopolar hook

order to create the pneumoperitoneum as already described.

The area assigned to the surgeon should first be carefully surrounded with a grasping forceps for better orientation. Then the skin is grasped at any place on the marking and elevated. Now the skin can be carefully incised and dissected with a unipolar hook electrode.

Care must be taken not to injure the subcutaneous muscle tissue with the hook probe! This can be avoided by grasping and elevating the tissue with the grasping forceps near to the area to be dissected (Fig. 17).

The approximately 4 × 4 cm isolated specimen can be removed from the Pelvi-trainer through an appendix extractor.

### Bipolar Dissection

With newly designed electrodes it is possible to execute the above-described unipolar dissection with a bipolar hook electrode. The procedure is similar to that above. It is not necessary to connect a neutral electrode.

As far as the different biophysical mechanisms are concerned, above all the risks when using a unipolar hook, we refer to the available literature [2, 6].

### Scissors Technique

The following dissection technique can be performed without using any electric current. The additional use of a mini-dissecting pledget can be very helpful, which is inserted into the simulation trainer through an appendix extractor. After the skin has been incised along the marking, the tissue can be grasped with the grasping forceps. Now the surgeon can dissect the individual tissue layers with the inserted dissecting pledget. This technique is nearly the same as the procedure used for laparotomy.

## Cholecystectomy

For trocar placement, see Fig. 5 (1, 2, 3). In this part of the exercise we use a pig's liver, including gallbladder and cystic duct. The specimen is placed on the neutral electrode in the right upper quadrant of the simulation trainer, mimicking the live situation.

### Unipolar Dissection

For trocar placement, see Fig. 5 (1, 2, 3). A dissecting scissors is inserted through point 2 and a grasping forceps through point 3.

First the cystic duct must be freed with the help of a scissors and a grasping forceps. If this succeeds, the vessels can be ligated with two endoclips and dissected.

The free end of the cystic duct or the infundibulum, located proximal to the gallbladder, is grasped with a grasping forceps and advanced to cranial. Afterwards, the gallbladder can be freed from the liver bed with a unipolar hook electrode inserted through point 2. In case of smoke production, the dissection should be interrupted for a few seconds.

It should be noted that the grasping forceps remains fixed near to the tissue to be prepared. The isolated gallbladder can be emptied after being punctured with a cannula approximately 2 mm in diameter that has been inserted into the Pelvi-trainer with a Roeder loop carrier through an additional 5.5 cm trocar. Afterwards, the gallbladder is grasped at the infundibulum with a grasping forceps that has been inserted through point 2; then the gallbladder is drawn into the appendix extractor sleeve and removed from the "abdominal cavity."

### Bipolar Dissection

For trocar placement, see Fig. 5 (1, 2, 3). The gallbladder is freed from the liver bed according to the above-described procedure. Only new bipolar hook electrodes are used (Fig. 18). It is not necessary to use a neutral electrode in this technique.

### Scissors Technique

As already described with regard to adhesiolysis, the scissors/dissecting pledget technique has the advantage of being a method of operation well known

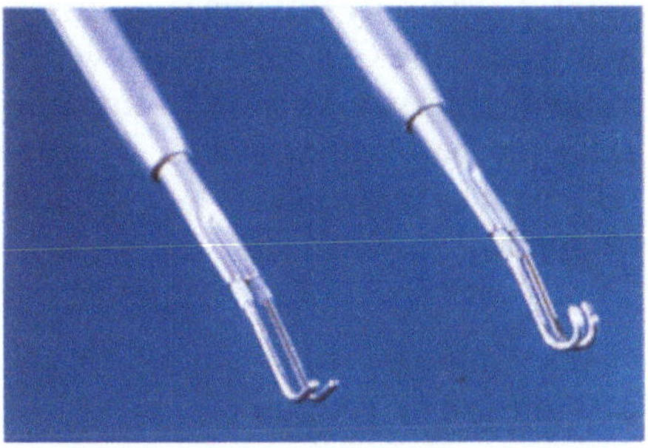

**Fig. 18.** Bipolar hook electrodes

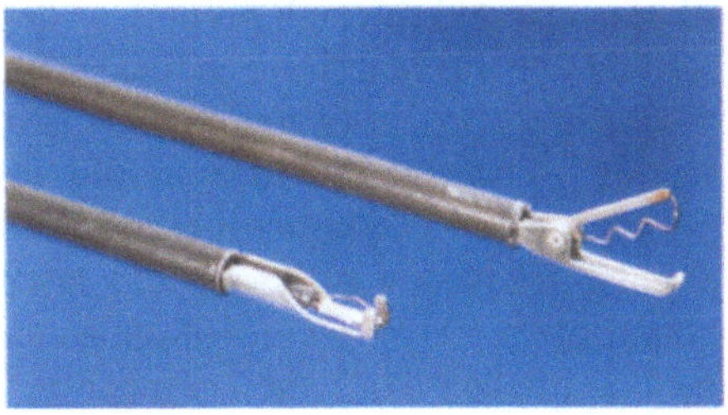

**Fig. 19.** Bipolar dissectors and hook electrodes

to the surgeon. After a circular dissection and ligature of the cystic duct, a mini-dissecting pledget is inserted through the appendix extractor sleeve at point 2 (paramedian).

The infundibulum is grasped with a grasping forceps (point 3) and pushed to cranial. The tissue layers are now well exposed and with the reciprocal use of dissecting pledget and dissecting scissors the gallbladder can be carefully detached. In human subjects prior bipolar coagulation of vascular tissue and a sharp dissection or dissection with a pledget is possible. The puncture of the gallbladder has already been described.

According to our method of operation, the gallbladder is removed from the abdomen through a 15–20 mm trocar that has been placed in the right anterior axillary line [5].

# Phase 5

## Use of Laser Instruments

Following is a description of the application of an Nd:YAG and a $CO_2$ laser instrument. Under consideration of biophysical properties, the $CO_2$ laser has a better and more precise cutting effect than the Nd:YAG laser. However, its coagulation effect is not as good. The Nd:YAG laser can be used both in the noncontact or in the contact method. Both laser instruments (Nd:YAG and $CO_2$) can be operated in continuous wave or pulsed mode.

A further distinguishing feature is the rigid transmitter system (mirror system) of the $CO_2$ laser. With the Nd:YAG laser the invisible laser light (300–600 nm) is brought into the operating field with a flexible glass fiber. For a thorough description of the properties of laser instruments we refer to the relevant literature [1, 8].

## Exercises Regarding the Functioning of Laser Instruments

### Nd:YAG Laser

For trocar placement, see Fig. 5 (1, 2, 3). Simple exercises to train the handling of a laser instrument and the mode of operation can be performed with an apple or with a steak placed in the simulation trainer. The apple is placed in the pit provided for the modules. In the meantime, the laser fiber (300–600 nm) has been adapted to the laser instrument and calibrated. It goes without saying that the preventive measures related to the use of laser instruments have to be observed!

The fiber optic can simply be inserted through a Roeder loop carrier and then introduced into the simulation trainer through point 3. With a grasping forceps (point 2) the tip of the fiber is fixed and the surgeon can now perform his first "operation" with the contact or the noncontact method and become acquainted with the mode of operation of an Nd:YAG laser using an apple. The same exercise can also be performed with a steak, using different output stages.

The beam of a $CO_2$ laser is brought directly into the operating field with a so-called coupler, through a sleeve or with the help of an infraguide. The correct aiming of the laser beam is ensured by a pilot beam. The further method of operation is analogous to that with the Nd:YAG laser.

### Adhesiolysis with the Nd:YAG Laser

As described above under Adhesiolysis, the exercise is performed on a chicken that has been placed in the right upper quadrant of the simulation trainer. We have achieved best results with the laser instrument in the contact method with a power of 15–25 W in the continuous-wave mode.

The Roeder loop applicator containing a 600-nm glass fiber is inserted at point 2 through the 5.5 mm trocar. A grasping forceps (point 3) grasps the tissue to be dissected and places tension on it so that the dissection can be performed carefully millimeter by millimeter. The fiber must be very carefully swept over the tissue vertically. If the tip of the laser is dirty it must be cleaned immediately; otherwise the fiber can burn up.

### Adhesiolysis with the $CO_2$ Laser

The $CO_2$ laser instrument is inserted into the simulation trainer through point 2. The only difference in dissection is that the $CO_2$ laser beam with a focus of approximately 0.5–1 mm provides the best cutting effect; if necessary, a pulse frequency of 1–2 s should be applied.

### Cholecystectomy with the Nd:YAG Laser

When the pig-liver specimen including gallbladder has been placed in the right upper quadrant, the Nd:YAG laser fiber is inserted through point 2, as described before. The tissue is fixed with a grasping forceps (point 3). The dissection of the gallbladder should be performed in the contact method with a pulse frequency of 1–2 s and laser power of 15–20 W. Hemorrhaging can be controlled in the noncontact method with laser power of up to 30 W.

Due to the biophysical properties of the Nd:YAG laser (lower absorption in water) it is – with this method of operation – possible to control bleeding which occurs during the irrigation of the liver bed. In this case, the hemostatic effect of a laser instrument is even stronger.

### Cholecystectomy with the $CO_2$ Laser

The dissection of the gallbladder with a $CO_2$ laser instrument has proven to be more difficult, as the contact method cannot be applied; however, this can be traind with the simulation trainer. We recommend dissection in a pulsed mode (1–2.5 s) with a maximal laser power of 40 W and a focus diameter (spot) of 0.5–1 mm. The laser light is inserted through point 2, a grasping forceps through point 3. The further dissecting procedure is analogous to that with an Nd:YAG laser as described before.

## Summary

The Pier/Götz simulation trainer which we are introducing here allows the training of all laparoscopic phases of operation. The possibility of creating a pneumoperitoneum seems to be of great importance to a beginner. According to our experience in the beginning most surgeons have difficulties with the correct placement of the Veress needle. If the Veress needle is not placed correctly, a preperitoneal emphysema can be created with this simulation trainer.

All trocars, including the safety trocars which have been introduced to the market, can be used. The torso form, the pneumoperitoneum, and the artificial "abdominal wall" allow a training that simulates reality.

We have deliberately not mentioned the method of operation with a direct view through the laparoscope, because with this technique the surgeon's posture is unphysiological, the OP team is excluded from the situs, and, above all, in case of difficult operations the decisive advantage of watching the operation together on a TV screen is lost.

## References

1. Berlien FP, Müller G (1990) Angewandte Lasermedizin – Lehr- und Handbuch für Praxis und Klinik. ecomed, Landsberg
2. Brinkmann K, Schaefer H (1978) Der Elektrounfall. Springer, Berlin Heidelberg New York
3. Frangenheim H (1958) Die Bedeutung der Laparoskopie für die gynäkologische Diagnostik. Fortschr Med 76:451
4. Palmer R (1974) Safety in laparoscopy. J Reprod Med 13:1–5
5. Pier A, Thevissen P, Ablaßmaier B (1991) Die Technik der laparoskopischen Cholecystektomie. Chirurg 62:323–331
6. Semm K (1984) Operationslehre für endoskopische Abdominalchirurgie. Schattauer, Stuttgart
7. Wickham JEA (1987) The new surgery. Br Med J 295:1581–1582
8. Pier A, Thevissen P (1992) Laser in der laparoskopischen Chirurgie. Chirurg (in press)

# Tuboscopy

D. Pollmann[1], D. Wallwiener, C. Sohn, S. Rimbach, and G. Bastert

The correct selection criteria for patients for either surgical revision of disturbances in tubal function or an IVF program will play an increasingly significant role in the future [2, 8–10]. Preoperative diagnostic hysterosalpingography and laparoscopy (to preclude a distal and proximal tubal occlusion respectively) ought to be performed on patients in whom occlusion of both tubes is suspected; they have an extremely poor prognosis as far as fertility subsequent to microsurgical intervention is concerned [3, 4]. If a double tubal occlusion is confirmed the patient is advised against surgical revision and in vitro fertilization is recommended. Considering the 20%–30% pregnancy rate after in vitro fertilization, tubal reconstruction is no longer performed when the following are present [7]:

- Chambered or thick-walled sactosalpinges
- Double tubal occlusions (proximal and distal)
- Re-sactosalpinges

Where preoperative hysterosalpingographical preclusion of a proximal tubal occlusion is not possible, an ascending chromopertubation is performed on the occasion of the laparoscopy. The patient is also advised against microsurgical revision if bilateral proximal and distal tubal occlusion is evident. A one-sided laser laparoscopic revision is carried out if partial or complete distal tubal occlusion is detected in proximally open tubes, or if other factors influencing the ovum pick-up mechanism (e.g., adhesions or endometrial infections) are present. In cases where the success rate is not particularly optimistic, a diagnostic laparoscopy is converted into a surgical endoscopic intervention without further burdening the patient or renewing anesthesia. Surgical endoscopic reconstruction is performed on patients with sactosalpinges, hydrosalpinges, or partial distal tubal occlusions.

To date, no satisfactory answer has been found regarding which patients ought to avoid tubal revision due to irreparable tubal mucosal damage [1, 5, 7]. Here, the introduction of intraluminal tuboscopy could present new aspects.

Henri-Suchet et al. were the first group who were able to visualize the tubal lumina from the ampullae to the isthmoampullary transition through thin rigid optics during microsurgical fimbrioplasty [6]. This enables determination of the intraluminal tubal status, especially as far as the intactness of the mucosa or the extent of mucosal damage is concerned. This was

[1] Universitäts-Frauenklinik, Abt. für Geburtshilfe und Gynäkologie, Voßstr. 9, W-6900 Heidelberg, FRG.

Lasers in Gynecology
Edited by G. Bastert and D. Wallwiener

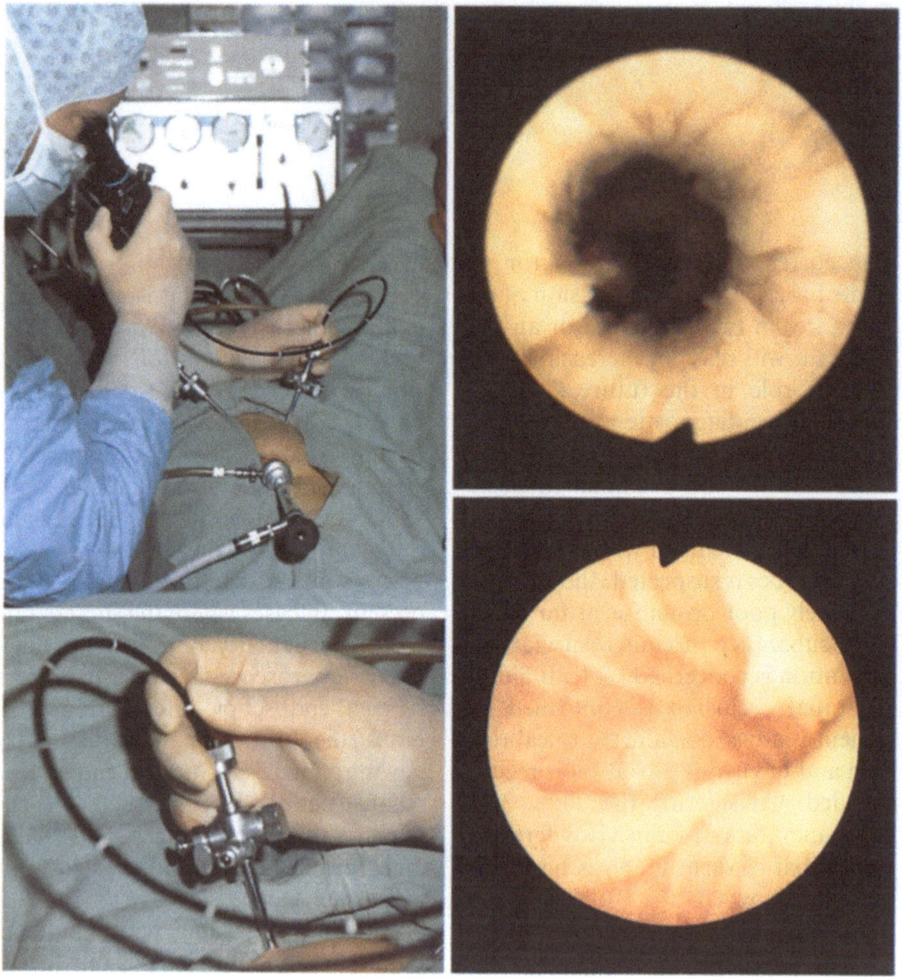

**Fig. 1 a–d.** Tuboscopy during a laparoscopic salpingostomy. **a–c** Insertion of the tuboscope into the neosalpingostoma via a second puncture; **d** mucosal status and view of the isthmoampullary transition

the first pre- or intraoperative semiquantitative documentation of mucosal damage. It was thus possible to correlate the extent of the fimbriotic areas with the extrauterine pregnancy rate.

Now that newly developed flexible endoscopes, which can be inserted via the working channel of the laparoscope or through a second or third puncture, are available, a tuboscopy can be performed during diagnostic or surgical laparoscopy (Fig. 1). The ampulla and the isthmoapullary transition can then be seen and the mucosal surface and rigidity of the wall assessed, and intraluminal changes such as polyps, septa, and synechiae can be detected [3].

## References


1. Boerrigter RM, Vemer HM, Willemsen WNP, Rolland R (1986) The differences between findings at laparoscopy and at subsequent fertility surgery. Eur J Obstet Gynecol Reprod Biol 23:181–185
2. Brockdorff U, Gräfin von (1987) Die operative Pelviskopie aus gesundheitsökonomischer Sicht. In: Mettler L, Weichert-von Hassel M (eds) Medizin, Gesundheit, Politik – operative Pelviskopie in Klinik und Praxis. Deutscher Ärzte-Verlag, Cologne, pp 369–373
3. Cornier E (1982) La fibroscopie en gynecologie: la fibro-hysteroscopie et la fibrotuboscopie. Nouv Presse Med 11:2841–2843
4. Cornier E, Feintuch MJ, Bouccara L (1984) La fibrotuboscopie ampullaire. J Gynecol Obstet Biol Reprod 1:49–53
5. Donnez J, Casanas-Roux F (1986) Prognostic factors influencing the pregnancy rate after microsurgical cornual anastomosis. Fertil Steril 46:1089–1092
6. Henri-Suchet J, Loffredo V, Tesquier L, Pez JP (1985) Endoscopy of the tube (= tuboscopy): its prognostic value for tuboplasties. Acta Eur Fertil 16:139–145
7. Karbowski B, Schneider HPG (1986) Stellenwert mikrochirurgischer Operationen der tubarbedingten Sterilität – Abgrenzung zur extrakorporalen Befruchtung. Geburtshilfe Frauenheilkd 46:825–828
8. Pittaway DE (1986) Tubal reanastomosis of a contralateral fallopian tube. Fertil Steril 46:976–977
9. Puttemans P, Brosens I, Delattin P, Vasquez G, Boeckx W (1987) Salpingoscopy versus hysterosalpingography in hydrosalpinges. Hum Reprod 2:535–540
10. Reich H (1987) Laparoscopic oophorectomy and salpingo-oophorectomy in the treatment of benign tubo-ovarian disease. Int J Fertil 32:233–236

# Non-malignant and Premalignant Diseases of Portio, Vulva, Vagina

# Preconditions and Indications for Laser Therapy of Benign and Precancerous Lesions of the Lower Female Genital Tract – An Overview

W. KÜHN [1] and D. WALLWIENER

Laser surgery has revolutionized the treatment of some diseases of the lower genital tract. Until recently, "bloody surgery" led to considerable tissue destruction, scar formation, often wound-healing problems, and to abandonment of sexual activity. Hospitalization times ranged from several days to weeks. In contrast, laser therapy – frequently without anesthesia – is often performed as an outpatient procedure, and thus costs are considerably reduced. The cosmetic result after laser vaporization is mostly excellent. A new squamocolumnar junction and a transitional zone form. No scars are generated on the vulva or the vagina, or perineally and perianally, which is a convincing argument in favor of laser therapy, especially for young women. One also has to consider that benign and premalignant lesions will occur more frequently in the future in sexually active women.

The indications for laser surgery in the lower genital tract are some dystrophic lesions, virus-associated benign tumors, and intraepithelial neoplasias. The latter indication, however, is not without risks. An intraepithelial neoplasia may cause invasive cancer and thus threaten the patient's life. Undertreatment by laser therapy may lead to subsequent mutilating surgery or even death. So the first maxim should be: No laser surgery without clinical experience and an exact diagnosis. No tumors, papules, hyperkeratosis, or eczematoid regions should be lased without a previous clinical diagnosis and cytologic as well as histologic examination. Cytology alone is not sufficient for therapy planning, since its estimated sensitivity is only 95%, and there is always a small rate of false-positive diagnoses. Moreover, cytology gives no reliable information on size and extent of the lesion. The laser surgeon thus has to be familiar with both cytologic and histologic terminology, as well as with the biology of the lesions to be treated. Several lesions can be visualized only by colposcopy and after function tests. So experience in colposcopy is a precondition. Using the laser colposcope the surgeon has the advantage of visualizing exactly the extent and depth of the lased defect. Cosmetic results can be improved by this management. Wound healing will be initiated by reepithelialization from intact cells of the basalis or the surrounding epithelial cells. On the other hand, all diseased cells have to be destroyd by the laser to avoid a recurrence. Because several of these lesions are associated with the presence of human papilloma virus, knowledge of the validity of oncogenic and nononco-

[1] Klinikum Steglitz, Abt. für Geburtshilfe und Gynäkologie, Hindenburgdamm 30, 1000 Berlin 45, FRG.

Lasers in Gynecology
Edited by G. Bastert and D. Wallwiener

genic virus types is mandatory. Condylomata accuminata and intraepithelial neoplasias often regress spontaneously, especially after pregnancy. An observation time of 3–4 months often demonstrates this biological behavior of the disease, without being of disadvantage for the patient. An expectant management should therefore sometimes be considered before laser therapy.

One of the most important preconditions for laser therapy is a reliable patient since, generally, there is no histologic control of the treatment. Only an examination several months after the laser procedure can reveal whether surgery has been successful or a repeat procedure has to be considered.

Treatment of CIN III is controversial. Most physicians advocate conventional conization for therapy of severe lesions, such as carcinoma in situ. Laser conization is a valuable alternative, being an elegant procedure allowing for exact histologic evaluation, but it should be performed only under optimal institutional conditions. A very small number of surgeons perform laser vaporization. Since the patient's safety is of first priority, however, the latter method is not to be recommended.

Complications after laser surgery are seldom. Occasionally, a larger vulvar edema or a local thrombosis, extending to the thighs, can be seen if laser therapy has been too extensive or too deep, or due to a preexistent infection. Prophylaxis of thrombosis or preoperative antiphlogistic, antieczematoid, antimycotic, or antidiabetic treatment is necessary.

Laser surgery is a modern concept of treatment for many lesions of the lower genital tract, but it will certainly not remain the only one. New drugs, so-called biological response modifiers, are already being investigated, alone or in combination with laser therapy, and will certainly lead to a further decrease in "bloody surgery".

# Laser Vaporization of Premalignant Diseases of the Vulva

W. KÜHN [1]

Only 1%–2% of the integument is vulvar skin, but in this area many more benign and malignant changes occur than elsewhere [6]. This is due to the fact that the vulvar skin has its origin in all three germinative layers [5, 7]. There are many premalignant lesions and potentially malignant degenerative changes of the vulvar skin, and while it is no problem to find an adequate therapy for precancerous lesions of the vulva, it is occasionally difficult to differentiate these premalignant lesions from benign or degenerative diseases, or from a carcinoma or a melanoma in situ.

Before planning a therapy it is necessary to have a diagnosis. The physical examination is followed by colposcopy and functional tests with toluidine blue. If necessary, a biopsy and histology should be done, combined with immunohistology for determination of different types of human papillomavirus. Cytology is not a very good method, because most lesions are accompanied by hyperkeratosis. But atypical cells are sometimes found in cases of red lesions – an erythroplakia. Before beginning treatment it is necessary to document the lesion by means of a photograph or a sketch.

[1] Klinikum Steglitz, Abt. für Geburtshilfe und Gynäkologie, Hindenburgdamm 30, 1000 Berlin 45, FRG.

The International Society of Vulvar Diseases has created a new classification for lesions of the vulvar skin. Today we call kraurosis vulvae "vulvar dystrophy." According to clinical and pathological criteria, we differentiate three types: atrophy or lichen sclerosus, hyperplastic dystrophy, and the mixed type, which has components of the two other types [1]. Hyperplastic dystrophy is potentially malignant. Sometimes it is very difficult to differentiate this type from a verrucous carcinoma with its extreme hyperkeratosis. Lichen sclerosus is not a premalignant lesion.

Not all forms of vulvar dystrophy are treated in the same way. The atrophic form, lichen sclerosus et atrophicus, is primarily not an epithelial disease; we find an atrophic, stratified squamous epithelium and diseased connective tissue with sclerosis and loss of collagenous filaments. From the clinical aspect this disease is very typical, with atrophic, small labia and a constricted introitus (Fig. 1). The diagnosis of this atrophic type can be made from the clinical examination; a biopsy should be avoided because the patient's discomfort is increased by this procedure. The retarded development of the epithelium is due to the lack of nutrition; therefore, a therapy which destroys the epithelium is not indicated. The use of keratolytic ointment or laser therapy should be avoided [3]. Recom-

Lasers in Gynecology
Edited by G. Bastert and D. Wallwiener

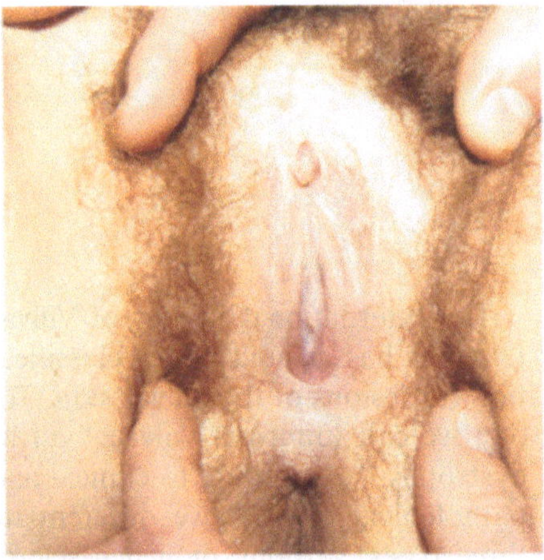

1

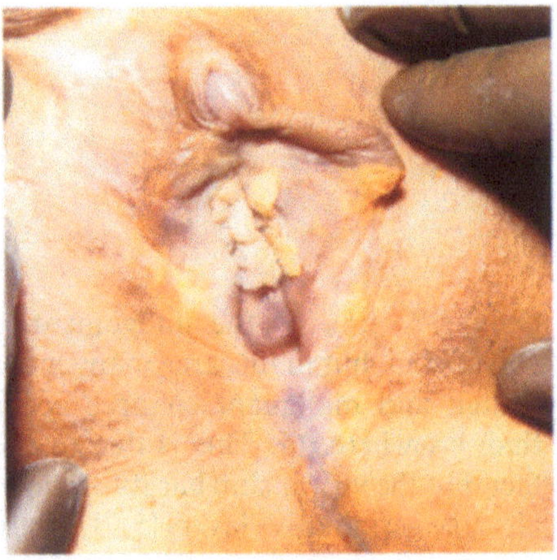

2

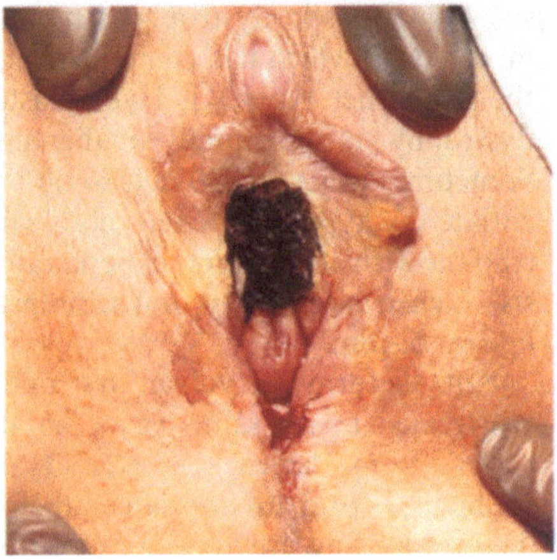

3

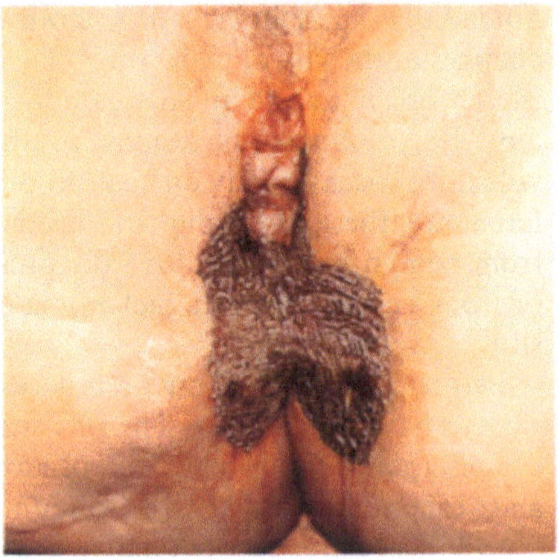

4

mended is a therapy with testosterone, which helps to build up the epithelium in these mainly young women.

In the hypertrophic type we find extreme hyperkeratosis, a significant thickening of the squamous epithelium, and less diseased connective tissue (Fig. 2). The diagnosis of hyperplastic dystrophy should be made from a biopsy of the suspicious area via colposcopy. This procedure is necessary to rule out a well-differentiated carcinoma, for example a verrucous carcinoma. For treatment keratolytic ointment was formerly used; nowadays, we recommend laser vaporization (Fig. 3), followed by application of a fat-containing ointment to relieve the pain and perhaps delay a relapse.

**Fig. 1.** Lichen sclerosus with atrophic small labia and a constricted introitus

**Fig. 2.** Hyperplastic dystrophy, with thickening and hyperkeratosis of the epithelium

**Fig. 3.** Laser vaporization of hyperplastic dystrophy

**Fig. 4.** Mixed type of vulvar dystrophy. Laser vaporization of the hyperplastic areas

Treatment of the mixed type is somewhat complicated: Only the areas with

5
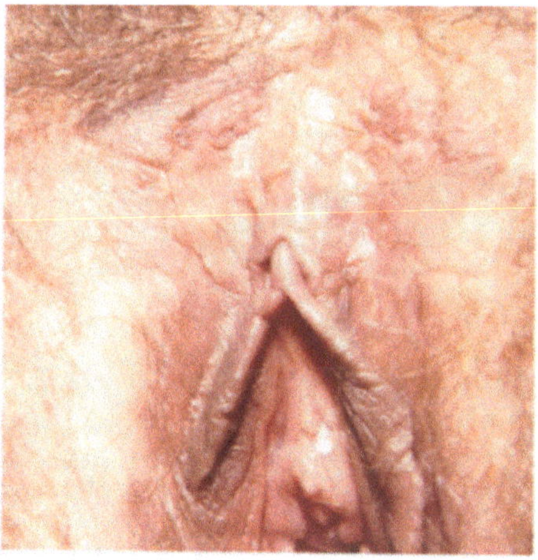

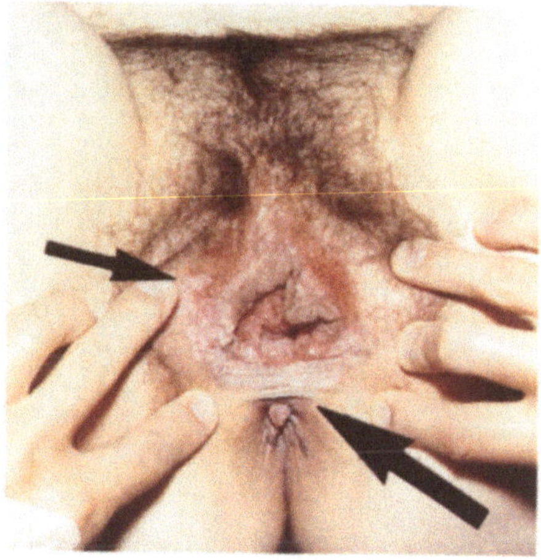
6

**Fig. 5.** Bowenoid papulosis, with several small brown papules

**Fig. 6.** Bowen's disease, with whitish hard plaques (*arrows*)

hyperplastic dystrophy should be lased (Fig. 4). Therapy with testosterone should follow to relieve the itching and help regenerate the atrophic epithelium.

In an attempt to arrive at one classification for all precancerous conditions, the International Society of Vulvar Diseases recommended the term "vulvar intraepithelial neoplasia" (VIN) [10]. Unfortunately, this classification is of no help in planning a regimen of therapy. There are many types of precancerous diseases of the vulva, and they can be differentiated quite well with modern clinical methods. The modern diagnosis of viral diseases and the fundamental results of the German Cancer Research Center in Heidelberg have led to a completely new view, not only of carcinomas but also of premalignant lesions [4]. "VIN" is now used for all precancerous lesions, although some of these lesions are associated with viral diseases and others are not.

One of the virus-associated precancerous lesions often seen is bowenoid papulosis, which can be diagnosed clinically on the basis of small papular changes [2] (Fig. 5). Histologically, the pattern is like that of Bowen's disease, so the pathologist cannot differentiate between the two. But the clinician can make the right diagnosis by using histology and modern diagnostic methods for viral diseases. The papules are human papillomavirus type 16/18 positive; they appear multiply, but they can disappear quickly and appear again in other vulvar areas [9]. The removal of these papules by laser is a very elegant and tissue-protecting method, but it is wise to wait some time before using laser surgery, as spontaneous regression occurs in many patients, similar to that of condylomata accuminata. Many of these patients are very young, so they should be advised to use condoms for sexual intercourse and have their partners checked. In addition, vaginal suppositories should be used to diminish vaginal discharge.

Bowen's disease, although similar to the above-described histological and

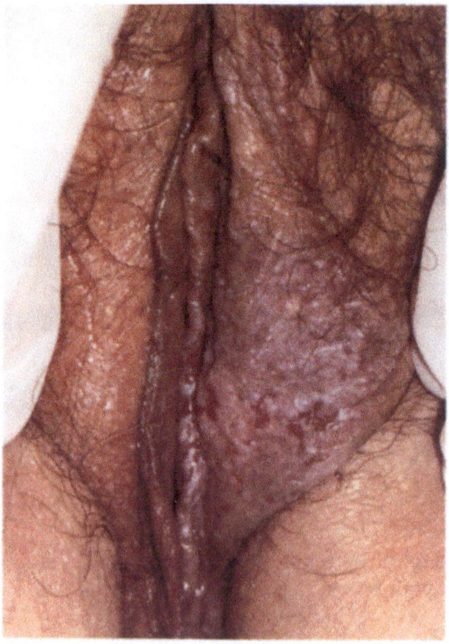

7

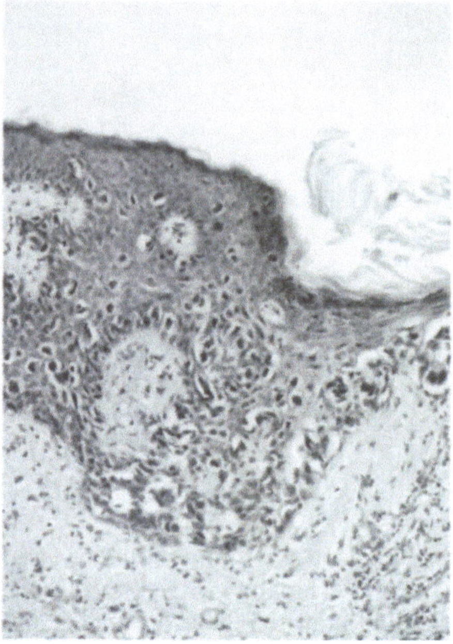

8

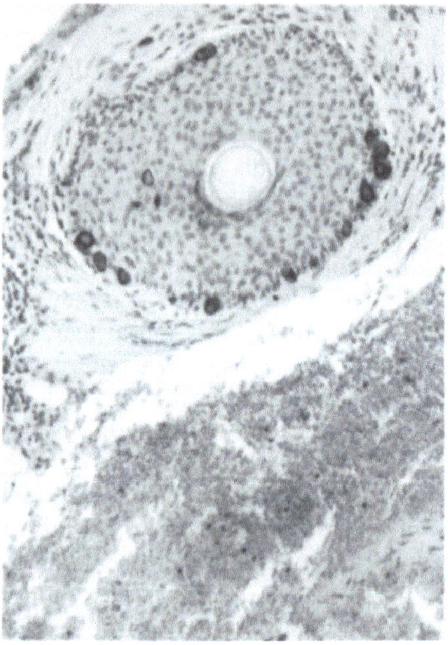

9

**Fig. 7.** Paget disease, with leukoplakia and erythroplakia

**Fig. 8.** Typical Paget cells in the superficial squamous epithelium

**Fig. 9.** Paget disease of the vulva: carcinoembryonic antigen-positive tumor cells in a hair follicle (immunohistochemistry)

virus pattern, is diagnosed clinically on the basis of whitish, slightly raised hard plaques [9] (Fig. 6). It is not possible to predict when the lesion will develop into Bowen's carcinoma, so it must be removed. For removal of the diseased superficial epithelium we recommend laser vaporization.

There are other precancerous lesions which are not caused by or associated with a human papillomavirus. One of these is Paget disease of the vulva, which clinically can be mistaken for an eczema [8]. But eczemas of the vulva are rare, and Paget disease is more likely. Sometimes a vulvar eczema has to be differentiated from a secondary eczematous change which is the result of a chronic psychogenic pruritus ani et vulvae. The short history makes the diagnosis of Paget disease unlikely. If a psychogenic itching is confirmed, a talk with the patient to clarify the situation often improves the symptoms without any medication. The classic Paget disease has some typical clinical signs: in addition to areas of leukoplakia we find areas of erythroplakia (Fig. 7). A biopsy is indicated to confirm the diagnosis. Histologically, typical cells can be seen in the superficial stratified epithelium (Fig. 8), but with immunohistochemistry, carcinoembryonic antigen-positive tumor cells can also be found in the integumentary system (Fig. 9). It must be added that about 25% of these patients have carcinomas in adjacent areas such as in Bartholin's gland or the perianal gland [8]. In cases of Paget disease laser therapy is not recommended; the lesion should be removed by surgery, not flat but deep.

There is a fourth group of VIN, lesions with neither virus particles nor the characteristics of Paget disease. The typical carcinoma in situ of the vulva spreads superficially (Fig. 10), but it cannot be diagnosed well clinically. After confirmation of the diagnosis histologically, laser therapy can be used, because the lesion is only superficial and the deep integumentary system is not affected.

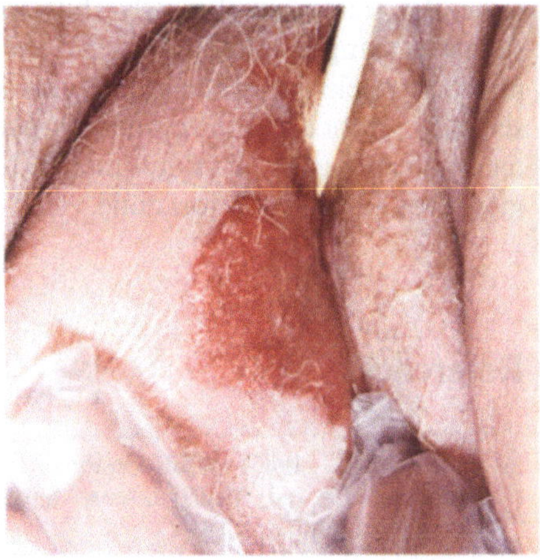

**Fig. 10.** Superficial spread of a carcinoma in situ of the vulva

Among all possible therapies for premalignant lesions, laser therapy should be emphasized, because it does not put too much strain on the patient, has good functional results, and allows the patient to continue to have sexual intercourse. The vulva can be saved. But it is essential to have a good clinical knowledge of vulvar diseases and to have a good pathologist as a co-worker. Diagnostic methods for viral diseases – nowadays immunohistochemistry – make sense for virus-caused or virus-associated lesions.

In treatment of vulvar lesions it is occasionally helpful to use a laser colposcope. With colposcopy we can better estimate the extent of the defect for vaporization. The thickness of the diseased skin is very variable; we can estimate the thickness clinically, but we suggest asking a pathologist to determine the thickness of the hyperkeratosis and acanthosis of the epidermis. We recommend using a micrometer to measure the vaporization zone.

## References


1. Friedrich E (1983) White lesions. In: Friedman EA (ed) Vulvar disease. Saunders, Philadelphia, p 131
2. Gross G (1986) Virusinfektionen der Vulva. In: Zander J, Baltzer J (eds) Erkrankungen der Vulva. Urban & Schwarzenberg, Munich, p 24
3. Heinzl S (1986) Anwendungen des Lasers bei Erkrankungen der Vulva. In: Zander J, Baltzer J (eds) Erkrankungen der Vulva. Urban & Schwarzenberg, Munich, p 113
4. Ikenberg H, Schwörer D, Pfleiderer A (1988) Nachweis von humaner Papilloma-Virus (HPV)-DNA in Vulvakarzinomen. Geburtshilfe Frauenheilkd 48:776–780
5. Kühn W, Rummel HH (1986) Gutartige Tumoren im Bereich des äußeren Genitales. In: Zander J, Baltzer J (eds) Erkrankungen der Vulva. Urban & Schwarzenberg, Munich, p 44
6. Kühn W, Rummel HH (1987) Benigne Tumoren der Vulva und Vagina. Arch Gynecol Obstet 241:152–160
7. Kühn W, Rummel HH (1987) Gutartige Tumoren an Vulva und Vagina. Schweiz Z Mod Ther Fortb 16:8–18
8. Kühn W, Kübler HC, Rummel HH (1989) Die zytologische Diagnose des M. Paget der Vulva. Geburtshilfe Frauenheilkd 49:67–69
9. Obalek S, Jablonska S, Beaudenom S, Walczak L, Orth G (1986) Bowenoid papulosis of the male and female genitalia: risk of cervical neoplasia. J Am Acad Dermatol 14:433–444
10. Stegner HE (1986) Das Carcinoma in situ der Vulva – keratinozytäre (squamöse) Praeneoplasie. In: Zander J, Baltzer J (eds) Erkrankungen der Vulva. Urban & Schwarzenberg, Munich, p 83

# Vulvar Intraepithelial Neoplasia (VIN): Laser Therapy

H. KOELBL [1], CH. DADAK, CH. KAINZ, G. GITSCH, and H. JANISCH

## Introduction

Vulvar cancer is uncommon, representing about 4% of malignancies of the female genital tract. Over the past decade, a number of significant advances have been made in the management of vulvar cancer. These changes include:

1. Individualization of treatment for patients with early disease, with particular emphasis on vulvar conservation, if appropriate.
2. A more rational approach to the management of the groin and pelvic lymph nodes.
3. The use of separate groin incisions for lymphadenectomy in patients without clinically suspicious groin nodes.

There has been confusion in the past over the nomenclature for vulvar intraepithelial neoplasia. Four major terms were used: erythroplasia of Queyrat, Bowens's disease, carcinoma in situ simplex, and Paget's disease. The International Society for the Study of Vulvar Disease decreed that the first three lesions were merely gross variants of the same disease process, and that all should be included under the umbrella term "squamous cell carcinoma in situ." As with squamous cell carcinoma of the cervix,there are varying degrees of intraepithelial abnormality, and these have been referred to as vulvar intraepithelial neoplasia (VIN) 1, 2, 3, the latter representing the most severe change. VIN is more frequently diagnosed in younger patients [3]. Moreover, the prevalence of VIN has increased over the past few years, often associated with HPV infection.

After the pioneering work of Taussig in the United States and Way in Great Britain, en bloc radical vulvectomy and bilateral dissection of the groin and pelvic nodes became the standard treatment for the most patients with operable vulvar cancer [7, 8]. If the disease involved the anus, rectovaginal septum, or proximal urethra, some type of pelvic exenteration was combined with the above dissection. Although the survival improved markedly with an aggressive surgical approach, several factors have led to modification of this "standard" treatment plan during the past decade. These factors may be summarized as follows:

1. An increasing proportion of patients with early-stage disease – up to 50% of patients in many centers – have vulvar intraepithelial neoplasia and stage-I disease.
2. There is concern about the postoperative morbidity and associated

[1] II. Frauenklinik Wien, Spitalgasse 23, 1090 Wien, Austria.

Lasers in Gynecology
Edited by G. Bastert and D. Wallwiener

long-term hospitalization common with en block radical dissection.
3. There is increasing awareness of the psychosexual consequences of radical vulvectomy.

Conventional surgical treatment is often mutilating, and recurrence rates have been reported of approximately 30%.

The modern approach to the management of patients with vulvar intraepithelial neoplasia and stage-I carcinoma of the vulva is based on the individualization of treatment. There is no "standard" operation applicable to every patient, and emphasis is on performing the most conservative operation that is consistent with cure of the disease.

Although radical vulvectomy has been regarded as the standard treatment for the primary vulvar lesion, significant postoperative complications of wound healing and infection are ascribed to the cold-knife technique, while vulvectomy by electrocauterization aims basically at secondary healing, resulting in gross scarring [5]. The significant disturbances of sexual function and body image are regarded as the major long-term morbidity associated with the treatment of vulvar cancer. Anderson and Hacker reported that sexual arousal was reduced to the eighth percentile and body image to the fourth percentile in women who had undergone vulvectomy, when compared with healthy adult women [1]. These facts prompted us to use laser applications in the treatment of vulvar lesions.

## Patients and Methods

Since 1983, radical vulvectomy as well as radical local excisions have been carried out in 50 patients. Three groups of patients underwent treatment: 12 women had gross condylomas, whereas VIN was present in 11 cases, and 27 women had invasive cancer of the vulva. Nicotine abuse and diabetes were similarily distributed in the three groups. Before laser treatment, biopsies were obtained to confirm the clinical diagnosis. For laser treatment a Sharplan Laser was used at 40–60 W continuous exposure. Radical vulvectomy comprised a 100% circumcision of the vaginal introitus, whereas radical local excision implied a wide and deep excision of the primary disease. Regardless of whether a radical vulvectomy or a radical local excision was performed, the aim was to obtain surgical margins of 2 cm, although this would not be possible posteriorly for perineal lesions. Radical local excision is most appropriate for lesions on the lateral or posterior aspects of the vulva, where preservation of the clitoris is feasible. Major bleeding was stopped by electrocoagulation, irrespective of the technique. After resection the wound bed was vaporized additionally in order to stop minimal or diffuse bleeding. The surgical defect was then closed in two layers and drained on both sides.

## Results

Radical vulvectomy was carried out in 35 patients only with invasive malignancy, whereas radical local excision was performed in 15 cases. Extensive resection requiring local flaps was necessary in three women with invasive cancer of the vulva and in one patient with vulvar intraepithelial neoplasia. VIN III was present in nine of 11 patients, and VIN I in the remaining two cases.

In order to obtain radical resection, toluidine was applied liberally to the vulva to better identify suspicious lesions before initiation of laser excision. However, histology revealed lack of radicality in one of 27 patients with invasive cancer and in two women with VIN. In two women with invasive malignancy and three patients with VIN III the histologist was uncertain about completeness of the neoplastic resection. In all of the cases of invasive cancer and VIN III, second-step laser treatment was successfully carried out for complete resection of these alterations. In cases of VIN I, short-term follow-up checks with regular biopsies have been performed. Secondary closure of wound healing defects was necessary in two patients who underwent radical vulvectomy. Catheter application lasted between 8 and 11 days. Patients left the hospital after 16–21 days.

To date, local recurrence of vulvar intraepithelial neoplasia has been ascertained in one patient, whereas six of 27 patients have had a relapse of invasive cancer of the vulva, with a mean duration of recurrent disease of 11.8 ± 5.2 months.

## Discussion

Various therapeutic modalities have been introduced in the treatment of VIN. The necessity for careful pretreatment, histological diagnosis, and vulvoscopic mapping of the various conditions has been stressed. The relative risks of malignant progression of the various conditions have been discussed, along with the various modes and methods of surgical therapy. The facts emerging are that traditional surgical therapy is giving ground to more modern techniques of ablation, particularly that of carbon dioxide laser photoevaporation therapy. Particularly in the treatment of VIN, laser therapy has begun to show many advantages. The general tendency of close surveillance and conservative treatment benefits the management of noninvasive genital tract intraepithelial neoplasia.

Laser treatment of vulvar intraepithelial lesions is a good therapeutic tool, resulting in satisfying cosmetic results. This is in contrast to previously reported complications of wound healing after cold-knife and electrocautery techniques [5]. Currently, the literature about laser treatment of VIN is rather scant. However, discussion about the significance of laser treatment in patients with vulvar lesions remains controversial.

Using laser techniques, Helmerhorst et al. have reported good postoperative results in ten patients with VIN III undergoing laser vaporization only. In their patient group one relapse of VIN III was observed. The authors conclude that laser vaporization appears to be an effective and nonmutilating therapy, and preferable for young VIN patients [4]. In a longitudinal study covering 10 years, Shafi et al. reported more frequent relapses in the laser-skinning group than in the vulvectomy group [6]. Bornstein et al. recommend the combined approach of surgical excision and carbon dioxide laser vaporization for multifocal vulvar intraepithelial neoplasia, resulting in a satisfactory cosmetic outcome and providing an adequate specimen for histopathologic diagnosis [2]. The carbon dioxide laser almost certainly has a role in the management of vulvar lesions. Although most of the data possibly do not reflect its full potential, they demonstrate a need for multiple controlled prospective and comparative studies.

## References


1. Anderson BL, Hacker NF (1983) Psychological adjustment after vulvar surgery. Obstet Gynecol 63:457
2. Bornstein J, Kaufmann RH (1988) Combination of surgical excision and carbon dioxide laser vaporization for multifocal vulvar intraepithelial neoplasia. Am J Obstet Gynecol 158:459
3. Gardner HL, Friedrich EG jr, Kaufmann RH, Woodruff JD (1976) The vulvar dystrophies, atypias, and carcinoma in situ. An invitational symposium. J Reprod Med 17:131
4. Helmerhorst TJ, van-der-Vaart CH, Dijkhuizen GH, Calame JJ, Kenemans P, Stolk JG (1990) $CO_2$-laser therapy in patients with vulvar intraepithelial neoplasia. Eur J Obstet Gynecol Reprod Biol 34:149
5. Kreienberg R, Beck T, Bartzke G, Henne M, Friedberg F (1990) Resultate der operativen Behandlung des Vulvakarzinoms. Geburtshilfe Frauenheilkd 50:375
6. Shafi MI, Luesley DM, Byrne P, Samra JS, Redman CW, Jordan JA, Rollason TP (1989) Vulvar intraepithelial neoplasia – management and outcome. Br J Obstet Gynecol 96:1339
7. Taussig FJ (1930) Leukoplakia and cancer of the vulva. Arch Dermatol Syphilol 21:431
8. Way S (1948) The anatomy of the lymphatic drainage of the vulva and its influence on the radical operation for carcinoma. Ann R Coll Surg Engl 3:187

# $CO_2$ Laser Therapy and HPV Typing in CIN – Follow-up Observations

A. GÖPPINGER [1], G. BIRMELIN, H. IKENBERG, M. HILGARTH, and H. G. HILLEMANNS

## Introduction

Former examinations showed that 60% of cervical intraepithelial neoplasias are associated with the papillomavirus types 6/11 and 16/18 [2]. Papillomaviruses, particulary type 16/18, are said to have an effect on the development of cervical carcinoma [3]. Thus, a mode of therapy for precursors of cervical carcinoma should also remove a possible carcinogen. The aim of our investigations was to evaluate the therapeutic success of laser treatment on the cervix, with particular regard to HPV findings and the degree of severity of the intraepithelial neoplasia being treated.

## Patients and Methods

Between 1981 and 1986 in the outpatient department of the University of Freiburg Department of Gynecology, 118 women with CIN were treated by vaporization with the Coherent $CO_2$ laser. The CIN in every case had been diagnosed by colposcopy, cytology, and target punch biopsy.

Of 108 women who had undergone the last laser therapy a year or more before, 104 appeared for follow-up examinations. The mean observation time after therapy was 27.7 months.

Up to five follow-up examinations were performed; on the average 2.0 examinations were carried out. The mean age of the women at the first laser therapy session was 28.7 years. None of them had had previous cervical treatment. The follow-up examinations consisted of colposcopy and cytology; from 1984 on HPV typing was also performed. The cervix was judged to be healthy when multiple cytologic smears were normal and when colposcopy showed no more than a transformation zone or uncharacteristic white epithelium. Patients with the latter findings were examined again. Mosaic and punctate lesions were histologically clarified by punch biopsy.

Beginning in 1984, 123 HPV findings were obtained from 83 patients. In situ hybridization of smears taken from the ectocervix and endocervix with a cotton swab was carried out according to the procedure described elsewhere [1]. In 23 patients the HPV typing was performed only prior to laser therapy, in 20 only after laser treatment. In 40 patients, hybridization with HPV was done before and at a mean of 17.6 months after therapy.

[1] Albert-Ludwig-Universität, Abt. für Geburtshilfe und Gynäkologie, Hugstetter Str. 55, W-7800 Freiburg i. Br., FRG.

Lasers in Gynecology
Edited by G. Bastert and D. Wallwiener

## Results

### Course of Therapy

Dysplastic changes were no longer detectable in 94 women during a mean observation period of 26.5 months after the last laser therapy (90.3%). An average of 1.25 treatment sessions were necessary for this therapeutic result. In 80 instances a single treatment was adequate (77%); in 14 cases more than one laser treatment had to be performed. For eight of the 104 patients, therapy for persisting or recurrent lesions after the first laser treatment was changed to conization or hysterectomy. In three cases this was because the patients had changed doctors, in another three cases because the patients themselves requested it, and in two cases therapy was changed because of an unfavorably located CIN III extending into the endocervix. In two women with recurrent or persisting lesions after a single laser treatment another therapy had not yet been administered at the end of our study.

### Diagnosis of Recurrent Lesions

In 80 of the 104 cases of CIN, no further growth or recurrence of the lesion was seen after a single laser treatment during an observation period of at least 1 year. In 24 cases of CIN, repeated treatment was necessary after the first laser session. Indications included:

In 15 cases – cytologically and/or histologically confirmed persistence ($n = 11$) or recurrence ($n = 4$) of the lesion.

In six cases – a second treatment had been planned right from the beginning because of the spread of the lesion.

**Table 1.** Cytologically and/or histologically proven failures, according to severity of CIN, after a single laser treatment

| Grade | Patients (*n*) | Failures (*n*) | Failures (%) |
|---|---|---|---|
| CIN I | 43 | 5 | 11.9 |
| CIN II | 32 | 3 | 9.4 |
| CIN III | 29 | 7 | 24.0 |

In three cases – a second treatment took place only on the basis of the colposcopic findings in the presence of normal cytologic smears.

Table 1 shows the cytologically and/or histologically proven failures after a single laser treatment, according to the severity of CIN. It is striking that the percentage of recurrent or persistent lesions is higher in CIN III than in CIN I and II. For six of the 15 failures we did HPV typing: four showed HPV 16/18, and two were negative.

### HPV Incidence and Course

Table 2 illustrates 123 findings before and after completed laser therapy in 83 of 118 patients who were treated by laser from 1981 to 1986. Not included are the eight women without HPV findings and those for whom the therapy was changed. The incidence of HPV findings before therapy (55.6%) was reduced to 13.3% after completed laser therapy; the percentage of HPV negative findings was twice as high.

The 52 patients with negative HPV findings after laser therapy had a Pap II diagnosis; in 44 cases colposcopy disclosed an original portio. There was a white epithelium or a transformation zone in eight cases. In the eight HPV-positive cases, condylomata could no longer be cytologically confirmed after laser therapy. Colposcopically, an un-

**Table 2.** HPV incidence in 110 patients before and after completed laser therapy

| HPV Type | Before laser therapy | | After laser therapy | |
|---|---|---|---|---|
| | Patients (*n*) | HPV incidence (%) | Patients (*n*) | HPV incidence (%) |
| 6/11 | 11 | 55.6 | 4 | 13.3[a] |
| 16/18 | 19 | | 4 | |
| 6/11 + 16/18 | 5 | | – | |
| Negative | 28 | | 52 | |
| ∅ HPV | 47 | | 50 | |
| Total | 110 | | 110 | |

[a] The reduction of the HPV incidence from 55.6% to 13.3% is statistically significant ($P \leq 0.001$). ∅, Not done

**Table 3.** Follow-up of 24 HPV-positive patients before ($t_0$) and after laser therapy ($t_1$) of CIN

| $t_0$ (CIN I–III) | | $t_1$ (∅ CIN) | | |
|---|---|---|---|---|
| HPV Type | (*n*) | HPV Type | (*n*) | |
| 6/11 | 9 | 6/11 | 3 | |
| 16/18 | 13 | 16/18 | 1 | 16.6% persistence[a] |
| 6/11 + 16/18 | 2 | 6/11 + 16/18 | 0 | |
| Negative | 0 | Negative | 20 | 83.3% negativity |
| Total | 24 | | 24 | |

[a] The reduction of the HPV infections to 16.6% is statistically significant

characteristic white epithelium was established in two cases at the time of the smear for HPV identification. In three cases no such change was noticed until the next follow-up.

HPV identification was performed in 40 patients before and after laser therapy. Of the 24 pretherapeutically positive cases 20 were HPV negative (83.3%) after completion of the laser therapy; in four (16.6%) the HPV findings persisted (see Table 3).

## Discussion

Within an observation period of at least 1 year and a mean observation time of 27.6 months after completed laser therapy, i.e., an average of 1.25 treatments, dysplastic changes were no longer detected in 90.3% of CIN cases. Recurrence after laser therapy is easy to detect and can be healed by another laser treatment or by any other form of therapy. The laser method is safe, as long as contraindications are ruled out. We observed no invasive tumor growth after laser therapy or as a result of a misinterpretation of the primary findings. Our rate of recurrence or continued growth was altogether 14.4% (15/104).

According to our studies, recurrence or continued growth of the lesion can be expected after laser therapy, especially among the severe grades of CIN with HPV 16/18 identification.

Laser therapy successfully reduced the prevalence of HPV, as well as the CIN itself. The HPV prevalence before laser therapy amounted to 56%. The HPV prevalence among patients in whom the CIN was cleared up by laser therapy was reduced to 13.3%. This corresponds to the incidence of HPV infections among the normal female population.

We consider laser therapy to be an adequate method of treating CIN grades I–III. A possibly causal agent, namely local HPV infection, cannot be removed totally. Because of the higher rate of relapses a reliable follow-up must be guaranteed, especially before the treatment of a CIN III by laser.

## References

1. Göppinger A, Birmelin G, Ikenberg H et al. (1987) HPV standardization and DNA cytophotometry in CIN. J Reprod Med 32:609–613
2. Wagner D, Ikenberg H, Böhm N (1984) Identification of human papillomavirus in cervical swabs by desoxyribonucleic acid in situ hybridization. Obstet Gynecol 64:767
3. zur Hausen H, Gissmann L, Schlehofer J (1984) Viruses in the etiology of human genital cancer. Prog Med Virol 30:170

# $CO_2$ Laser Excisional Conization: Five Years' Experience Treating Cervical Intraepithelial Neoplasia

U. HECKMANN [1]

## Introduction

In the past decade, $CO_2$ laser microsurgery has been tested as a new, effective modality for the treatment of cervical intraepithelial neoplasia (CIN) [1–3]. Excisional cervical conization was the standard procedure in patients with CIN. It was suggested that by studying the margins of the cone specimen, one could predict with reasonable accuracy the presence or absence of residual disease in the uterus.

Laser conization is performed with an operating microscope, and the exact boundaries of the abnormal transformation zone can therefore be visualized accurately. This microsurgical technique is intended to tailor the defect, incorporating only the diseased tissue. Thus, in accordance with the geometry of the cervical gland field and the endocervical clefts, a cylindrical specimen of the endocervix will preserve more connective tissue and reduce the stroma defect [4].

[1] Frauenklinik am Knappschaftskrankenhaus, Wieckesweg 27, W-4600 Dortmund, FRG.

## Material and Methods

Between 1985 and 1990, 220 patients underwent excisional laser conization for CIN. In 193 patients the follow-up ranged from 6 months to 5 years; in 16 patients it was less than 6 months. Eleven patients were lost to follow-up.

A histological specimen was needed based on the endocervical location of the lesion, on disparity between results of cytology, colposcopy, and histology, and to rule out early stroma invasion. Intraepithelial neoplasia of the endocervix was seen in 93% of the patients. The $CO_2$ laser excisional cone biopsy was performed in the operating room, using either general anesthesia or – when the patient's compliance made it possible – local anesthesia.

For local anesthesia Xylocaine 1% with Ornipressin (POR 8) 1:30 was subepithelially infiltrated in aliquots of 1 ml into the four quadrants of the ectocervix.

The laser speculum incorporates into its anterior blade one smoke evacuator tube and a second tube for nitrogen insufflation into the operating field, in order to minimize carbonization in the absence of oxygen. Prior to preparation of the specimen, the length of the cervical canal is evaluated by means of a cannula with a millimeter ruler on its surface. The apical third of this cannula is perforated for instillation of methylene blue 1% into the cervical canal.

Lasers in Gynecology
Edited by G. Bastert and D. Wallwiener

1 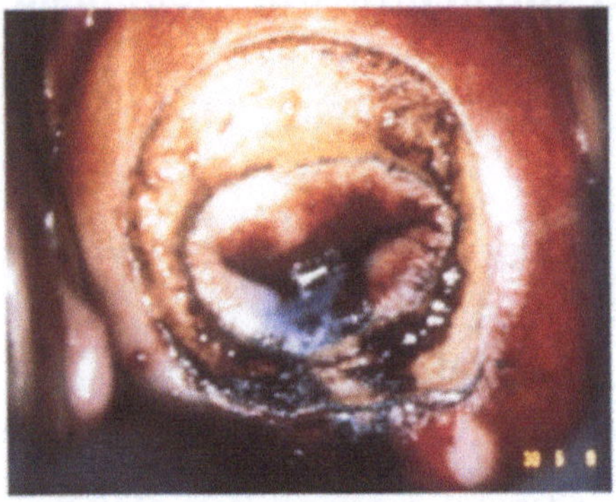

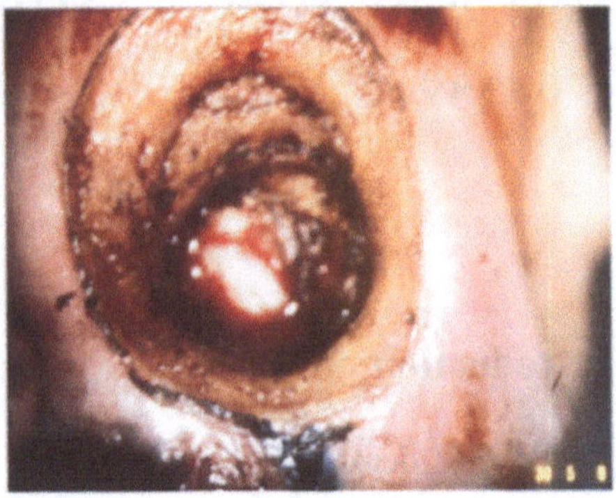 2

**Fig. 1.** Preparation of the cone

**Fig. 2.** Cervical defect following excision of the cone

Thus, the methylene blue solution penetrates into the endocervical clefts. When the laser beam hits the mucus of a gland crypt at the periphery of the cone margin, a blue bubble forms. This is a signal for the surgeon to incorporate this lesion in the preparation figure.

The laser equipment used was a Coherent 450 model, with a maximum power output of 30 W, an objective focal length of 300 mm, and an effective laser imprint of 1.2 mm (spot size). The power density for the cutting procedure ranged from 1400 to 1600 W/cm$^2$. The $CO_2$ laser beam is delivered to the cervix by a micromanipulator coupled to a colposcope. The preoperative colposcopy with 3% acetic acid and Lugol's solution visualizes the ectocervical lesion. Approximately 2–3 mm of the normal cervix should be included in the peripheral margin around the abnormal transformation zone. The periphery of the specimen is then outlined by imprinting a series of circular spots on the portio (Fig. 1). The initial circular cut is made with a penetration depth of 8 mm. For manipulating the cervix, a laser hook is placed on the margin of the specimen. When traction is placed on the hook, moving around the periphery of the specimen, the laser beam is directed into the stroma toward the upper portion of the endocervical canal until the specimen is freed. The configuration of an excisional cone will depend on the ectocervical lesion and the length of the cervical canal as measured prior to surgery.

The endocervical margin is cut with a scalpel or with the superpulse laser to minimize thermal artifacts. To evert the endocervix another neat 2-mm margin is created by clipping the tissue surrounding the canal, preventing stenosis after the healing process (Fig. 2). A combined excisional and vaporization procedure was performed where large lesions encompassed more than 50% of the surface of the ectocervix, or included the vaginal fornices. A cylindrical excisional laser conization is followed by vaporization of the remaining ectocervix to a depth of 8 mm. The vaginal wall can be vaporized to a depth of only 2–3 mm.

Minimal residual bleeding can be controlled by a defocused spot > 2 mm. Any arterial bleeding can be sutured under microscopic guidance if necessary. Hemostasis can be difficult to achieve when cervicitis or multifocal or early invasive neoplasia is present.

## Results

Of 193 patients who underwent laser excisional conization for CIN of all grades, 182 (94%) had no evidence of disease (NED) during an observation time of at least 6 months to 5 years with careful cytologic and colposcopic studies. The persistence rate was for four of 117 patients (4%) CIN grade 2, and seven of 61 patients (11%) for CIN grade 3 during follow-up examination. Persistent or recurrent abnormal cytology prompted subsequent and repetitive conization.

Twenty-eight patients of the 193 (14%) underwent hysterectomy for positive margins of the cone specimen, multifocal disease, and other gynecologic indications. The hysterectomy specimens revealed no evidence of disease in 12 of these 28 (43%) patients. In 16 of the 28 cases (57%) persistence of CIN, mostly of a lower grade, was seen. In elderly women hysterectomy was performed for multifocal disease; only two of ten were found to be free of CIN. As a consequence of these data, careful follow-up by colposcopy and cytology studies at quarterly intervals is mandatory.

The geometry of the biopsy specimen was determined in 134 patients. The maximum diameter, the axial length, and the volume varied according to the endocervical and ectocervical lesion. The average specimen of the cylinder-shaped cones measured 15.78 mm (range 10–25 mm) in diameter and 20.45 mm (range 10–34 mm) in length, with a mean volume of 2.09 ml (range 1.00– > 5.00 ml). When a cylinder dissection was used rather than a conical or semiconical excision, the stroma defect was reduced by 40%.

## Complications

When $CO_2$ laser conization was first used, bleeding was the most common complication observed during the excision procedure. Suturing of an arterial pumping hemorrhage under colposcopic guidance was necessary in 2% (four of 220 patients). Light bleeding was easily controlled by defocusing the laser beam, coagulating the vessel, and sealing it off before its entrance into the crater. Delayed bleeding occurred 6–12 days after the operation in 3% (seven of 220 patients); the application of a vaginal gauze pack was an effective treatment on an outpatient basis. Blood-tinged discharge lasted as long as 2 weeks after laser excision procedures. The discharge gradually diminished after the first 7 days.

No pelvic infection was observed in the entire series. No case of an incompetent cervix was seen. The organ mass was restored after 3 weeks.

## Conclusion

Microsurgical techniques of $CO_2$ laser excisional conization for the treatment of CIN allow the surgeon to tailor the operation to fit the geographic extent of the disease. A cylindrical biopsy specimen preserves more cervical stroma.

The efficacy of $CO_2$ laser treatment of CIN was documented by the absence of disease in 94% of the patients.

The persistence rate of 6% (CIN grade 2: 4%, grade 3: 11.5%) indicates that the compliance of the patient is mandatory. Regular cytologic and colposcopic follow-up at quarterly intervals will eliminate persistence and recurrence of disease by allowing early application of more radical laser surgery.

## References

1. Baggish MS, Dorsey JH, Adelson M (1989) A ten-year experience treating cervical intraepithelial neoplasia with the $CO_2$ laser. Am J Obstet Gynecol 161:60–68
2. Dorsey JH, Diggs ES (1979) Microsurgical conization of the cervix by carbon dioxide laser. Obstet Gynecol 54:565
3. Toaff R (1976) The carbon dioxide laser in gynecological surgery. In: Kaplan I (ed) Laser surgery. Jerusalem Academic Press, p 64
4. Wright VC, Davies E, Riopelle MA (1983) Laser surgery for cervical intraepithelial neoplasia. Am J Obstet Gynecol 145:181

# Malignant Diseases

# Surgery of Primary Breast Cancer with the $CO_2$ Laser: Report on 449 Operations

K. DINSTL [1]

Since 1975, 1600 operations with the $CO_2$ laser (Sharplan 791) have been performed at the First Department of Surgery, KA Rudolfstiftung, Vienna. Most of the operations were done for diseases of the breast.

The use of the $CO_2$ laser in breast surgery is based on the following characteristics [1]:

1. Reduced bleeding
2. Reduced postoperative pain
3. Reduced postoperative wound infections
4. Possibly reduced local recurrence (but there are no significant results as compared with the general survival time)

Reports on the results of laser surgery in cancer patients have been controversial. The use of the $CO_2$ laser in tumor surgery is based on the theory that lymph vessels can be sealed to prevent the spread of tumor cells during the operation. Published papers show not only positive but also negative results [2].

Therefore, the following points have been investigated:

1. Experimental use of the $CO_2$ laser compared with the scalpel or the electric knife (with regard to lymph flow and cancer)
2. Treatment of breast cancer with and without the $CO_2$ laser in selected patients
3. $CO_2$ laser surgery treatment of breast cancer in all patients
4. The value of using the $CO_2$ laser for extensive malignant tumors of the breast

## Results

An experimental study on $CO_2$ laser surgery in mice with Lewis lung carcinoma was done at our department to address the following questions:

1. Can the $CO_2$ laser improve survival in comparison to the conventional scalpel?
2. Is the rate of local recurrence lower?
3. Are there morphological findings specific to the method applied?
4. Is there a lower incidence of implantation metastasis?
5. Does the tumor size have any influence on the results?

The study provided the following conclusions:

1. $CO_2$ lasers are not capable of improving overal survival rates significantly, except where small tumors are concerned; here the survival rates differ significantly in favor of the laser method.
2. The local recurrence rate is lower with laser excision.

[1] Krankenhaus Rudolfstiftung, 1. chirurgische Abt., Juchgasse 25, 1030 Wien, Austria.

Lasers in Gynecology
Edited by G. Bastert and D. Wallwiener

3. Histological and cytological examinations done after the laser is used reveal tissue and cell destruction that lowers the rate of local tumor growth.
4. The incidence of tumor growth was much lower in mice whose wounds had been exposed to $CO_2$ laser before tumor cells were swabbed into them.
5. The tumor size has a great influence on the results, especially when small tumors are treated by the $CO_2$ laser [2].

A prospective controlled study of female patients with breast cancer was started in 1981. The selection criteria for 135 patients were:

1. Age less then 70 years
2. Clinical tumor stage 1 or 2
3. Pre- or postmenopausal state

Randomization to type of surgery was done by computer, and a radical mastectomy was performed in all patients. Postoperative therapy was standardized according to hormone receptor state (either tamoxifen or chemotherapy). We do not yet have definitive results because the period of investigation has been too short (Tables 1 and 2).

The follow-up of 449 patients operated on because of breast cancer from 1975 to 1985 showed reduced incidence of recurrence only when a radical mastectomy had been done (Table 3). The incidence of tumor recurrence depended on the kind of procedure used, but it seems that $CO_2$ laser surgery in breast cancer gives an advantage over conventional methods relating to local recurrence. The standardized axillary lymph node dissection could not be done by laser because it would have endangered axillary vessels or nerves.

**Table 1.** Randomization to surgery with and without the $CO_2$ laser

| Age of patient | Type of surgery | |
|---|---|---|
| | Laser (*n*) | Without laser (*n*) |
| ≤ 50 years | 20 | 20 |
| 51–69 years | 47 | 48 |
| Total | 67 | 68 |

**Table 2.** Results of follow-up (6.5 years)

| Type of surgery | Treatments | Metastases | | Mortality | |
|---|---|---|---|---|---|
| | (*n*) | (*n*) | (%) | (*n*) | (%) |
| Laser | 67 | 3 | 4 | 9 | 14 |
| Without laser | 68 | 5 | 8 | 8 | 12 |

**Table 3.** Postoperative local recurrence according to type of procedure

| Procedure | Treatments | Recurrences | |
|---|---|---|---|
| | (*n*) | (*n*) | (%) |
| Lumpectomy | 18 | 6 | 33 |
| Modified radical mastectomy | 15 | 6 | 40 |
| Radical mastectomy | 416 | 20 | 4.8 |
| | 449 | 32 | 7.1 |

**Fig. 1.** Extensive breast cancer: skin incision with scalpel

**Fig. 2.** Radical operation with $CO_2$ laser, after removal of tumor

**Fig. 3.** Primary skin grafting after surgery with the $CO_2$ laser

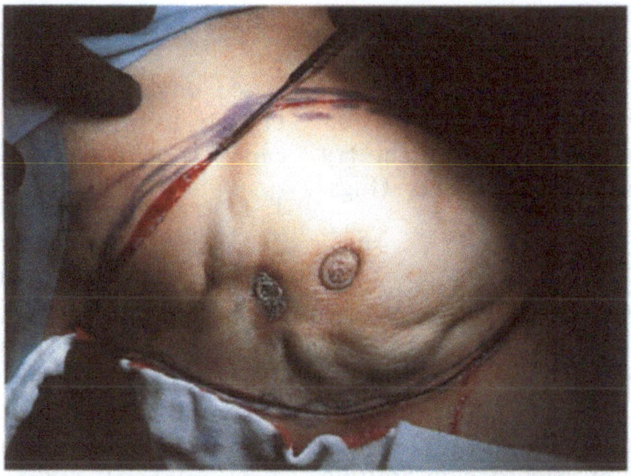

1

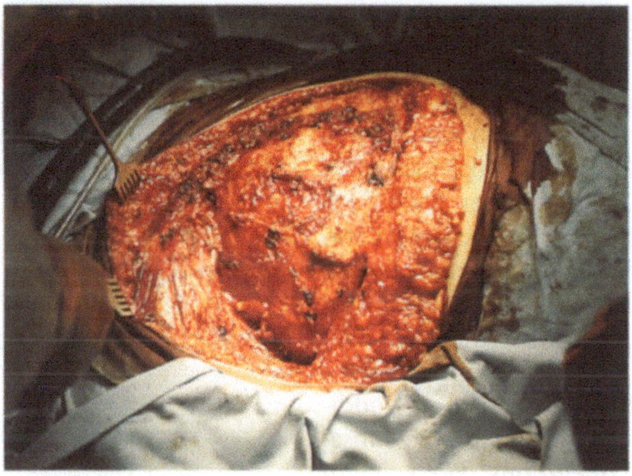

2

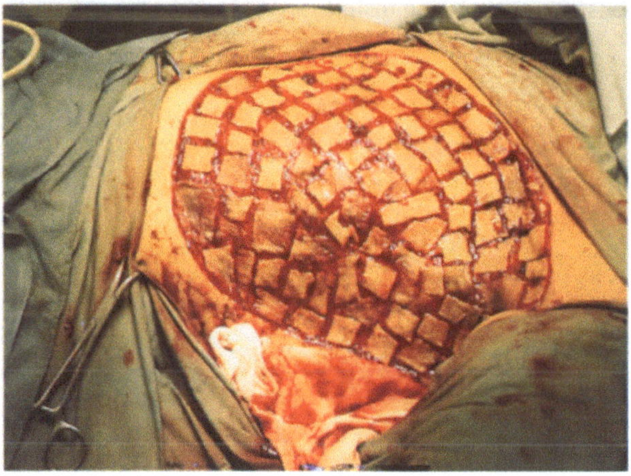

3

Undeniable are the advantages of the $CO_2$ laser in the excision of extensive tumors of the breast. Our experience with 39 patients showed the following advantages of laser surgery over conventional procedures (Figs. 1–3):

1. Less bleeding
2. Possibility of reducing the excision of skin to a minimum
3. Significant reduction of local recurrences
4. Less postoperative pain
5. Possibility of primary skin grafting in ulcerated or infected tumors

A definitive statement regarding survival time cannot be made; the number of patients treated is too small for a prospective study. Recently, such cases of extensive breast cancer have been treated preoperatively by chemotherapy. We found a significant reduction of the tumor size in patients responding to preoperative chemotherapy after three cycles. We hope to prolong the survival of these patients by the additional use of $CO_2$ laser for the subsequent radical operations. The value of this procedure is currently being tested.

## Conclusion

Based on our experience, the advantages of using the $CO_2$ laser in breast cancer surgery are

1. Operative:
   a) cutting without pressure
   b) reduced bleeding
   c) less postoperative pain
   d) fewer wound infections (and seromas)
2. Oncological: reduced local recurrence (survival time?)

The disadvanges are prolonged operating time, prolonged wound healing of skin incisions, and the possibility of thermic lesions.

## References


1. Dinstl K, Fischer PL (1981) Der Laser. Springer, Berlin Heidelberg New York
2. Tuchmann A, Bauer P, Plenk H jr, Braun O, Dinstl K (1986) Comparative study of conventional scalpel and $CO_2$ laser in experimental tumor surgery. Res Exp Med 186:375–386

# Laser in Gynecologic Oncology? Laser Palliation of Malignoma Recurrences

D. WALLWIENER[1], S. RIMBACH, M. KAUFMANN, H. SCHMID, G. BASTERT and D. VON FOURNIER

## Introduction

The so-called local recurrence of carcinomas of the breast and genitals after a complete course of treatment still poses a therapeutic challenge. Palliative local therapy, particularly in the sense of treatment of the wound, is aimed for when ubiquitous distant metastases exist or when there is progressive local tumor development during systemic hormone therapy or chemotherapy. After maximum radiation of the affected area a second course of irradiation is generally not possible. Complex plastic surgery is often out of the question – even in less unfavorable situations – due to e.g. internistic contraindications, advanced age, or the patient's poor general condition.

Nevertheless, an improvement in the quality of life must be the therapeutic aim, particularly during these chronic stages [5]. Local tumor control, even well into the advanced stage, thus plays an exceptionally important role.

Laser treatment is a possible palliative procedure, by which the tumorous area is vaporized or coagulated, in an attempt to sanitize, at least temporarily, the often ulcerating recurrent tumor and to reduce the local pain it causes. Such an intervention is also indicated to combat the extreme physical and psychic strain the patient must endure as a result of the uninterrupted confrontation with the visible manifestation of her tumor and the restriction in movement it causes.

## Material and Methods

### Patient Population

Palliative laser treatment was carried out on 45 patients in a pilot study. The indications for treatment and age distribution of our patient population are shown in Table 1.

### Surgical Procedure

The surgical procedure was to a large extent uniform during laser application, despite the diverse forms of primary tumors and their varied localization.

The clinical diagnosis was confirmed histologically with excised specimens. Successive thermal destruction or coagulation of the tumorous tissue using a $CO_2$ laser, an Nd:YAG laser, or a combination of the two was then carried out. An electrosurgical tumor reduction was performed when large tumor masses were present. Tumor de-

[1] Universitäts-Frauenklinik, Abt. für Geburtshilfe und Gynäkologie, Voßstr. 9, W-6900 Heidelberg, FRG.

Lasers in Gynecology
Edited by G. Bastert and D. Wallwiener

**Table 1.** Indications for laser treatment and age distribution of 45 patients

| Indication | $n$ | Age | (average) |
|---|---|---|---|
| Mammary carcinoma (recurrence on the thoracic wall) | 29 | 38–86 | (54) |
| Vulvar carcinoma (local recurrence) | 11 | 67–84 | (78.5) |
| Uterine sarcoma (growing vaginally) | 2 | 64, 66 | (65) |
| Carcinoma of the clitoris (local recurrence) | 1 | 45 | – |
| Carcinoma of the rectum (vulvar and vaginal metastases) | 1 | 83 | – |
| Carcinoma of the body (suburethral recurrence) | 1 | 78 | – |
| Total | 45 | | |

struction was carried out until macroscopically unaffected tissue layers were visible. Healing was attained by granulation.

## Results

In all 45 patients the removal of visible and palpable surface recurrences and tumor parts was attained with palliative laser treatment. Multiple evidence with up to 25 isolated coin-sized lumps was present in all cases of recurrent breast carcinomas on the thoracic wall. In ten patients – i.e., in approximately 1/3 of the cases – the tumor had a diameter of over 10 cm. A third (i.e., in 4 patients) of the locally recurrent vulvar carcinomas were multifocal. Whereas multiple surface evidence was vaporized with the $CO_2$ laser, the combination of $CO_2$ laser vaporization and Nd:YAG laser coagulation was a new approach in the surgical treatment of ten cases of recurrences on the thoracic wall and in three cases of keratinizing vulvar carcinoma recurrences, as well as in both cases of vaginally growing uterine sarcomas. The removal of tumors thicker than 1 cm using a laser is an extremely lengthly procedure; hence, a diathermal loop was initially used to attain rapid and extensive tumor reduction, followed by tumor bed coagulation using the Nd:YAG laser.

Re-infections were just as rare as other complications during the wound-healing process. Within 3–5 days, when the pain of the wound had eased, the subjective condition of all patients had improved. The removal of the mostly ulcerating visible parts of tumor, led to a marked improvement in the subjective quality of life; no influence was evident however, as far as objective parameters, such as frequency and extent of the metastatic spread and recurrence rate, or even the survival time were concerned. Nevertheless, the aim of the laser treatment was exactly this improvement in the quality of life, i.e., palliation, and not a curative influence. The procedure is illustrated below by case reports.

### Case 1

A 75-year-old patient with an 18-year history of a mammary carcinoma suffered a second local recurrence which was removed by therapy and was in poor general condition. She had a 10 × 10 cm immovable axillary tumor and more than 25 small cutaneous metastases, severe edematic swelling of the arm with permanent adduction posture, extreme pain, and commencing ulceration of the tumor (Fig. 1).

The axillary recurrence and numerous cutaneous metastases were vapor-

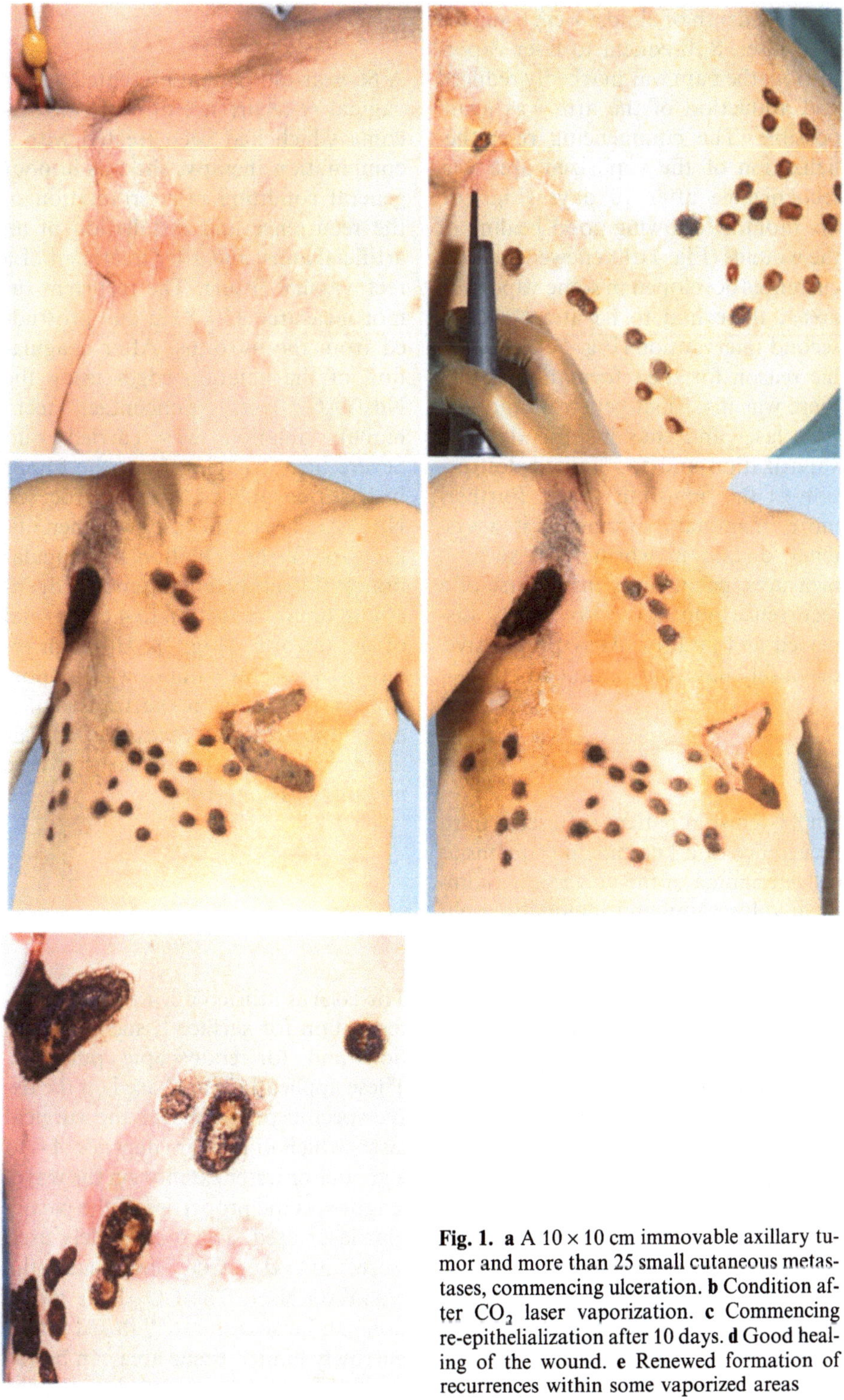

**Fig. 1.** **a** A 10 × 10 cm immovable axillary tumor and more than 25 small cutaneous metastases, commencing ulceration. **b** Condition after $CO_2$ laser vaporization. **c** Commencing re-epithelialization after 10 days. **d** Good healing of the wound. **e** Renewed formation of recurrences within some vaporized areas

ized to a depth of almost 1 cm using the $CO_2$ laser. Subsequent to laser vaporization, the pain was markedly reduced and adduction of the arm was again possible. The commencing re-epithelialization of the vaporized area was recognizable after 10 days (Fig. 1 c). Six months following good healing of the wounds (Fig. 1 d), renewed local recurrences developed in some vaporized areas; these had to be attacked in a second intervention (Fig. 1 e). A possible reason for their formation is that there was insuffient penetration of the $CO_2$ laser and thus insuffiently deep vaporization and too limited destruction of the tumorous tissue. Furthermore, mammary carcinoma is a systemic disease in which local surgical measures, particularly in the case of a recurrence which is also widely dispersed and multilocularly progressive, inevitably have only a limited effect.

**Case 2**

An 84-year-old patient suffered a third recurrence of a keratinizing squamous cell carcinoma of the vulva; after complete vulvectomy and inguinal lymphonodectomy, she was in good general health.

Although the left side of the tumor was satisfactorily vaporized to a depth of approximately 1 cm, the right side of the tumor had such a thick corneous plaque that $CO_2$ laser vaporization was not possible. However, tumor coagulation using the Nd:YAG laser presented no problems.

**Case 3**

A 66-year-old patient presented with a vaginal re-recurrence of a uterine sarcoma which had been treated with a combination therapy; she was in poor general condition after irradiation of the recurrence and installation of an artificial anus, and after treatment of a rectovaginal fistula. The recurrent tumor measured $15 \times 8$ cm and protruded from the introitus. After coagulation of the ablation edges using the Nd:YAG laser (noncontact technique), ablation was carried out. Severe hemorrhaging from the tumor bed was coagulated with a defocused Nd:YAG laser. The laser was then utilized for further ablation by coagulation of the intravaginal tumor bed, enabling tumor reduction to 4 cm cranial of the introitus (Fig. 2). Although removal of the entire tumor was not possible, a reduction in the transvaginal exophytic area was necessary for palliative reasons, particularly where nursing was concerned.

## Discussion

The laser is utilized during surgical intervention for surface tissue vaporization and for endoscopic purposes. These applications are based on the tissue-specific properties of the surgical laser, which in turn are dependent – to a greater or lesser extent – on the wavelength-specific properties of the particular laser used. The most widely used lasers are the $CO_2$ laser and the Nd:YAG laser. The $CO_2$ laser radiation can be so accurately dosed that a narrowly limited tissue area can be vaporized by the transformation of light

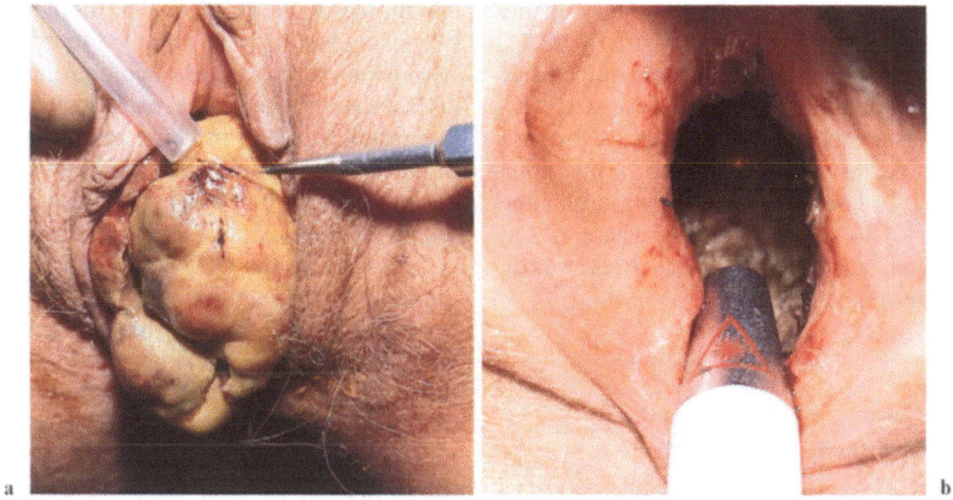

**Fig. 2. a** Vaginal re-recurrence of a uterine sarcoma, 15 × 8 cm, protruding from the introitus. **b** Condition after combination treatment with laser coagulation of the intravaginal tumor bed and tumor reduction to 4 cm cranial of the introitus

energy into thermal energy when it is sharply focused, thus enabling extremely precise and thin tissue incisions.

On the other hand, it can be utilized to vaporize surface tissue when applied in a less sharply focused state. The vaporization zone is surrounded by a thermally damaged tissue zone in which lymphatic clefts and blood vessels up to a thickness of 0.5 mm are sealed. The Nd:YAG laser requires an extremely high energy output before it is capable of tissue vaporization. It has a greater tissue penetration depth than the $CO_2$ laser, its beam being less absorbent in water. One millimeter-deep tissue coagulation and extensive coagulation of the blood vessels results during tissue irradiation with the Nd:YAG laser. The respective range of application depends upon these wavelength-specific tissue properties and the resultant penetration depth.

## Experimental Surgical Basis

Initially, the influence of the various $CO_2$ laser operating techniques on tumor behavior was tested experimentally on C 56 BL/6 mice with Lewis-Lung carcinoma. The wound-healing process, local recurrence rate, metastatic spread, and the survival rate after tumor excision using either a laser or a scalpel were investigated and compared. Wounds resulting from laser damage healed without complications, but the process was delayed compared with that after scalpel excision.

As fas as the local recurrence and distant metastatic spread rates were concerned, the laser was found to be distinctly superior to the scalpel: the average local recurrence rate after lasing and scalpel excision was 34.4% and 78.6% respectively. Scalpel excision was also found to be the statistically significantly poorer method as far as the occurrence of pulmonary metastases was concerned, with a 92.8% rate, while the average rate of metastatic spread was only 17.8% after $CO_2$ laser excision. An average post-operative survival rate of 65.5% was noted after laser excision; this was also superior to scalpel excision, where an animal survival rate of only 7.1% was determined [6].

## Laser-assisted Treatment of Primary Carcinomas

The laser was used for surgical treatment of primary tumors, including mammary carcinomas, relatively early, due to the small thermally damaged zone attained during tissue surgery with simultaneous sealing of the smaller blood vessels and thus a proportionate reduction in hemorrhaging from the wound edges [3].

A slight delay in the healing of the cutis incision was observed, however; this was also the case after primary treatment of vulvar carcinomas. The cutting speed (a maximum speed of 2 cm/s) was reduced compared with that of the scalpel or electroscalpel.

Recognition of the tissue layer, which is of exceptional importance in axillary lymphonodectomy, for example, is more difficult during laser preparation than during preparation using sharp instruments.

The largest study of the primary treatment of carcinomas of the breast using the laser is presently being conducted by Dinstl in Vienna. His prelimary results appear to confirm the trend that there are no advantages to be expected concerning the local recurrence rate and metastatic spread after primary treatment of the mammary carcinoma with the laser. Although promising results were obtained in experimental surgical studies comparing laser excision with scalpel excision [6], laser application for primary treatment of carcinomas of the breast and vulva has not proved to be successful. This is due to (a) the lack of coagulation ability – at least of the $CO_2$ laser – which makes electrocoagulation necessary during preparation, (b) the delayed wound healing, and (c) the increased danger of wound dehiscence as a result of secondary sequestration of the coagulated edges. Furthermore, the expectation that the survival and recurrence rates would be improved due to the sealing of lymph clefts and smaller blood vessels (which does indeed occur) has not been fulfilled. Whereas laser treatment of the primary carcinoma does not offer any advantages over conventional excision (K. Dinstl, this volume), laser reduction of tumor tissue is a highly promising approach for palliative surgery [4].

## Laser-assisted Palliative Treatment of Locoregional Recurrences of Carcinomas

The clear advantage of laser-assisted treatment of malignant growths lies in the palliative therapy of recurrences and soft-tissue metastases which are no longer accessible by conventional surgery. On the basis of the illustrated cases, it becomes evident that palliative laser treatment is a last possible solution when therapeutic alternatives are lacking. The basic advantages of the laser over conventional surgery using a scalpel lie in the relative bloodlessness of the surgical area, in the extensive conservation of the unaffected surrounding tissue, and in the repeatablity of the intervention – which is often performed with the patient under local anesthesia – without any great impairment of the patient [2].

In contrast to the vaporization of carcinomas in situ, during which surgery is performed conserving as much tissue as possible using a high energy output for a short exposure time, palliative removal of recurrent tumors requires a high output – in optimum cases up to 100 W/s – and long exposure periods to attain maximum tissue destruction. The results of our pilot study appear to indicate that the $CO_2$ laser effect does not suffice for such palliative interventions, as the tissue effect extends over only fractions of millimeters; renewed recurrences were also often observed in already vaporized zones, suggesting that tumorous tissue remains at a greater depth. A new approach is to be found in the utilization of the Nd:YAG laser, with its penetration depth of up to 7 mm. The coagulation effect of the Nd:YAG laser when using a focusing handpiece and the non-contact technique is clearly recognizable in Fig. 3. One can differentiate between the coagulated tumorous tissue and the surrounding healthy tissue due to the yellow discoloration of the former and its secondary sequestration during the wound-healing process. The combination of surface vaporization with the $CO_2$ laser and the subsequent destruction of tumor parts located deeper in the tissue by means of Nd:YAG laser coagulation also appears to be a good policy and thus presents a new approach to be studied.

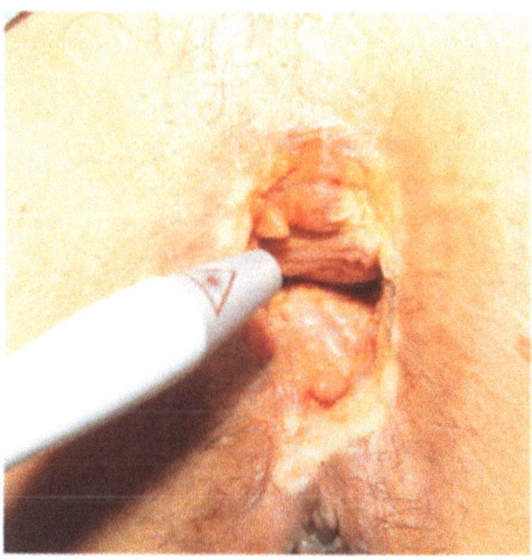

**Fig. 3.** Coagulatory effect of the Nd:YAG laser using the focusing handpiece and non-contact technique

Thus the indications for palliative laser treatment in gynecological oncology are:

- Advanced age and poor general health with high surgical risk
- Limited possibilities for plastic surgery due to internistic risks
- Wide spread of the tumor and/or multiple evidence
- Failure of systemic hormone therapy and/or chemotherapy
- Condition after maximum utilization of irradiation

In summary, one can say that laser application as a palliative measure to treat locoregional recurrences and soft-tissue metastases in carcinomas of the breast and malignant genital tumors, as well as vulvar carcinomas, constitutes an efficient supplement to the currently available therapeutic measures. Under no circumstances should laser palliation be viewed isolated from the oncological master plan. In future, one must examine which other therapeutic procedures (e.g., topical chemotherapy) can be usefully combined with laser surgery and which position the latter holds in the oncological overview.

## References


1. Giebel GD, Jaeger K (1990) Die Versorgung des fortgeschrittenen Mammakarzinoms, des Lokalrezidivs und des Strahlenschadens. In: Giebel GD, Jaeger K, Stark GB (eds) Brustrekonstruktion nach Mammakarzinom. Springer, Berlin Heidelberg New York Tokyo Rome, pp 75–85
2. Jako GJ (1987) New strategies and initiatives in cancer treatment. Laser Surg Med 7:717–718
3. Kaplan I, Giller S (1984) $CO_2$ laser surgery. Springer, Berlin Heidelberg New York Tokyo, pp 20–23
4. Leuchter RS, Townsend DE, Hacker NF, Pritorius RG, Lagasse LD, Wade ME (1984) Treatment of vulvar carcinoma in situ with the $CO_2$ laser. Gynecol Oncol 19:314–322
5. Schmidt-Matthiesen H, Bastert G (1987) Gynäkologische Onkologie. Schattauer, Stuttgart
6. Wallwiener D, Rimbach S, Pollmann D, Klinger U, Stolz W, Block M, Bastert G, von Fournier D (1990) Laser surgery in gynecological oncology. Part I. Experimental results: relapse and metastatic behavior of the Lewis lung carcinoma subsequent to $CO_2$ laser surgery of the primary tumor. An experimental model to compare the effectiveness of various $CO_2$ laser application techniques. Eur J Gynaecol Oncol 11:331–341

# Technical and Experimental Research

# Biophysical Fundamentals for Laser Application in Medicine

F. FRANK [1]

## Introduction

The term "laser" is an acronym for "light amplification by stimulated emission of radiation." The possibility of laser action was first suggested by Albert Einstein in 1917 [7]. All the preliminary work culminated in 1960, when Theodore H. Maiman constructed the first working laser using a rod of crystalline ruby excited by a coaxial helical flash lamp. In rather rapid succession, other lasers were built [1,3,13, 14,18,19,22,23].

Light amplification by stimulated emission of radiation means that we want to achieve light amplification, and that stimulated emission of radiation is the way to achieve it. A light amplifier is a lasing medium, i.e., a material with which stimulated emission is possible, enclosed between two parallel mirrors, the so-called optical resonator. The lasing medium, which may be ions, atoms, or molecules in a solid, liquid, or gas phase, is excited by a pump source that uses light, electrical, or chemical energy. Reflection of photons by the mirrors amplifies the number of photons generated by stimulated emission. One of the mirrors is a partial transmitter which permits some of the photons to be emitted from the laser resonator, creating a beam of laser light (Fig. 1).

Laser light can be focused to form beams of very small diameter. Thus, its intensity can be increased by many orders of magnitude. Striking a match produces an energy of 200 J of incoherent light. With only 1 J of coherent light from a ruby laser it is possible, by

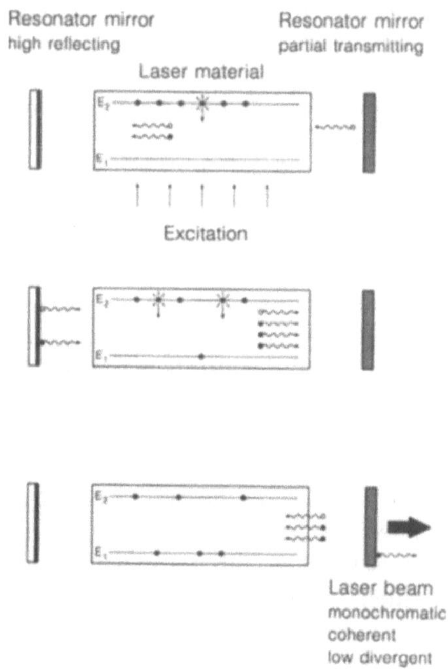


**Fig. 1.** Principle of laser with the optical resonator, excitation of the laser medium, and light amplification

[1] MBB-Medizintechnik, Applikationsforschung, Postfach 801168, W-8000 München 80, FRG.

Lasers in Gynecology
Edited by G. Bastert and D. Wallwiener

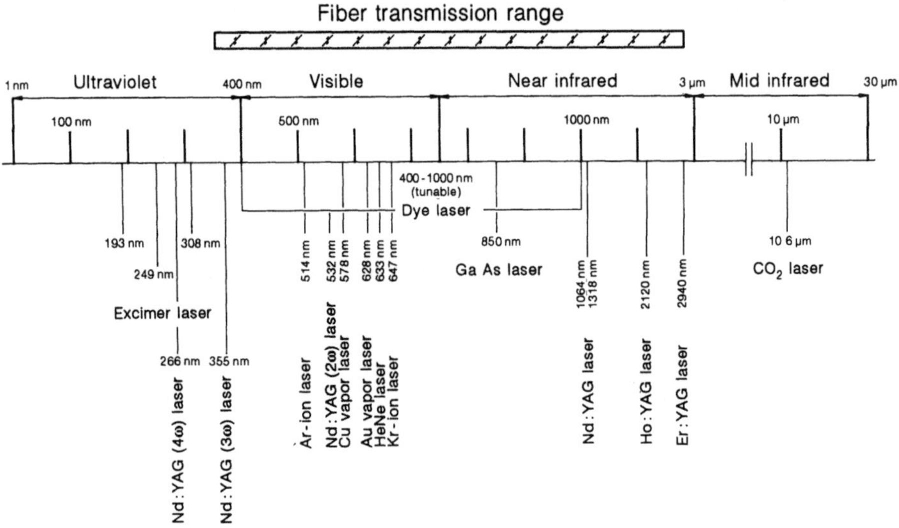


**Fig. 2.** Different types of lasers and their spectral wavelengths

focusing with a simple lens, to drill a hole through a metal blade. Sunlight has an average power density of 0.1 W/cm$^2$, as compared with power densities of 100 000 W/cm$^2$ which are easily obtained with surgical laser systems.

By using different lasing media, laser systems emitting light of different wavelengths can be constructed, beginning in the ultraviolet region at about 200 nm and reaching into the infrared up to a wavelength of 10 µm. In the region of wavelengths between 300 nm and 2.2 µm it is possible to transmit laser light through very thin but still mechanically robust quartz glass fibers. This permits the use of lasers in all endoscopic techniques (Fig. 2).

Laser systems differ also with regard to their features concerning duration and power of the emitted laser radiation. In continuous-wave lasers (cw mode) with power outputs of up to $10^3$ W, the lasing medium is excited continuously. With pulsed lasers, excitation is effected in a single pulse or in on-line pulses (free-running mode). Peak powers of $10^5$ W can be developed for a duration of 10 ms – 100 µs. Storing the excitation energy and releasing it suddenly (q-switch mode or mode-locking) leads to a peak power increase of up to $10^{10}$–$10^{12}$ W and a pulse duration of 100 ns – 10 ps.

## Laser-Tissue Interactions

In order to determine the interaction between laser light and biologic tissues quantitatively, the physical parameters of the biologic object must be related to the parameters of the laser light. The degree and extent of the effect depend on the properties of the tissue, which are determined by the structure, water content, and blood circulation, as well as on the geometry of the laser beam (Fig. 3).

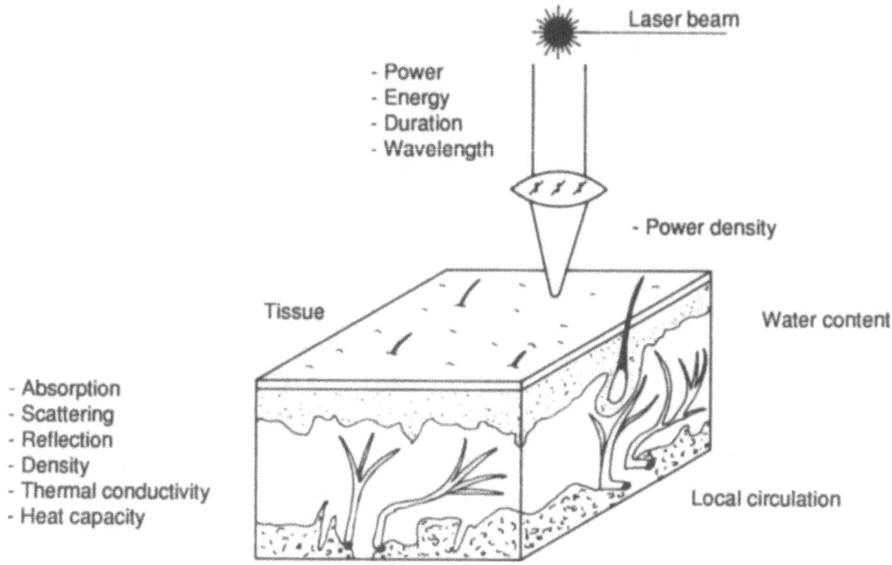


**Fig. 3.** Physical parameters which determine the effect of laser light in tissue

Depending on the duration of the laser irradiation on the tissue (interaction time) on the one hand, and on the laser irradiance in surface or volume interaction with tissue (effective power density) on the other, three types of tissue interaction can be distinguished [2, 21]:

Photochemical effects
(10 s–1000 s; $10^{-3}$–1 W/cm$^2$)
Photothermal effects
(1 ms–100 s; 1–$10^6$ W/cm$^2$)
Photoionizing effects
(10 ps–100 ns; $10^8$–$10^{12}$ W/cm$^2$)

The photochemical and photothermal processes can be achieved with quasi-continuous-wave irradiations. The photoablative and electromechanical processes are induced with short pulses in the high power regions (Fig. 4).

## Photochemical Effects

With extremely long interaction times and low power densities photochemical transformation occurs by absorption of light with no primary heating of the tissue (Fig. 4). The most important example of photochemical effects is photosensitized oxidation. The combined use of laser light and an injected photosensitizer, today mainly hematoporphyrin derivatives (HPD), initiates a cytotoxic process. Most of the tissue is destroyed after excitation of the photosensitizer by laser light. The stimulated sensitizer undergoes a series of intramolecular chemical reactions that lead to the oxidation of various cellular components [5]. The fact that the residence time of HPD is longer in pathologic tissue than in healthy tissue permits selective tumor eradication. In photodynamic therapy, use is made of argon-pumped dye lasers (1 W at 630 nm, cw) and gold vapor lasers (10 W at 628 nm, pulsed). HPD has significant side effects, so other photo-

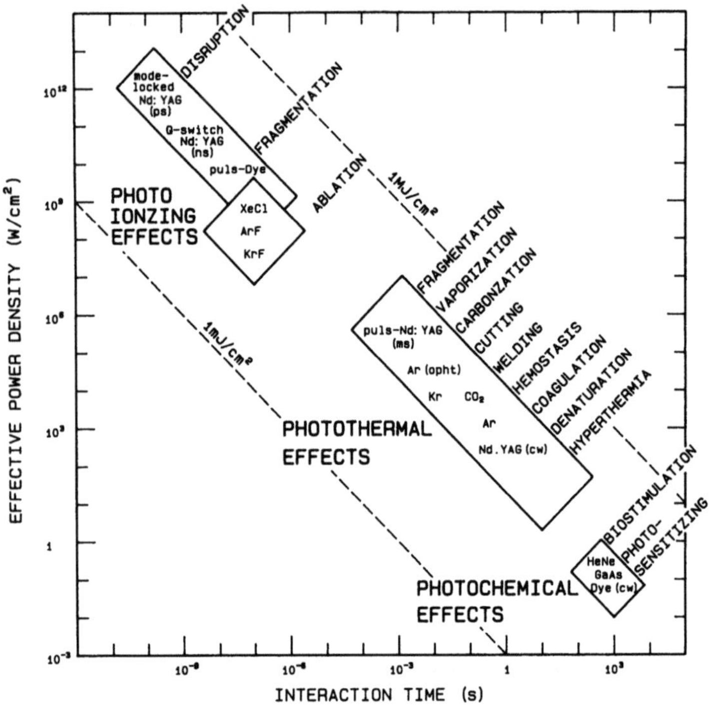


**Fig. 4.** Photochemical, photothermal, and photoionizing tissue effects of different laser systems depending on interaction time and effective power density (from [2])

sensitizers, maybe even special ones for the different kinds of tissue, have to be developed. At present pheophorbide-a (Ph-a) is being investigated. This photosensitizer can be excited by infrared Nd:YAG laser light. Biostimulation, mainly for wound healing or pain relief, with stimulation of microcirculatory effects and other features must be mentioned as well. From what is known so far, the difference in the photon statistics of a laser light source as compared with a thermal light source is of decisive importance. Systematic studies have not yet given reasonable explanations for the clinically observed improvements [20]. Use is made of He-Ne lasers (1–5 mW at 633 nm) and Ga-As laser diodes (5 mW at 850 nm).

**Photothermal Effects**

With decreasing interaction time and higher power density, the transition to photothermally induced effects begins. The main surgical applications for lasers are based on the conversion of laser light into heat (Fig. 4). The optical and thermal properties of the tissue and the laser beam geometry and energy of the incident light influence the degree and extent of the thermal action. The most important optical parameter is the wavelength-dependent absorption of biologic molecules. Since the building blocks of living systems – amino acids, proteins, and nucleic acids – in spite of their great variety are made up of only a few basic

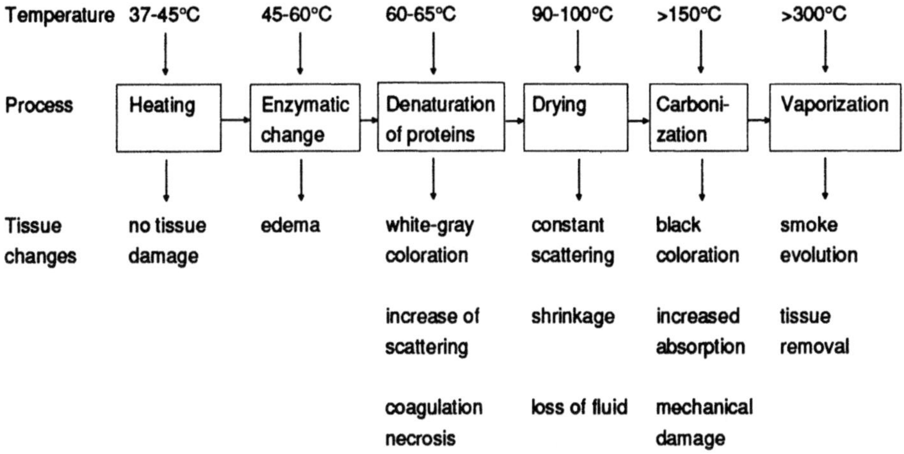


**Fig. 5.** Thermal tissue alterations following laser irradiation

elements, a few fundamental rules can be formulated for the absorption of optical radiation.

The main absorption of biologic molecules occurs within the range of wavelengths shorter than about 280 nm (ultraviolet). The far more molecule-specific vibrational and rotational absorption bands are all in the range of wavelengths longer than 1 μm (infrared). Visible laser radiation is hardly absorbed by biologic material. Two of the most important exceptions to this rule are the hemoglobin in the red blood corpuscles and melanin, which is stored as a pigment in the skin and also in large quantities in the pigment epithelium of the retina. A strong absorption in the green spectrum occurs in both substances. Due to the high water content (60%–80%) of most tissue there is an extensive absorption of infrared radiation. This leads to a very efficient energy transfer and heating of the tissue when lasers of these wavelengths are used for irradation.

In addition to absorption, scattering must be considered as a further optical tissue parameter. Tissue is a highly structured medium, so that directed optical radiation is completely altered in its spatial distribution due to reflection, refraction, and diffraction. This scattering effect is relevant mainly when absorption is weak. The thermal properties of tissue, the heat capacity and conductivity, can be taken in the first approximation to be the same as those of water. However, estimation of the spread of energy by thermal conductivity is often very difficult when tissue layers of strongly differing structure and complicated geometry are involved, such as the stomach wall, the retina, and the bladder wall, or when blood vessels give rise to a very nonhomogeneous removal of energy due to the usually irregular blood flow.

The temperature increase and temperature distribution in tissue exposed to laser radiation depend on the energy absorbed by the volume of tissue and on the thermal properties of the tissue. According to the temperature in the tissue, changes such as discoloration, coagulation, shrinkage, carbonization, and vaporization occur (Fig. 5).

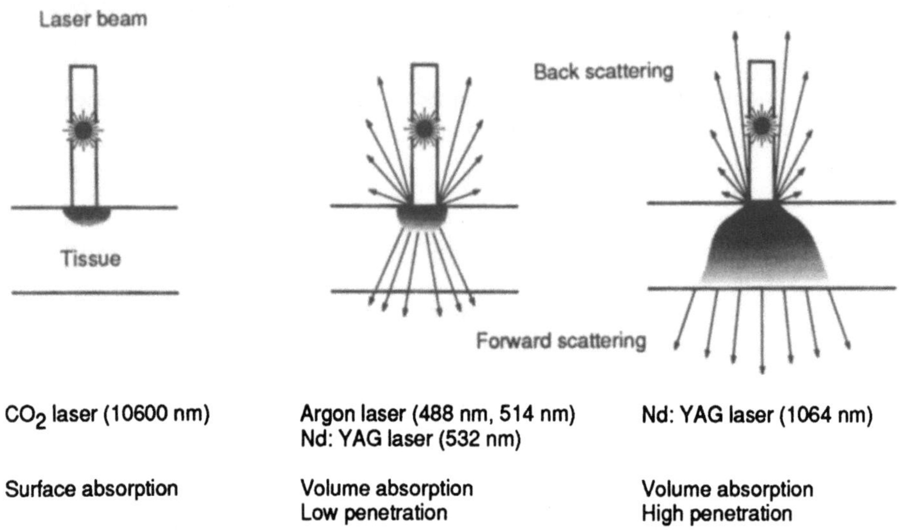


**Fig. 6.** Different distribution of laser light in tissue for the $CO_2$ laser, Nd:YAG laser, and argon laser

As a matter of fact, the processes involved in the thermal interaction between laser radiation and biologic tissue are quite complex. This makes it difficult to develop suitable interaction models that make it possible to approximate at least qualitatively the optimal laser parameters for an intended and expected interaction. The choice of wavelength determines the depth of penetration according to the kind of tissue and therefore influences the interplay between the different tissue reactions. Thus, a consideration of no more than the dependence of absorption and scattering in the tissue on the wavelength used leads to a simple explanation of the various thermal effects of laser systems applied in surgery (Fig. 6). The photobiologic processes based on thermal effects are broadly applied and have gained significant importance for the routine clinical application of laser in surgery and the adjacent endoscopic specialties using the argon laser, the $CO_2$ laser, and the Nd:YAG laser.

With the $CO_2$ laser (up to 100 W at 10.6 μm), absorption by tissue is the strongest and scattering is negligible. The light energy is therefore completely converted into heat at the tissue surface. For this reason, the $CO_2$ laser as a cutting tool is well suited for removing tissue, leaving a narrow edematous border.

The absorption of argon laser radiation (up to 10 W, 488 nm, 514 nm) is weaker than that of $CO_2$ laser radiation. The scattering of the argon laser emission is less pronounced. Penetration is limited due to the selective absorption in hemoglobin and melanin. The same applies for the radiation of the frequency-converted Nd:YAG laser at 532 nm. The application is thus restricted to indications in which removal of the tissue with simultaneous limited coagulation is desired. A penetration depth of 0.5 mm is characteristic. The Nd:YAG laser emits light in the near infrared range (up to 100 W at 1064 nm and up to 30 W at 1318 nm). At the wavelength of 1064 nm, absorption in tissue is very low. Scattering is therefore very pronounced, resulting in a uniform distribution of the radiation in the tissue. Slow heating of a large tissue volume around the point of impingement of the radiation occurs, followed by a deep coagulation, that progresses slowly. Finally, protoplasm vaporizes at the tissue surface, leading to a marked shrinking, although the tissue surface itself is hardly damaged. The surface is covered with a thin layer of fibrin only. The shrinking of the tissue combined with uniform coagulation results in the sealing of blood and lymph vessels. Owing to the shrinkage, the coagulation layer causes mechanical compression of the vessels, which results in hemostasis. The sealing off of the source of bleeding is assisted by the gluelike consistency of the denatured tissue. Arteries of up to 2 mm and veins of up to 3 mm in diameter are closed rapidly and reliably.

The low absorption leads to coagulation depths of down to 6 mm. The tissue volume covered by the laser light is heated, and this results in the delayed death of the tissue, with no noticeable structural damage. The denatured volume or damage zone increases according to intensity. After coagulation, carbonization occurs at the surface, and the tissue is vaporized and removed [9].

Of essential importance for the form and structure of tissue damage by Nd:YAG laser irradiation is surface cooling. When the surface of the tissue is cooled with gas the extent of the necrosis is clearly marked at the surface. With water cooling, a deep drop-shaped necrosis develops and the surface of the tissue is hardly injured.

A comparison of the coagulation effect of electric current and Nd:YAG laser light reveals marked differences, which are particularly noticeable in multilayered tissue. With comparable surface necrosis, the coagulation zone produced by electrocoagulation is shallow, with unsharp, irregular lateral and deep boundaries. Nd:YAG lasing results in a laterally sharply delimited necrosis involving all the layers.

The absorption coefficient of water and saline is approximately ten times higher using the Nd:YAG laser at a wavelength of 1318 nm than it is at 1064 nm. This means that energy is more efficiently converted into heat in tissue at 1318 nm. The extinction coefficient (which depends on scattering and absorption) in blood at 1318 nm is only one third that at 1064 nm [26]. This means less heat dissipation by blood and deeper penetration in tissue at 1318 nm. Particularly in tissue with a high water content, a marked loosening of the necrotic tissue is obtained with laser light at the 1318 nm wavelength. The focused application (5–10 kw/cm$^2$) has a certain ablative effect. However, it is by no means to be compared with the well-known precise incision obtained with a 10.6 μm $CO_2$ laser. Unlike the $CO_2$ laser incision, tissue ablation with the Nd:YAG laser at 1318 nm is distinguished by a clear and effective coagulation zone along the borders of the incision. With a significantly lower power density (100–200 W/cm$^2$) and a shorter interaction time (0.1 s) the resulting sharply defined coagulation can be used for tissue welding [10, 24].

For microsurgery and the surgery of tumors in neurosurgery, gynecology, otorhinolaryngology, thoracic surgery and plastic surgery, different $CO_2$ laser systems with power levels between 20 W and 100 W are available. The use of lasers in surgery is determined by the contact-free technique; with $CO_2$ lasers the beam is transmitted over mirror systems (articulated arms) to the operative site and applied by means of a focusing handpiece. For microsurgery the $CO_2$ laser is adapted directly to the microscope. The laser beam is guided over a movable mirror controlled by a joystick.

According to the medical indications and applications surgical Nd:YAG lasers are graded in their output power. The systems can generate up to 120 W of power in the cw mode and are used in gastroenterology, urology, gynecology, pulmonology, neurosurgery, vascular surgery, dental surgery, and dermatology.

For the medical application in dermatology, otorhinolaryngology, and ophthalmology argon laser and frequency-converted Nd:YAG laser systems generating power levels of up to 10 W are available. With Nd:YAG and argon lasers, power transmission is possible through thin, flexible quartz-glass light guides, thus permitting their use in endoscopy. In order to achieve a better control of the irradiation geometry and power density on tissue, ceramic tips are mounted on the light guide and are in direct contact with the tissue. With the diverse shapes of ceramic tips different distributions of power density can be achieved to produce either coagulation, vaporization, incision, or merely a local warming.

The bare fiber can be used in contact to prepare and incise tissue. Very small and precise cuts are possible. Due to the thermal and mechanical sensitivity of the fiber tip a power feedback system is necessary. By the detection of heat radiation the fibertome system (manufacturer: MBB-Medizintechnik GmbH) makes it possible to keep the temperature at the fiber-tissue interface constant. In order to obtain optimal cutting effects, the laser power can be set according to the different kinds of tissue (see pp 247–250: G. Hauptmann, F. Frank, Nd:YAG Laser for Contact Surgery with Tissue-Effect Control). A considerable extension of the thermal application is interstitial thermotherapy (ITT). Here, a special light guide is pierced into the pathologic tissue, controlled by magnetic resonance imaging and other modern imaging methods [11]. Bare fibers, diffuse emitting fibers, and sapphire tips can be used for interstitial application. Problems regarding the carbonization of adjacent tissue due to high power density on the one hand and no effective coagulation of larger areas on the other hand have led to the development of a special ITT guide. The configuration of the fiber tip, combined with the circumferential beam characteristic, makes it possible to heat a relatively large tissue volume, which is associated with a homogeneous coagulation. The extension of the necrosis depends on the exposure time. Multifiber systems can be applied to extend the coagulation necrosis and to model its expansion according to the tumor geometry by varying the exposure time and power output.

The thermal application of laser in gynecology comprises the $CO_2$ laser for cutting and vaporization, the Nd:YAG laser for noncontact coagulation and vaporization of tissue, and the Nd:YAG laser with contact tips or special bare fibers for scalpel-like cutting and preparation with an optimal sealing effect (Fig. 7).

| | |
|---|---|
| **$CO_2$ Laser**<br>cutting<br>vaporization | – adhesiolysis<br>– fimbrioplasty, salpingostomatoplasty<br>– salpingotomy (oviduct pregnancy)<br>– tubal sterilization (severing)<br>– endometriosis<br>– conization (CIN)<br>– condylomata acuminata<br>– vulvectomy<br>– mammoplasty |
| **Nd:YAG Laser**<br>coagulation<br>vaporization | – endometriosis<br>– tubal sterilization (coagulation)<br>– ablation of the endometrium<br>– intrauterine polyps and fibromas<br>– intrauterine septa and synechiae<br>– condylomata acuminata<br>– cancer of the vulva |
| **Nd:YAG Laser**<br>with<br>contact tips | – adhesiolysis<br>– tumor excision<br>– intrauterine septa and synechiae<br>– polyps<br>– conization<br>– condylomata acuminata<br>– vulvectomy<br>– mammoplasty |

**Fig. 7.** Laser applications in gynecology

## Photoionizing Effects

When a power density of $10^7$ W/cm$^2$ is exceeded nonlinear effects result. The high irradiance generates strong electric fields, which lead either to a breaking of the intracellular structures or to a dissociation or ionization of the tissue material involved. Thus, laser light is converted into kinetic energy (Fig. 4).

The so-called photoablative effect occurs only with energy densities above a certain threshold. Below this threshold the affected tissue volume, determined by spot diameter and penetration depth, is not removed spontaneously but the absorbed energy is converted into heat which results in linear thermal effects. With energy densities exceeding the threshold, the photoablation effect starts and increases. Far above the ablation threshold, the ablation rate is saturated due to the plasma shielding of the laser beam [4] (Fig. 8). The threshold of the energy density necessary for photoablation can, due to high absorption, be achieved in the ultraviolet region using ArF-, KrF-, and XeCl-excimer lasers ($10^8$ W/cm$^2$ with 10 ns at 193 nm, 249 nm, and 308 nm) and with solid-state lasers of wavelengths between 2 μm and 3 μm in the mid infrared region such as the Er:YAG laser at 2.94 μm and the Ho:YLF or the Ho,Tm:YAG laser at 2.10 μm [4] (Fig. 9).

Photoablation is a combination of tissue evaporation and expulsion of liquified material by hydrodynamic mechanisms [12]. The high intensity of the laser radiation evaporates tissue and the arising pressure ejects the molten tissue material out of the irradiated area. The extent of this process depends on the elastic properties of the tissue and on its viscosity. The laser energy is used particularly for phase transitions such as vaporization and liquefaction, as well as for the kinetic energy of the expulsed tissue particles. Due to the small amount of remaining energy, only minimal thermal damage occurs at the margins (Fig. 10). Photoablation is a thermal effect with almost no thermal injury.

Photoablation has been exploited in the ultraviolet region with ArF-, KrF-, and XeCl-excimer lasers. With low ablation rates, precise, non-necrotic cuts (smaller than 50 μm) can be produced. In ophthalmology first applications have been in keratotomy. Experimen-

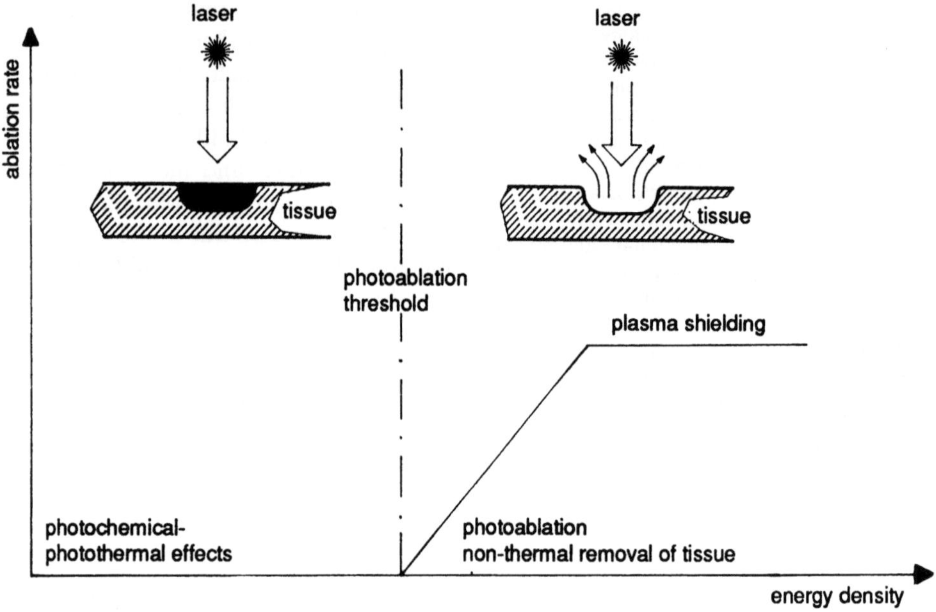


**Fig. 8.** Photoablation depending on the energy density on tissue and the threshold for nonthermal ablation (from [4])

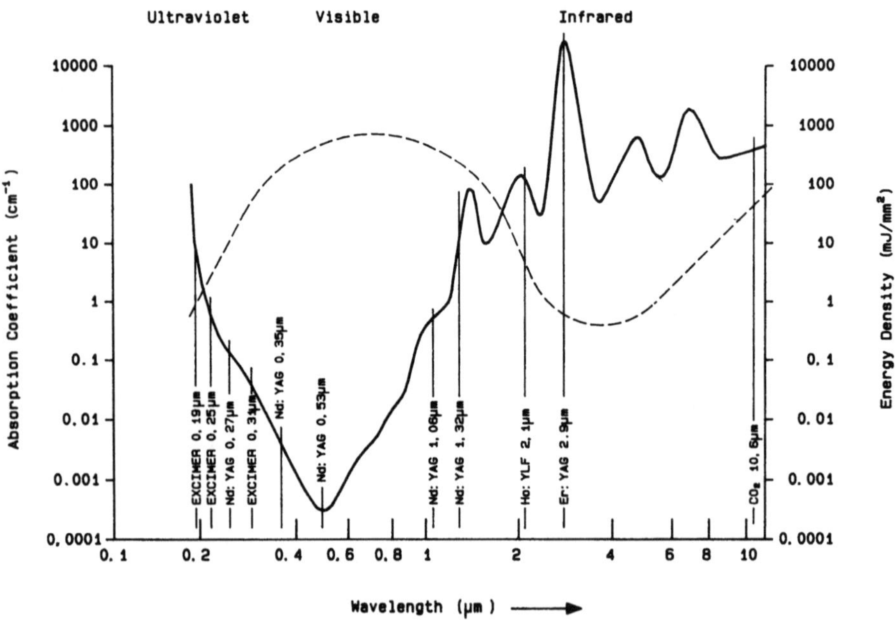


**Fig. 9.** Absorption coefficient of tissue and threshold energy density for nonthermal photoablation depending on wavelength (from [4]). ——, Absorption coefficient of $H_2O$ ($cm^{-1}$), - - - approx. threshold energy density for tissue ablation ($mJ/mm^2$)

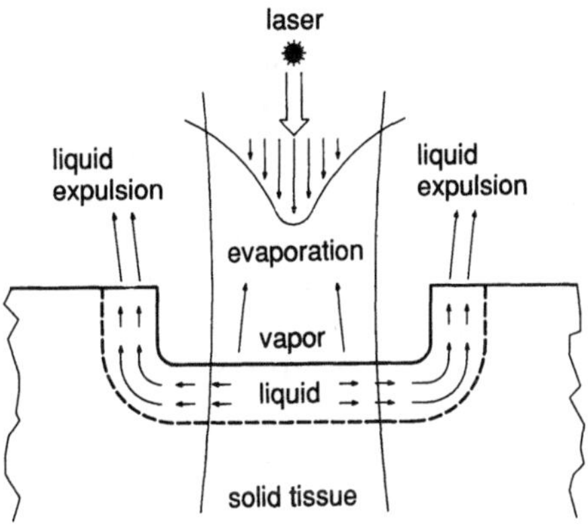


**Fig. 10.** Different phase division and mechanisms for photoablation of tissue (from [12])

tal treatments in orthopedics have been successful. First clinical results in vascular recanalization have been obtained; however, they are confined to wavelengths larger than 300 nm, because only for this range are fiber transmission systems available.

With mid infrared solid-state laser systems an average power of 15 W with pulses of 1–3 J energy and a pulse duration of 250 μs can be achieved. So the ablation rates for hard tissue are definitely higher than those achieved with ultraviolet lasers. The Ho-laser can be used with fiber transmission systems. With the Ho-laser, incisions in cartilage and bone can be produced with only slight secondary damage. Calcified plaque, too, can be destroyed effectively. Thus, the Ho-laser is successful not only in the endoscopic and open ablation of tumor treatment but also in arthroscopic orthopedics and the recanalization of vessels.

Application with the Er:YAG laser is still performed using articulated arms. The Er:YAG laser can work as a dental drill. With suitable application parameters and by cooling the surface of the tooth, cracks in the enamel can be avoided.

The focusing of even shorter high peak-power laser pulses creates power densities ($10^{10}$ W/cm$^2$ for ns pulses and $10^{12}$ W/cm$^2$ for ps pulses) which generate such powerful electric fields ($10^6$–$10^7$ V/cm) that spontaneous ionization to free electrons and ionized atoms (plasma) is induced. When a certain degree of ionization has been reached, a rise in plasma temperature follows. Thereupon, the plasma undergoes a sudden expansion accompanied by a mechanical-acoustic shockwave. The shockwave ruptures the tissue structure (photodisruption) or disintegrates the targeted material (photofragmentation).

Use is made of photodisruption in ophthalmology for microsurgical interventions within the eye without any damage to the healthy anatomic structure. For the laser-induced intracorporeal shock wave lithotripsy photofragmentation using pulsed dye lasers (100 mJ with 2 μs at 590 nm), q-switched Nd:YAG lasers (60 mJ with 20 ns at 1064 μm), or q-switched alexandrite lasers (80 mJ with 200 ns at 750 nm) is applied. All laser systems under investigation for laser lithotripsy differ with regard to wavelength and pulse energy, and especially with regard to pulse length. With the dye and the alexandrite laser it is possible to transmit the high power pulses through 300 μm or 200 μm quartz-glass fibers and to obtain a plasma formation at the surface or within the stone. The stones are gradually destructed. First clinical applications to fragment kidney stones, ureter stones, and gallstones in patients have been performed [6, 8, 15–17, 25, 27].

## Prospects

In addition to the various tested and approved applications of the laser in medicine, further promising applications in tumor surgery, vascular surgery, and stone fragmentation, as outlined, will certainly be developed. Time and again, new laser systems will be investigated with the aim of finding potential new applications in medicine. Particularly photoablation effects resulting from excimer lasers at wavelengths from 193 nm to 308 nm (UV) as well as from pulsed irradiation with solid-state lasers with wavelengths between 2 μm and 3 μm (IR) as well as frequency-converted Nd:YAG lasers with wavelengths of 266 nm and 355 nm (UV) are being investigated. Certainly, free-electron lasers are not very competitive, due to their enormous dimensions and high technical expenditure. The free-electron laser is characterized by the fact that it can be tuned to different wavelengths and operate at extremely high power levels with excellent optical beam quality. The efficiency of the systems currently in use still lies below 1%. More promising are laser diodes. Although they are small, power levels of up to 1 W cw can be achieved. Efficiency is extremely high, lying between 25% and 60%.

Laser application is always indicated where it proves superior to other methods. Its use in endoscopy in particular means less strain for the patient and avoids long hospitalization. For Theodore Maiman, who was the first to realize the laser technically, the fascination lies in the fact that the laser light, which initially was dismissed as a ray of death, turned out to have a life-preserving function. The trend is towards a reasonably priced integration of laser units and application instruments, such as the combination of laser and operation microscopes, and towards effect-controlled dose systems in replacement of the subjective irradiation control by the surgeon, which render the application safer and easier and benefit the patient.

Totally new prospects open up with interstitial laser application. Noncontact application stood at the beginning of laser therapy. This was succeeded by endoscopic application, and today interstitial thermotherapy is being launched, involving the potentialities of magnetic resonance imaging and other modern imaging methods for control.

## References

1. Bassov NG, Prokhorov AM (1958) Molecular oscillators and amplifiers. Priroda 7:24
2. Boulnois JL (1986) Photophysical processes in recent medical laser developments: a review. Laser Med Sci 1:47–66
3. Bridges WB (1969) Laser oscillation in singly ionized argon in the visible spectrum. Appl Phys Lett 4:128
4. Dörschel K, Müller G (1989) Dosimetry for photoablation technique. SPIE Institute Series, vol. IS 5:226–237
5. Dougherty TJ (1983) Photoradiation therapy. Clinical and drug advances, porphyrin photosensitisation. Plenum, New York
6. Dretler StP (1988) Laser lithotripsy: a review of 20 years of research and clinical applications. Lasers Surg Med 8:341–356
7. Einstein A (1917) On the quantum theory of radiation. Physikalische Z 18:121
8. Ell C, Wondrazek F, Frank F, Hochberger J, Lux G, Demling L (1986) Laser-induced shockwave lithotripsy of gallstones. Endoscopy 18:95–96
9. Frank F, Hofstetter AG, Keiditsch E (1981) Experimental investigation and new instrumentation for Nd:YAG laser treatment in urology. In: Bellina JE (ed)

Gynecologic laser surgery. Plenum, New York, pp 345–356
10. Frank F, Beck OJ, Hessel St, Keiditsch E (1986) Comparative investigations of the effects of the Nd:YAG laser at 1.06 µm and 1.32 µm on tissue. Lasers Surg Med 6:546–551
11. Frank F, Hessel St (1990) Technische Voraussetzungen für die interstitielle Thermotherapie mit dem Nd:YAG Laser. Lasermedizin 0:36–40
12. Frenz M, Romano V, Zweig AD, Weber HP, Chapliev NI, Silenok AV (1989) Instabilities in laser cutting of soft media. J Appl Phys 66:4496–4503
13. Geusic JE, Marcos HW, van Uitert LG (1964) Laser oscillations in Nd: doped yttrium aluminium, yttrium gallium, and gadolinium garnets. Appl Phys Lett 4:182
14. Gordon JP, Zeiger HJ, Townes CH (1954) Molecular microwave oscillator and new hyperfine structure in the microwave spectrum of $NH_3$. Phys Rev 95:282
15. Hofstetter AG, Frank F, Keiditsch E, Wondrazek F (1985) Intracorporale laserinduzierte Stoßwellen-Lithotripsie. Laser 1:155–158
16. Hofstetter AG, Schmeller N, Pensel J, Arnholdt H, Frank F, Wondrazek F (1986) Harnstein-Lithotripsie mit laserinduzierten Stoßwellen. Fortschr Med 104:32–34
17. Hofstetter AG, Thomas S (1989) Intrakorporale, laserinduzierte Stoßwellenapplikation zur Zerstörung von Harnsteinen. Urologe (A) 28:145–147
18. Javan A, Bennett WR jr, Herriot DR (1961) Population inversion and continuous optical maser oscillation in a gas discharge containing a helium-neon mixture. Phys Rev Lett 6:106
19. Maiman TH (1960) Optical and microwave experiments in ruby. Phys Rev Lett 4:564
20. Mester E (1980) Laser application in promoting wound healing. In: Koebner HK (ed) Lasers in medicine. Wiley, Chichester, pp 83–95
21. Müller GJ, Berlien P, Scholz C (1986) Der Laser in der Medizin. Umschau 4:233–240
22. Patel CKN (1964) Selective excitation through vibrational energy transfer and optical maser action in $N_2$-$CO_2$. Phys Rev Lett 13:617
23. Schawlow AL, Townes CH (1958) Infrared and optical masers. Phys Rev 112:1940
24. Schober R, Ulrich F, Sander T, Dürselen H, Hessel St (1986) Laser-induced alteration of collagen substructure allows microsurgical tissue welding. Science 232:1421–1422
25. Simon W, Hering P (1987) Laserinduzierte Stoßwellenlithotripsie an Nieren- und Gallensteinen (in vitro). Laser Optoelektronik 1:33–34
26. Stokes LF, Auth DC, Tanaka D, Gray JL, Gulacsik C (1981) Biomedical utility of 1.32 µm Nd:YAG laser radiation. IEEE Trans Biomed Eng BME 28:297–299
27. Thomas S, Pensel J, Engelhardt R, Meyer W, Hofstetter AG (1988) The pulsed dye laser versus the q-switched Nd:YAG laser in laser-induced shock-wave lithotripsy. Lasers Surg Med 8:363–370

# Thermal Preparation Techniques and Thermal Tissue Destruction: Attempt at a Synopsis

D. WALLWIENER[1], D. POLLMANN, S. RIMBACH, D. HEBERLING, and G. BASTERT

Thermal preparation includes photothermal techniques, such as with the thermal laser, and electrothermal techniques using high-frequency electropreparation.

Contrary to sharp preparation or the cutting of tissue with athermal lasers, tissue preparation with thermal techniques is always connected with more or less extensive thermal damage of the wound edges. The greater the coagulatory potential of the chosen technique for simultaneous hemostasis, the greater the associated thermal tissue damage. It thus stands to reason that a minimization of thermal damage is accompanied by a minimization of the hemostatic effect (Fig. 1). It is therefore imperative to compare and contrast the common preparation methods which aim at a minimization of the thermal tissue damage according to the morphometrically determined thermally damaged zone, and to base such a study on enzyme-histochemical techniques.

The entire thermally damaged zone including the wound edges is composed

**enzyme free zone**
**twice as wide**
**as necrotic zone with van Gieson**

average width of necrosis by $CO_2$ laser*
= 200 μm by van Gieson stain
=0.4 mm by LDH enzyme-free zone

(* 10 W/s, cw, focus diameter 0.22 mm)

**Fig. 2.** LDH-enzyme-free zone = thermal damage

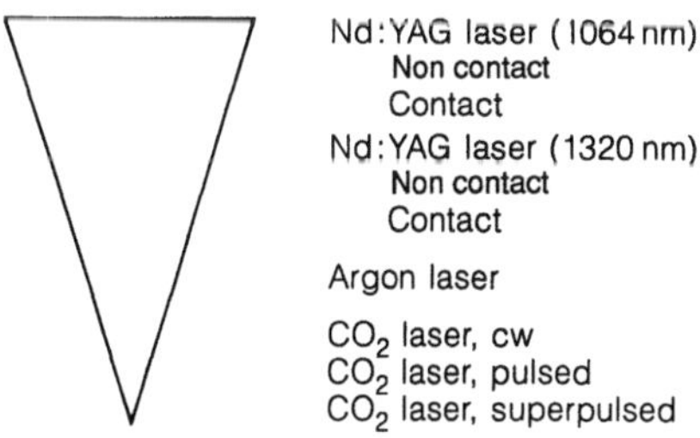


**Fig. 1.** Laser and hemostasis: coagulation effects

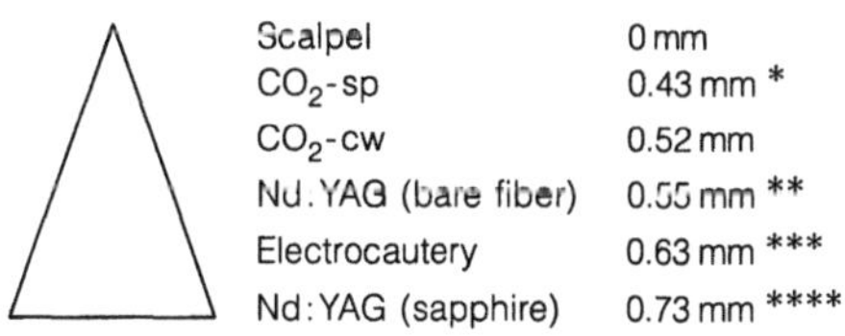


* 10 W/s; focus 0.22 mm
** 14 W/s; cw, 400 μm
*** 50 W/s; monopolar
**** sapphire tip, diameter 400 μm, 14 W/s; cw (1.06 μm/1.32 μm)

**Fig. 3.** LDH-enzyme-free zone = thermal damage

[1] Universitäts-Frauenklinik, Abt. für Geburtshilfe und Gynäkologie, Voßstr. 9, W-6900 Heidelberg, FRG.

Lasers in Gynecology
Edited by G. Bastert and D. Wallwiener

of the light-microscopically visible damaged tissue zone (necrotic or carbonization zone) and a damaged zone which cannot be identified with a light microscope and cannot be distinguished from the undamaged tissue. Histochemical presentation (Fig. 2) is necessary to show that the actual thermally damaged tissue zone after the utilization of thermal preparation techniques is in fact twice as large as the area detectable light microscopically.

If one now attempts an enzyme-histochemical morphometrical evaluation of the entire thermally damaged zone of the various preparation techniques based on a META-analysis and against the background of our own results, the synopsis shown in Fig. 3 is obtained.

## Conclusion

If one bears the statistical uncertainties of such a combination in mind when critically evaluating these parameters and, furthermore, takes the numerous variables which exert diverse influences on the morphometrically measured results (diverse and in part not standardized tissue lesions) into consideration, it is nevertheless amazing to find that the differences in the thermally destructive action of the various preparation techniques lie only fractions of millimeters apart under optimum conditions. These differences lie in the decimal unit range. This fact should be borne in mind when one is discussing different thermal preparation techniques!

# Technical Background of Nd:YAG Laser Contact Technique

D. Pollmann [1], D. Wallwiener, H. Hauptmann, M. Kappler, U. Felderhoff, S. Rimbach, and G. Bastert

## Nd:YAG Laser Contact Technique

Noncontact application of the Nd:YAG laser results in a homogeneous coagulation zone of the tissue of 4–6 mm in depth [1,2]. Effective cutting by vaporization can be attained only at an extremely high density and at a low application speed; however, a large coagulation zone cannot be avoided. Fine cuts/preparation, as with the $CO_2$ laser, cannot be achieved with the noncontact technique [1,2,4–7].

With development of the contact technique, the application of the Nd:YAG laser as a scalpel became possible: the laser beam is applied under direct tissue contact [3]. First, various types of coniform sapphire crystals were distally attached to protect the thermolabile quartz glass fibers and to adapt the radiation geometry of the emitted laser beam for different application purposes (contact coagulation, contact vaporization) [1,11,17]. The tip is adapted using a metal connector on the flexible quartz fibers or with a screw thread on a hand applicator to which the quartz fibers are affixed [6,7]. Later, the possibility of tissue preparation by the contact technique without a sapphire tip, with so-called bare quartz glass fibers, was examined.

The output loss due to reflection is considerably lower with the contact technique than with the noncontact technique. Furthermore, destruction of the quartz glass fibers by overheating after contamination with tissue or blood is prevented using this method. Conical formation of the sapphire crystal ensures that the laser radiation leaves the tip of the crystal after total reflection within it (Fig. 1). The exceptionally high light density thus attained leads to immediate vaporization of a minute tissue volume at the point of contact with the tip, enabling utilization of the Nd:YAG laser as a scalpel [13,16]. Exceptionally fine prepara-

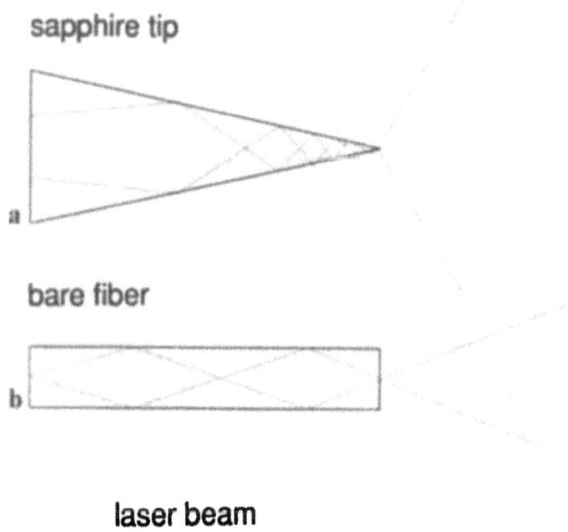


**Fig. 1 a, b.** Total reflection of the laser beam within (**a**) a sapphire tip; (**b**) a bare fiber

[1] Universitäts-Frauenklinik, Abt. für Geburtshilfe und Gynäkologie, Voßstr. 9, W-6900 Heidelberg, FRG.

Lasers in Gynecology
Edited by G. Bastert and D. Wallwiener

tions can be attained by guiding the contact tip over the tissue and slightly touching it [4, 17].

The tissue effect of Nd:YAG contact laser preparation depends upon the shape of the particular contact tip utilized. It was thus necessary to test the various sapphire tips in standard energy ranges to determine the ideal contact tip for surgery. A further preclinical problem was whether or not the contact tip presented a danger to the patient.

The following detailed technical questions regarding the laser were posed:

1. Is there a danger of explosion and/or breaking off or melting of the sapphire crystal during use, and would toxic substances be released during these processes?
2. Can sapphire tips and/or quartz glass fibers be optimized with respect to the expected effect?
3. What technical differences are discernible between Nd:YAG laser contact preparation with bare fibers and contact preparation using sapphire tips, especially as far as attrition is concerned?

## Physicochemical Examination of the Sapphire Tips

Phase-analytical and radiological examinations (Mineralogical Institute of the University of the Saarland) of both new and used sapphire tips disclosed that the contact tip utilized (Osada), purportedly made of 100% pure synthetic sapphire, in fact included a high percentage (approximately 20%) of silica (Quartz) [14]. This impurity accounts for the vitrification of the tip, as evidenced by a spherical opaque melting at the point of deflection of the beam when the tip attains the melting temperature of the crystal – approximately 2050 °C. The release of toxic substances was precluded by phase-analytical investigation of both a new and a used sapphire tip [14].

## Radiation Geometry and Deflection Characteristics

### Sapphire Tips

Calculation of the radiation geometry within the sapphire tips (Fig. 1) shows that the length of the sapphire tip is limited: the angle of incidence on the interface, as a function of the angle of the cone of the sapphire probe, becomes steeper with every total reflection (Table 1).

If this angle exceeds the threshold angle, determined by the difference in the angle of refraction of the two media, the ray is emitted laterally prior to reaching the tip. This loss of intensity is due to the fact that the decreasing angle of incidence is smaller than the angle of total reflection; it is calculated

**Table 1.** Data on the sapphire tips for geometric construction and resultant measurements

| Length of tip | Diameter of tip | Last measured angle | Number of reflections |
|---|---|---|---|
| 40 mm | 0.8 mm | 68° | 6 |
| 40 mm | 0.6 mm | 32° | 15 |
| 30 mm | 1.0 mm | 77° | 3 |
| 30 mm | 0.8 mm | 76° | 3 |
| 30 mm | 0.4 mm | 60° | 8 |
| 20 mm | 0.8 mm | 62° | 4 |
| 20 mm | 0.4 mm | 56° | 7 |

with the following formula, using Snell's law:

$$\sin\beta(\text{incidental})/\sin\beta(\text{fractional}) = n(\text{air})/n(\text{sapphire})$$

$$\sin\beta(\text{T}) = \frac{n(\text{air})}{n(\text{sapphire})} = \frac{1}{1.75} = 0.571$$

$\beta(\text{T})$ = angle of total reflection = 34.85°

$n$ = refractive index, i.e.

$$\sin\beta(\text{T}) = n(\text{water or tissue})/n(\text{sapphire}) = 1.33/1.75 = 0.76$$

$\sin(\text{T})$ = angle of total reflection = 49.46°

Total reflection enables transmission of all the laser rays; hence, the majority are not reflected at the crystal/water interface, being emitted from the crystal instead. The transmitted beam is then broken according to Snell's refractive index (see above). Those rays which have left the crystal prior to reaching the end of the tip lead to a reduction in the energy density at the tip; this, in turn, results in a reduction in the desired cutting property.

If the maximum length of the tip is exceeded (dependent upon the diameter of the tip and the conical angle) a radiation loss results [8–10]. The hand applicator utilized (Osada) thus includes an additional optical system which focuses the laser beam into the sapphire crystal; long tips can thus be employed.

The angle of deflection is thus dependent upon numerous factors:
- The length and diameter of the probe (Table 1)
- The shape of the end of the tip
- The difference in the refractive indices of sapphire and tissue

## Quartz Glass Fibers (Bare Fibers)

The transmission of laser radiation via light-guiding fibers is based upon its repeated total reflection on the interface of the more highly refractive nucleus and the surrounding medium with a lower refractive index. The nucleus of the utilized light-guiding fibers consists of fused silica. It is surrounded with a so-called cladding, consisting of a layer of silicone or specially developed quartz glass with a lower refractive index to maintain the total reflection. Stability and flexibility of the fibers is attained by an exterior coating of synthetic material (Table 2) [14].

The rays emitted from the laser resonator are focused on a spot with a smaller diameter than the nucleus of the utilized fibers to enable their enclosure in a fiber. If thin fibers are used, this is possible only with a lens with a short focal distance and thus is attainable only over a larger enclosure system. This, in turn, leads to a greater divergence of the rays emitted at the distal end of the fiber.

Table 2 shows the composition of the fibers utilized with nuclei of 200, 360, and 600 μm in diameter. The theoretical emission angle, dependent upon the type of enclosure system, is also shown. The cladding of the 200- and 360-μm fibers is constructed of special quartz glass which cannot be insulated; hence, the diameter of the fibers which actually come into contact with the tissue is 280 and 400 μm respectively.

Extremely high temperatures, lying within the range of the heat of transformation of quartz glass (i.e., + 1300 °C), are attained on the fiber tip during vaporization with direct tissue contact. Heating of the fiber tip is locally restricted, and the heating and cooling times are limited to approxi-

**Table 2.** Technical data on the quartz fibers

| Diameter | | | Material | | Theoret. emission angle | Manu-facturer |
|---|---|---|---|---|---|---|
| Nucleus (μm) | Cladding (μm) | Total (μm) | Nucleus | Cladding | | |
| 600 | 830 | 1050 | Quartz | Silicone | 8° | EOTec |
| 360 | 400 | 570 | Quartz | Silicone | 19° | Schott |
| 200 | 280 | 450 | Quartz | Quartz | 19° | Quartz & Silice |

mately 0.1 s due to the poor heat capacity and thermal conductivity of quartz glass [7].

## Radiation Profile

Integral procedure for determination of laser light deflection by "bare fibers" or sapphire tip: A cylinder with a slit diaphragm rotates in front of a photodetector. The distance between photodetector and light-guiding fiber must be so determined, that the light cone is entirely focused on the photodetector. The rotation of the cylinder is transmitted to an operation recorder that measures the density of the light-guiding fiber, which depends on the angle of incidence (Fig. 2, left).

Local procedure for determination of laser light deflection by "bare fibers" or sapphire tip: A photodetector installed behind a pin diaphragm describes a well-defined circular course around the light-guiding fiber. The distance recorded is transposed into the angle of the photodetector. Applying the amplitude established on the distance covered, the light density geometry can be established, which is dependent on the reflection angle (Fig. 2, right).

The deflection curve (Figs. 3, 4) of a new sapphire tip or bare fibers (Table 3) is comparable to a normal distribution curve, i.e., a maximum point with a continuous decrease on both sides. The maximum point in this measurement is the point of maximum radiation intensity. The radiation intensity measured at a 90° angle approximated 0.

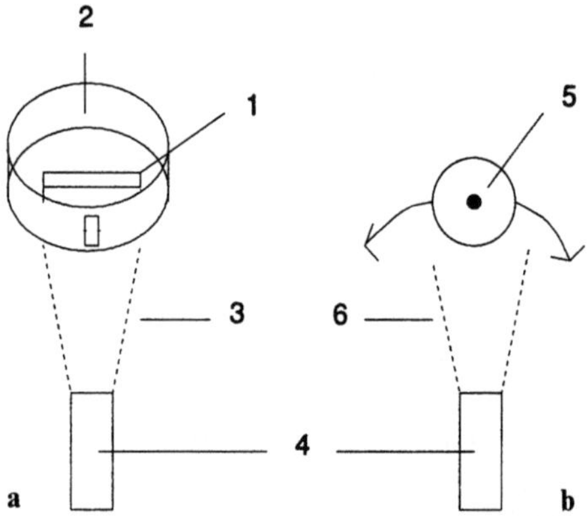


**Fig. 2 a, b.** Test design for measurement of the deflection by (**a**) integral determination and (**b**) local determination. *1*, Photodetector; *2*, rotating cylinder with slit diaphragm; *3*, distance 0.2 cm; *4*, object to be measured (bare fiber); *5*, pin diaphragm with photodetector; *6*, distance 1.2 cm

The conical deflection curve of a new sapphire tip or bare fibers (Figs. 3a, b; 4a, b) changes with the degree of wear of the tip to form a multipointed, uneven flatter curve (Figs. 3c, d; 4c, d). The radiation intensity measured at 90° also changes, i.e., increases.

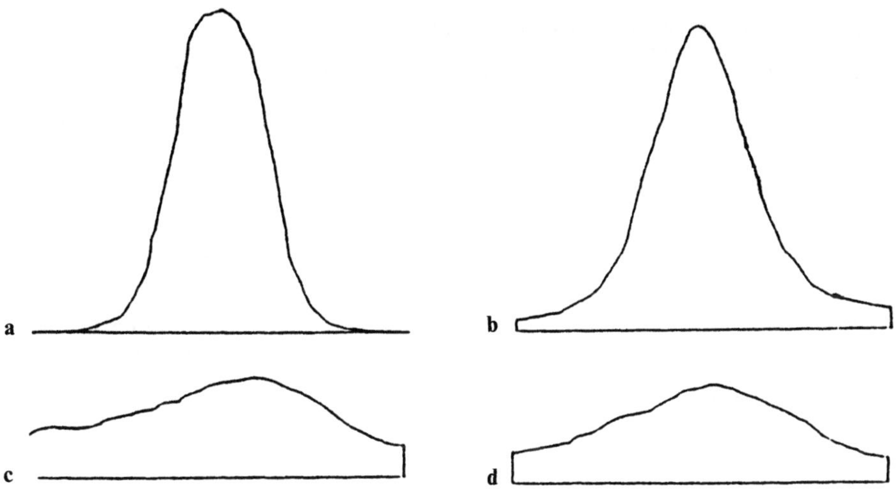


**Fig. 3.** Deflection characteristics determined by the local procedure. **a** New fiber; **b** new tip; **c** used fiber (20 applications); **d** used tip (20 applications)

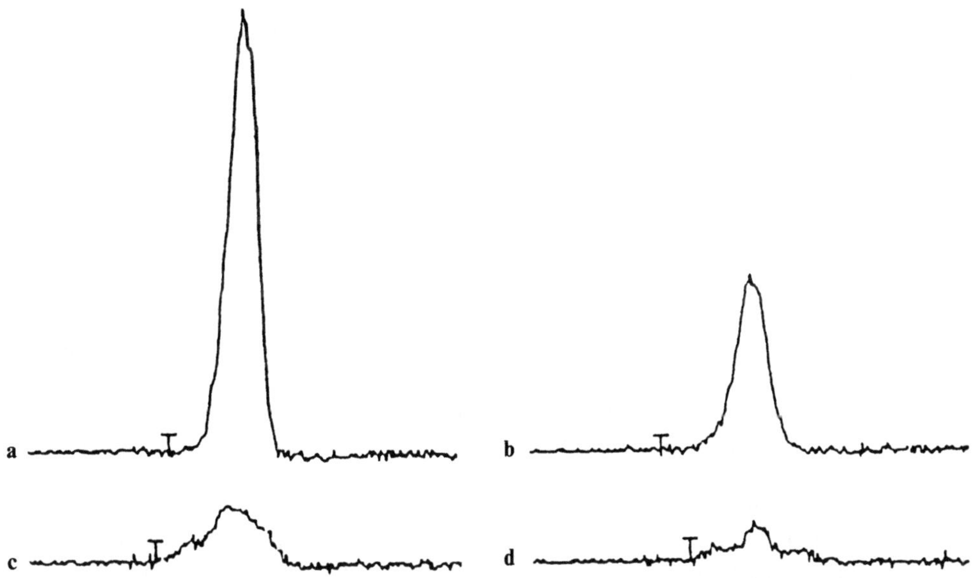


**Fig. 4.** Deflection characteristics determined by the integral procedure. **a** New fiber; **b** new tip; **c** used fiber (20 applications on liver tissue); **d** used tip (20 applications on liver tissue)

**Table 3.** Type and condition of examined tips and bare fibers

| Application method | Length (mm) | Diameter (mm) | Condition |
|---|---|---|---|
| Hand applicator | 20 | 0.2 | New/little used |
| | 16.5 | 0.4 | New |
| | 40 | 0.2 | Coniform melting |
| | 30 | 0.4 | Very worn |
| Endoscope | 5.5 | 0.4 | New/little used |
| | 7.5 | 0.4 | New/little used |
| Bare fibers | – | 0.2 | Often used/damaged |
| | – | 0.4 | New, used, damaged |
| | – | 0.6 | New, used |

In cases of coniform "vitrification" (e.g., of a sapphire tip, Figs. 3c, d; 4c, d) the radiation intensity is identical at all points of the semicircle, amounting only to a fraction of the maximum radiation intensity of a new tip or fiber. Utilization of oft-used sapphire tips or fibers (Table 3) leads to a reduction in the energy density within the axis of the tip and to an increased deflection to the periphery (Figs. 3c, d; 4c, d). Hence, the sum of the applied energy is just as great as in a new tip, but the tissue effect exerted by the tip decreases due to the reduced energy density. The deflected laser energy in the periphery simultaneously results in a larger thermally damaged zone.

This is due to the gradual attrition of the end surface of the tip. The initially smooth surface becomes rough; it can be compared to an "undulating countryside" when viewed under a microscope. The laser beam now strikes this uneven surface. Thus, the angle of incidence of the laser beam is subject to greater variation: it can vary from 0° to 90°. This, in turn, causes the multipeaked character of the radiation intensity in oft-used sapphire tips or fibers (Fig. 4). The output density, which is directly dependent upon the applied energy, thus decreases at 0°, increasing as the angle approaches 90°.

During determination of the radiation profile of identically constructed tips it was shown that the deflection characteristics change with changing length of the probe and diameter of the tip; however, a linear correlation is not present. This can be accounted for by the radiation geometry within the sapphire tip: the probability that the laser beam is emitted from the crystal prior to reaching the end surface of the tip increases, the longer the tip and the smaller its diameter, due to the change in the angle of incidence (Table 1). This results in a relatively flat radiation intensity curve and a smaller maximum radiation intensity.

Thus, a longer sapphire tip does have better cutting properties, due to its higher output density as a result of the radiation concentration within the tip. In contrast, it was shown that the output density in the tissue is directly dependent upon the shape of the tip utilized. It is thus important that the producers supply information on the radiation profile of each sapphire tip they manufacture so that the user is able to assess the distribution of the laser intensity.

*Conclusions:* The deflection curve of "bare fibers," even at a certain degree of wear, is still steeper than that of the sapphire tip. The use of "bare fibers" results in a maximal radiation intensity along the axis of the fiber (Figs. 3,4).

## Determination of the Output Density

During our test series for the determination of the output density of the laser the 20-W maximum output recommended by the manufacturer of the sapphire tips was not exceeded. Energy outputs of 7, 10, and 14 W were selected for cutting purposes. The resultant output density on the tissue surface could be approximately calculated with the following formula:

$$\text{output density} = \text{laser output} \times \frac{1}{\text{contact surface of the tip}}$$

whereby the diameter of the tip approximates a focused noncontact laser beam and therefore the circular surface of the sapphire tip or bare fiber and the resulting radiated area [6].

Furthermore, one must bear in mind that the affect of the Nd:YAG laser is also dependent upon other parameters. The penetration depth of the laser beam into the tissue depends upon both the output density and the tissue absorption, which in turn is dependent upon the wave length of the laser beam. The $CO_2$ laser beam with a wavelength of 1060 nm has a penetration depth of only 0.1 mm; hence, the laser output is almost entirely absorbed over these short distances, whereas the Nd:YAG laser beam has a penetration depth of 2–4 mm [4,7,10]. Frank [5] even reported a coagulation depth of 6 mm when utilizing a sapphire tip which was not blood soiled; i.e., the laser output is absorbed by a tissue volume 60 times as large as that which absorbs the $CO_2$ laser. This is also evident when one considers the various tissue absorption coefficients for different laser wavelengths: 10/cm for 1060 nm and 1000/cm for 10600 nm [5]. Even though the output density decreases exponentially to the tissue depth, the Nd:YAG laser energy absorbed per tissue volume on the tissue surface is smaller than that of a $CO_2$ laser with identical initial output and focus. It would thus be incorrect to directly compare the output density of various lasers.

Within the framework of preliminary tests it was shown that a sapphire tip must be soiled with coagulated blood before it can be utilized with small outputs. This indicates that the output density cannot simply be calculated from the laser output and the tissue surface when using the contact technique; a large proportion of the output is absorbed by the tip and changed into heat before being applied to the tissue. An additional affect component of the Nd:YAG contact technique is thus present, which also supplies an explanation for the limited penetration depth of this technique – the reduced laser output, which does not suffice for tissue defects at a depth of 6 mm [6,16,18].

## Attrition of Bare Fibers and Sapphire Tips

The energy loss and deflection characteristics in relation to various energy ranges were determined by tissue cuts (in vitro tests) to enable comparison of

**Table 4.** Advantages ( + ) and disadvantages ( – ) of laser preparation using bare quartz glass fibers compared with laser application with the sapphire probe

| Sapphire probe | Bare fibers |
|---|---|
| + High melting point (2050 °C) | – Lower melting point (1300 °C) |
| – Cooling time 2 s | + Cooling time 0.1 s |
| + High mechanical durability | – Low mechanical resistance |
| – Limited flexibility | + Highly flexible |
| – Adaptation system necessary | + Adaptation system unnecessary |
| – Cooling system necessary | + Cooling unnecessary |
| – Increased diameter of system | + Diameter of system equal to fiber cross section |
| – Subject to attrition | – Subject to attrition |
| – Repairs are not possible | + Routine repair |
| – Expensive individual construction | + Inexpensive mass product |

the attrition of the sapphire tips with that of the bare fibers.

Numerous factors influence the rate of attrition of the sapphire tips: The smooth semicircular end surface of a new sapphire tip becomes increasingly rough with increasing use and cleaning; it can even be broken when carelessly handled (Table 4) [14].

Coniform vitrification of the tip has been observed when utilizing a 40-mm sapphire tip with a diameter of 0.2 mm for 1 min at 15 W without cooling it with gas or liquid. Continuous utilization of this laser output level thus suffices for the melting point of the crystal to be attained at its tip, thus making it useless if not cooled; the maximum radiation intensity of this tip is no longer adequate for cutting purposes [5, 14, 16].

The bare fibers are damaged immediately during vaporization if the synthetic coating (which would already burn at a temperature of + 200) is not removed from the fibers for a length of 3–5 mm, due to the heat conductivity and thermal capacity of the fibers mentioned above. The soft silicone cladding should be removed from the 600-µm fibers (diameter of the nucleus) for the same reason. If these precautionary measures are not carried out, the divergence of emission of the fibers is substantially changed by the melting and transformation processes. The theoretical angle of emission of the various fibers utilized (Table 2), which is of exceptional importance during noncontact application, determining the size of the irradiated tissue, thus cannot be assumed for the contact technique and is of only secondary importance for the cutting effect. Furthermore, fibers "soiled" by the melting processes are no longer suitable for noncontact application. They must be newly sharpened for such purposes (Table 4).

The laser beam emitted from the distal end of the fibers can be fully utilized to obtain the desired tissue effect during preparation with bare fibers, while the laser energy is reduced to a maximum of 85% of the initial output after transmission through a sapphire probe [14, 16]. The loss in energy is mostly the result of reflection on the optic interface and leads to heating of the metal setting of the contact probe, thus necessitating cooling of the light guide (constant gas flow or liquid irrigation during contact preparation; Table 4) [14, 18].

On the other hand, the thermal conductivity of sapphire and its high melting temperature ( + 2000 °C) prevent premature damage of the tip. Incorrect utilization (noncontact application, insufficient cooling), however, can lead to such extreme heating of the sapphire tip that the melting point of sapphire is exceeded and the sapphire tip thus useless for further operations [16, 17].

The additional thermal capacity of the sapphire tip (heated by thermal conduction), due to its relatively large volume, results in the cooling phase being considerably longer subsequent to switching off the laser radiation (2 s) than is the case with bare fibers (0.1 s). During this time interval there is a greater danger of secondary thermal damage, such as the adhesion of the tissue to the hot sapphire tip or its metal setting. The initially observed adhesion of the tissue to the quartz glass fibers is avoided by breaking the tissue contact shortly before ending the laser application.

No cooling of the "bare fiber" is necessary, as it has a reduced thermal capacity and conductivity. For endoscopic application this means that systems with smaller diameters (e.g., tuboscopes) can be used and laser beam application in organs with small lumina becomes possible.

However, an approximately 40% energy loss was noted after only ten ($10 \times 2\,s \times 14\,W/s$) in vitro tissue cuts with bare fibers without quartz cladding (Table 2). The energy loss and deflection characteristics were even more unsuitable after more than ten preparations, leaving these bare fibers useless for contact preparation if the tips are not suitable refurbished.

Standardizable, representative deflection characteristics were not determined when fibers were used for contact preparation and only broken off distally when repaired. It is thus recommended that bare fibers be smoothly cut distally for contact preparation and recut after use. This is the only method which ensures deflection characteristics with a maximum energy radiation.

Correct application is imperative with the contact technique. Improper handling consists in:

1. A lack of cooling of the sapphire tip
2. No tissue contact during preparation
3. Too high energy levels
4. Not suitably refurbished bare fiber

Destruction or breaking of the sapphire tip was not observed in any of the tissue preparation tests. The observed vitrification was avoided by adequate cooling. A lesion of the connector is the only danger present during insufficient cooling. The hot sapphire tip can then fall off and damage intraperitoneal organs.

Damaged bare fibers must also be punctually replaced, as they influence the stability of the distal fiber ends with increasing thermal strain.

Preparation with bare fibers is superior to sapphire tip preparation where flexibility is concerned, as the flexibility of the former is not hampered by the metal connector. Because the high flexibility of the "bare fiber" makes tissue preparation difficult, a special prototype of an applicator was designed for this type of contact preparation, the "Heidelberger bare fiber endless system". The length of this applicator can be varied, and thus it can be used in laparoscopic as well as extracorporeal tissue preparation.

In summary, safe use of the contact preparation technique is possible with both the sapphire tip and bare fibers, insofar as the laser technical details determined are taken into consideration. Optimization of the configuration of the sapphire tip and a careful watch for early attrition of the bare fibers are imperative, however. Table 4 summarizes the advantages and disadvantages of both techniques. The superiority of the bare fiber technique, at least as far as endoscopy of small lumina is concerned, is clearly discernible.

## References


1. Daikuzono N, Joffe SN (1985) Artificial sapphire probe for contact photocoagulation and tissue vaporization with the Nd:YAG laser. Med Instrum 19:173–178
2. Ell C, Hochberger J, Lux G (1986) Clinical experience of non-contact and contact Nd:YAG laser therapy for inoperable malignant stenoses of the oesophagus and stomach. Lasers Med Sci 1:143–146
3. Fasano VA, Ponzio RM (1986) Nd:YAG laser contact: a coming of the microperformances. In: Oguro Y, Atsumi K, Joffe SN (eds) Nd:YG laser in medicine and surgery. Professional Postgraduate Services, Tokyo, pp 358–362
4. Fasano VA, Urciuoli R, Lombard GF, Benech F, Ponzio RM, Lanotte MM (1985) Advanced technologies in neurosurgery. Laser Med Surg 1:44–49
5. Frank F (1986) Biophysical basis and technical prerequisites for the endoscopic and surgical use of the Nd:YAG laser. Laser Med Surg 3:124–132
6. Frank F (1987) Der Laser in Chirurgie und Endoskopie: Einführung in die physikalischen, biophysikalischen und technischen Grundlagen. Dtsch Arztebl 84: 2238–2243
7. Frank F, Hessel S, Ischinger T (1988) Einsatzmöglichkeiten des Nd:YAG-Lasers bei der Gefäßrekanalisierung. Laser Med Surg 2:66–71
8. Godlewski G, Ginoves P, Chincholles J, Viel E, Bureau JP, Rouy S, Mion H, Dubois A, Fesquet J (1983) Hepatic resection with an Nd:YAG laser in pig. Laser Med Surg 3:217–224
9. Godlewski G, Rouy S, Bureau JP, Dubois A, Mion H, Fesquet J, Romieu M, Giraudon M (1983) Résections hépatiques au laser YAG Nd – résultats anatomiques et biologiques d'une étude expérimentale. J Chir (Paris) 120:443–448
10. Hofstetter A, Frank F, Kreiditsch E, Bowering R (1981) Endoscopic Nd:YAG laser application for destroying bladder tumors; clinical papers. Eur Urol 7:278–282
11. Joffe N (1986) Contact neodymium-YAG laser surgery in gastroenterology: a preliminary report. Laser Med Surg 6:155–157
12. Keckstein J, Finger A, Steiner R (1988) Laser application in contact and noncontact procedures: sapphire tips in comparison to bare fiber, argon laser in comparison to Nd:YAG laser. Laser Med Surg 4:158–162
13. Meyer HJ, Haverkampf K (1982) Experimental study of partial liver resection with a combined $CO_2$ and Nd:YAG laser. Laser Med Surg 2:149–154
14. Pollmann D (1989) Der Nd:YAG Kontakt-Laser – lasertechnischer Background und Gewebeeffekte. Dissertation, Homburg/Saar
15. Verdaasdonk R, Cross F, Borst C (1987) Physical properties of sapphire fibre-tips for laser angioplasty. Laser Med Surg 2:183–188
16. Wallwiener D, Pollmann D, Morawski A, Bastert G, Krampe C (1988) Die Nd:YAG-Laser-Kontakttechnik mit Saphirschneidespitzen – lasertechnischer Background und Gewebeeffekte. Laser Med Surg 4:75–85
17. Wallwiener D, Stolz W, Pollmann D, Spiller M, Bastert G (1988) Neodym YAG contact-Laser (Kontakttechnik mit Saphirspitzen) in der operativen Laparoskopie. Laser Med Surg 4:172–179
18. Wallwiener D, Pollmann D, Kappler M, Bastert G (1989) Die Nd:YAG-Laser-Kontakttechnik mit nackten Glasfasern – lasertechnischer Background und Gewebeeffekte im Vergleich zur Präparation mit Saphirschneidespitzen. Laser Med Surg [Suppl] 1:5–10

# Tissue Effects by Different Laser Techniques ($Nd:YAG$ 1.06 and 1.32, $CO_2$ Laser)

D. POLLMANN [1], D. WALLWIENER, U. FELDERHOFF, S. RIMBACH, W. KÜHN *, and G. BASTERT

## Standardization of Tissue Lesions

The tissue effect exerted by the laser had to be standardized, as the interaction between the laser and the tissue is of an extremely complex nature. Minute alterations in the water content of, or the circulation within the tissue result in marked alterations in the thermal penetration as well as in the absorption of the laser energy. Furthermore, the pressure the sapphire tip or the bare fiber exerts on the tissue must be continuously measured and regulated, as the tissue surface is not entirely flat; unevenness can result in a pressure increase or decrease. However, this would result in this model losing its clinical relevance, as no surgeon can possibly hope to fulfill these conditions. An increased standard deviation was thus deliberately accepted to enable the application of the test results to the clinical situation.

The beam of the $CO_2$ laser can be applied with a micromanipulator with no problem [3]. Linear uterotomies were carried out with a beam length of 2 cm (1 cm/s) on the uterine horn of rats to enable histological determination of the incision depth and the thermally damaged zone resulting from preparation using the various laser techniques.

The rat has become the animal of choice for experimental microsurgical preparation in fertility surgery, as the size of the uterine horn of a mature rat corresponds well with that of the human fallopian tube. In this experiment, sexually mature female Wistar rats were used as test animals (Table 1) [17].

Subsequent to induction of anesthesia with ether, the animals were intraperitoneally injected with single doses of a 3.6% chloral hydrate solution at a rate of 1 ml/100 g body weight. The laparotomy was carried out over a median longitudinal section once the animal had been shaved and disinfected, and the abdomen was closed with a double-layered suture subsequent to the manipulation.

The main objective of this study was the morphometric determination of the relationship between the tissue defect observed and the methods of laser application (wavelength, mode of application, i.e., continuous or superpulsed wave, sapphire tips, quartz fibers, and handpieces) and the applied energy (Table 2). The following parameters were measured:

- Incision depth
- Incision width
- Thermally damaged zone
- Healing procedure

[1] Universitäts Frauenklinik, Abt. für Geburtshilfe und Gynäkologie, Voßstr. 9, W-6900 Heidelberg, FRG.

* Berlin.

Lasers in Gynecology
Edited by G. Bastert and D. Wallwiener

**Table 1.** Sample model design (tissue effects exerted by bare fibers and the DT-10 sapphire tip on the uterine horn of the rat)

| Fiber sapphire-tip | 200 µm | 400 µm | 400 µm | 600 µm | 200 µm |
|---|---|---|---|---|---|
| Energy | 14 W | | | | 20 W |
| 0 days | 6 animals | 6 animals | 6 animals | 6 animals | 6 animals |
| 7 days | 6 animals | 6 animals | 6 animals | 6 animals | / |
| 14 days | 6 animals | 6 animals | 6 animals | 6 animals | / |
| | | | | Total | 78 animals |

**Table 2.** Morophometric test groups (bare fiber versus sapphire tip)

| Group | Dependent upon | Tip/fiber | W | W/cm | $n$ | $n^a$ | $n^b$ |
|---|---|---|---|---|---|---|---|
| | Length and | DT 15 | | 2500 | 5 | 20 | 29 |
| | diameter | DT 14 | | 5600 | 5 | 20 | 27 |
| | (tip) | DT 13 | | 22000 | 5 | 20 | 28 |
| I | | DT 11 | 7 | 2500 | 5 | 20 | 29 |
| | | DT 10 | | 5600 | 5 | 20 | 29 |
| | | DT 9 | | 22000 | 5 | 20 | 28 |
| | | DT 7 | | 2500 | 5 | 20 | 27 |
| II | Type | Endoscope | 7 | 5600 | 5 | 20 | 29 |
| III | Energy | DT 10 | 7 | 5600 | 30 | 90 | 112 |
| | | | 14 | 11 200 | 30 | 78 | 156 |
| IV | Degree of wear | DT 10 | 10 | 8000 | 30 | 71 | 121 |
| | Diameter | 600 | | 5000 | 20 | 102 | 115 |
| V | (bare fiber) | 400 | 14 | 11200 | 20 | 75 | 82 |
| | | 200 | | 44500 | 20 | 79 | 80 |

$n$ Number of animals (III–V); $n^a$ number of incisions (average); $n^b$ number of individual measurements/parameter (average)

A special model was developed to standardize the examinations when the Nd:YAG contact technique was utilized. For this purpose, the tissue to be lased was affixed on a specially designed operating table which was connected to an electrical motor, allowing it to be moved in an axial direction. The tissue was stretched slightly in this axial direction to compensate for the muscular curvature.

Hand applicators for the sapphire tips and quartz fibers were mounted on a stand. Under microscopic observation the sapphire tip or bare fiber was then brought into contact with the tissue by lowering the stand.

Linear uterotomies 2 cm in length were carried out by commencing the laser application simultaneous with the electrical motor of the operating table; a constant velocity of 1 cm/s and a laser pulse duration of 2 s were maintained.

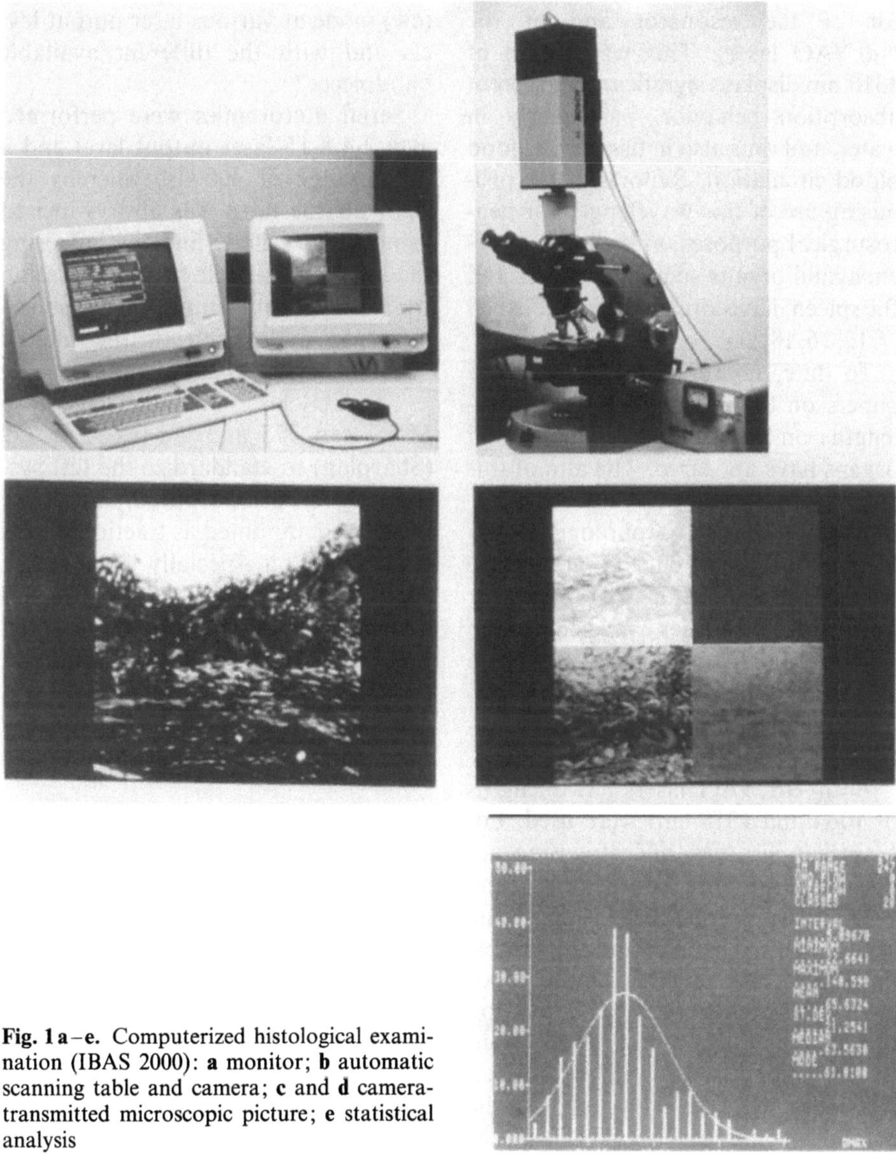

Fig. 1 a–e. Computerized histological examination (IBAS 2000): **a** monitor; **b** automatic scanning table and camera; **c** and **d** camera-transmitted microscopic picture; **e** statistical analysis

## Nd:YAG Laser with a Wavelength of 1.06 and 1.32 μm

The Nd:YAG laser with a wavelength of 1064 nm is already well established for clinical purposes. It can be used for coagulation purposes with a focusing applicator or with flexible quartz glass fibers for noncontact irradiation and with "bare fibers" or a sapphire contact tip for contact surgery [5, 8–10, 14, 15, 19–21].

A second wavelength of 1318 nm can be emitted by the Nd:YAG laser after appropriate modification of the resonator (crystal coating of the reflector of the resonator and of the Nd:YAG laser). This wavelength of 1318 nm displays significantly different absorption behavior, particularly in water, and thus also in tissue with good blood circulation. Reports of the promising use of this wavelength for neurosurgical purposes, as well as on parenchymal organs such as the liver and the spleen, have already been published [7,12,16,18].

To date, no fundamental scientific papers on the use of these two wavelengths on the internal female genital organs have appeared. The aim of this comparative study was the determination of the basic morphological data (coagulatory and cutting ability) as well as the applicatory parameter for both Nd:YAG laser wavelengths on this specific tissue. Special attention was paid to the comparison of the different application methods (noncontact and contact techniques).

Both Nd:YAG lasers (wavelengths of 1064 and 1318 nm) were used. Focusing applicators, microfocusing applicators, sapphire contact tips, and "bare fibers" were utilized as transmission systems. The liver and the uterine horn of the rat served as animal models. Linear surgery was performed on the liver and the uterus under standardized conditions and the tissue effects obtained were evaluated with the help of a computer (Fig. 1) [11,12].

### $CO_2$ Laser, Continuous-wave Mode and Superpulsed-wave Mode

The following question was to be answered: Is there a significant difference in the tissue effect (i.e., depth of vaporization and thermally damaged zone) exerted when using the $CO_2$ laser in the superpulsed- (sp) and continuous-wave (cw) mode at various laser output levels and with the different available handpieces?

Serial uterotomies were performed at a 2-4-8-15 laser output level and a pulse range of 400 Hz, whereby the right uterine horn was always incised using the cw mode and the left using the sp mode. Histological preparation was carried out immediately subsequent to extirpation of the uterine horns.

The $CO_2$ laser was connected to a Microscan 775 and a Microslad 779 (Sharplan) to standardize the test system (Fig. 2). Once exposed, the uterine horns were mounted as traction free as possible on a specially constructed Plexiglas operating table (Fig. 3). This system, especially the scanner, ensured precise control of the laser beam at a constant velocity of 1 cm/s for an incision length of 2 cm.

## Histological Examinations

The uterine horn segments which had been surgically manipulated were fixed in formalin (5%). For this purpose the uterine horns of the Wistar rats were placed on a stylet directly after extirpation to prevent curvature due to subsequent muscle contraction. This enabled vertical fixation of the uterine horns during the embedding, which in turn enabled histological assessment of the entire horn.

The fixed tissue segments were embedded in synthetic hydroxymethacrylate, because laser-treated tissue segments embedded in paraffin tore when sections were produced, and were thus no longer assessable. The synthetic embedding enabled hardening of the tissue damaged by laser manipulation; 2-

2 a 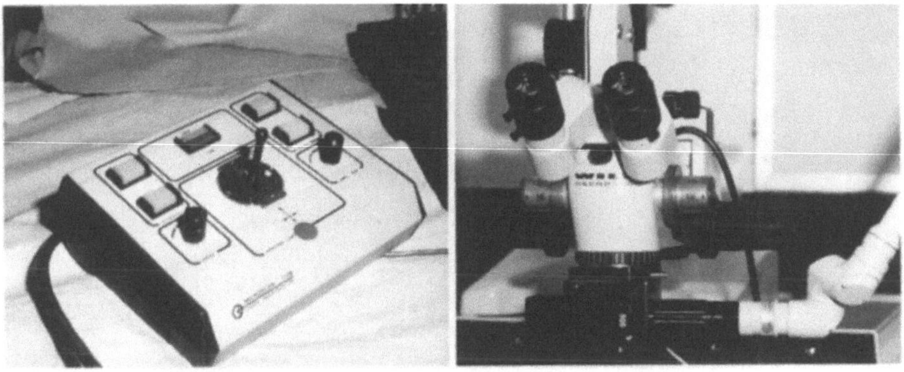 b

3 

**Fig. 2a, b.** Standardization of the laser uterotomies. **a** Microscan 775; **b** Microslad 779

**Fig. 3.** Operating table

to 3-μm segments could thus be produced without structural changes being induced.

The sections were stained either with a modified Romanovsky-Giemsa stain, developed and standardized by Wittekind (Anatomic Institute of the University of Freiburg) [22] or with hematoxylin-eosin.

Histological assessment of the laser-induced tissue effects was carried out subsequent to an extremely simplified subdivision of the tissue lesions under a light microscope (Fig. 1). We differentiated between tissue defects and the following thermally damaged zones [11]:

- Carbonization zone
- Irreversibly thermally damaged zone
- Reversibly thermally damaged zone

However, it is extremely difficult to demarcate the edges of reversible and irreversible tissue damage, as they are subject to numerous influencing factors and can really be determined only by means of a time-consuming enzyme histochemical examination. Hence, the vaporization depth and width as well as the coagulation zone were used in this study to enable comparison of the tissue effects attained. The coagulation zone, by definition, includes all areas of tissue damage which can be detected under a light microscope, i.e., tissue carbonization, increased staining of the intercellular substance, and decreased staining of the nucleus.

These parameters were measured using the IBAS 2000 image analysis system [11]. Statistical programs attached to an image analysis procedure enabled

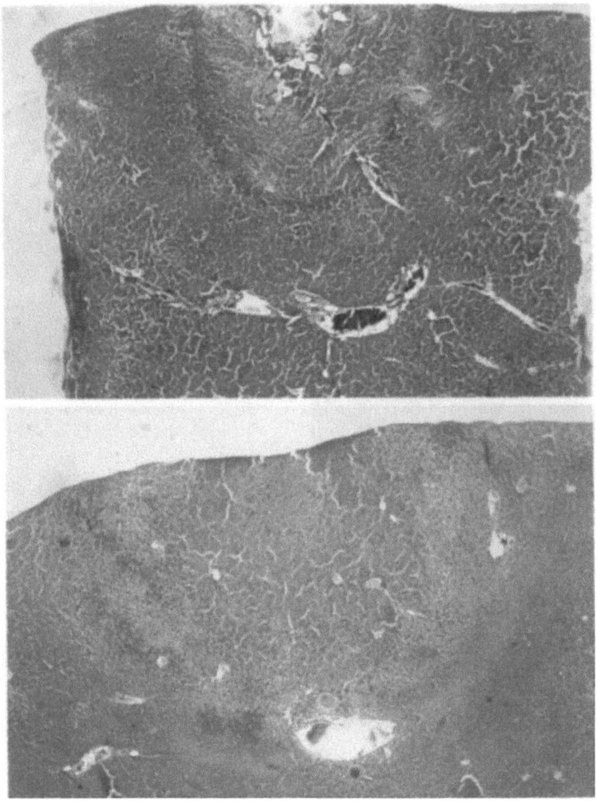

a

b

Fig. 4 a, b. Tissue effects of Nd:YAG laser – noncontact technique on rat liver, 15 W, H & E, × 25. **a** Nd:YAG laser, 1.32 μm; **b** Nd:YAG laser, 1.06 μm

the printing of all histological parameters as classified histograms with simultaneous statistical evaluation (maximum and minimum values, mean, median, standard deviation, Gaussian distribution curve) (Fig. 1). The results of the laser tissue preparation are summarized in Tables 3–10.

## Evaluation of Results and Conclusions

### Nd:YAG Laser Noncontact Technique

When the noncontact method was used on the liver, the cutting ability of the 1318-nm laser was found to be far better than that of the 1064-nm laser. This can be explained by the fact that the beam is ten times more absorbable in water. A more efficient transformation of the energy from light energy into heat within the tissue results. Tissue vaporization thus commences considerably earlier, which leads to a greater incision depth when using the 1318-nm laser. Approximately three times as much energy is required to attain the same cutting effect on the liver with the 1064-nm laser. The necrosis obtained

**Table 3.** Tissue effects in relationship to the various types of tips, power 7 W

| Tip | Length (mm) | Diameter (mm) | Cutting width (mm) | Cutting depth (mm) | Thermally damaged zone Vertical (mm) | Thermally damaged zone Horizontal (mm) |
|---|---|---|---|---|---|---|
| DT 15 | 40 | 0.6 | 0.656 ± 0.103 | 0.238 ± 0.078 | 0.101 ± 0.021 | 0.104 ± 0.044 |
| DT 14 | 40 | 0.4 | 0.618 ± 0.073 | 0.183 ± 0.069 | 0.104 ± 0.027 | 0.132 ± 0.052 |
| DT 13 | 40 | 0.2 | 0.371 ± 0.107 | 0.537 ± 0.107 | 0.158 ± 0.037 | 0.129 ± 0.050 |
| DT 11 | 30 | 0.6 | 0.631 ± 0.068 | 0.263 ± 0.032 | 0.100 ± 0.024 | 0.120 ± 0.033 |
| DT 10 | 30 | 0.4 | 0.453 ± 0.060 | 0.288 ± 0.053 | 0.103 ± 0.018 | 0.105 ± 0.017 |
| DT 9 | 30 | 0.2 | 0.440 ± 0.067 | 0.120 ± 0.059 | 0.150 ± 0.023 | 0.120 ± 0.048 |
| DT 7 | 20 | 0.6 | 0.685 ± 0.071 | 0.161 ± 0.020 | 0.140 ± 0.018 | 0.207 ± 0.050 |
| Endoscope | 7.5 | 0.4 | 0.429 ± 0.074 | 0.136 ± 0.065 | 0.149 ± 0.033 | 0.168 ± 0.020 |
| DT 10 7 W | 30 | 0.4 | 0.578 ± 0.111 | 0.081 ± 0.038 | 0.063 ± 0.026 | |
| DT 10 14 W | 30 | 0.4 | 0.745 ± 0.110 | 0.126 ± 0.034 | 0.066 ± 0.013 | |
| DT 10 10 W | 30 | 0.4 | 0.511 ± 0.164 | 0.085 ± 0.034 | 0.042 ± 0.013 | |
| Fiber 14 W | – | 0.6 | 1.281 ± 0.233 | 0.217 ± 0.051 | 0.032 ± 0.006 | 0.029 ± 0.006 |
| Fiber 14 W | – | 0.4 | 0.560 ± 0.21 | 0.187 ± 0.107 | 0.046 ± 0.019 | 0.032 ± 0.007 |
| Fiber 14 W | – | 0.2 | 0.231 ± 0.079 | 0.112 ± 0.050 | 0.022 ± 0.004 | 0.020 ± 0.004 |

with the 1318-nm laser shows a structure with three distinct zones. This structure is composed of a central vaporization zone, surrounded by a white zone of thermal denaturation (coagulation and hemostasis are clearly evident) and a hyperemic border as an outer ring (Fig. 4a). However, the necrosis is distinctly more dispersed and broken up than that attained using the 1064-nm wavelength laser. Only two zones were discerned when the latter was used: a central vaporization zone and a thermal denaturation zone (Fig. 4b).

## Nd:YAG Laser Contact Technique

1. The DT-10, -13, and -9 sapphire tips cut significantly less widely than the DT-15, -11, -7, and -14 tips and are thus to be recommended as far as incision width is concerned (Tables 3 and 4).

2. The DT-10 sapphire tip cut significantly deeper than the DT-14, -9, and -7, and the DT-13 significantly deeper than all the above-mentioned tips (Table 5; Fig. 5a, b).

**Table 4.** Incision width – sapphire tips (7 W)

| Student $t$-test ($t, f, p$) | | DT 15 | DT 11 | DT 7 | DT 14 |
|---|---|---|---|---|---|
| Incision width ($n$= 5) | | 0.656 ± 0.103 | 0.613 ± 0.068 | 0.685 ± 0.071 | 0.618 ± 0.073 |
| DT 10 | 0.453 ± 0.060 | $t = 3.81$ $p < 0.01$ | $t = 3.95$ $p < 0.01$ | $t = 5.58$ $p < 0.001$ | $t = 3.90$ $p < 0.01$ |
| DT 13 | 0.371 ± 0.107 | $t = 4.29$ $p < 0.01$ | $t = 4.27$ $p < 0.01$ | $t = 5.47$ $p < 0.001$ | $t = 4.26$ $p < 0.01$ |
| DT 9 | 0.440 ± 0.067 | $t = 3.93$ $p < 0.01$ | $t = 4.05$ $p < 0.01$ | $t = 5.61$ $p < 0.001$ | $t = 4.02$ $p < 0.01$ |

$f$= 8 for all $t$-tests

**Table 5.** Incision depth – sapphire tips (7 W)

| Student $t$-test Incision depth | ($t, f, p$) ($n$ = 5) | DT 10 0.288 ± 0.053 |
|---|---|---|
| DT 15 | 0.238 ± 0.078 | $t = 1.19$ n.s. |
| DT 14 | 0.183 ± 0.069 | $t = 2.70$ $p < 0.05$ |
| DT 13 | 0.537 ± 0.107 | $t = 4.66$ $p < 0.01$ |
| DT 11 | 0.263 ± 0.032 | $t = 0.90$ n.s. |
| DT 9 | 0.120 ± 0.059 | $t = 4.74$ $p < 0.01$ |
| DT 7 | 0.161 ± 0.065 | $t = 5.01$ $p < 0.01$ |

$f$ = 8 for all $t$-tests

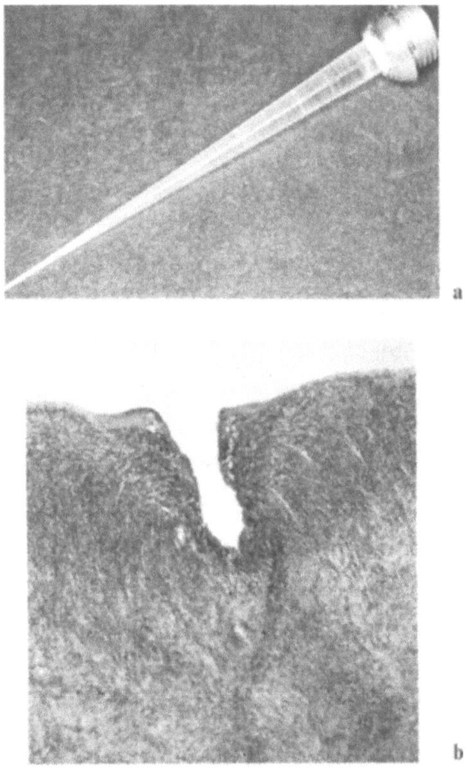

**Fig. 5. a** Sapphire tip, DT 13. **b** Tissue effect of sapphire tip DT 13. H & E, ×25.

3. The DT-7 sapphire tip causes significantly more extensive horizontal tissue damage than the other sapphire tips do. Other differences between the sapphire tips as far as horizontal damage is concerned were not observed. The DT-10 has the lowest value of the DT-10, -13, and -9 tips, and is thus to be recommended as far as incision width and horizontal damage are concerned (Table 6; Fig. 6).

4. Significantly deeper vertical tissue damage is attained with the DT-13, as well as the DT-9 and DT-7 tips, however.

Recommendation of the DT-10 sapphire tip is justified due to its significantly smaller incision width and its least extensive horizontal damage. Hence the following results were obtained for the questions posed:

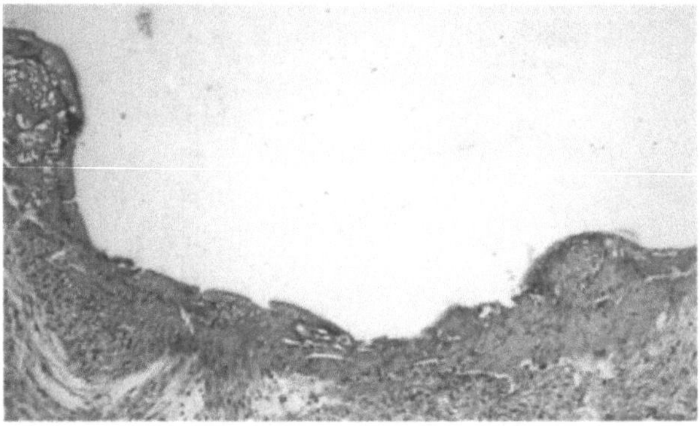

**Fig. 6.** Tissue effect of sapphire tip DT 10. H & E, ×100

- Incision depth and width are dependent upon the length, diameter, and configuration of the sapphire tip. The best cutting action is obtained with the DT-13 sapphire tip (length = 40 mm, diameter = 0.2 mm; Fig. 5a).
- The thermally damaged zone increases with increasing diameter of the end surface of the tip.
- An increase in energy output from 7 to 14 W results in a 1.4-fold increase in the tissue effect.
- The healing procedure is marked by the rapid development and degeneration of granulation tissue.

5. The DT-10 tip was selected for comparison of the tissue effects attained with bare fibers and a sapphire tip by reason of the findings noted in 1–4 above.

Significantly better results were attained with the bare fibers for all four parameters (Table 7; Fig. 7). The relatively high standard deviations can be explained by the difficulties associated with the contact technique (i.e., standardization, see above).

The results presented show that tissue preparation with bare quartz fibers is possible with all fiber diameters (200 μm, 400 μm, 600 μm; Table 3).

Fibers with a 400-μm diameter of the nucleus were found to be superior to the 600-μm fiber: the tissue damage they caused was basically of the same

**Table 6.** Horizontal thermal damage – sapphire tips (7 W)

| Student $t$-test ($t, f, p$) | | DT 15 | DT 11 | DT 7 | DT 14 |
|---|---|---|---|---|---|
| Horizontal damage ($n = 5$) | | 0.104 ± 0.044 | 0.120 ± 0.033 | 0.207 ± 0.050 | 0.132 ± 0.052 |
| DT 10 | 0.105 ± 0.017 | $t = 0.05$ n.s. | $t = 0.90$ n.s. | $t = 4.32$ $p < 0.01$ | $t = 1.10$ n.s. |
| DT 13 | 0.129 ± 0.050 | $t = 0.84$ n.s. | $t = 0.34$ n.s. | $t = 2.47$ $p < 0.05$ | $t = 0.09$ n.s. |
| DT 9 | 0.120 ± 0.048 | $t = 0.55$ n.s. | $t = 0.00$ n.s. | $t = 2.81$ $p < 0.05$ | $t = 0.38$ n.s. |

$f = 8$ for all $t$-tests

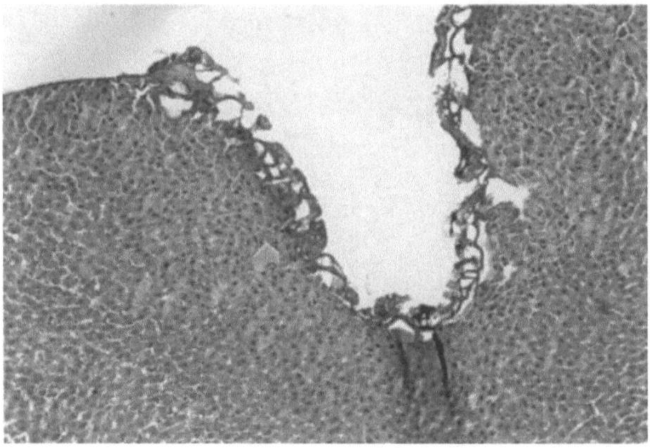

**Fig. 7.** Typical tissue effect of bare fiber. H & E, ×100

depth as that observed with the 600-µm fiber, yet only some 50% as wide. A comparison of the damage caused by the 200-µm fibers at various energy levels showed than an increased output led to a consistent increase in the width and depth of the damaged tissue, whereas the thermally damaged zone did not increase significantly.

The tissue lesion was in fact characterized by minimal thermal damage as a result of the divergence of the laser energy into the surrounding tissue. The thermally damaged zone measured a maximum of 46 µm when the 600-µm fibers were used; a minimum of 20 µm was attained with the 200 µm fiber (Table 3).

This leads to the following conclusions for the questions posed as far as contact preparation using bare fibers in concerned (Table 8):

- Incision depth and width are dependent upon the diameter of the bare fiber.
- The thermally damaged zone increases with increasing diameter of the bare fiber.
- There was no linear relationship for an increased output.

The healing procedure is characterized by the rapid development and degeneration of granulation tissue. Full recov-

**Table 7.** Tissue effects – bare fibers versus sapphire tip: identical output (14 W), identical diameter (0.4 mm)

| Type mm | Incision width | Horizontal damage | Incision depth | Vertical damage |
|---|---|---|---|---|
| Sapphire tip (n = 30) | 0.679 ± 0.040 | 0.066 ± 0.021 | 0.126 ± 0.052 | 0.066 ± 0.021 |
| Bare fiber (n = 20) | 0.560 ± 0.210 | 0.032 ± 0.007 | 0.187 ± 0.107 | 0.046 ± 0.019 |
| Student's *t*-test ($t, f, p$) | $t = 3.04$ $f = 48$ $p < 0.01$ | $t = 6.97$ $f = 48$ $p < 0.001$ | $t = 2.69$ $f = 48$ $p < 0.05$ | $t = 3.42$ $f = 48$ $p < 0.01$ |

**Table 8.** Tissue effects – sapphire tip versus bare fiber

| 10 W<br>110 W/cm | Sapphire tip<br>(diameter 0.4 mm) | Bare fiber<br>(diameter 0.4 mm) |
|---|---|---|
| Damage | Trough shaped (endoscope)<br>Oval or wedge shaped<br>(applicator) | Wedge shaped |
| Incision depth (vaporization) | 0.126 mm | 0.187 mm; 1/3 > sapphire vaporization |
| Incision width | 0.679 mm | 0.560 mm; 1/3 < sapphire incision width |
| Thermal damage | 0.066 mm | 0.035 mm; up to 50% less |

ery was not noted during the observation period (21 days). Fibers 0.4 mm in diameter gave the best correlation between all parameters.

On the one hand, it was shown that tissue lesions approximating those obtained using the contact technique with sapphire tips of the same diameter can be attained with extremely thin fibers. An identical laser output does lead to a minimization of thermal damage with a comparable incision depth, however.

Furthermore, a cooling or cleaning system is not necessary during contact utilization of bare fibers [21]. Tissue contamination leads to a heating of the fiber tip, resulting from the reduced deflection of the laser light due to the minimal thermal conduction of quartz glass. The melting procedure causes a deformation of the fiber tip. This alteration in material forms the basis of the locally restricted cutting action during the contact technique; it, in turn, is dependent upon the tissue vaporization resulting from the strong heat development in the fiber tips. On the other hand, the depth of penetration characteristic of the Nd:YAG laser and the associated coagulation are reduced. It is thus obvious that the tip with its modified light guide can be utilized only for contact laser preparation. The noncontact technique prevents the diversion of heat into the tissue; this in turn leads to an overheating and subsequent melting of the fiber tip.

During practical use one must ensure that the laser beam is applied only after tissue contact has been established by the fiber tip.

The fact that the light guide can be restored to its original condition by shortening the tip by 1–2 mm and peeling off the surrounding cladding is extremely important for practical clinical use. The quartz-glass light guide is thus re-usable.

The results shown in Table 8 allow the following conclusions to be drawn for bare quartz-glass fibers: If one takes the tissue effect exerted by the sapphire tip as a reference value, the incision depth of the bare 400-μm-thick fiber is greater (1.5-fold), while the incision width and the thermally damaged zone are reduced (0.8- and 0.05-fold respectively); i.e., use of bare fibers results in an improvement in the cutting action with a simultaneous reduction in thermal damage. There is no significant difference in the healing procedure between the two application methods (Table 9).

**Table 9.** Healing procedure: sapphire tip/bare fiber 7, 14, 21 days

| Group | Tip (DT)/ fiber | Length (mm) | Diameter (mm) | Regenerative tissue layer 7 days (mm) | 14 days (mm) | 21 days (mm) |
|---|---|---|---|---|---|---|
| III | DT 10 7 W | 30 | 0.4 | 0.175 ± 0.046 | 0.166 ± 0.038 | 0.141 ± 0.028 |
| | DT 10 14 W | 30 | 0.4 | 0.275 ± 0.103 | 0.206 ± 0.052 | 0.170 ± 0.035 |
| IV | DT 10 10 W | 30 | 0.4 | 0.290 ± 0.067 | 0.176 ± 0.032 | 0.149 ± 0.031 |
| | fiber 14 W | – | 0.6 | 0.113 ± 0.034 | 0.135 ± 0.036 | – |
| V | fiber 14 W | – | 0.4 | 0.093 ± 0.022 | 0.105 ± 0.015 | – |
| | fiber 14 W | – | 0.2 | 0.064 ± 0.017 | 0.085 ± 0.016 | – |

Miyasaka et al. [15] used sapphire tips with lengths of 30 and 40 mm. The tissue defects they described in association with the various sapphire tips utilized differed from those observed in our study [20, 21]. All details concerning penetration depth and the extent of the damage are missing, however; hence, direct comparison of the results is not possible.

The thermally damaged zone resulting from Nd:YAG laser application is comparable to that of the $CO_2$ laser in a range of 150–750 μm [1]. Wakita and Fasano attained similar results [5, 6, 19].

Keckstein et al. investigated the cutting and coagulatory action of different laser transmission systems [9]. The MRTP 5 (manufactured by SLT) sapphire tip, a 400-μm-diameter bare fiber (MBB), and a 600-μm-diameter bare fiber were tested. During contact utilization the sapphire tip was pressed about 0.5 mm into the tissue. The depth of vaporization and the coagulation of the incision borders were used as criteria for the effects attained. Smaller coagulation zones were noted when the sapphire tip was used than with the bare-fiber technique. In contrast to the presented study, Keckstein undertook in vitro tests; i.e., he examined dead tissue. This probably accounts for the contrasting observations, as dead tissue has entirely different absorption characteristics because ot the lack of blood circulation.

## Comparisons of the Two Different Wavelengths of the Nd:YAG Laser

The difference is not as distinct on organs with a poorer blood circulation, such as the uterine horn of the rat used as model is this study. A slight increase in the vaporization depth was observed only with the microfocusing applicator on the 1318-nm laser. The 1318-nm wavelength does not offer an evident advantage when the contact technique is employed.

However, this is attributable exclusively to the property of this technique, i.e., the direct transformation of laser energy into heat at the point of contact of the fiber with the tissue.

At present, we thus see a profitable advantage in this new wavelength, primarily for noncontact preparation purposes. Further clinical studies are necessary, however, to enable a more accurate evaluation.

## $CO_2$ Laser

The results of the tissue effects exerted by the $CO_2$ laser are summarized in Table 10. A linear correlation between the tissue defects measured and the particular laser method could not be ascertained.

Comparison of the different application methods, cw or sp, revealed that a greater incision depth was attained with the superpulsed mode in all energy ranges. A factor of 1.25 (1.1–1.4) was determined from the data gained.

The incision width was less extensive in all energy ranges with the sp mode. This reduction in incision width is negligible, however, with a factor of 1.09 (1.01–1.13).

In an energy range exceeding 4 W the microscopically definable thermally damaged zone was 1.3 (1.1–1.5) times smaller with the superpulsed mode than with the continuous-wave mode. The reason for the exception when using a power output of 2 W, where the thermally damaged zone was found to be greater with sp utilization, will be investigated in a further study in which superpulsed and low power ranges are to be compared.

Within the framework of an experimental study, Bellina was also able to establish that there is no liner reduction in tissue necrosis when using a pulse rate of 650 Hz [4]. He found no statistical difference in thermal necrosis for the 350 pps superpulsed and cw mode.

Lanzafame found a 23% reduction in the necrosis zone using a 25 W sp mode and a spot size of 0.2 mm. He did

**Table 10.** Comparison of the tissue effects attained using the two laser techniques (cw versus sp) – day 0 –

| Mode | Wattage (W) | Width of incision (μm) | Depth of incision (μm) | Thermally damaged zone (μm) |
|---|---|---|---|---|
| cw | 2 | 337.4 ± 67.8 | 95.4 ± 21.7 | 19.8 ± 3.9 |
| sp | 2 | 332.4 ± 50.1 | 133.7 ± 16.7 | 25.6 ± 2.4 |
| cw | 4 | 564.0 ± 35.5 | 166.5 ± 13.0 | 59.6 ± 4.3 |
| sp | 4 | 522.0 ± 46.4 | 200.2 ± 7.2 | 41.9 ± 3.2 |
| cw | 8 | 794.4 ± 54.7 | 194.1 ± 13.9 | 123.1 ± 4.9 |
| sp | 8 | 700.9 ± 45.7 | 257.1 ± 14.8 | 112.0 ± 4.6 |
| cw | 15 | 767.3 ± 35.0 | 257.2 ± 15.3 | 134.8 ± 5.9 |
| sp | 15 | 681.9 ± 28.3 | 279.6 ± 23.2 | 90.1 ± 5.5 |

*cw*, continuous wave; *sp*, superpulsed wave

not attempt to quantify the reversible wound area. Our tests in this field are not yet completed [13]. The 2.5 factor (sp/cw) for thermal necrosis, in favor of the superpulsed technique, reported by Baggish and El Bakry [2] during partial transection of uterine horns of rats, thus could not be confirmed within the framework of this study (see above).

The peak output obtained when utilizing relatively high energy ranges ( > 20 W/s) results in extremely deep incisions. This accounts for the discrepancy between the histological results obtained in this study and the shape of the damaged zone as reported by Baggish and El Bakry [2]. They described the damaged zone typical of superpulsed utilization as "funnel-shaped"; this was not histologically detected in this study. One may thus assume that this characteristic shape occurs only in higher energy ranges, which we did not use in this study, and which are not required in microsurgery.

The advantages of the sp mode are to be found in the deeper incisions due to the higher output density, and in the minimization of the tissue defect obtainable during correct handling. This technique does require greater experience, however, as incorrect application (incorrect angle of incidence of the laser beam, incorrect distance from the working surface, and an incalculable penetration effect when using too high a power output) can lead to more extensive tissue damage.

In summary, in agreement with the literature, it can be said that a minimization of tissue damage can be attained by the correct application of the $CO_2$ superpulsed-wave laser. The tissue effects attained by the application of different laser techniques to uterine tissue show that the tissue effect is minimalized and the incision depth greater with the $CO_2$ superpulsed-wave mode.

## References


1. Badawy S, El Bakry MM, Baggish MS (1986) Comparative study of continuous and pulsed $CO_2$ laser on tissue healing and fertility outcome in tubal anastomosis. Fertil Steril 46:21–27
2. Baggish MS, El Bakry MM (1986) Comparison of electronically superpulsed and continuous-wave $CO_2$ laser on the rat uterine horn. Fertil Steril 45:120–127
3. Baggish MS, Sze E, Badawy S, Choe J (1988) Carbon dioxide laser laparoscopy by means of a 3.0-mm-diameter rigid wave guide. Fertil Steril 50:419–424
4. Bellina JH (1984) Analysis of electronically pulsed versus quasi-continuous wave carbon dioxide lasers in an animal model. Am J Obstet Gynecol 150:934–940
5. Fasano VA, Ponzio RM (1986) Nd:YAG laser contact: a coming of the microperformances. In: Oguro Y, Atsumi K, Joffe SN (eds) Nd:YAG laser in medicine and surgery: fundamental and clinical aspects. Professional Postgraduate Services, Tokyo, pp 358–362
6. Fasano VA, Urciuoli R, Lombard GF, Benech F, Ponzio RM, Lanotte MM (1985) Advanced technologies in neurosurgery. Laser Med Surg 1:44–49
7. Frank F, Beck OJ, Hessel S, Keiditsch E (1987) Comparative investigations of the effects of the Nd:YAG laser at 1.06 microns and 1.32 microns on tissue. Laser Med Surg 6:546–551
8. Frank F, Hessel S, Ischinger T (1988) Einsatzmöglichkeiten des Nd:YAG Laser bei der Gefäßrekanalisierung. Laser Med Surg 2:66–71
9. Keckstein J, Finger A, Steiner R (1988) Laser application in contact and noncontact procedures: sapphire tips in comparison to "bare-fiber", argon laser in comparison to Nd:YAG laser. Laser Med Surg 4:158–162
10. Keckstein J, Hepp S, Schneider V, Sasse V, Steiner R (1990) The contact Nd:YAG laser: a new technique for conservation of the fallopian tube in unruptured ectopic

pregnancy. Br J Obstet Gynecol 97:352–356
11. Köditz W (1986) Mehr Komfort durch Joystick zum Scanningtisch-Steuern. Image – Kontron Bildanalyse. Kontron Werke & Druck GmbH, Eching
12. Kraus M, Beer M, Bauer E, Jocham D, Permanetter W, Ruprecht L, Staehler G, Unsöld E (1988) Die Bedeutung der geometrischen Bestrahlungsbedingungen für die Lasersysteme Nd:YAG 1318 nm, Nd:YAG 1064 nm und Argon 488 nm. Laser Med Surg 4:152–157
13. Lanzafame RJ, Nahu JO, Rogers DW, Hinshaw JR (1988) Comparison of continuous-wave, chop-wave and superpulsed laser wounds. Laser Med Surg 8:119–124
14. Mecke H, Riedel HH, Semm K (1989) Die 3-Stufen-Therapie bei Endometriose als Kombination von medikamentöser Therapie und operativer Pelviskopie. In: Bastert G, Wallwiener D (eds) Der Einsatz neuer Lasertechniken in der gynäkologischen Fertilitätschirurgie. Laser Med Surg [Suppl] 1:71–75
15. Miyasaka M, Morita T, Yamaguchi H, Saito H, Nishimura M, Tanino R, Osada M, Wakaki M, Takeyama N, Yokota H (1986) Basic researches of Nd:YAG laser with contact type scalpel. In: Oguro Y, Atsumi K, Joffe SN (eds) Nd:YAG laser in medicine and surgery: fundamental and clinical aspects. Professional Postgraduate Services, Tokyo, pp 643–637
16. Rolle A, Unsöld E, Ruprecht L, Permanetter W, Frank F (1988) Morphologische Aspekte der Nd:YAG-Laseranwendung (Wellenlängen: 1064 nm und 1318 nm) am Lungenparenchym. Laser Med Surg 4:10–14
17. Scheidel P (1982) Mikrochirurgie in der Gynäkologie – Tierexperimentelle und klinische Untersuchungen zum Einsatz mikrochirurgischer Techniken in der rekonstruktiven Tubenchirurgie. Habilitationsschrift, Homburg/Saar
18. Stokes LF, Auth DC, Tanaka D, Gray JL, Gulacsik C (1981) Biomedical utility of 1.34 microns Nd:YAG laser radiation. Trans Biomed Engineering 28:297–299
19. Wakita K, Kuramoto H, Sasaki N, Arai M (1986) YAG laser incisional conization for cervical neoplasia. In: Oguro Y, Atsumi K, Joffe SN (eds) Nd:YAG laser in medicine and surgery: fundamental and clinical aspects. Professional Postgraduate Services, Tokyo, pp 593–597
20. Wallwiener D, Pollmann D, Morawski A, Bastert G, Krampe C (1988) Die Neodym:YAG Laser Kontakttechnik mit Saphirschneidespitzen – Lasertechnischer Background und Gewebeeffekte. Laser Med Surg 4:75–85
21. Wallwiener D, Pollmann D, Kappler M, Bastert G (1989) Die Nd:YAG-Laser-Kontakttechnik mit nackten Glasfasern – Lasertechnischer Background und Gewebeeffekte im Vergleich zur Präparation mit Saphirschneidespitzen. Laser Med Surg [Suppl] 1:5–10
22. Wittekind D, Schulte E (1987) Die Bedeutung der Standardisierung der Zell- und Gewebspräparation für bildanalytische Operationen. In: Quantitative und strukturelle Bildanalyse in der Medizin. GIT Verlag, pp 5–12

# The $CO_2$ Superpulsed Laser in Microsurgical Tubal Preparation: Experimental Results – $CO_2$ Superpulsed Laser Versus Microelectrode

D. WALLWIENER [1], D. POLLMANN, S. RIMBACH, E. KELLER, and G. BASTERT

The development of the $CO_2$ superpulsed technique enabled preparation with substantially less tissue damage [10]. The question thus arose as to whether the results of surgical revision of tubal occlusion could be improved by utilizing the $CO_2$ laser technique. Presently these problems cannot be clinically clarified due to the lack of standardization [17]; hence, an experimental model (the minipig) was selected.

The minipig was selected as an animal model, as the size relationships of the minipig correspond best with those of the human body, thus enabling realistic comparison of the effect exerted by the laser beam on the tubal tissue [21, 22]. Furthermore, the practicability of this test series is also ensured by the unproblematic utilization of the laser laparoscopic and electrical [20] instruments, also the handling of the pneumoperitoneum.

The possibility of surgical laparoscopy on the minipig using clinical instruments has already been described by Daniell and Brown [6]. In this study the minipig was once more shown to be an exceptionally suitable animal model for the development of laser operation methods. Despite all fundamental arguments against animal experiments, one must bear in mind that preclinical testing and improvement of laser laparoscopic methods are not possible without animal experiments; this must be taken into consideration when evaluating this experimental model. Paterson et al. also used the minipig as an experimental model [15].

A better analysis of the surgical success than would be the case in clinical practice is made possible by the fact that the minipig is a polyovulatory species, thus enabling more accurate determination of the nidation index due to the plurality of the implantations. The nidation index, i.e., the quotient of the number of implantations to the number of corpora lutea on the ipsilateral side can be seen as a good indicator of tubal function.

Isthmo-isthmic end-to-end anastomosis subsequent to the coagulation of a 2-cm-long isthmic tubal segment during laparoscopic sterilization was chosen as the best standardizable model. Disturbing influences such as differing residual tubal segments and lumen differences were thus precluded.

The aim of this comparative study was to clarify whether or not microsurgical preparation, i.e., resection of occluded tubal segments using the newly developed $CO_2$ superpulsed laser is superior to conventional microsurgical preparation using monopolar high-frequency electrodes.

[1] Universitäts-Frauenklinik, Abt. für Geburtshilfe und Gynäkologie, Voßstr. 9, W-6900 Heidelberg, FRG.

Lasers in Gynecology
Edited by G. Bastert and D. Wallwiener

The following aspects were considered:

1. The possibility of tubal preparation using the $CO_2$ superpulsed laser within the framework of microsurgical refertilization with regard to:
   - mucosa agglutination due to laser preparation
   - surgical handling and the time factor during microsurgical preparation
2. The pregnancy rate and nidation index after microsurgical preparation using the $CO_2$ superpulsed laser or microelectrode
3. Adhesions with regard to the preparation method

## Experimental Method

Isthmo-isthmic reanastomosis of the fallopian tube of the minipig, subsequent to sterilization by laparoscopic coagulation of a 2-cm-long isthmic segment with consecutive transection, served as the experimental model for microsurgical tubal preparation. Laparoscopic sterilization was performed with the laparoscopic instruments utilized in human medicine. The endocoagulatory instrument developed by Semm (Wisap Company) was used for laparoscopic sterilization [18]. A laparotomy, during which a one-sided salpingo-oophorectomy and a contralateral microsurgical refertilization were performed, was carried out on the animals after an interval of approximately 8 weeks. After a further 8 weeks the animals were mated or inseminated to test the success of the operation. Mating ensued with a boar of the same species, the animals being placed together as soon as the commencing estrus was evident.

Ejaculation was confirmed by semen in a vaginal smear. A second-look laparotomy followed 6 weeks later if no further estrus developed. If another estrus occurred, however, the animals were remated. If pregnancy ensued, the animal was relaparotomized to appraise the embryos within the remaining uterine horn. If no pregnancy ensued after the two matings, artificial insemination was carried out and repeated a maximum of two times. The animals were injected with 2.5 ml suigonan i.m. simultaneous with the commencement of estrus to enable the technical execution of insemination (5 ml suigonan contains 6.8 mg dry substance, equivalent to 200 I.U. chorionic gonadotropin, 400 I.U. serum gonadotropin, and 0.5 mg methyl hydroxbenzoate as preservative). Five days later, upon onset of the estrus, instrumental insemination was carried out with 250 ml living semen from the "Pietrain" breed. Laparotomy was performed on those animals which were not pregnant after two inseminations. The number of implantations and number of corpora lutea were noted at second-look laparotomy. A transuterine chromopertubation was performed on those animals in whom neither a pregnancy nor an implantation were discernible. The site of anastomosis was examined in every case.

The animals were randomly divided into two groups of ten animals each, 6 weeks after the laparoscopic sterilization. One animal, which was not included in the study, was used to determine the optimum laser energy level by performing tubal transections at various $CO_2$-laser energy levels (laser output) using the superpulsed mode. A laser output of 10 W/s with a focal site 200 μm in diameter was found to be the most practicable.

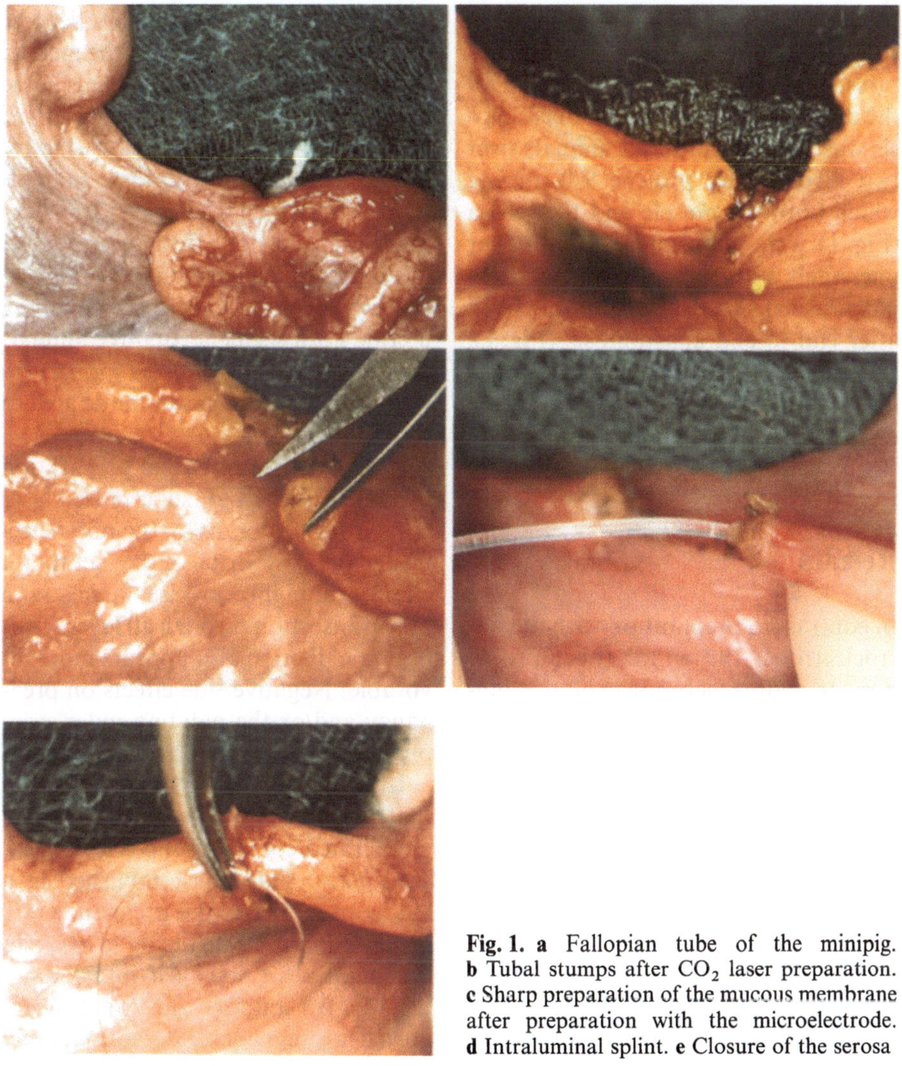

a b c d e

**Fig. 1. a** Fallopian tube of the minipig. **b** Tubal stumps after $CO_2$ laser preparation. **c** Sharp preparation of the mucous membrane after preparation with the microelectrode. **d** Intraluminal splint. **e** Closure of the serosa

A left-sided salpingo-oophorectomy was performed after presentation of the internal genitals, after which the residual tube was exposed (Fig. 1). The animals wer refertilized by microsurgical preparation, by means of either a $CO_2$ laser (Fig. 1) or a microelectrode from Martin, type 151 B 200 (Fig. 1), depending upon which group they belonged to.

All microsurgical manipulation was carried out using an operating microscope with 15-fold magnification. The tubal stumps of all animals were prepared with a laser or a microelectrode, under meticulous hemostasis, as a layer resection, until intact tubal lumen was macroscopically discernible. A trocar was used to insert a splint from distal to proximal from the tubal lumen through the wall of the uterine horn (Fig. 1). Two mesosalpinx fixation sutures were applied for adaptation of the dissected tubal stumps. Anastomosis was carried out in all animals over two layers using prolene 9/0, during which both the muscle and the serosa (Fig. 1) were approximated with four one-stitch sutures each. Mucosa was not grasped. The splint was removed after end-to-end anastomosis (Fig. 1), and the abdomen was closed layer by layer with a peritoneum muscle suture and one-stitch cutaneous sutures.

## Results

In contrast to preparation with the microelectrode, the entire tubal dissection, including severance of the mucous membrane, was performed with the $CO_2$ superpulsed laser. The accurate layer resection of the mucosa was carried out with surgical scissors (Fig. 1) during preparation with a microelectrode. Mucosa agglutination was not microscopically discerned during any of the laser preparations. There was no difference in hemostasis between laser and microelectrode preparation. Bipolar coagulation was necessary in all animals in whom the vascular connection between the ovarian and tubal arteries had been struck.

Pregnancy was attained in seven animals. An occluded tube was found in one animal in each group. Two animals in each group did not become pregnant due to the restricted mating and insemination opportunities. Statistical analysis showed that there was no significant difference between the nidation indices (chi-square test).

There was no discernible reason for the lack of conception in the two animals in each group that did not become pregnant. Adhesive band in the median suture were found in two animals in the laser group; these did not inhibit the ovum pick-up mechanism. Adhesive bands were also seen in three animals prepared with the microelectrode; these also did not affect the tube, yet extended on the left side of the salpingo-oophorectomy to the remaining uterine horn segment. In three animals in each group adhesive formations were seen from the anastomotic area to the pelvic wall.

It was shown that tubal preparation by means of the $CO_2$ superpulsed laser is superior to that using the microelectrode, alone by virtue of the fact that it enables a highly precise, quick, and easy transection through all layers. Adhesion induction also appears more favorable. Negative side effects on pregnancy and/or the nidation index were precluded.

It must be borne in mind, however, that the normal pregnancy rate in an unoperated control group does not exceed 70%. This was established in a study on minipigs, the pregnancy rate being found to be distinctly lower than the number of so-called intact tubes [12, 15]. All authors thus far have reported a pregnancy rate considerably lower than that of the intact tubes.

The present study showed that there was no significant difference in nidation index and/or pregnancy rate, with the same number of permeable tubes present after microsurgical preparation utilizing the microelectrode and the superpulsed $CO_2$ laser (Table 1).

The limited statistical validity, due to the relatively small number of animals, must be borne in mind when interpreting these results. It was evident, however, that there is no significant difference between the preparation

**Table 1.** Refertilization of minipigs – pregnancy rate and nidation index (NI[a]) using $CO_2$ laser/microelectrode ($n = 20$)

| Group | Preparation technique | Pregnant animals *n* | Permeable tubes *n* | Pregnant animals | | | | |
|---|---|---|---|---|---|---|---|---|
| | | | | Corpora lutea | | Implantations | | NI[a] |
| | | | | Total | Per animal | Total | Per animal | |
| I | $CO_2$ laser superpulsed | 7 | 9 | 88 | 11 | 53 | 6.6 | 60 |
| II | Micro-electrode | 7 | 9 | 72 | 9 | 56 | 7 | 77 |

NI[a], percentage implantations with reference to the number of corpora lutea. Statistics: No significant difference (chi-square test)

techniques as far as the permeability rate, and thus ultimately pregnancy, is concerned.

The results are nevertheless especially important where adhesion induction is concerned. It can be assumed that, given a larger collective and/or a clinical situation, this adhesion problem is ultimately the most relevant question as far as functional integrity, i.e., the integrity of the ovum pick-up mechanism, is concerned.

In a comparative study of the superpulsed and continuous-wave modes Badawy was able to attain a 20% higher pregnancy rate using the superpulsed made after uterine horn transection in rats with subsequent reanastomosis [1].

Table 2 shows the advantages of laser tubal preparation [11, 14, 16, 19, 23]. The nidation indices Paterson attained within two groups subsequent to end-to-end anastomosis with resection of the isthmoampullary transition are comparable to those attained in the present study [15].

The greatest difficulty in comparing various laser studies lies in the diversity of the technical differentiation of the

**Table 2.** Advantages of $CO_2$ superpulsed laser preparation

1. No tissue and/or parenchyma loss due to minimalized tissue destruction
2. Exceptionally rapid preparation even when including mucosa preparation due to the absence of a destructive coagulation effect

instruments. In 1983, using a weaker $CO_2$ laser with a relatively large spotsize diameter of 2 mm and a basic laser energy density of 700 W/cm$^2$, Choe attained poorer experimental surgical results as far as postoperative adhesion formation and the success of tubal preparation were concerned [4, 7, 8].

In 1984, Bellina carried out adhesiolyses using a laser energy ranging from 800 to 4000 W/cm$^2$, emphasizing that the adhesion insertion sites were prophylactically relasered with a 300–600 W/cm$^2$ laser energy [3]. He carried out distal tubal laser surgery using the same energy range, however, performing the eversion with a laser energy of 100–300 W/cm$^2$.

Fayez reported mucosa coagulation, as well as postoperative fibrosis and relatively extensive tissue necrosis, after laser tubal preparation [9]. However, he also utilized a 900 $W/cm^2$ power density and a focal spot diameter of 2 mm.

The statements made concerning the damaged tissue zone are just as diverse. In 1982, Daniell reported a maximum laser penetration depth of 1 mm when appropriately focused [6]. In 1985 he described a damaged zone of less than 500 μm and a necrotic area under 1000 μm [5]. On the other hand, Baggish demonstrated that necrotic zones of 0.2 mm (17 W superpulsed laser) and 0.5 mm (20 W continuous-wave mode) can be attained using the newest type of $CO_2$ laser instruments [2].

## References


1. Badawy S, El Bakry MM, Baggish MS (1987) Comparative study of continuous and pulsed $CO_2$ laser on tissue healing and fertility outcome in tubal anastomosis. Fertil Steril 47:843–847
2. Baggish MS, El Bakry MM (1986) Comparison of electronically superpulsed and continuous-wave $CO_2$ laser on the rat uterine horn. Fertil Steril 45:120–127
3. Bellina JH (1984) Analysis of electronically pulsed versus quasi-continuous wave carbon dioxide lasers in an animal model. Am J Obstet Gynecol 150:934–940
4. Choe JK, Dawood MW, Bardawil WA, Andrews AH (1984) Clinical and histological evaluation of laser reanastomosis of the uterine tube. Fertil Steril 41:754–760
5. Daniell JF (1985) Operative laparoscopy for endometriosis. Semin Reprod Endocrinol 3:353–359
6. Daniell JF, Brown D (1982) Carbon dioxide laser laparoscopy: initial experience in experimental animals and humans. Obstet Gynecol 59:761–764
7. Daniell JF, Diamond MP, McLaughlin DS, Martin DC, Feste J, Surrey MW, Friedman S, Vaughn WK (1986) Clinical results of terminal salpingotomy with the use of $CO_2$ laser: report of the intra-abdominal laser study group. Fertil Steril 45:175–178
8. Diamond MP, Daniell JF, Martin DC, Feste J, Vaughn DS, McLaughlin DS (1984) Tubal patency and pelvic adhesions at early second-look laparoscopy following intra-abdominal use of the carbon dioxide laser: initial report of the intra-abdominal laser study group. Fertil Steril 42:717–723
9. Fayez JA, McComb JS, Harper MA (1983) Comparison of tubal surgery with the $CO_2$ laser and the unipolar microelectrode. Fertil Steril 40:476–480
10. Foulot H, Lefebvre G, Jagueux M, Darbois Y (1987) Experimental study of $CO_2$-laser-induced histological effects on human fallopian tube: determination of $CO_2$-laser parameters to be used in microsurgery. Laser Med Surg 7:202–206
11. Frantzen C, Schlösser H, Beck L (1979) Die mikrochirurgische Refertilisierung nach Tubensterilisation. Dtsch Arztebl 42:2715–2719
12. Hulka JF, Ulberg LC (1975) Reversibility of clip sterilization. Fertil Steril 26:1132–1133
13. Kaplan I (1976) Laser surgery. Jerusalem Academic Press, Jerusalem
14. Mage G, Bruhat MA (1983) Pregnancy following salpingostomy: comparison between $CO_2$ laser and electrosurgery procedures. Fertil Steril 40:472–475
15. Paterson PJ, Downing B, Trounson OA, Cumming IA (1981) Fertility and tubal morphology after microsurgical removal of segments of the porcine tube. Fertil Steril 35:209–213
16. Rock J, Bergquist CA, Zacur HA, Parmley TH, Guzick DS, Jones HW (1982) Tubal anastomosis following unipolar cautery. Fertil Steril 37:613–618
17. Scheidel P (1982) Mikrochirurgie in der Gynäkologie – tierexperimentelle und klinische Untersuchungen zum Einsatz mikrochirurgischer Techniken in der rekonstruktiven Tubenchirurgie. Habilitationsschrift, Homburg/Saar
18. Semm K (1976) Endocoagulation: a new field of endoscopic surgery. J Reprod Med 16:194–203
19. Silber SJ, Cohen R (1989) Microsurgical reversal of female sterilization: the role of tubal length. Fertil Steril 33:598–601

20. Swolin K (1967) 50 Fertilitätsoperationen. Acta Obstet Gynecol Scand 46:234–267
21. Tadir Y, Kaplan I, Zuckerman Z, Edelstein T, Ovadia J (1984) New instrumentation and technique for laparoscopic carbon dioxide laser·operations: a preliminary report. Obstet Gynecol 63:582–585
22. Tadir Y, Kaplan I, Zuckerman Z, Ovadia J (1986) Actual effect of $CO_2$-laser power on tissue in endoscopic surgery. Fertil Steril 45:492–495
23. Wheeless CR (1977) Problems with tubal reconstruction following laparoscopic sterilization using the electrocoagulation and resection technique. Fertil Steril 28:723–727

# $CO_2$ Laser or Microelectrode for Adhesiolysis? Experimental Results

P. SCHEIDEL [1], G. WALLWIENER, and H. HEPP

## Introduction

Adhesions in the genital region and neighboring organs are found in a high percentage of sterile women. They are considered the primary cause of sterility in approximately 10% of all cases [3]. Schlösser et al. [5] reported that in 83 cases of microsurgical reconstruction of the oviduct, 49 were combined with an adhesiolysis. In 14 of those cases, adhesiolysis proved to be the only microsurgical procedure performed. Altogether, 63 of 83 (75.9%) patients required the removal of adhesions. The renewed postoperative appearance of adhesions is considered responsible for the failure of more than 40% of all procedures [3]. A multitude of attempts have been made to prevent the postoperative development of adhesions. Despite the abundance of scientific research, it is still not possible to explain the exact pathologic-genetic mechanism of adhesion formation.

Local ischemia undoubtedly plays a significant role in the development of adhesions. Ischemia generally appears after a trauma of the serosa. Specifically, there seems to be a subsequent decrease in fibrolytic activity (plasminogen activator activity, PAA) [1]. It is quite possible that a relationship exists between the reduction of fibrolytic activity and the development of adhesions [2]. Animal experiments have shown that the trauma defect was closed by subperitoneal fibroblasts after the initial cleaning phase of phagocytizing cells was completed. It remains to be seen whether this new mesothelium is a direct result of fibroblast activity, or whether it is the result of monocyte and macrophage activity from the exudate of the surrounding peritoneum [4].

The development of microsurgery and its application in reconstructive tubal surgery strives to reduce the operative trauma by providing exact hemostasis and a minimum of operative preparation. The number of postoperative adhesions has been substantially reduced through preparation under the operative microscope with such special instruments as Swolin's [6] microelectrode.

Nevertheless, it is inevitable that adhesions will reappear despite scrupulous preparation and a minimum of trauma. The combination of microsurgical procedures with intraoperative cortisone therapy in reconstructive tubal surgery was an attempt to improve on hitherto disappointing results. Toaff [8] initiated the use of the $CO_2$ laser in reconstructive tubal surgery, using it primarily for adhesiolysis. He noted an obvious improve-

[1] Marienkrankenhaus, Abt. für Geburtshilfe und Gynäkologie, Alfredstr. 9, W-2000 Hamburg 76, FRG.

Lasers in Gynecology
Edited by G. Bastert and D. Wallwiener

ment over conventional methods after eight operations. A direct comparison with microsurgical techniques could not be conducted. We have therefore attempted to compare the advantages of the $CO_2$ laser with those of the microelectrode in experimental animals using standardized operative procedures.

## Materials and Methods

The initial traumatization of the serosa was performed on 60 white, female rats of approximately 300 g each. After anesthesia with chloral hydrate (3.6% intraperitoneal), the animals were laparotomized with a midline incision along the abdomen. After sufficient exposure of the uterus, a hard Perlon brush was stroked ten times on the anterior and ten times on the posterior side of the uterus. This procedure served to induce adhesions. Thereafter, the uterus was severed along the middle portion. The distal fragment was attached to the lateral pelvic wall with an 8/0 nylon suture. The severing of the uterus was conducted in order to combine the adhesiolysis with a reconstructive procedure (end-to-end anastomosis of the uterine horns) during the second laparotomy 14 days later. This second laparotomy also served as an opportunity to classify the degree of adhesions in the individual animals. Grades 0 and + were not considered significant adhesions, whereas grades + + and + + + were seen as significant adhesions. Two experimental groups were created by randomly selecting a rat for adhesiolysis by alternating the $CO_2$ laser method with the microelectrode method. Both procedures were conducted under an operative microscope. During this second laparotomy, the sealed ends of the uterine horns were opened and reanastomosed under the operative microscope with 9/0 nylon suture material. The $CO_2$ laser utilized in the adhesiolysis was from the Coherent Radiation Co. It was attached to a Zeiss OPMI 1 operative microscope and controlled by a micromanipulator through a mirror-reflex system. The microelectrodes used in this study were from the Gebrüder Martin Co., along with the high frequency device which powers them.

No antibiotics, corticosteroids, or other drugs were administered after relaparotomy. A third laparotomy was conducted 14 days later to inspect the operative results, and to repeat the classification of the adhesions according to the above-mentioned categories.

## Results

Severe to very severe adhesions (+ + and + + +) were induced in 42 of 60 animals by the method described previously. The remaining 18 animals also showed obvious evidence of adhesions. Not one of the treated animals was completely free of adhesions. The adhesion status was clearly improved in both experimental groups after microsurgical adhesiolysis. Altogether, it was possible to obtain a marked reduction of adhesions in 32 animals. Considering that a pharmaceutical prophylaxis was dispensed with, the results of microsurgical adhesiolysis may be seen as satisfactory. Regarding the difference between the two experimental groups, the laser-adhesiolysis group showed 23 of 30 animals (76.7%) without adhesions, whereas the microelectrode group showed 17 of 30 animals

(56.7%) in whom only a minimal degree of adhesions was found. These figures consider only the overall results and not the individual status of the animals prior to the experiment.

The laser-treated group also showed a distinct trend towards better results when a reduction in the degree of adhesions was considered (for example: from a preoperative severe degree of adhesions (+ +) to a postoperative adhesion-free status (0) is a reduction of two grades. A reduction of three grades was found in seven (23.3%) of the laser-treated animals. A reduction of one grade was found in ten (33.3%) animals of the laser-treated group. The postoperative result was not different from the initial adhesion status in only four (13.3%) of the laser-treated animals. Only two (6.6%) animals in this group showed more severe adhesions postoperatively than they had preoperatively.

A reduction of three grades was achieved in only two (6.6%) of the animals treated with the microelectrode method. A reduction of two grades was found in six (20%) animals and a reduction of one grade was found in seven (23.3%). The number of rats that demonstrated no significant change from their preoperative status was as high as 13 (43.3%) in the microelectrode-treated group. Just as in the laser-treated group, only two (6.6%) animals showed a more severe postoperative adhesion status. The results were statistically evaluated using the chi-square test and the U-test according to Mann-Whitney ($p = 0.05$). A significant difference was obtained between the two groups.

## Discussion

The principal problem of any study dealing with the formation or prevention of adhesions lies in the unavoidable subjective classification of the degree of adhesion. An additional difficulty is encountered when several subgroups are judged. The classification in this study was always made by the same researcher, who had no previous knowledge of the operative procedure (laser or microelectrode) performed. Nevertheless, the problem of subjectivity in this study remains. Although a significant difference was obtained between to two groups, this may not be drawn upon in attempting to prove the superiority of the $CO_2$ laser due to the subjectivity of the classification. The absence of secondary adhesion bands after $CO_2$ laser adhesiolysis of peritoneal adhesions was particularly conspicuous. These secondary adhesion were often found in the microelectrode group.

The $CO_2$ laser appears to require more familiarization than any other operative instrument, not only with its manipulation, but also with the correct determination of wattage (0–30 W) and the duration of application (pulsating or continuous). In a preparatory series of ten animals which were not included in this study, three serious injuries were inflicted by the use of the laser. In one case, the vena iliaca was ruptured during adhesiolysis. Two further cases involved injuries to the ureter and the intestines. All three cases undoubtedly involved errors in the manipulation of the $CO_2$ laser. Consequently, our operative results clearly improved with increased experience in laser manipulation.

In this study, the $CO_2$ laser proved superior to the microelectrode particularly for very severe adhesions. A conclusive explanation for this phenomenon cannot be given because the effect of the $CO_2$ laser on different tissues has not been sufficiently investigated. The assumption can be made that improved hemostasis on the one hand, and the minimal depth of penetration (when using an appropriate energy setting) on the other, will result in laser superiority over the microelectrode. In comparison, the microelectrode with its high-frequently current is difficult to steer, and clumsy when attempting to gauge the depth of penetration. The result of this experimental animal study are not sufficient to prove the possible superiority of the $CO_2$ laser in adhesiolysis. The disadvantage of the laser lies not only in its high initial cost, but also in its hazard to neighboring organs. The procedure must therefore be conducted with the utmost caution. The ensuing time loss will be more than compensated by the ease and rapidity of laser adhesiolysis. The laser is particularly timesaving when one is dealing with vessels up to 0.5 mm, which then require no additional hemostasis in the form of bipolar coagulation. This would allow the microsurgeon a clear field of operation with fewer complications. The remaining reconstructive steps may be implemented in the usual manner after completion of the laser adhesiolysis.

The $CO_2$ laser could play an increased role in gynecological macro- and microsurgery, if it were possible to solve some of the instrumental problems (reducing the focal-point diameter, simplifying the reflex-mirror-arm manipulation and the angular projection of the laser beam). It is especially noteworthy that an independent microsurgical research team has been able to confirm the results of this study [7].

## References

1. Benzer H, Blümel G, Piza F (1963) Zusammenhänge zwischen fibrinolytischer Aktivität des Blutes und aseptischen Wundheilungsstörungen im Tierexperiment. Langenbecks Arch Clin Chir 302:463
2. Buckman RF, Buckman PD, Hufnagel HV, Gervin AS (1976) A physiologic basis for the adhesion-free healing of deperitonealized surfaces. J Surg Res 21:67
3. Gordij M (1975) Pelvic adhesions and sterility. Acta Eur Fertil 6:279
4. Raftery AT (1973) Regeneration of parietal and visceral peritoneum: a light-microscopical study. Br J Surg 60:293
5. Schlösser HW, Frantzen C, Beck L (1979) Über den Stellenwert mikrochirurgischer Techniken bei der operativen Therapie der tubaren Sterilität. Geburtshilfe Frauenheilkd 39:545
6. Swolin K (1975) Electromicrosurgery and salpingostomy: long-term results. Am J Obstet Gynecol 121:418
7. Tadir Y, Margara R, Ovadia T, Winston RML (1980) Laser microsurgery for lysis of intraperitoneal adhesions. X. World Congress on Fertility and Sterility, Madrid
8. Toaff R (1975) The carbon dioxide laser in gynecological surgery. In: Kaplan I (ed) Laser surgery. Jerusalem Academic Press, Jerusalem, p 129

# Is the Nd:YAG Contact Laser as Suitable for Adhesiolysis as the $CO_2$ Laser?

D. WALLWIENER [1], D. POLLMANN, S. RIMBACH, and G. BASTERT

## Introduction

The $CO_2$ laser has been used as the instrument of choice in the adhesiolysis of intraperitoneal adhesions in female reproductive organs for many years [1,3,5]. It was thus of interest to investigate whether intraperitoneal adhesiolysis is also possible using the neodymium:YAG contact method, and how the results thus attained compare with those achieved using the $CO_2$ laser.

## Adhesiolysis Studies on the Uterine Horn of the Rat

Adhesions were artificially induced in rats according to a standard procedure [2] to enable comparison of the various surgical methods of adhesiolysis. After the abdomen had been opened and the uterine horns exposed, the central area of both uterine horns was traumatized by ten hard strokes with a Perlon brush (Fig. 1). The adhesions were classified within the framework of a second laparotomy (Table 1), after which adhesiolysis was carried out. The remaining and/or newly developed adhesions were classified within the framework of a third laparotomy.

Forty female Wistar rats were randomized into two groups of 20 each and were tested. Laser adhesiolysis of the previously induced adhesions was carried out using the Nd:YAG contact laser with sapphire tip in group I and the $CO_2$ laser in group II.

## Adhesiolysis

Microsurgical adhesiolysis using the $CO_2$ laser and the Nd:YAG contact laser was performed using microsurgical security measures according to Swolin [6]. The adhesions or the affected neighboring organs were exposed

**Table 1.** Classification of adhesions

| | |
|---|---|
| Grade 0 | No adhesions whatsoever |
| Grade I | Isolated adhesions in the visceral peritoneum |
| Grade II | Wide adhesion bands between the visceral parts of the peritoneum and/or the parietal peritoneum (only genital organs involved) |
| Grade III | Wide adhesion bands extending beyond the genitals and involving the intestines and/or bladder |

[1] Universitäts-Frauenklinik, Abt. für Geburtshilfe und Gynäkologie, Voßstr. 9, W-6900 Heidelberg, FRG.

Lasers in Gynecology
Edited by G. Bastert and D. Wallwiener

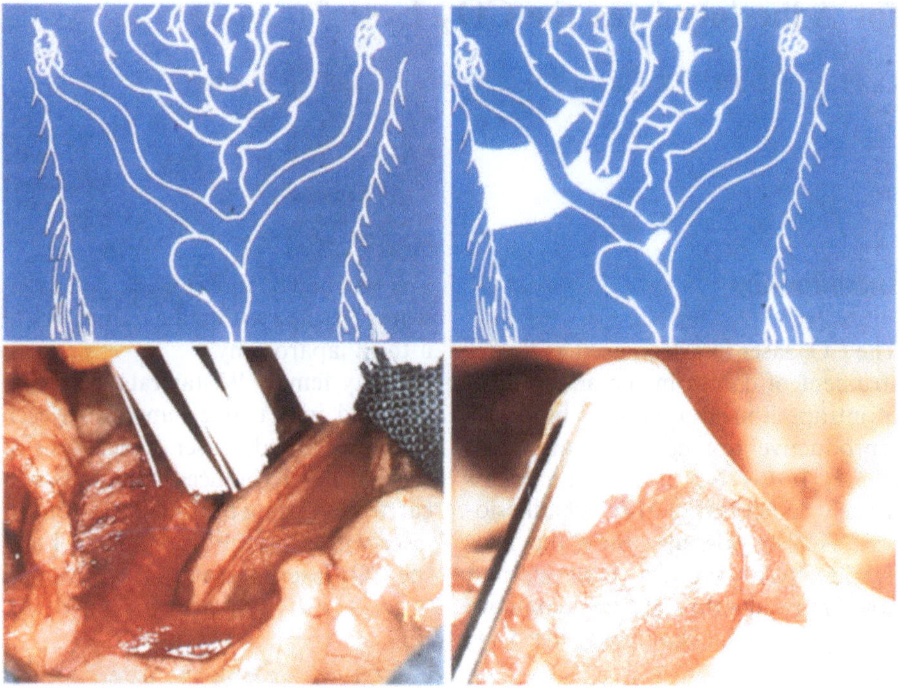

a b

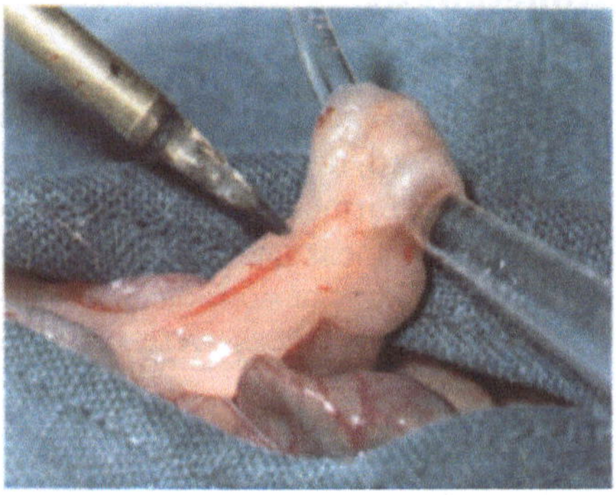

c

**Fig. 1a–c.** Situs of the rat. **a** Adhesion induction with a brush; **b** adhesion grade + +; **c** adhesion grade + + + (affecting the intestine and bladder): adhesiolysis using the sapphire tip

with specially prepared quartz rods and the organs elevated, so that the adhesion filaments remained tense during laser adhesiolysis. An 8-W output, equivalent to an engery density of 5000 $W/cm^2$ at the tip, was selected for contact adhesiolysis. Microsurgical $CO_2$ laser adhesiolysis was carried out using the continuous-wave mode with an energy density of 10000 $W/cm^2$.

The severity of the adhesions induced during the first laparotomy was termed the preoperative adhesion status. There was no difference in the distribution of the severity of the preoperative adhesions within the two groups; 13 animals had extremely severe, 14 severe, and 13 minor adhesions. No rat was free of adhesions. The improvements in the degree of ad-

**Table 2.** Effects of adhesiolysis (improvement in adhesion grades)

| Group | Improvement in adhesion grade | | | No improvement | Deterioration |
|---|---|---|---|---|---|
| | 3 | 2 | 1 | 0 | −1 |
| Group I | 5 | 5 | 7 | 2 | 1 |
| Group II | 5 | 6 | 6 | 3 | 0 |
| Adhesiolysis results according to point score | | | | | |
| Group I | 15 | 10 | 7 | 0 | −1 |
| Group II | 15 | 12 | 6 | 0 | 0 |

hesion attained within the two groups were subdivided to enable comparison of the effectiveness of the two adhesiolysis methods (Table 2).

The adhesiolysis results were represented by the improvement attained in the degree of adhesion to illustrate the results of the operation. An improvement of three adhesion grades, for example, represented a decline in adhesion from grade III to grade 0. Improvement of grade 0 represented no success in adhesiolysis and a deterioration in adhesion grade −1. The number of animals in each group on which a particular grade of improvement in adhesion had been attained was multiplied by the improvement. To illustrate: five animals in group 1 showed an improvement in adhesion of three grades – this gives a point score of 15 (Table 2). Where there was no improvement, 0 points were scored, while a deterioration resulted in a minus score. Tables 3 and 4 show the preoperative and postoperative adhesion status as well as the difference within the two groups. Comparable operation results were attained in both groups as far as the improvement in the degree of adhesion observed in the individual animals is concerned. Hence, the results of Nd:YAG contact laser adhesiolysis were not worse than those of $CO_2$ laser adhesiolysis.

**Table 3.** Adhesion reformation after Nd:YAG (I) and $CO_2$ (II) laser adhesiolysis

| | Grade | | | |
|---|---|---|---|---|
| | 3 | 2 | 1 | 0 |
| Group I | 7 | 7 | 6 | 0 |
| Group II | 6 | 7 | 7 | 0 |

**Table 4.** Table to calculate the Fisher exact test

| | Improved | Not improved | Total |
|---|---|---|---|
| Group I | 31 | 10 | 41 |
| Group II | 33 | 6 | 39 |
| Total | 64 | 16 | 80 |

Table 3 shows two groups of 20 animals each with adhesions of grade i ($i = 0, 1, 2, 3$). These grades are added to determine the adhesive state (Table 4). The adhesive state of group I is: $3 \times 7 + 2 \times 7 + 1 \times 6 + 0 \times 0 = 41$; that of group II is: $3 \times 6 + 2 \times 7 + 1 \times 7 + 0 \times 0 = 39$. Group I was treated by means of Nd:YAG laser adhesiolysis, whereby an improvement of 31 grades was attained. Group II was treated by means of $CO_2$ laser adhesiolysis, whereby an improvement of 33 grades was attained. This results in a quality factor of $31/41 = 0.76$ for group I and of $33/39 = 0.85$ for group II.

The statistical significance of these quality factors calculated using the Fisher's exact test were $p > 0.135$. This implies that there is no significant difference between the two techniques as far as this method of comparison is concerned ($p > 0.05$).

The experimental adhesiolysis results obtained using the Nd:YAG contact technique are therefore in no way inferior to those attained with conventional $CO_2$ laser adhesiolysis.

In a comparative study of lysis of previously induced adhesions in rabbits, Mecke et al. [4] were able to show significantly fewer re-adhesions after preparation using sapphire tips.

# References


1. Bellina JH (1981) Reconstructive microsurgery of the fallopian tube with the carbon dioxide laser – procedures and preliminary results. Reproduction 5:1–17
2. Gazzangia AB, James JM, Shobe JB, Oppenheim EB (1975) Prevention of peritoneal adhesions in the rat. Arch Surg 110:429–432
3. Joffe N (1986) Contact neodymium:YAG laser surgery in gastroenterology: a preliminary report. Laser Med Surg 6:155–157
4. Mecke H, Riedel HH, Semm K (1989) Die 3-Stufen-Therapie bei Endometriose als Kombination von medikamentöser Therapie und operative Pelviskopie. In: Der Einsatz neuer Laser-Techniken in der gynäkologischen Fertilitätschirurgie, ed by G. Bastert and D. Wallwiener. Laser Med Surg [Suppl] 1:71–75
5. Pittaway DE, Maxson WS, Daniell JF (1983) A comparison of the $CO_2$ laser and electrocautery on postoperative intraperitoneal adhesion formation in rabbits. Fertil Steril 40:366–368
6. Swolin K (1967) Fertilitätsoperationen. Acta Obstet Gynecol Scand 46:234–267

# Linear Salpingotomy Without Suture? Laser Techniques in Laparoscopy of the Minipig

D. WALLWIENER[1], S. RIMBACH, D. POLLMANN, and G. BASTERT

## Introduction

To test newly developed laser techniques and the possibilities offered by their application in operative laparoscopy, the following points were considered:

1. The healing process after linear salpingotomy, with respect to organ-preserving laparoscopic treatment of tubal pregnancies without endosuture
2. The possibility of developing an experimental model for laser laparoscopy in fertility surgery

## Material and Methods

The mediLas 40 N (manufactured by MBB) neodymium YAG laser and the Sharplan 1060 $CO_2$ laser were used. The optimum laser energy range was determined by performing tubal transections on a minipig (which was not included in the study) in different $CO_2$ and Nd:YAG laser energy ranges (laser output) [3]. The laser outputs shown in Table 1 were found to be the most favorable in the respective test groups; surgery was carried out on five animals in each group. A primary laparotomy was performed on the animals in group I, as the laser equipment initially had to be tested on the open abdomen. Laparoscopic salpingotomy was performed on the other groups.

### Surgical Laparoscopy

In laser laparoscopy of the minipig, puncture was carried out subumbilically with 10-mm optics, subsequent to the application of a pneumoperitoneum of approximately 2 l $CO_2$ with a Verres needle via a Z-puncture using an 11-mm trocar. The second puncture and/or the fixation forceps were then inserted some 20 cm subumbilical to the left of the midline using a 5-mm trocar. Both tubes were exposed and the isthmic segments distended with the fixation forceps prior to further surgical manipulation. After surgery, the intra-abdominal blood dryness was controlled, the laparoscopic instruments were removed, and the incisions were closed with single knotted sutures.

### Operation Procedure Within the Individual Groups

Laparotomy was carried out over a median section in group I. The tubes

[1] Universitäts-Frauenklinik, Abt. für Geburtshilfe und Gynäkologie, Voßstr. 9, W-6900 Heidelberg, FRG.

Lasers in Gynecology
Edited by G. Bastert and D. Wallwiener

**Table 1.** Laser laparoscopy/laparotomy isthmic (segmental and linear) salpingotomy

| Group (*n*) | Method | Mode | Energy (W/s) | Output density (W/cm$^2$) |
|---|---|---|---|---|
| I 5 | $CO_2$ laser | cw* | 20 | |
| II 5 | $CO_2$ laser | cw | 20 | 50000 |
| III 5 | $CO_2$ laser | sp | 15 | |
| IV 5 | Nd:YAG laser | contact | 15 | 12000 |

a

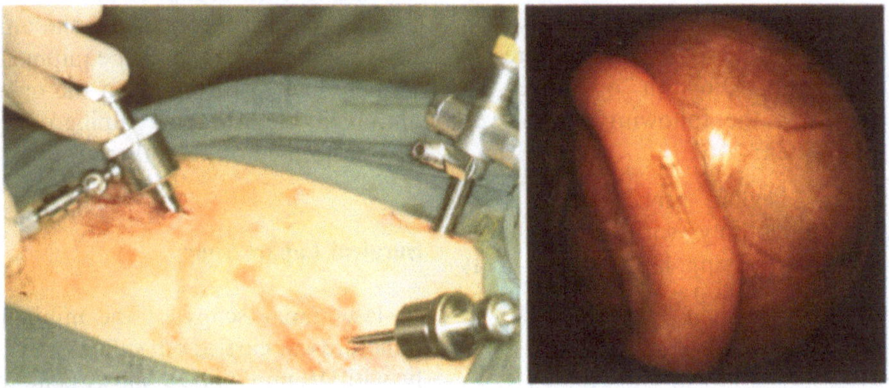

b

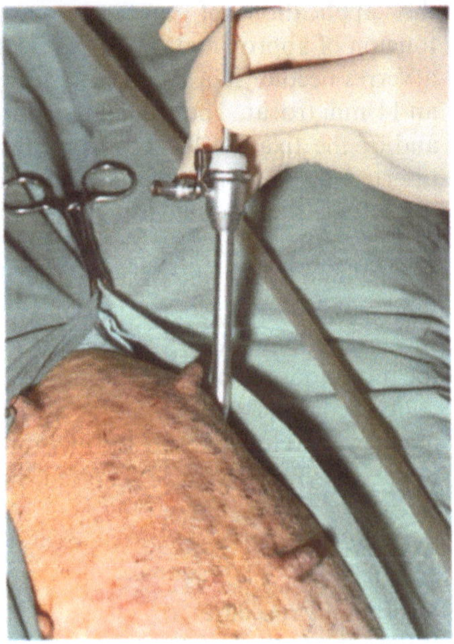

c

**Fig. 1. a** Laser laparoscopy on the minipig. **b** Double puncture trocar. **c** Salpingotomy

**Table 2.** Preparation possibilites by laparoscopic laser application ($CO_2$ laser, continuous wave, superpulsed wave, and Nd:YAG contact technique)

| Criterion | Test groups | | |
|---|---|---|---|
| | II<br>cw | III<br>sp | IV<br>contact |
| Accessibility of organ parts | With shield extremely bad; without shield good | | Very good with rigid and flexible application |
| Tissue destruction (macroscopic) | Scarcely any | None | A little, due to coagulation |
| Preparation or cutting speed | Rapid | Slow | The quickest |
| Coagulability | Energy dependent | Scarcely any | Very good |
| Damage to neighboring organs | Without shield: one ovary injury<br>With shield: two damaged uterine horns due to heated shield | | None |

were placed on synthetic plates and fixed in a suitable position with glass probes subsequent to the exposure of the uterine horns and the appendages. Laser salpingotomy was performed on the isthmic segment. Transection was carried out by continuous-wave $CO_2$ laser vaporization using a focusing handpiece (125 mm) with an output of 20 W/s. Salpingotomy was carried out laparoscopically in the other groups (Fig. 1). In groups I and II the laser laparoscope was intra-abdominally inserted via a second puncture trocar; the tubal segment was distended and the transection carried out at 20 W/s. The superpulsed-wave mode, again with an output of 15 W/s, was used for the transection in group III. In group IV the laparoscopic rigid handpiece for the sapphire contact tip was also inserted via a second puncture trocar. The Nd:YAG laser beam was transmitted over a flexible light guide from the rigid handpiece to the sapphire contact tip and the distended tubal segment was transected with an output of 15 W/s. A linear salpingotomy over 2 cm was performed in the same manner on the contralateral tube (Fig. 1) [2].

The possibilities of laparoscopic preparation were to be tested as a model on the fallopian tubes of the minipig by performing segmental isthmic salpingotomy using the various techniques. Group V served as a control group. It was known that entirely blood-free transection subsequent to tissue coagulation is possible. The laser equipment and instruments were tested in the operations on the animals in group I; these preparations also served as training for the surgical procedure.

Second-look laparotomy was performed after a 7–8 week interval. Special attention was paid to the healing of the laser laparoscopic linear salpingotomy, due to the much-discussed question of whether or not salpingotomy wounds ought to be closed by endosutures after organ-preserving treatment of tubal pregnancy by laparoscopic salpingotomy. A good and correctly layered approximation of the wound margins was visible under the operation microscope in all 20 salpingotomies carried out over 2 cm in the isthmic section (Fig. 1). A very small fibrous scar only a few millimeters wide was discernible in three cases. A fistula had not developed. Chromopertubation showed all tubes to be permeable.

Within the framework of this laser laparoscopic study on the minipig, the necessity of a suture after linear laser salpingotomy was investigated. In all animals on whom a laser laparoscopic linear salpingotomy was performed, the resulting wound healed within 6 weeks without fistulization and without the need of a suture [1,2]. Even permeability of the prepared tubes was evident in all cases.

This study not only enabled experimental handling of surgical problems, but also ensured that the technical difficulties could be solved and improvements made prior to the clinical utilization of the laser equipment, thus avoiding serious complications.

## References

1. Nayel SA, Habashy E, Osman AM, Gawad AA (1985) The effect of electro-cutting and diathermy coagulation on wound healing in gynecological surgery. Asia Oceania J Obstet Gynecol 11:65–67
2. Semm K (1976) Endocoagulation: a new field of endoscopic surgery. J Reprod Med 16:194–203
3. Wallwiener D, Pollmann D, Morawski A, Bastert G, Krampe C (1988) Die Nd: YAG-Laser-Kontakttechnik mit Saphirschneidespitzen – Lasertechnischer Background und Gewebeeffekte. Laser Med Surg 4:75–85

# Superoxide Dismutase: Prophylaxis of Intraperitoneal Adhesion Reformation

S. Rimbach [1], D. Wallwiener, D. Pollmann, L. Stehling, W. Stolz, and G. Bastert

## Introduction

Irrespective of the etiopathology, intraperitoneal adhesions can lead to a complex spectrum of symptoms which are of utmost importance in reproductive surgery [1]. The main problem consists in dysfunction of the ovum pick-up mechanism [2].

Especially drug prophylaxis of recurrent adhesions is of focal interest as a supplement to operative adhesiolysis. To date, only intraperitoneally administered cortisone preparations and instillation of high-molecular-weight dextran solutions as artificial ascites have been accepted [7]. However, the problem remains essentially unsolved [6].

The adiministration of superoxide dismutase (SOD) as an antiphlogistic agent in other fields [3] gave rise to the question whether SOD could be successful in the prophylaxis of postoperative intraperitoneal adhesion formation. An animal experimental study was done to evaluate whether SOD displays an antiadhesive action comparable to that of cortisone.

[1] Universitäts-Frauenklinik, Abt. für Geburtshilfe und Gynäkologie, Voßstr. 9, W-6900 Heidelberg, FRG.

## Material and Methods

Intra-abdominal adhesions were induced in 102 adult female rats (Wistar strain) by brushing the uterine horns. After an interval of 20 days a second-look laparotomy was performed. Adhesions were classified according to a grading score (Table 1). This score differentiates four grades: extremely extensive (grade III), extensive (grade II), distinct (grade I), and no adhesions (grade 0) (Figs. 1–4). Then surgical adhesiolysis was done using a carbon dioxide laser.

When the surgical adhesiolysis was completed, re-adhesion prophylaxis with either SOD, cortisone, or the dilution medium (NaCl 0.9%) as a control followed. Thus, the animals were randomized to three groups. Group 1 (control group, $n = 30$): 4 ml NaCl

**Table 1.** Classification of adhesions

| Grade | Description |
|---|---|
| 0 | Absolutely no adhesions |
| I | Single adhesions in the visceral peritoneum |
| II | Broad, bandlike adhesions in the visceral parts of the peritoneum and/or the parietal peritoneum (including only genital organs) |
| III | Broad, bandlike adhesions involving not only the genital organs but also the intestines and/or the bladder |

Lasers in Gynecology
Edited by G. Bastert and D. Wallwiener

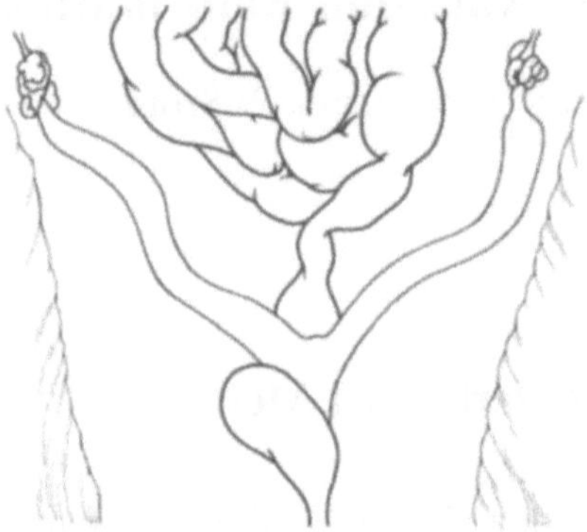

1

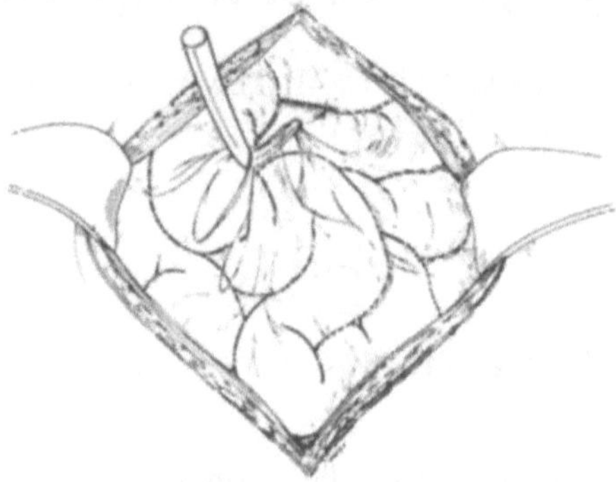

4

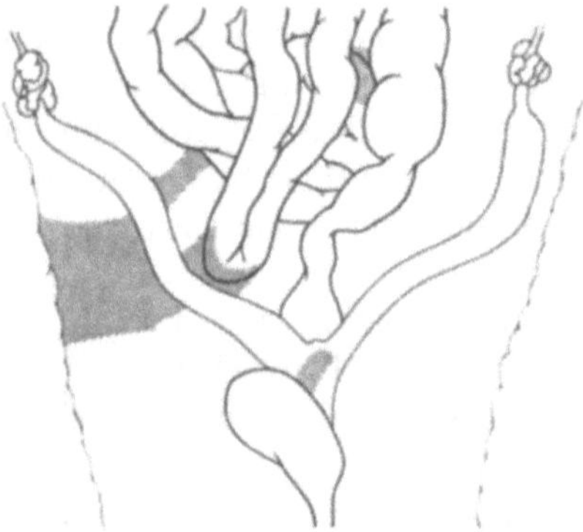

2

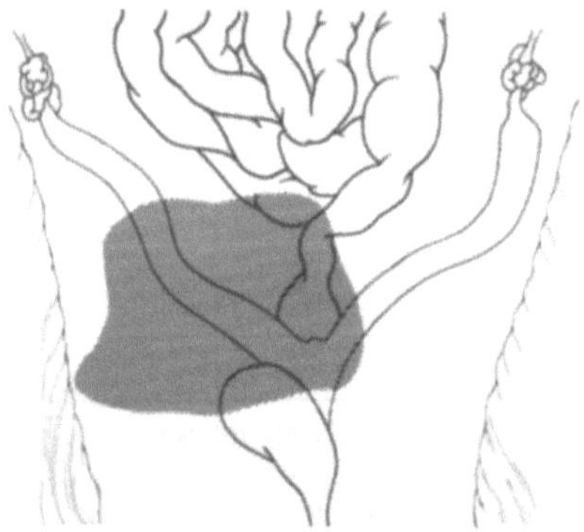

3

**Fig. 1.** Extremely extensive adhesions (grade III)

**Fig. 2.** Extensive adhesions (grade II)

**Fig. 3.** Distinct adhesions (grade I)

**Fig. 4.** No adhesions (grade 0)

0.9% (dilution medium) was administered intraoperatively; group 2 ($n = 30$): cortisone (100 mg/kg body weight) was administered intraoperatively; group 3 ($n = 42$): SOD was administered in varying patterns. In 38 animals SOD was given in a dosage of totally 80 mg/kg body weight. Thirty of these 38 animals received 8 mg intraoperatively and 4 mg after 24 h or 48 h, resulting in 16 mg/animal (= 80 mg/kg body wt.). Four animals received 8 mg intraoperatively and 2 mg every 12 h, up to 15 mg total, and four other animals received 16 mg intraoperatively as a single dose. In four animals a double dose was administered, resulting in 160 mg/kg body weight. Each animal received 16 mg intraoperatively, and 8 mg after 24 h or 48 h.

Three weeks later a third-look laparotomy was performed. In all animals classification of recurrent adhesions was carried out, thus controlling the efficiency of the drug prophylaxis. In order to compare the different drugs and administration patterns, the results according to the grading score were statistically analyzed using the chi-square and Fisher's exact tests.

## Results

As a result of the induction of adhesions, the second-look laparotomy showed extremely extensive adhesions (grade III) in 23% of the animals operated on. Extensive adhesions (grade II) were found in 35% and distinct adhesions in 42%. None of the animals was entirely without adhesions.

The subsequent randomization ensured statistically the same initial adhesion state for every group prior to surgical lysis of adhesions and pharmacological prophylaxis of recurrent adhesions.

As a result of adhesiolysis and prophylaxis of re-adhesion formation, the third-look laparotomy revealed in group 1 ($n = 30$, dilution medium only) nine animals with grade III, five with grade II, 11 with grade I, and five with no re-adhesions.

In the cortisone group ($n = 30$, group 2) four animals had extremely extensive (grade III) re-adhesions. Six animals had extensive re-adhesions (grade II), seven had distinct (grade I), and 13 had no re-adhesions.

Surgical adhesiolysis in combination with SOD prophylaxis of re-adhesions (group 3, $n = 30$) resulted in four animals with grade-III (extremely extensive) re-adhesions, two animals with grade II (extensive), eight animals with grade I, and 16 animals with no re-adhesions.

Statistically, (chi-square test) both cortisone and SOD proved to be more efficient in the prophylaxis of re-adhesions than the dilution medium alone ($p < 0.05\%$), but there were no significant differences between the cortisone and the SOD groups. Thus, cortisone, as well as SOD, seems to be efficient in the prophylaxis of re-adhesion formation. On the other hand, statistical analysis did not show any differences between the two drugs. Therefore, SOD had statistically the same efficiency in re-adhesion prophylaxis as cortisone.

Still, the re-adhesions formed in the SOD group appeared more transparent than in the other groups. The consistency was softer and breaking strength was reduced. Neither an increase in dosage nor a change in the administration pattern increased efficiency. Statistical analysis (Fisher's exact test) revealed no significant differences.

## Discussion

The antiadhesive action of the cortisone administered is well established [8]. The selected dosage corresponds to the average recommended dosage mentioned in the literature [7]. Side effects in the usual dosage range are not yet known. Nevertheless, it is becoming increasingly evident that such steroid prophylaxis brings satisfactory results only in cases of less extensive adhesions.

Based on the above, and due to the successful experimental and clinical anti-inflammatory administration of SOD, we tested the efficacy of this drug as a prophylactic agent for intraperitoneal re-adhesions, comparing it with that of the cortical steroids.

SOD prophylaxis was shown to be at least as effective as steroid prophylaxis. There are presently no reference values available in the literature for the reduction in re-adhesion formation attained in this study. The majority of authors reporting basic experimental prophylaxis studies using cortical steroids commenced drug adhesion prophylaxis after adhesion induction, and not only after adhesiolysis [4, 5, 8–10]. When interpreting the results obtained in the question posed, one may assume that SOD broadens the spectrum of antiphlogistic drugs available for intraperitoneal adhesion prophylaxis. The antiadhesive power of SOD is particularly evident when comparing the functionally serious adhesions, e.g., the extensive adhesions within the groups. A reduction in such functionally serious adhesions was attained using SOD, as well as with cortisone treatment.

Adhesion prophylaxis using SOD appears to be as successful a therapeutic concept as steroid prophylaxis and is possibly even better, at least as far as a particular range of indications is concerned. Whereas lysis of induced adhesions is, as introduced by Swolin [9], the best method of testing the adhesion prophylactic action of a drug, administration after adhesiolysis of the experimentally induced adhesions appeared, in this case, to be the most objective manner of comparing the re-adhesion prophylactic power of the drugs under consideration.

The mechanism of action of SOD still appears to be hypothetical. Whether or not the reduction in dangerous radicals influences the fibrinolytic activity of the serosa or decreases the ischemic zone remains to be elucidated in further studies. A direct enzymatic action on the fibrinolytic structure is also feasible.

The SOD dosage range was selected in accordance with the details on antiphlogistics found in the literature. Nevertheless, the dosage used in this study is, as in experimental models described in the literature, relatively high compared with the clinical range of administration. For clinical administration, e.g., 8 mg of Orgotein is recommended as a weekly unit dose, with treatment continuing over numerous weeks (basic project of the Grünenthal Company; SOD is Orgotein).

To summarize, SOD exerted an antiadhesive action comparable to that of cortisone in our experimental study, thus possibly broadening the spectrum of drugs for re-adhesion prophylaxis also in clinical fertility surgery. The details for use in human beings must be established in controlled clinical studies.

## References

1. Bogart LW (1937) Intra-abdominal adhesions: a experimental study. Arch Surg 34:129
2. Diamond E (1979) A comparison of gross and microsurgical techniques for repair of cornual occlusion in infertility: a retrospective study 1968–1978. Fertil Steril 32:370
3. Edsmyr F, Meander-Huber K (1981) Orgotein efficacy in ameliorating side effects

due to radiation therapy. Eur J Rheumatol Inflamm 4:228
4. Ehlers PN (1960) Adhäsionsverhütung durch intraabdominale Instillation von Prednisolon-Acetat. Langenbecks Arch Klin Chir 293:612
5. Gazzangia AB, James JM, Shobe JB, Oppenheim EB (1975) Prevention of peritoneal adhesions in the rat. Arch Surg 110:429
6. Jackson BB (1958) Observations on intraperitoneal adhesions. An experimental study. Surgery 44:507
7. Kastendieck E (1981) Tierexperimentelle Untersuchungen zur intraperitonealen Adhäsionsprophylaxe bei Ratten. 4. Arbeitstagung "Mikrochirurgie der Tube", Oberlech, Austria, 8 January
8. Lehmann V, Memmel L, Freudrich H (1978) Tierexperimentelle Untersuchungen zur medikamentösen Adhäsionsprophylaxe. Geburtshilfe Frauenheilkd 38:203
9. Swolin K (1966) Experimentelle Studien zur Prophylaxe von intraabdominalen Verwachsungen. Acta Obstet Gynecol Scand 45:473
10. Thomaschek G (1959) Experimentelle Untersuchungen über Adhäsionseinschränkungen im Bauchraum der Ratte durch intraperitoneale Hydrocortisongaben. Arch Gynäkologie 191:377

# Experimental $CO_2$ Laser Surgery on the Lewis Lung Carcinoma Tumor Model in C56BL/6 Mice

S. RIMBACH [1], D. WALLWIENER, D. POLLMANN, U. KLINGER, W. STOLZ, M. BLOCK, G. BASTERT, and D. VON FOURNIER

## Introduction

The superiority of the laser over conventional surgical techniques, especially where tumor surgery is concerned, has been well illustrated in numerous experimental studies in which tumors were excised using a scalpel, electrosurgery, and $CO_2$ laser surgery [6]. Laser surgery resulted in a reduction in (a) the intraoperative propagation of tumor cells, (b) the manifestation of metastases [10, 15], and (c) the local recurrence rate, as well as in an increase in the survival time of the treated animals [7, 9]. $CO_2$ laser surgery is no longer restricted to experimental models, being firmly established in clinical oncology. Nevertheless, hardly any intramethodic comparisons have been made to investigate the influence of various application techniques on tumor behavior.

Two modes of $CO_2$ laser application – the continuous wave (cw) and the superpulsed wave (sp) techniques – were thus compared and contrasted in this study on Lewis lung carcinoma subcutaneously implanted in C 57 B1/6 mice [14]. The comparisons were also undertaken in combination with vaporization of the wound bed utilizing the defocused cw technique, and the vaporization effect alone was investigated. These results were then compared with those obtained by conventional scalpel excision.

The following questions were to be answered by observing the healing procedure, local recurrences, and metastatic spread, as well as the survival rate subsequent to excision of the tumors using different techniques:

1. Does laser excision have advantages over conventional scalpel excision?
2. Is the superpulsed technique superior to the continuous-wave technique in this oncological model?
3. What is the significance of additional vaporization of the wound bed for subsequent tumor behavior?
4. Does vaporization alone possibly suffice?

## Material and Methods

The tests were carried out with a $CO_2$ laser (1060-μm wavelength, continuous wave or superpulsed with a pulse repetition rate of 250–300 Hz, maximum output density approximately 460000 $W/cm^2$) [3, 11, 12].

[1] Universitäts-Frauenklinik, Abt. für Geburtshilfe und Gynäkologie, Voßstr. 9, W-6900 Heidelberg, FRG.

Lasers in Gynecology
Edited by G. Bastert and D. Wallwiener

**Table 1.** Test groups for the examination of the recurrence rate and metastatic spread of Lewis Lung carcinomas after $CO_2$ laser surgery of the primary tumor

| Group | Technique | No. of animals |
|---|---|---|
| 1: sp | superpulsed wave | 17 |
| 2: cw | continuous wave | 14 |
| 3: vap | vaporization | 14 |
| 4: spvap | sp + vap | 17 |
| 5: cwvap | cw + vap | 10 |
| 6: scalpel | | 14 |
| 7: control | | 20 |

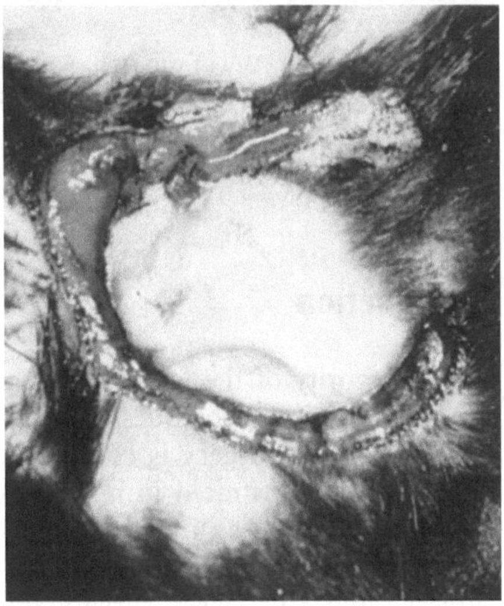

Fig. 1. Tumor excision – circular incision of the skin and subcutaneous tissue

## Experimental Design

Tumor pieces of 1–2 mm from host animals (BDF/1 mice, original tumor substance supplied by Höchst AG, FFM) were subcutaneously transplanted into the backs of 106 B6D2F1 mice (C57B1/6 × DBA/2 hybrids).

The tumor-bearing animals were divided into seven groups 6 days after inoculation. Of the 106 tumor-bearing mice, 86 were subsequently operated upon, 72 using the laser in five different application techniques and 14 using the scalpel. Twenty animals served as an untreated control group (Table 1).

## Tumor Operation Procedure

All operations were performed while the animals were under i.p. chloral hydrate (0.35 mg/g body wt.) anesthesia. In groups 1, 2, 4, and 5 the tumor was excised by incising the skin und subcutaneous tissue around the tumor (Fig. 1; power output: 10 W cw or 7 W sp), prior to removing the tumor mass from the underlying muscle. In groups 4 and 5 the wound bed was additionally vaporized using a defocused 5-W cw setting. The animals in group 3 were treated by vaporization alone commencing with a 10-W cw output. Both the tumor and a surrounding area of macroscopically healthy-appearing tissue (spot size diameter 3–5 mm) were removed, the output being reduced to 5 W cw subsequent to removal of the tumor mass. Tumor excision was carried out with a scalpel in group 6.

## Evaluation

The vitality and general condition of the animals were controlled, and the healing procedure and development of local recurrences were examined on a daily basis. Cardiopulmonary organs were removed from animals that died during the course of the observation period or after a maximum of 72 days, and the tumor mass, including local recurrences, was excised and histologically prepared (paraffin sections, hematoxylin and eosin staining method).

**Table 2.** Survival time, metastatic spread, and local recurrence in correlation with the operation technique

| | Group | | | | | | |
|---|---|---|---|---|---|---|---|
| | sp | cw | vap | spvap | cwvap | scalpel | control |
| *n* | 17 | 14 | 14 | 17 | 10 | 14 | 20 |
| Survivors | 10 | 11 | 7 | 12 | 7 | 1 | 0 |
| (%) | 58.2 | 78.6 | 50.0 | 70.6 | 70.0 | 7.1 | 0 |
| Local recurrences | 7 | 3 | 7 | 5 | 3 | 11 | – |
| (%) | 41.1 | 21.4 | 50.0 | 29.4 | 30.0 | 78.6 | – |
| Pulmonary metastases | 3 | 1 | 4 | 1 | 3 | 13 | 20 |
| (%) | 17.6 | 7.1 | 28.5 | 5.8 | 30.0 | 92.8 | 100 |

This enabled investigation of the success of the radical surgery as well as of the incidence and nature of both local recurrences and pulmonary metastases. The incidence of local and distant recurrences and the survival rate of the animals within the individual groups were statistically analyzed using Fisher's exact test (significance level of $p < 0.05$).

## Results

### Healing Procedure

Fewer complications arose during the healing of laser-damaged tissue than after granulation, but healing was retarded compared with scalpel excision. A complete restoration of epithelial cells with only minor scar formation was observed during the second and third postoperative week, vaporization, however, leading to a slight delay in the healing procedure. There was no discernible difference between the healing procedure after utilization of the cw and sp modes (Table 2).

### Tissue Effects of Laser Treatment

Cellular examination of laser-cut surfaces disclosed method-dependent morphological differences. A three-zone construction of the thermally damaged area, with formation of a functional barrier zone, was found to be typical after cw application. Utilization of the sp technique led to an irregular and narrower necrobiotic zone with distinct vacuolation, interspersed with intact cells (Figs. 2 and 3).

### Local Recurrences

The local recurrence rate was lowest after utilization of the cw technique (21.4%), followed by the two techniques with additional vaporization (spvap and cwvap, 29.4% and 30% respectively). Somewhat poorer results (41.4%) were attained using the sp technique. Statistical analysis showed the results obtained using a scalpel (recurrence rate of 78.6%) to be significantly worse than those obtained using the laser techniques.

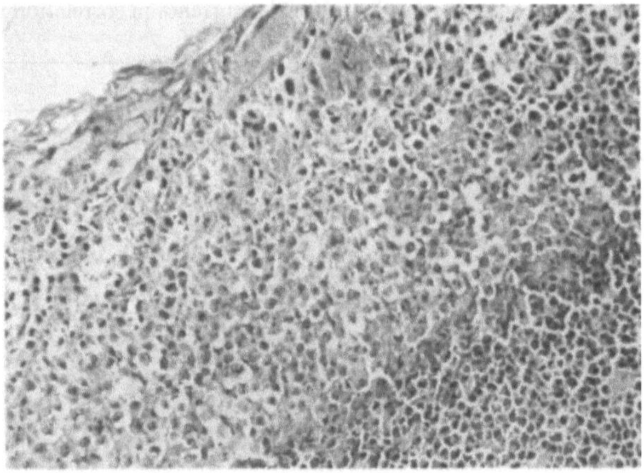

2

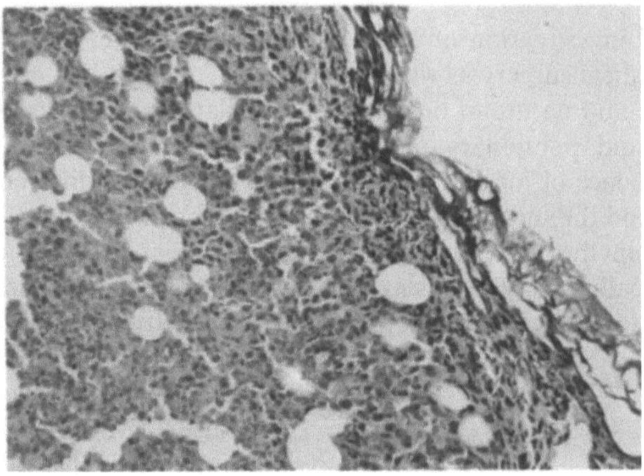

3

**Fig. 2.** Detail of the edge of the incision of a tumor transected using the 10-W cw mode with three damaged tissue zones: an outer carbonization zone with cell proliferation and vacuolation, a necrotic zone with altered stain reaction and indistinct cell structure, and an edematous zone with definite signs of damage. H & E, ×80

**Fig. 3.** Detail of the incision edge of a tumor transected using the 7-W sp mode showing an irregular carbonization zone with a subjacent necrotic zone. This tissue is partially heat fixed with signs of intracellular dehydration, which has led to a compression of the outer cell layers, thus looking like squamous epithelia. An inhomogeneous zone with multiple vacuolation (popcorn effect) is discernible underneath. H & E, ×100

## Radicality of Surgery

Histological preparation of the wound margins of the excised tumors showed that all local recurrences ensued after nonradical tumor excision. Bands of tumor cells, or the fibrous capsules surrounding the carcinoma, extended into the necrobiotic zone. In sano operations were attained in significantly more animals treated with the $CO_2$ laser than in those in which the tumor was excised with a scalpel; this is in accordance with the observed local recurrence rate. No statistical difference was found between the various laser techniques as far as radicality of the surgical procedure is concerned, as was also evident from the local recurrence rates. The techniques with additional vaporization as well as the cw mode were more favorable, however, vaporization alone being found to be disadvantageous. A correlation between tumor size prior to surgery and the local recurrence rate was not established.

### Pulmonary Metastases

Scalpel excision also led to significantly worse results as far as the development of pulmonary metastases is concerned, i.e., 92.8% compared with a rate between 5.8% and 30% after $CO_2$ laser excision. Although there was no statistically significant difference between the various application methods, the metastatic spread rate was lowest after utilizing the cw and spvap modes (7.1% and 5.8% respectively), followed by the sp mode (17.6%), with poor results being attained after vaporization alone and the cw and vaporization combination technique (28.5% and 30% respectively). All animals in the control group had pulmonary metastases.

### Postoperative Survival Rate

None of the control animals survived for longer than 34 days post intervention (d.p.i.) (mean survival time 30.5 ± 1.002 d.p.i.). The mean survival time of those animals with recurrences (local and/or metastases) was comparable to that of the control group. None of the animals still alive ($n = 48$) at the end of the observation period (72 days) showed signs of either local recurrences or pulmonary metastases, and they were thus defined as cured. The numbers of healed mice within the different groups varied greatly. A statistically significantly ($p = 0.010$) greater number of long-term survivors was registered in the purely cw-treated and vaporization combination groups (78.6% using the cw, 70.6% the spvap, and 70% the cwvap technique) than in the scalpel group (7.1% survival rate). The survival rates within the other two groups lay between those mentioned above: 50% (vap) and 58.2% (sp).

## Discussion

Laser application has been tested in various modes by numerous authors with the aim of combining the advantages offered by laser surgery with those of conventional scalpel excision [2,4,8,13,15]. In experimental studies electronic superpulsing with a relatively low impulse frequency (300 to max. 500 pps), short impulse duration (0.3–0.7 ms), and consistently high peak output was found to produce the best results [1].

Minimal tissue traumatization is attained using the sp mode. Whether or not, and to what degree this reduction in tissue traumatization is desirable in tumor surgery remains to be clarified. It is uncertain whether tissue traumatization as negligible as that attainable using the sp technique can still ensure sufficient local cell destruction on the edge of the tumor and an adequate occlusion of lymph tracts with subsequent reduction in the intraoperative dislocation of tumor cells. This problem was optimally investigated in the chosen model: the Lewis Lung carcinoma is characterized by constant tumor behavior with a 100% attainability, high aggressivity, early and inevitable metastatic spread, and death of all untreated animals. Investigation of the influence exerted by different application methods, such as the superpulsed and continuous-wave techniques with or without additional vaporization of the wound, on tumor behavior was thus possible. A delay in the healing procedure of laser incisions when compared with conventional scalpel application, as has already been described in the literature [4, 5], was observed. However, the postoperative survival time and metastatic spread were notably better after $CO_2$ laser application than after scalpel utilization; the possible difference between the various laser application modes was thus worth investigating. The accelerated healing process and reduced inflammatory reaction after sp compared with cw utilization, as postulated in a comparative study undertaken by Baggish and El Bakry [1], was not verified. The local recurrence rate and incidence of pulmonary metastases noted in this study did not show the sp mode to have any fundamental advantages over the other application modes. The cw and the sp mode with additional vaporization appear to offer slight advantages, whereas vaporization alone does not suffice for the prevention of local recurrences. Pulmonary metastases were discernible in all tumor-bearing animals, although method-dependent differences were not established. This observation is in keeping with those made by both Mahn et al. [9] and Tuchmann et al. [15]. The incidence of pulmonary metastases is probably determined by the tumor cells already in circulation at the time of surgical intervention and independent of the operation technique utilized.

The postoperative survival rate confirms the results in favor of the cw mode and the sp or cw and vaporization combination techniques; the survival rate is considerably higher in these three groups than where the sp mode or vaporization was utilized alone. The survival rate subsequent to one of these application methods is interpretable only as animal and model specific, however, and is not directly transferable to clinical conditions.

In summary, the various $CO_2$ laser application methods utilized in this chosen model were shown to be superior to the scalpel, only minor differences being discernible between the various techniques. Vaporization alone did not prove to be successful, not resulting in a lower recurrence rate than the other techniques, and leading to a retarded healing process. The superpulsed technique remains oncologically inferior to the continuous-wave mode as long as the wound bed is not additionally vaporized. Although the precise incision and tissue preservation attainable when using the sp mode make it advantageous from a purely technical point of view, it does not yet ensure sufficient tumor cell destruction, as is indeed possible with the cw technique. Additional vaporization (spvap) does enable the fulfillment of these requirements, however, and thus leads to results comparable to those attainable with the continuous-wave mode, both with or without vaporization, at least as far as the local recurrence rate and metastatic spread are concerned. The cw mode and the combination techniques (i.e., cwvap and spvap) lead to a better survival rate than the sp mode and can hence be considered superior.

# References


1. Baggish MS, El Bakry MM (1986) Comparison of electronically superpulsed and continuous wave $CO_2$ laser on the rat uterine horn. Fertil Steril 45:120–127
2. Bellina JH (1984) Analysis of electronically pulsed versus quasi-continuous wave carbon dioxide laser in an animal model. Am J Obstet Gynecol 150:934–940
3. Dagan J (1984) The Sharplan family of $CO_2$ lasers for surgery. Neurosurg Rev 7:113–121
4. Herrmann U (1986) Lasertherapie bei essentiellem Pruritus vulvae. Laser Med Surg 3:144–145
5. Kaplan I, Goldman J, Ger R (1973) The treatment of erosions of the uterine cervix by means of the $CO_2$ laser. Obstet Gynecol 41:795–796
6. Kaplan I, Giller S (1984) $CO_2$ laser surgery. Springer, Berlin Heidelberg New York
7. Lanzafame RJ (1985) The effect of $CO_2$ laser excision on local tumor recurrence. In: Proc Sixth Congress of the International Society for Laser Surgery, Jerusalem, 13.–18. October
8. Larsson G (1982) Laser conization versus cold-knife conization. Surg Gynecol Obstet 154:59–61
9. Mahn HR (1982) Tierexperimenteller Vergleich der therapeutischen Effektivität einer Tumorexcision mittels Skalpell und $CO_2$ laser am s.c. implantierten LLC. Z Exp Chir 15:38–47
10. Peled I, Kaplan I, Shohat B, Gassner S (1976) Excision of epithelial tumors: $CO_2$ laser versus conventional methods. Cancer Lett 2:41–46
11. Samoilov S (1985) Sharplan $CO_2$ surgical lasers – summary of temporal operating modes. Laser Industries Ltd, Tel Aviv
12. Schwabel A, Samoilov S (1985) Superpulse. Laser Industries Ltd, Tel Aviv
13. Strong MS, Jako GJ, Polanyi TG (1973) Laser surgery in the aerodigestive tract. Am J Surg 126:529–533
14. Sugiura K, Stock CC (1985) Studies in a tumor spectrum. III. The effect of phosphoramides on the growth of a variety of mouse and rat tumors. Cancer Res 15:38–51
15. Tuchmann A (1985) Surgical knife and $CO_2$ laser in experimental tumor surgery. In: Proc. Sixth Congress of the International Society of Laser Surgery. Jerusalem

# Laser-assisted Diagnosis and Therapy of Photosensitized Tumors *

E. UNSÖLD [1], R. BAUMGARTNER, D. JOCHAM, and H. STEPP

Lasers have proved their feasibility and value in many fields of invasive and operative medicine. Surgical techniques using the laser for coagulation, denaturation, ablation, or cutting of tissue generally can be summarized from a physical view point as *thermal* methods. The incident laser radiation is mainly converted to heat during interaction with the target. The application is guided by the physical properties of the laser radiation (wavelength, power density, and energy density) and of the tissue (absorbance, scattering, and thermal conductivity).

In contrast, the diagnostic and therapeutic methods to be described here are based on primarily *nonthermal* phenomena of fluorescence excitation [2] and light-induced biochemical reactions [14] of photosensitizing drugs. These are systemically applied, tumor-selectively stored, and provide a transient marking of malignant tissue. The ratio of concentration of sensitizer in tumorous versus normal tissue ranges between 2 and 30, depending on the individual organ. Those involved in metabolizing the sensitizer, e.g., liver and kidney, do not show significant contrast.

The drug most commonly used today is a hematoporphyrin derivative (HPD) which to a great degree consists of ether- and/or ester-linked hematoporphyrin molecules such as dihematoporphyrin ether and ester (DHE) (Fig. 1) [6]. Due to their imperfect synthesis up to about 25 further porphyrins can be detected in HPD fractions commercially available. Synergism of these many compounds cannot be excluded. Several other sensitizers, e.g., of the group of phthalocyanins [3, 8, 14], purpurines [4], or antibody- [15] and liposome-bound dyes [9], are under investigation, but all substances commercially available today are not admitted for clinical use, only for experimental application.

## Detection of Fluorescence for Diagnosis of Tumors

The power density for exciting DHE fluorescence is – due to the onset of photobleaching – limited to values below about 50 mW/cm$^2$ (DHE, excitation at 407 nm cw, $Kr^+$ laser). The corresponding fluorescence emission between 600 and 700 nm is rather weak and – especially from usually subvisual

* This work was supported by the *Bundesministerium für Forschung und Technologie*, Bonn, under Grant no. MMT 51.

[1] GSF – Zentrales Laserlaboratorium, Ingolstädter Landstr. 1, W-8042 Neuherberg, FRG.

Lasers in Gynecology
Edited by G. Bastert and D. Wallwiener

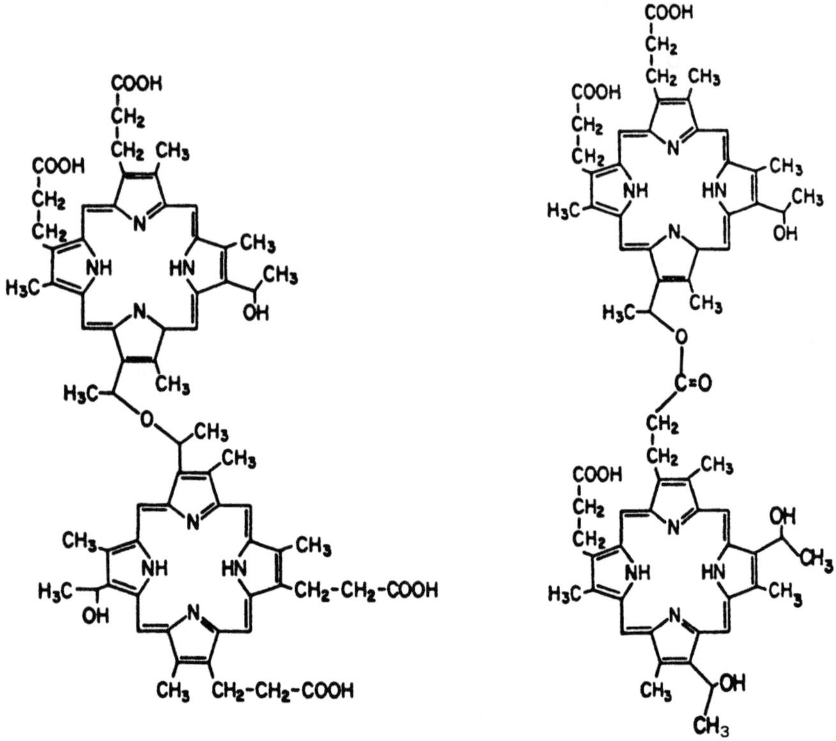

**Fig. 1.** Structural formulas of dihematoporphyrin ether (*left*) and ester (*right*). (After D. Kessel et al. 1985)

tumor foci – has to be detected with high sensitivity. Sophisticated detection techniques are necessary since most tissues show intrinsic autofluorescence which is simultaneously excited and superposes the DHE fluorescence in the same spectral range. In order to discriminate both parts of fluorescence a new method has been developed. It is based on laser excitation of DHE alternately at two wavelengths. Fluorescence emission is detected with an intensified charge-coupled device (ICCD) camera combined with a real-time image processing unit [1,13]. Due to the high sensitivity of the system the sensitizer DHE can be injected at doses of only 0.2 mg/kg body wt., about 1/10 of the dose usually applied for photodynamic therapy. General photosensitization, especially of the skin, a serious side effect of DHE application, may be reduced considerably.

The ICCD camera can easily be coupled to all conventional endoscopes, e.g., hysteroscopes or laparoscopes, to extend the applicability of this new diagnostic method to the various fields of medicine.

## Photodynamic Therapy

Photoactivation of singlet oxygen is claimed to be the basic mechanism underlying the photodynamic therapy

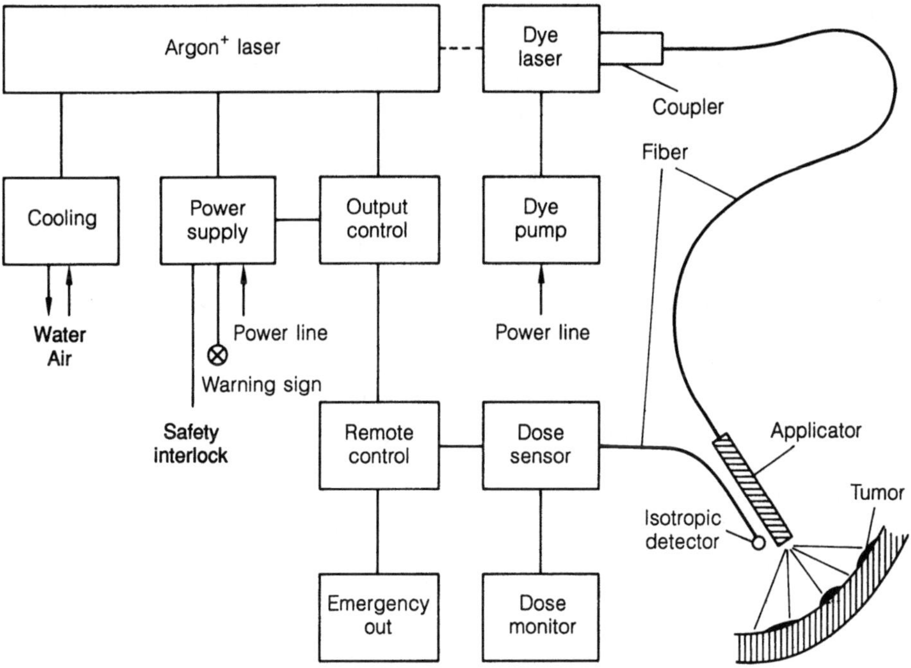


**Fig. 2.** A laser system for photodynamic therapy

(PDT) of tumors. This effect can be induced only by light sufficiently absorbed by the photosensitizer but little screened by the overlying tissue in order to obtain a remarkable effect in depth.

A wavelength of 630 nm has been chosen as the best compromise for DHE excitation. The $Ar^+$ laser-pumped dye laser is well established as a light source and allows for irradiation of hollow organs also endoscopically (Fig. 2). Also the Au vapor laser, the flashlamp-, the excimer laser-, and the Cu vapor laser-pumped dye laser are in use experimentally. Even incoherent light sources seem applicable under certain circumstances for PDT. At present, fully developed irradiation systems with satisfying and safe performance do not exist. In particular, the light applicators so far available ensure sufficient irradiation of a tumor-bearing area or the whole organ in only a few fields of application [14]. For example, the use of a light-scattering medium and special twin-balloon catheters for introduction and fixation of the laser light fiber seems to be a good approach for homogeneous irradiation of the entire inner surface of the bladder [5] (so-called integral PDT). But the development of adequate equipment for intracorporeal gynecologic treatments is only at the beginning. Several treatments of cutaneous and subcutaneous malignancies of breast cancer and its metastases have been reported, however [7].

## Dosimetry

The light dose absorbed by the sensitizer, and consequently the quantity of photo-oxidant produced, seems to be essential for the success or failure of PDT. Several factors, especially of the irradiation system, have to be taken into account to ensure reliable dosimetry. Control of the output power of the laser, i.e., "light mode control", or monitoring of the light power transmitted to the area to be irradiated does not guarantee sufficient treatment. Small isotropic detectors coupled to separate sensor fibers and placed at different sites of the area to be treated with PDT may serve to reliably control the light dose applied [10] (R. Baur 1987, GSF-Zentrales Laserlaboratorium, Munich, unpublished work). These detectors should, of course, be integrated in the light applicators and devices for irradiation wherever possible to provide a compact system for safe and easy handling.

## Side Effects

General photosensitization, especially of the skin and the retina, is the most striking side effect after application of DHE. One has to bear in mind that 12–15 min of sun irradiation provides about the same dose of light as therapeutic irradiation to understand the importance of light protection for patients treated with DHE [13]. The skin remains sensitive for up to 3 months after treatment, and this hinders the use of DHE for routine application in tumor diagnosis. Other sensitizers, e.g., aluminum sulfonated phthalocyanine (AlSPc), seem to overcome the problem of skin photosensitization, and this enhances the possibility of their future use in PDT [11].

## Concluding Remark

The problems briefly outlined here must be satisfactorily solved before fluorescence diagnosis and photodynamic treatment can be applied in daily clinical work. But PDT, which is "reversible" and can be applied repeatedly, promises to supplement classical "final" surgical treatment of tumors.

*Acknowledgement.* The authors thank Hannelore Fißlinger and Pauline Zandl for their helpful assistance.

## References

1. Baumgartner R, Unsöld E (1987) High-contrast fluorescence imaging using two-wavelength laser excitation and image processing. J Photochem Photobiol [B] Biology 1:130–132
2. Baumgartner R, Feyh J, Götz A, Jocham D, Schneckenburger H, Stepp H, Unsöld E (1986) Experimental study on laser-induced fluorescence of hematoporphyrin derivative (HpD) in tumor cells and animal tissue. Laser Med Surg 2:4–9
3. Biolo R, Spada A, Reddi E, Milanesi C, Jori G (1987) TPPS and Zn-phthalocyanine as photodynamic agents in vivo: pharmacokinetic and toxicological studies. 2nd Congr Europ Soc Photobiol 6–10 Sept 1987. Padua, Italy, Abstracts, Elsevier Sequoia, Lausanne C 163, 148
4. Garbo GM, Keck RW, Selman SH, Morgan AR (1987) Comparison of two delivery systems for a hydrophobic photosensitizer in photodynamic therapy of rat bladder tumors. 2nd Congr Europ Soc Photobiol, 6–10 Sept 1987. Padua, Italy, Abstracts, Elsevier Sequoia, Lausanne, C 124, 126
5. Jocham D, Schmiedt E, Baumgartner R, Unsöld E (1986) Integral laser photodynamic treatment of multifocal bladder

carcinoma photosensitized by hematoporphyrin derivative. Eur Urol 12 (Suppl 1):43–46
6. Kessel D, Chang CK, Musselmann B (1985) Chemical, biologic, and biophysical studies on hematoporphyrin derivative. In: Kessel D (ed) Methods in porphyrin photosensitization. Plenum, New York, pp 213–227
7. McCaughan JS (1984) Photoradiation of malignant tumors presensitized with hematoporphyrin derivative. In: Doiron DR, Gomer CJ (eds) Porphyrin localization and treatment of tumors. A. R. Liss, New York, pp 805–827
8. Milanesi C, Cavallini L, Reddi E, Biolo R, Jori G (1987) TPPS and Zn-phthalocyanine as photodynamic agents in vivo. Studies on the mechanism of photoinduced tumor necrosis. 2nd Congr Europ Soc Photobiol, 6–10 Sept 1987. Padua, Italy, Abstracts, Elsevier Sequoia, Lausanne, C 164, 148
9. Reddi E, Valduga G, Jori G (1985) Photosensitizing properties of liposome-bound hematoporphyrin. In: Jori G, Perria C (eds) Photodynamic therapy of tumors and other diseases. Libreria Progetto, Padua, pp 67–69
10. Star WM, Marijnissen JPA (1988) Quantitative light dosimetry in photodynamic therapy. In: Siomos K (ed) Lasers in medicine. Pelekanakis, Chania (Crete), Greece, pp 74–87
11. Tralau CJ, Young AR, Wacker NPJ, Brown SB, Brown SG (1987) A comparison of the skin photosensitizing potentials of dihematoporphyrin ether (DHE) and aluminum sulphonated phthalocyanine (AlSPc) in mice. 2nd Congr Europ Soc Photobiol, 6-10 Sept 1987. Padua, Italy, Abstracts, Elsevier Sequoia, Lausanne C 129, 129
12. Unsöld E, Baumgartner R, Jocham D, Stepp H (1987) Application of photosensitizers in diagnosis and therapy. Laser Med Surg 3:210–214
13. Unsöld E, Jocham D (1988) Grundlagen photodynamischer Laser-Therapieverfahren. Chirurg 59:76–80
14. van Lier JE, Ali H, Rousseau J (1984) Phthalocyanines labeled with gamma-emitting radionuclides as possible tumor-scanning agents. In: Doiron DR, Gomer CJ (eds) Porphyrin localization and treatment of tumors. A. R. Liss, New York, pp 315–319
15. Wat CK, Mew D, Levy JG, Towers GHN (1984) Photosensitizer – protein conjugates: potential use as photoimmunotherapeutic agents. In: Doiron DR, Gomer CJ (eds) Porphyrin localization and treatment of tumors. A. R. Liss, New York, pp 351–359

# Photodynamic Therapy of Gynecological Tumors

S. SCHMIDT [1], U. WAGNER, and D. KREBS

The results as yet achieved in the surgical and chemotherapeutic treatment of gynecological tumors and their recurrence are still unsatisfactory. It, therefore, seems appropriate to develop complementary methods to improve the treatment modalities of these complaints [4, 7]. Photodynamic therapy (PDT) has been suggested as an additional tool in cancer therapy in other fields. Cell-devitalization during PDT is induced by the interaction of light of a defined wavelength and a photosensitizer accumulated in the target object. The phototoxic effect is assumed to be achieved through the augmentation of oxygen radicals during radiation with laser light of low energy, thereby avoiding side effects of thermal origin.

Hematoporphyrin derivatives (HPD) have been proved to be an effective photosensitizing agent. HPD is accumulated in tumor tissue and shows a selective and prolonged retention [2, 5]. However, experience of the use of HPD in the field of gynecology is only very limited. Up to now, various research institutions have reported the use of HPD on a total of 56 patients. The results relating to partial and/or complete remission in these studies were most favorable. The majority of patients were treated for cervical carcinomas ($n = 23$), and experience was also gained in the therapy of vaginal ($n = 20$) and endometrial ($n = 4$) carcinomas, as well as in carcinomas of the vulva and the ovaries ($n = 2$) [1]. Although the effectiveness of using HPD has been evaluated in about 3000 cases, its potentially harmful side effect, namely the phototoxic reaction induced by daylight exposure, has limited its being widely used [5]. Thus the synthesis of new photosensitizers with a different absorption characteristic such as phthalocyanines has been proposed to reduce the daylight phototoxicity [3, 6]. Furthermore, the penetration depth during radiation is improved, as phthalocyanines are absorbed at a larger wavelength (670 nm).

We have developed a new technique of antibody-targeted photodynamic laser therapy using Zn(II) phthalocyanine as a photosensitizer in order to increase the selectivity of PDT [5]. During the procedure, the photosensitizer is chemically bonded to a tumor-selective antibody against a tumor-associated antigen. The systemic dosage of phthalocyanine can be reduced by using antibody-targeted PDT, while providing the same accumulation concentration and effectiveness in tumor tissue. For example, while a dose of 5 mg/kg was proposed to be effective in in vivo experiments in mice, with the

[1] Universitätsfrauenklinik, Sigmund-Freud-Str. 25, W-5300 Bonn 1, FRG.

Lasers in Gynecology
Edited by G. Bastert and D. Wallwiener

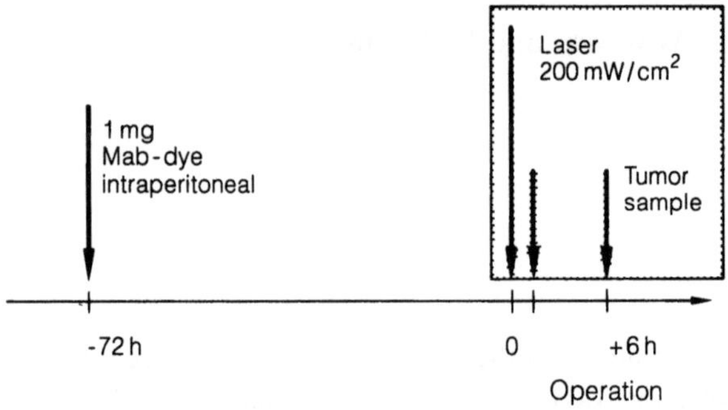


**Fig. 1.** Protocol of antibody-targeted PDT using Zn(II) phthalocyanine as a photosensitizer. The antibody-dye complex (*Mab-Dye*) is applied 72 h before the operation. Laser radiation starts after laparotomy

antibody-targeted type of photodynamic therapy the dose can be reduced by a factor of 1000. The risk of potential side effects is decreased by the same degree.

The first clinical use of the antibody-targeted photodynamic cancer treatment was carried out in three patients, all of whom were suffering from an advanced ovarian carcinoma (FIGO III). All patients had elevated serum concentrations of the tumor-associated antigen CA 125. One milligram of an antibody-dye complex directed against CA 125 was diluted in 1 l Ringer's solution and was instillated intraperitoneally 72 h before the interventional laparotomy for tumor reduction was performed (Fig. 1). Immediately after laparotomy, the intraperitoneal cavity was rinsed to remove unbounded antibody complexes. During surgery, laser radiation was performed at 670 nm, using an argon-pumped dye laser system providing 50 J/cm$^2$. The devitalization of tumor cells was proved by sampling after 6 h. Ultramicroscopical analysis showed damaged mitochondria matrices after exposure to laser radiation in comparison to control tissue.

The results suggest that tumor devitalization is possible using antibody-targeted photodynamic cancer therapy with Zn(II) phthalocyanine as a bonded photosensitizer. The scope of photodynamic laser therapy reaches far beyond the treatment of ovarian carcinomas and relates basically to all tumor types where tumor-associated antigens are available and which can be treated in a similar way. PDT may be indicated in patients with advanced carcinomas, where conventional forms of therapy have proved ineffective, to improve radical surgery for tumor reduction, or as supplementary treatment modality. By applying targeted PDT, it is expected that all superficial tumor tissue is completely destroyed. In advanced carcinomas, a palliative effect should be feasible, due to decompression and pain reduction.

## References

1. Corti L, Maluta S, Tomio L (1989) Photodynamic therapy of gynecologic cancer. Lasers Med Sci 4:155–158

2. Dougherty TJ, Potter WR, Weishaupt KR (1984) Porphyrin localisation and treatment of tumors. Progr Clin Biol Res 170: 301–314
3. Eckhauser M et al. (1990) Biodistribution of Fe(II) phthalocyanin tetrasulphonate. Lasers Med Sci 5:21–24
4. Käser O (1989) Die Entwicklung radikaler Operationstechniken in der Gynäkologie am Beispiel des Ovarial- und Zervixkarzinoms. Geburtshilfe Frauenheilkd 49: 1025–1030
5. Schlag P, Hünerbein M, Stern J, Gahlen J, Graschew G (1991) Photodynamische Therapie gastrointestinaler Karzinome. DMW 116, 16:619–623
6. Schmidt S, Wagner U, Schultes B, Oehr P, Decleer W, Lubaschowski H, Kindermann D, Krebs D (1990) Photodynamic laser therapy targeted by antibody linked dyes against gynecologic tumors. J Cancer Res Clin Oncol 116:324
7. Zander J (ed) (1982) Ovarialkarzinom – Fortschritte für das diagnostische und therapeutische Handeln. Urban and Schwarzenberg, Munich

[illegible]

# Significance of Laser Safety

C. PHILIPP [1], H. ALBRECHT, B. HUG, H.-P. BERLIEN, and G. MÜLLER

## Introduction

The following information regarding the safe use of class-3 B and -4 medical laser systems is based on the lectures held at the First Plenary Workshop of the European Community Medical Laser Concerted Action Program, 1989, in Berlin [15]. This survey is not complete; it focuses on some aspects and is not supposed to give directive advice to the reader. The appendix provides a survey of the existing regulations and standards in the EC countries as well as in the USA.

The reader should keep in mind that an awareness of possible dangers and training in the procedures offer the best protection. Any medical device should be used for the benefit of the patient, and all precautions should be taken to protect the patient and the staff as well as the operator from possible hazards.

## Current Situation

The current situation is described by an incident report of the FDA [1], which analyzed 134 incidents involving lasers that occurred in US hospitals between 1984 and March 1989. The number of published incidents may represent only the tip of the iceberg, but the description of the injuries and their probable causes provides valuable information about the type of incident that is associated with the application of lasers in medicine.

It should first be pointed out that medical lasers are relatively safe instruments which lack some "hidden" problems, such as the adverse current paths of electrocautery. Some problems they do have are typical for the medical use of electrical equipment, but others are laser specific. According to the report, 78 % of the incidents resulted in injury; in seven cases death occurred, and 45 cases are labeled serious (Table 1). The

**Table 1.** Incidents involving lasers between 1984 and March 1989 in the USA [1]

| Injury | Number of incidents | Percentage |
|---|---|---|
| Injury resulted | 78 | 58 |
| No injury resulted | 56 | 42 |
| **Extent of Injury** | **Number of incidents** | **Percentage** |
| Death | 7 | 5 |
| Serious | 45 | 34 |
| Minor | 26 | 19 |
| None | 56 | 42 |

[1] Laser Medizin Zentrum, Krahmerstr. 6–10, 1000 Berlin 45, FRG.

Lasers in Gynecology
Edited by G. Bastert and D. Wallwiener

**Table 2.** Nature of injuries and persons involved [1]

| Type of injury | Number of incidents | Percentage |
|---|---|---|
| Inappropriate or excessive eye exposure | 42 | 31 |
| Burn | 19 | 14 |
| Airway fire | 7 | 5 |
| **Person injured** | **Number of incidents** | **Percentage** |
| Patient | 56 | 72 |
| Medical team member | 8 | 10 |
| Maintenance person | 8 | 10 |
| Other | 6 | 8 |

**Table 3.** Potential causes of laser incidents [1]

| Potential cause | Number of incidents | Percentage |
|---|---|---|
| Laser malfunction | 74 | 48 |
| Lapse in safety precautions | 16 | 10 |
| Surgical error | 8 | 5 |
| Medical complication | 14 | 9 |
| Inconclusive/ unknown | 26 | 17 |
| Maintenance deficiency | 12 | 8 |
| **Situation** | **Number of incidents** | **Percentage** |
| During treatment | 90 | 67 |
| During testing or maintenance | 34 | 25 |

**Table 4.** Differences between use of lasers for material processing and in medicine

| Lasers in industry | Lasers in medicine |
|---|---|
| Closed system | Open system |
| One process | Several different treatment steps |
| No person in laser area | Several persons present: surgeon, nurse, anesthetist |
| Standardized uniform process; constant distance, power | Hand-held applicator; variation of distance, laser parameters |
| Constant conditions and properties of work-piece | intra- and inter-individual variations |
| Protection aim: Personal protection of operator | Personal protection of patient and OR team |
| Defective products are disposed of (only costs) | Not possible! |

types of injuries and persons injured are given in Table 2. The potential causes are difficult to determine (Table 3). Hints for prevention may be gained from an analysis of each situation.

Safety standards and regulations for clinical application in most countries are based on industrial norms, but there are differences between industrial and clinical laser application which require additional measures. Table 4 presents some of these differences.

Typically, the operating table is surrounded by many different devices that are used during treatment, and several procedures may be carried out by different persons simultaneously. Most treatment steps are performed without the laser. Laser applicators are normally hand-held, and during laser application the laser parameters are changed to optimize the treatment.

# Practical Aspects

The discussion of the potential risks or hazards of laser application must distinguish between radiation-related hazards, explosion and fire hazards, the dangers of toxic substances, and electrical hazards. The patient, the operating team, or the environment may be at risk.

## Radiation-related Hazards

Radiation may cause damage to the eyes or to the skin, it may result in failure of treatment, and it may be cancerogenic or mutagenic.

### Damage to the Eye

The human organ most vulnerable to laser exposure is the eye. Different parts are endangered, depending on the wavelength, the mode, and the power density of the laser beam (Figs. 1 and 2). Ultraviolet light rays shorter than 400 nm and infrared (IR) rays longer than 1400 nm are completely absorbed by the cornea and by intraocular media, such as the lens. Exposure can result in damage to these structures. The possible mutagenicity of several UV wavelengths must also be considered [6].

An incident beam of visible light (400–700 nm) will be focused by the optical apparatus to the retina, mostly to the fovea, causing severe retinal burn. Because of the collimation of the laser beam, laser entering the eye can be focused to a point image of 10–20 μm in diameter. This results in a very high power density.

A further source of possible damage is the type of laser, i.e., whether the laser is in the continuous-wave mode or pulsed. The radiation of a Q-switched Nd:YAG laser is stopped in the case of an optical breakdown almost by plasma shielding. If the energy density is below the threshold for optical breakdown, no plasma is produced and the beam passes the focus and strikes the area behind, which can cause damage.

Not only the direct beam of a laser is hazardous; the reflected beam can also seriously damage the eye. In the environment of the operating room there are many items that can cause specular or – less dangerous – diffuse reflection (Fig. 3). Shiny instruments, metal surfaces, and ceramics can act as mirrors. If an instrument has a concave surface

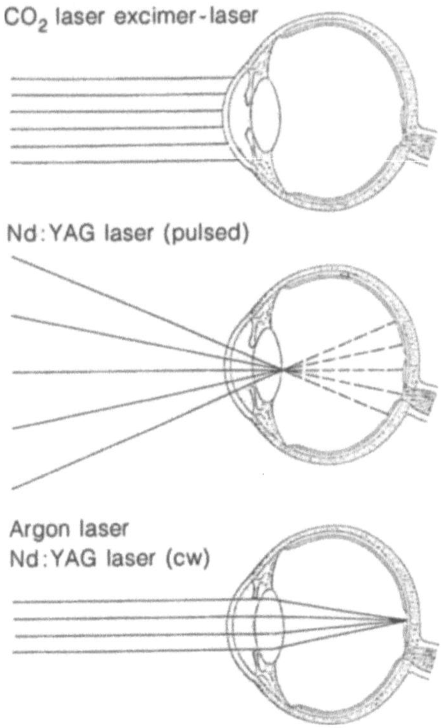


**Fig. 1.** Effects of different lasers on the eye and on plasma formation in the eye; *cw*, continuous wave

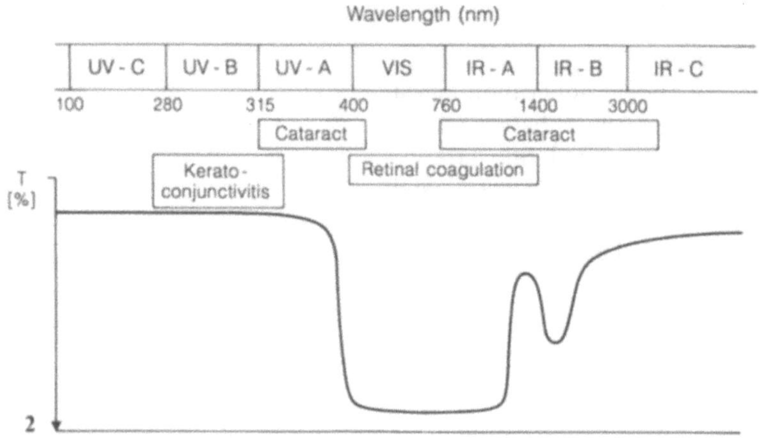


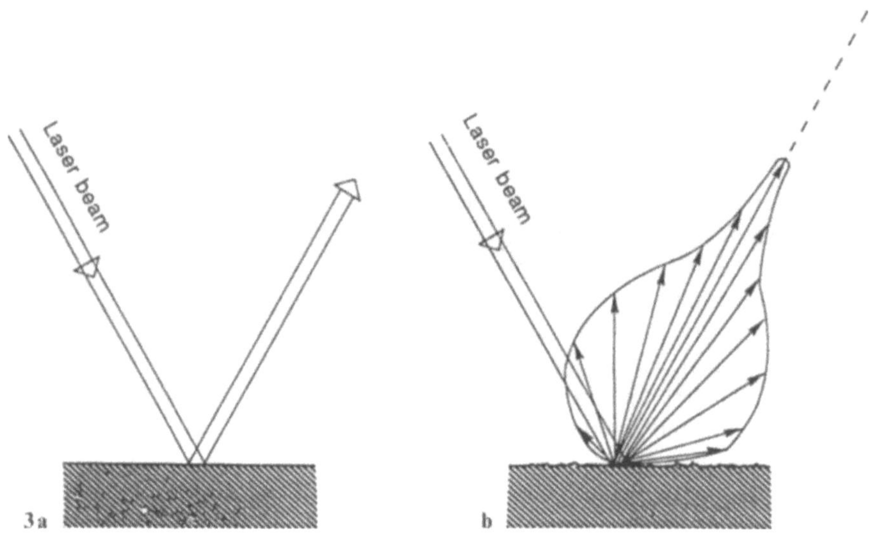


**Fig. 2.** Transmission behavior of the optical elements of the eye and possible hazards depending on the wavelength

**Fig. 3 a, b.** (**a**) Specular (polished metal surface) and (**b**) diffuse (rough metal surface) reflection

which is exposed even to a divergent laser beam, there may be a second focal point, as with a parabola-like mirror. Even materials that appear rough in visible light are not necessarily rough in longer wavelengths, such as IR. Great efforts have been made to develop instruments that reflect the laser beam only as diffuse radiation, but with a certain amount of roughness a surface presents considerable problems with cleaning and sterilization. For visible wavelengths, black surfaces act as absorbers, but this possibility is limited, because if exposed, the instrument will heat up and represent a thermal danger if it comes into contact with the patient's tissue. Although special instrumentation for laser surgery cannot eliminate all danger, it can help to diminish it where reflection cannot be avoided completely.

## Safety Glasses

Fortunately, there is broad agreement among laser users about the necessity for protective eyewear, and of protecting the patient adequately with eye covers. Unfortunately, the quality of eyewear varies according to the safety regulations of different countries. A new European standard is therefore under discussion, the PrEN 207 (Personal Eye Protection: Filters and Eye Protectors Against Laser Radiation). It is based on the German standard DIN 58215, which has been in effect since 1985. The new standard requires the filter to effectively resist beam penetration or shattering for at least 100 s or 1000 pulses. The frame must also resist penetration and be fire resistant. Furthermore, a polycarbonate laminate applied to the inner surface of the glass serves to retain its integrity when the glass itself shatters. The draft is likely to be adopted in 1992 by the European Community [7].

Although this new standard exceeds all present national regulations, it offers certain problems. It covers only the emitted wavelength of the laser, not secondary radiation in other spectral bands. Intelligent safety eyewear is being developed, which recognizes by sensors these other wavelengths and emits an audible alarm if preset maximum radiation limits are exceeded. According to the present state of the art, it is further possible to design filters that switch to a higher optical density in case of exposure. These dynamic eye protection filters could offer sufficient visible transmission for low interference with color perception.

In the endoscopic use of lasers, it is common to use filters which are fitted to the eye. Thus, the other eye, which is usually very close to the laser inlet, is completely unprotected. If the transmission system is damaged by manipulation the risk of accidental exposure to the unprotected eye is great. In video endoscopy as well, an optical leak can occur, emitting possibly hazardous light, so it is recommended that adequate eye glasses be worn for any laser procedure.

## Damage to the Skin

The skin is much less sensitive than the eye to laser irradiation. Thus, the threshold values are much higher. Depending on the amount of laser energy absorbed, injuries may vary from mild reddening to blistering and charring. Under normal conditions, the reaction of the "target" will lead to a rapid interruption of the exposure, but unnoticed exposure of the patient under general anaestesia, e.g., from a reflected beam, can cause severe damage.

## Failure of Treatment

In cases where tissue behind the intended target might accidentally be exposed to the laser beam (as with endoscopic use of the carbon dioxide laser) the use of a beam stop is highly recommended. Otherwise, this hidden tissue will be exposed unintentionally when the target tissue is vaporized. Safety measures should cover the accidental irradiation of tissue as well as the possible hazards of intended irradiation (Table 5).

**Table 5.** Hazards of intended tissue irradiation

| | |
|---|---|
| Overdosage | Increase in side effects<br>Failure of intended treatment<br>Maltreatment of adjacent healthy tissue/organs |
| Underdosage | Failure of intended treatment<br>Increase in side effects |

### Cancerogenity and Mutagenity

Further studies should be carried out to evaluate the cancerogenity and/or mutagenity of laser irradiation, e.g., of ultraviolet Excimer laser radiation and the multiphoton effects of high-energy pulses of IR lasers. To prevent late sequelae of treatment, the use of lasers with possibly hazardous wavelengths should be avoided.

## Explosion and Fire-related Hazards

These hazards are not specific to lasers; they exist with electrocautery as well. All inflammable materials in the operating room are potentially dangerous, including organic solvents and disinfectant solutions containing alcohol. Although most solvents do not ignite when directly radiated with a laser beam, when they are present as a thin film the ignition temperature is reached faster. Other inflammable materials are drapes, tissues, and swabs, most plastic articles such a tubes, cuffs, hoses, and catheters, and gases such as oxygen, nitrous oxide, and halothane. There is an urgent need for noninflammable materials, but suitable products are not readily available on the market, except for special laser tubes and very useful tube covers made of a compound (metal and plastic) foil, which adheres to the tube. As with instrumentation, no laser tube can be recommended for use with all laser types. Most are safe for carbon dioxide lasers, some for Nd:YAG radiation, and only a few for both. The recommendations of the manufacturer as well as the FDA regulations regarding these tubes and covers should be studied before use.

Even with proper use, exposure of the tube must be avoided. The laser operator must keep in mind that most accidents which lead to the death of a patient [1] are related to ignition of the tube, burning of the airway, or intoxication due to decomposition of plastic material from the tube, releasing hydrogen chloride, hydrogen fluoride, hydrogen cyanide, or various plastic monomers.

Additional safety can easily be achieved with the $CO_2$ laser by covering endangered areas (the cuff of the tube, surrounding tissue) with saline solution-soaked gauze or drapes. As the IR wavelength of this laser is absorbed by the water, it will not penetrate until the tissue is dry. For argon or Nd:YAG lasers this "water shield" is not appropriate, as their beams penetrate water easily.

## Hazards of Toxic Substances

### Laser Material

As dye lasers contain solvents of carcinogenic and/or mutagenic dyes, they can be hazardous when the equipment is damaged, incorrectly used, or manipulated (Table 6). Maintenance of these lasers (e.g., changing of the solutions) should be carried out by authorized personnel who are aware of the

**Table 6.** Hazards of toxic substances

| | |
|---|---|
| – Laser material | |
| Laser gases: | Excimer laser |
| Laser dyes: | Dye laser, exchange during maintenance, disposal of wastes |
| Laser tubes: | Beryllium oxide, disposal |
| Laser optics: | IR lenses and fibers: zinc selenide, cadmium telluride |
| Beam stop: | fired clay, asbestos, cadmium |
| – Laser plume: | Toxic substances in: – infectious and contagious material – disposal of filters – disposal of "filtered" air |

special hazards. Even better, the procedure itself should make contact with these substances impossible. It should also be mentioned that argon-laser tubes may contain beryllium oxide and must be disposed off appropriately, i.e., returned to the manufacturer.

## Laser Plume

Currently, there is far more concern about the products of tissue vaporization. Much research has been carried out in the past several years into the hidden dangers of laser plume [16]. In 1987, Baggish showed the acute effect of laser smoke on the lungs of rats [4, 5]. Moreover, it is an established fact that the products of tissue pyrolysis contain carcinogenic/mutagenic substances such as benzene, toluene, formaldehyde, acrolein, and PAHs [8, 10, 12, 13] (J. Kinzel, H. Städter, R. Hagemann, and G. Müller, manuscript in preparation). It is also possible that the laser plume may contain intact viral DNA [9, 11].

The danger of becoming infected from laser plume is still unsubstantiated, but the user of laser systems (and any other systems that produce smoke by the pyrolysis of tissue, e.g., electrocautery) should be aware of the possible risks and should take all possible precautions, such as using a vacuum system correctly. Many of these vacuum systems are promoted nowadays, but there are still unsolved problems [2, 18]. The first is to get all of the plume into the suction tube. The closer the inlet is held to the operation site, the more smoke is taken into the filter. New inlets, e.g., in the shape of a circle to surround the operation site, are currently being tested by the FDA.

Another problem is that as many evacuation systems are loud, some surgeons will not use them. However, the most serious concern is about the filters themselves, and whether those most commonly used are truly sufficient for removing particles from the airstream. The largest concentration of particles in the smoke are about 0.3 μm in diameter; some are smaller. The most common filters are only 95% efficient for 0.5-μm particles. High-efficiency particulate air filters (HEPA) with a 99.997% efficiency for 0.3-μm particles are available, and ultra low penetrating air filters (ULPA) with an efficiency of 99.9997% for all particles of 0.12 μm and larger are the standard filters for the newest generation of smoke evacuators now in the USA [14]. Therefore, every laser user should check the evacuation system for its efficiency and, whenever possible, upgrade it with the new types of filters. The maintenance of the evacuation system is also very important; only when the exchange intervals are carefully observed and the filters changed often they can maintain their efficiency. To date, there is no satisfactory solution to the disposal of the filters or of the "filtered" air.

### Electrical Hazards

As with all electrically energized equipment, the possible dangers should not be underestimated, as most lasers operate at 10 kV or more. Most countries have legal standards related to the IEC standards for electrical safety of laser equipment and installation (IEC 820) and for safety of medical electrical equipment (IEC 601.1). Where no national standards are given, the installation should correspond to the IEC standards.

## Objectives of Further Development

### Hazards of Equipment and Installation

*Classification and Maintenance of the Laser.* As all regulations are based on the classification of lasers into different classes (I–IV; Fig. 4), the laser installation should be labeled clearly. The protective devices mentioned in IEC standard 825 (protective housings, key controls, emission warning devices, beam stops, etc.) must be provided by the manufacturer. Laser systems may not emit radiation inadvertently. Maintenance and regular safety checks should be performed by specially trained personnel according to the regulations of the manufacturer and to national regulations, at least once a year.

*Controlled Laser Areas.* The nature of a laser implies that its operation can present a hazard not only to the user, but also to other people over a considerable distance. Therefore, controlled laser areas should be marked with significant symbols, and access to these areas should be limited to authorized personnel. No hazardous radiation should be present at any time outside these areas.

Although many problems appear to have been solved, further research and international cooperation are needed in the following areas to improve safety:

- Development of standard interfaces
- Harmonization of control elements and procedures (for the use of different laser systems)
- Development of safety standards
  - Safety functions
  - User eye protection (PrEN 207 [DIN 58215])
  - Inflammability of OR material
  - Marking of laser area (separation, warning signs)
- Development of laser-safe instruments
- Development of laser-safe OR material
- Development of dynamic eye protection
  - Fast sensors for transmission control
  - Filters with controllable transmission (i.e., laser radiation: primary; secondary radiation: fluorescence, plasma)
- Toxic and infectious material
  - Maintenance of procedures and standards for disposal of toxic laser material
  - Development of substitutes for toxic material

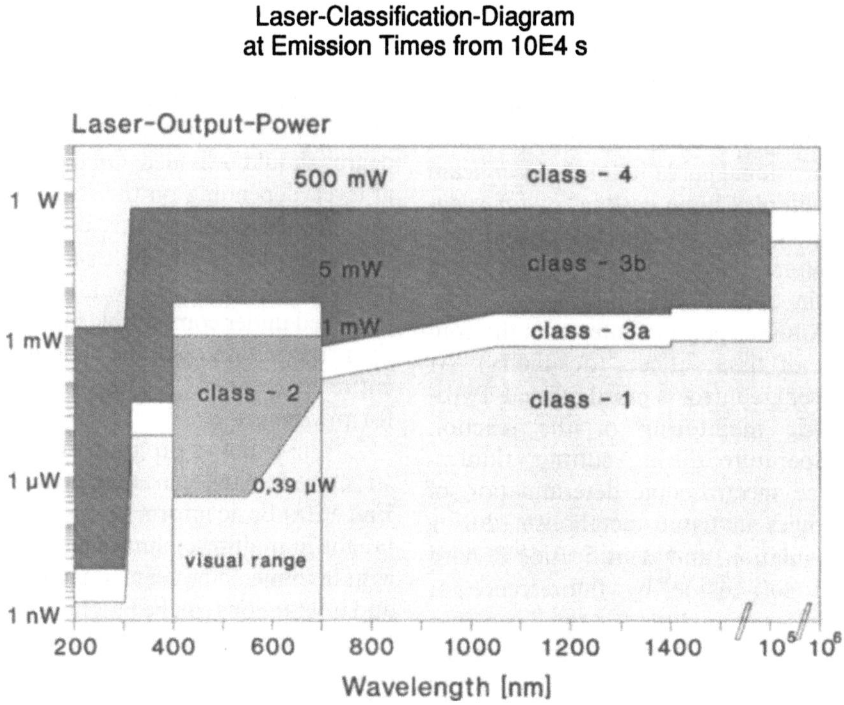


**Fig. 4.** Classification of laser systems. All medical laser systems are 3 b or 4

- Laser plume
  - Further evaluation of production of toxic material depending on process prarameters
    a) Test procedures
    b) Identification of significant substances
    c) Variation of parameter for process control
  - Evaluation of conditions for infectious material in laser plume
    a) Test procedures
    b) Type and survival time of living material
  - Evaluation of evacuators
    a) Specifications
    b) Test plans and procedures
    c) Efficiency during clinical use
    d) Disposal of contaminated parts (filters)
  - Development of improved filters and evacuators
    a) Air flow at OR site
    b) Filters
    c) Disposal of "filtered" air

## Hazards of Application

Several improvements in application have been made or are being worked on at present, to support the operating team in their attempt to increase the safety of treatment. The use of small instruments (handpieces with gas or liquid flushing, evacuator nozzle) that will not block visualization of the operation site, and of dynamic eye protection glasses with sufficient visible transmission for low interference with color perception, erroneous settings on laser console, and color changes of tissue, could improve the visual control.

A distal in-line measurement of the transmitted power or energy, to determine defects of the transmission line or to apply constant power for tissue welding, and computer-assisted multidimensional guidance of the laser beam in complex bone cutting, or for scanning systems in dermatology and neurosurgery, would offer better control of the applied radiation.

Another point of interest is the control of tissue effects (dosimetry). At present, control is mostly visual. Pyrometric monitoring of the reaction temperature during cutting, fluorescence spectroscopic determination of changes in tissue metabolism during coagulation, and identification of hard and soft tissue by fluorescence or shockwave analysis and intelligent systems to control laser power and energy could lead to better instruments and improved safety of application.

## Precautions for Personnel

Just as safe and reliable instruments are essential for any laser treatment, so are an understanding of laser/tissue interactions and the use of proper parameters. Therefore, the authorized personnel should be adequately trained in laser performance and safety precautions. The OR personnel must work as a team, and all must be familiar with laser use and must know their particular duties. The operator of the laser system must be familiar with the therapeutic procedures, with the bioeffects of lasers on irradiated tissue, the eye, the skin, and with the secondary effects of lasers and the procedure in case of an accident [3].

As training is the best insurance against dangers, a training program is proposed based on the standards of the German Laser Association and the European Laser Association [15]:

1. Independent of the respective professional field, a postgraduate type of degree should be issued. On the national level, depending on the structure of the medical training, this could be a university degree or one issued by the respective medical society. If training is provided under comparable guidelines, then recognition could be valid for the entire EC area, as is already the case for physicians.

2. The training program should be structured in two parts, as detailed in Fig. 5: (a) Basic information should be taught in multidisciplinary basic classes at laser medicine centers; laser safety and interactions can be taught, since all types of lasers are available. (b) Following the basic information classes, advanced classes covering the respective professional fields and their problems should be taught. This faculty-oriented special training should be conducted at hospitals where the clinical use of laser systems is part of the routine program. The basic training is already available, and for some disciplines (e.g., gynecology, otorhinolaryngology) advanced classes are now being conducted in Germany.

## Laser Safety Officer

A competent laser safety officer should be named for each device, with the basic training mentioned above. His or her duties should include supervision of compliance with safety rules and training of staff members for safe operation. The laser safety officer's training should be certified according to the regulations of the health authorities or – in Germany – the Berufsgenossenschaft. The officer should be indepen-

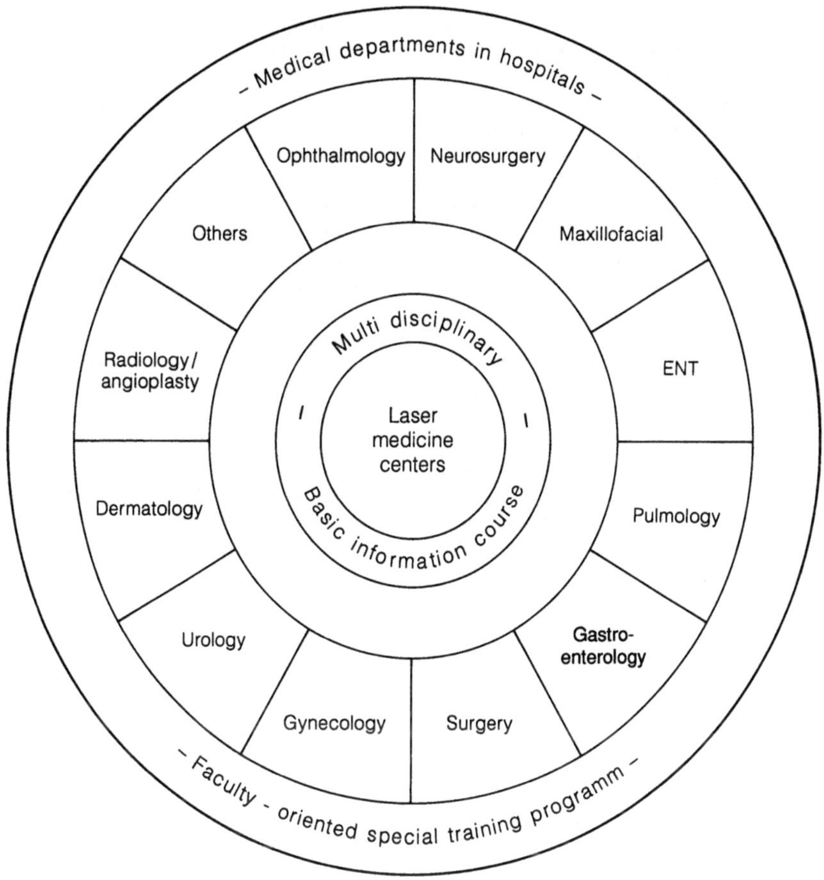


**Fig. 5.** Structure of a certification program: basic course provides a survey of all common laser systems, with minimal information for a laser safety officer. For medical doctors, a faculty-oriented training program is then necessary

dent of the authorities of a hospital or clinic and should have the power to give safety directives to technical and medical staff as well as to the physicians.

## Registration, Accident Report and Investigation

The installation of a laser system should (for Germany: must) be registered with the occupational safety authorities. Any accident which has caused injury or which could cause injury to persons should it reoccur should be reported to the laser safety officer, the manufacturer, and the occupational safety authorities. A close investigation of the circumstances should be made and a report submitted. All efforts must be made to prevent any reoccurrence of an accident.

# References

1. Anonymous (1989) FDA incident report: a sobering reminder of laser perils. Clin Laser Monthly 7:97–108

2. Anonymous (1989) Effectiveness of smoke evacuation systems questioned. Clin Laser Monthly 11:121–128
3. Anonymous (1990) Preventing and reacting to fires: steps the surgical team must take. Clin Laser Monthly 8:123–125
4. Baggish MS, Elbakry M (1987) Effects of laser smoke on lungs of rats. Am J Obstet Gynecol 156:1260–1265
5. Baggish MS, Baltoyannis P, Sze E (1988) Protection of the rat lung from the harmful effects of laser smoke. Lasers Surg Med 8:248–253
6. Bende T, Seiler T, Wollensak J (1988) Side effects in excimer corneal surgery. Graefes Arch Clin Exp Ophthalmol 226:277–280
7. Dance B (1989) Europe sets standards for safety goggles. Laser Focus World 10:33–34
8. Dierkesmann R, Huzley A (1985) Side effects of endobronchial laser treatment. Endoscopy 17:49–53
9. Edinger MN, Matchette LS (1989) In vitro production of viable bacteriophage in a laser plume. Lasers Surg Med 9:296–299
10. Fisher WR (1987) Laser smoke in the operation room. Biomed Technol Today 11/12:191–194
11. Garden JM, O'Bannion MK, Shelnitz LS, Pinski KS, Bakus AD, Reichmann ME, Sandburg JP (1988) Papillomavirus in the vapor of carbon dioxide laser-treated verrucae. JAMA 259:1199–1202
12. Kinzel J (1989) Analyse der Abbrandprodukte laserbestrahlten Gewebes. Diplom-Arbeit, Technische Universität Berlin, Institut für technischen Umweltschutz, Fachbereich Luftreinhaltung, Umweltchemie und Laser-Medizin-Zentrum Berlin
13. Kokosa JM, Eugene J (1989) Chemical composition of laser-tissue interaction smoke plume. J Laser Appl 1:59–63
14. Liu B, Rubow KL, Pui D (1988) Performance of HEPA and ULPA filters. Proc Institute of Environmental Sciences
15. Müller GJ, Berlien HP (eds) (1989) Safety and laser-tissue interaction. Advances in laser medicine, II. ecomed, Landsberg
16. Müller GJ, Berlien HP (Eds) (1989) Angewandte Lasermedizin – Ein Handbuch für Klinik und Praxis, ecomed, Landsberg
17. Nezhat C, Winer WK, Nezhat F, Nezhat C, Forrest D, Reeves WG (1987) Smoke from laser surgery: is there a health hazard? Lasers Surg Med 7:376–382
18. Smith JP, Moss E, Bryant CJ, Fleeger AK (1989) Evaluation of a smoke evacuator used for laser surgery. Lasers Surg Med 9:276–281

# Appendix

## EC and International Standards and Their National Equivalents

| Standard | | |
|---|---|---|
| | EN 29000<br>= ISO 9000 | Quality management and quality assurance |
| | EN 29001<br>= ISO 9001 | Model for quality assurance in design/development, production, and servicing |
| | Complement EN 29001 | For active medical devices; guide to application of quality systems to the manufacture of medical devices (EUROM/COCIR) |
| | EN 29002<br>= ISO 9002 | Model for quality assurance in production and installation |
| | EN 29003<br>= ISO 9003 | Model for quality assurance in final inspection and test |
| | EN 29004<br>= ISO 9004 | Quality management and quality system elements – guidelines |
| | pr EN 207 | Personal eye protection; filters and eye protectors against laser radiation |
| | pr EN 208<br>= DIN 58219 | Personal eye protection; laser adjustable goggles |
| | EC Directive | Active medical devices |
| | IEC 601.1<br>= HD 395.1<br>vgl. DIN VDE 0750 Teil 1<br>= BS 5724 Part 1 | Safety of medical electrical equipment; part 1: General requirements |
| | IEC 820<br>= HD 194<br>= DIN VDE 0836<br>= NF C 74-311<br>= NBN C 79-700 | Electrical safety of laser equipment and installation |
| | IEC 825<br>= HD 482<br>= DIN VDE 0837<br>= NF C 74-312 | Radiation safety of laser products; equipment classification, requirements, user's guide |
| | IEC 76(Sec)12<br>= DIN VDE 0835 | Power and energy measuring equipment for laser radiation |

## National Standards and Regulations

| Country | Standards | | | National regulations |
|---|---|---|---|---|
| Belgium | NBN C 79-700 | = IEC 820 | | |
| Denmark | – | | | |
| Federal Rep. of Germany | DIN VDE 0835 | = IEC 76(Sec)12 | | Medizingeräteverordnung<br>Unfallverhütungsvorschrift VBG 93 "Laserstrahlung" |
| | DIN VDE 0837 | = IEC 825 | | |
| | DIN 58215 | Laserschutzfilter und Laserschutzbrillen; Sicherheitstechnische Anforderungen und Prüfung | Filters and eye protectors against laser radiation; safety requirements and testing | |
| | DIN 58219 | = prEN 208 | | |
| | DIN VDE 750 Teil 1 | = IEC 601.1 | | |
| | VDE 750 Teil 226 | Medizinische elektrische Geräte<br>Diagnostische und therapeutische Lasergeräte<br>Besondere Festlegungen für die Sicherheit | Medical electrical equipment<br>Diagnostic and therapeutic laser equipment<br>Particular requirements for safety | |
| | DIN 31000 | Allgemeine Leitsätze für das sicherheitsgerechte Gestalten technischer Erzeugnisse | General guide for designing of technical equipment to satisfy safety requirements | |
| | DIN 31000 Teil 2 | Begriffe der Sicherheitstechnik Grundbegriffe | Concepts of safety technology basic concepts | |
| France | NF C 74-311<br>NF C 74-312 | = IEC 820<br>= IEC 825 | | Homologation |
| Greece | in preparation | | | in preparation |
| Ireland | – | | | – |
| Italy | in preparation | | | User's Guide for the Clinical Use of Laser Equipment |
| Portugal | – | | | – |

| | | | |
|---|---|---|---|
| Spain | – | | Real Decreto no. 2603/85 Especificaciones tecnicas de los aparatos mobiles radiocirgurgicos y su homologation por el Ministrio de Industria y Energia |
| The Netherlands | – | | Richtlijnen Laserveiligheid voor Research en Onderwijs Veiligheidsmaatregelen voor het gebruik van lasers medische toepassingen |
| United Kingdom | BS 4803 Part 1 | Radiation safety of laser products and systems | Guidance on the Safe Use of Laser |
| | BS 4803 Part 2 | Specification for manufacturing requirements for laser products | |
| | BS 4803 Part 3 | Guidance for users | |
| | BS 5724 Part 1 | = IEC 601.1 | |

# Subject Index

MIX
Papier aus verantwortungsvollen Quellen
Paper from responsible sources
FSC® C105338

If you have any concerns about our products,
you can contact us on
**ProductSafety@springernature.com**

In case Publisher is established outside the EU,
the EU authorized representative is:
**Springer Nature Customer Service Center GmbH**
**Europaplatz 3, 69115 Heidelberg, Germany**

Printed by Libri Plureos GmbH
in Hamburg, Germany